Mechanical Geometry Theorem Proving

Shang-Ching Chou

Institute for Computing Science,
University of Texas at Austin, Texas, U.S.A.

Mechanical Geometry Theorem Proving

D. Reidel Publishing Company

A MEMBER OF THE KLUWER ACADEMIC PUBLISHERS GROUP 

Dordrecht / Boston / Lancaster / Tokyo

Library of Congress Cataloging in Publication Data

Chou, Shang-Ching, 1942–
Mechanical geometry theorem proving / Shang-Ching Chou.
p. cm. — (Mathematics and its applications)
Bibliography: p.
Includes indexes.
ISBN 90–277–2650–7
1. Automatic theorem proving. I. Title. II. Series: Mathematics and its applications (D. Reidel Publishing Company)
QA76.9.A96C48 1987
516'.0028'5—dc 19 87–30387
CIP

ISBN 1-4020-0330-7
Transferred to Digital Print 2001

Published by D. Reidel Publishing Company,
P.O. Box 17, 3300 AA Dordrecht, Holland.

Sold and distributed in the U.S.A. and Canada
by Kluwer Academic Publishers,
101 Philip Drive, Assinippi Park, Norwell, MA 02061, U.S.A.

In all other countries, sold and distributed
by Kluwer Academic Publishers Group,
P.O. Box 322, 3300 AH Dordrecht, Holland.

SERIES EDITOR'S PREFACE

Approach your problems from the right end and begin with the answers. Then one day, perhaps you will find the final question.

'The Hermit Clad in Crane Feathers' in R. van Gulik's *The Chinese Maze Murders*.

It isn't that they can't see the solution. It is that they can't see the problem.

G.K. Chesterton. *The Scandal of Father Brown* 'The point of a Pin'.

Growing specialization and diversification have brought a host of monographs and textbooks on increasingly specialized topics. However, the "tree" of knowledge of mathematics and related fields does not grow only by putting forth new branches. It also happens, quite often in fact, that branches which were thought to be completely disparate are suddenly seen to be related.

Further, the kind and level of sophistication of mathematics applied in various sciences has changed drastically in recent years: measure theory is used (non-trivially) in regional and theoretical economics; algebraic geometry interacts with physics; the Minkowsky lemma, coding theory and the structure of water meet one another in packing and covering theory; quantum fields, crystal defects and mathematical programming profit from homotopy theory; Lie algebras are relevant to filtering; and prediction and electrical engineering can use Stein spaces. And in addition to this there are such new emerging subdisciplines as "experimental mathematics", "CFD", "completely integrable systems", "chaos, synergetics and large-scale order", which are almost impossible to fit into the existing classification schemes. They draw upon widely different sections of mathematics. This programme, Mathematics and Its Applications, is devoted to new emerging (sub)disciplines and to such (new) interrelations as exempla gratia:

- a central concept which plays an important role in several different mathematical and/or scientific specialized areas;
- new applications of the results and ideas from one area of scientific endeavour into another;
- influences which the results, problems and concepts of one field of enquiry have and have had on the development of another.

The Mathematics and Its Applications programme tries to make available a careful selection of books which fit the philosophy outlined above. With such books, which are stimulating rather than definitive, intriguing rather than encyclopaedic, we hope to contribute something towards better communication among the practitioners in diversified fields.

This is a book about automatic theorem proving. I have no difficulty in picturing sceptical smiles on the faces of many colleagues upon hearing this phrase. Indeed, if the phrase is taken in the sense of replacing most of the activities of the modern professional mathematician by computer programs, if might suffice to look into a mirror. But then there would still be a limit of uneasiness. Frankly, I have no idea how far most of our reasoning can be automated. Certainly much of what we do is based on sweat and routine and not on continuous creative sparks. The only way to see how far one can go is to try it. And certainly Chou's (theorem) prover will astonish many.

I see a great future and great usefulness of such provers and symbolic calculation expert systems in near and more distant future mathematics. After all, the 'talents' of computers and humans are vastly different and should be used for reinforcement rather than competition. This 'reinforcement idea', by the way, seems to be gaining headway in such areas as decision-support systems and NP-hard optimization problems.

In any case, provers such as the one described in this book will be of inestimable value in checking (verifying) all kinds of conjectural (differential) geometric statements such as are, for example, important in engineering contexts dealing with complex geometric (accessibility and mover) problems in robotics. Chou's methods have in fact been extended/implemented in algebraic-geometric directions precisely with a view to such applications.

The unreasonable effectiveness of mathematics in science ...

Eugene Wigner

Well, if you know of a better 'ole, go to it.

Bruce Bairnsfather

What is now proved was once only imagined.

William Blake

As long as algebra and geometry proceeded along separate paths, their advance was slow and their applications limited.

But when these sciences joined company they drew from each other fresh vitality and thenceforward marched on at a rapid pace towards perfection.

Joseph Louis Lagrange.

Bussum, November 1987

Michiel Hazewinkel

Table of Contents

Foreword by Larry Wos

When computers were first conceived, then designed, and finally implemented, few people (if any) would have conjectured that in 1987 computer programs would exist capable of proving theorems from diverse areas of mathematics. Even further, if a person at the inception of the computer age had seriously predicted that computer programs would be used to occasionally answer open questions taken from mathematics, that person would have received at best a polite smile.

The justifiable hesitancy to make such apparently rash conjectures and predictions rests mainly with two factors. The first factor focuses on the nature of theorems and questions from mathematics. Indeed, the study of some area of mathematics presents a most formidable challenge, for deep reasoning is almost always required. The second factor focuses on the automation of reasoning, and in particular on the automation of deep reasoning. The obstacles to be overcome, if the objective is the design and implementation of a computer program that can apply deep reasoning to given problems or questions, are many and varied. Some obstacles concern the presentation of information, some the inference rules for drawing conclusions, and some the strategy for controlling the application of the inference rules.

Fortunately, despite the formidable challenge presented by mathematics and despite the obstacles to the automation of reasoning, computer programs do exist that reason and reason so effectively that they can occasionally be used as a colleague. This monograph focuses on such a program, an automated theorem-proving program designed and implemented by Chou. Chou's program is remarkable because of its capacity to easily prove one theorem from classical geometry after another and, of greater significance, to occasionally answer an open question.

As one reads the material presented in this volume and examines the proofs obtained with Chou's program, one is likely to first feel some surprise, then delight, and finally astonishment at the uniformity of the success and the approach taken to achieve that success. One sees a long parade of theorems of varying difficulty, each proved without the need to choose from an array of parameter settings or supply appropriate lemmas. Selecting randomly from the theorems presented in this book and attempting to supply a proof quickly leads one to an appreciation of the excellence and scope of Chou's work. Although in no way should one assume that the problem of proving theorems in geometry is completely solved, Chou has demonstrated that a very large number of theorems can be proved by a single program relying on a well-defined paradigm. For theorems in this class, the burden of proof is, metaphorically, not on the mathematician but instead solely on the program-a most satisfying situation indeed.

We applaud Chou for choosing wisely the basic method for automating the proof of theorems from geometry, namely, that of W.-T. Wu. We also congratulate him for his powerful extension of Wu's approach to automated theorem proving. Most of all, we thank Chou for providing a standard that other paradigms for the automation of theorem proving might seek to emulate. Chou's program and monograph provide further evidence that we have in fact entered a new era – the era in which various computer programs function as effective and useful automated reasoning assistants.

Larry Wos
January 1987

Preface by the Author

This monograph consists of two parts: an introduction to mechanical geometry theorem proving using algebraic methods and a collection of 512 theorems proved by the computer program described in [11], which was based on the pioneering work of Wu Wen-Tsün [42], [46], [43], [45]. It has grown out of a collection of 360 theorems proved mechanically ([14], hereafter referred to as the *Collection*).

During the three years of my thesis work, I used the prover and its predecessors to prove many theorems from a variety of sources, including N. Altshiller-Court's textbook [1]. By May, 1985, at least 200 theorems had been mechanically proved. My thesis advisors, R. S. Boyer and J S. Moore, encouraged me to collect the theorems into a single technical report, believing that such a report would be of some interest to the theorem proving and geometry communities simply because it would show the power of the prover.

However, the job of collecting the theorems was much more tedious than I expected. Because the 200 theorems were proved over a period of several years by different versions of the prover, there were many replications in the list, the statements of the theorems were in a variety of different formats and were spread over a dozen files. In the end I gave up the task of collecting those theorems that had been proved during my thesis work. Instead I selected a single text, namely [1], and set out to prove a representative sample of theorems from it. Within two weeks I proved the 360 theorems listed in the *Collection*.

To make the *Collection* a monograph, I wrote Part I and chose about 300 theorems from the *Collection* and about 200 theorems from other sources to form Part II.

In Part I, I give a detailed presentation of those algebraic methods in geometry theorem proving that are most successful in practice: Wu's method and the Gröbner basis method. Many elegant examples were used to illustrate Wu's method, including Morley's trisector theorem and a conjecture of V. Thébault which was not confirmed until 1983. All these examples have been proved by my prover. Chapter 1 provides sufficient materials for those who wish to implement a prover (on a personal computer) to prove geometry theorems whose traditional proofs need enormous amounts of human intelligence (e.g., Feuerbach's theorem). The only background necessary for this chapter is high school algebra.

Most of Part II was generated mechanically. A typical entry for this part includes an identification number, a diagram, an exact statement of the theorem in the form of a construction, and a note on the "nondegenerate" conditions of the theorem. The latter are unstated constraints on the original construction that are necessary to insure that the statement is a theorem, e.g., that certain points are not collinear. Some entries include a conventional (informal) statement of the theorem in English. The diagrams and exact constructions are generated mechanically from the input to my theorem prover. The nondegenerate conditions are generated from the theorem-prover's output. I personally wrote only the informal statements of the theorems and the occasional remarks. All such text is printed in *italics*.

During the period of my thesis work and subsequent work, many people gave me help, encouragement and guidance, especially my thesis advisors, Professor Robert S. Boyer and Professor J Strother Moore. In fact, they are still guiding my work now. I am deeply grateful for their valuable help and guidance.

Dianne King of the Institute was the editor of this monograph. I am deeply indebted to her for the help, which was vitally important for the completion of this monograph and many other related papers.

Of course, the content of this work suggests the important role played by Professor Wu Wen-Tsün since our first correspondence in 1982. Professor W.W. Bledsoe's initial encouragement and support were important to me. I wish to thank Professor W.F. Schelter for his many forms of help.

During the writing of the monograph, discussions with Dr. J. G. Yang and Dr. H. P. Ko helped me clarify many points and speeded up the writing. I also thank Dr. N. Shankar, Dr. Richard Cohen and the other members of the Institute for their help and encouragement. I am deeply indebted to my wife for her patience and support during a long and *unusual* period of more than 15 years.

The work reported in this monograph was supported in part by NSF Grants DCR-8503498 and CCR-8702108.

Shang-Ching Chou
The Institute for Computing Science and Computer Applications
The University of Texas, Austin
January 1987

Part I

Methods in Mechanical Geometry Theorem Proving

Chapter 1. An Introduction to Wu's Method

In the area of proving geometry theorems using computer programs, the earliest successful work was done by H. Gelernter [20] and his collaborators. Their method was based on the Euclidean traditional proof method.

A. Tarski, on the other hand, gave a decision procedure for what he called elementary geometry, based on algebraic method, [38]. In spite of the subsequent improvements by A. Seidenberg [34], G. Collins [18] and others along this line, methods of Tarski's variants still seemed far away from mechanically proving nontrivial geometry theorems in practice.

It was really a surprise when Chinese mathematician Wu Wen-Tsün introduced an algebraic method in 1977 which can be used to prove many geometry theorems whose traditional proofs need enormous amounts of human intelligence.

1. The Defects in Traditional Proofs

In order to carry out proofs of geometry theorems mechanically, we have to follow some rules and axioms strictly. In this preliminary section, we shall point out difficulties in mechanically carrying out traditional proofs in geometry.

1.1. The Traditional Euclidean Proof

One of the main defects in the traditional Euclidean proof is its almost complete disregard of such notions as the *two sides of a line* and the *interior of an angle*. Without clarification of these ideas, absurd consequences result. (See [2].)

Examples (1.1). Let $ABCD$ be a parallelogram (i.e, $AB \parallel CD$, $BC \parallel AD$), E be the intersection of the diagonals AC and BD. Show $AE \equiv CE$ (Figure 1).

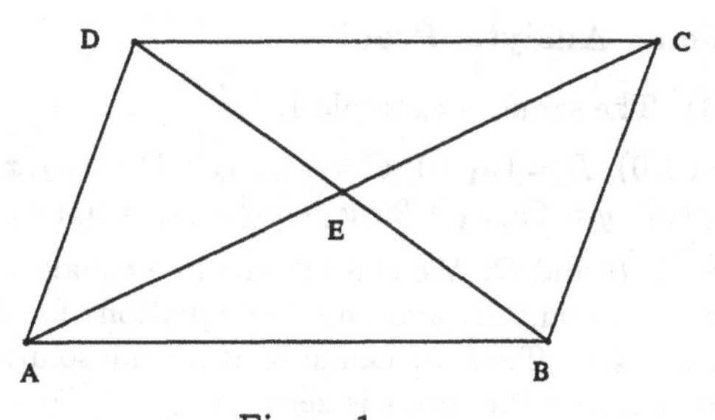

Figure 1

The traditional proof of this theorem is first to prove $\Delta ACB \cong \Delta CAD$ (hence $AB \equiv CD$), then to prove $\Delta AEB \cong \Delta CED$ (hence $AE \equiv CE$). In proving the congruence of these triangles, we have repeatedly used the fact $\angle CAB \equiv \angle ACD$. This fact is quite evident because the two angles are the alternative angles with respect to parallels AB and CD. However, here we have implicitly assumed the "trivial fact" that point D and B are on either sides of line AC. The last fact is harder to prove than the original statement. (Please try it!)

This extremely simple example reveals difficulty in implementing a powerful and sound geometry theorem prover based on traditional proofs. Of course, one can develop an interactive prover so that the user can input some trivial facts such as the one in the previous paragraph. These facts can be stored in a data base by programs. Then one would face a much more severe problem of the consistency of proofs.

Example (1.2). Every triangle is isosceles. Let ABC be a triangle as shown in Figure 2. We want to prove $CA \equiv CB$.

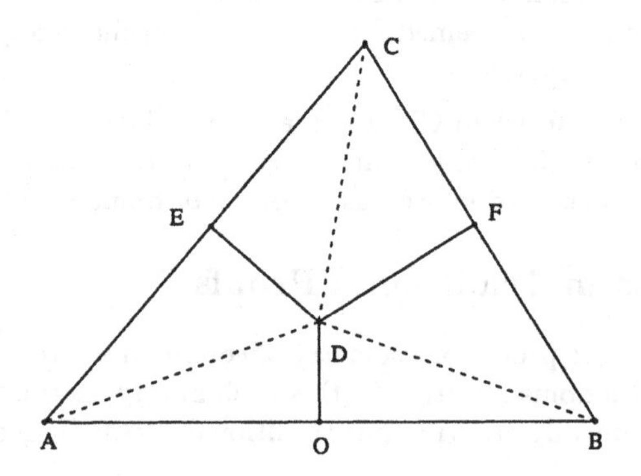

Figure 2

Proof. Let D be the intersection of the perpendicular bisector of AB and the internal bisector of angle ACB. Let $DE \perp AC$ and $DF \perp CB$. It is easy to see that $\Delta CDE \cong \Delta CDF$ and $\Delta ADE \cong \Delta BDF$. Hence $CE + EA = CF + FB$, i.e., $CA \equiv CB$. .QED.

We leave the reader to solve this paradox.

1.2. The Traditional Analytic Proof

Examples (1.3). The same as example 1.

We can let $A = (0,0)$, $B = (u_1, 0)$, $C = (u_2, u_3)$, $D = (x_2, x_1)$, and $E = (x_4, x_3)$. Then we want to prove $g = 2u_2x_4 + 2u_3x_3 - u_3^2 - u_2^2 = 0$, i.e., $AE \equiv CE$.

Let us fix points A, B and C. We can use the line equations for AD and CD to solve the coordinates of point D, and the line equations for BD and AC to solve the coordinates of point E. Then we can substitute the solutions in the conclusion polynomial g to see whether the result is zero.

Thus we have two line equations for x_1 and x_2:

$h_1 = u_1x_1 - u_1u_3 = 0$ — AB is parallel to DC

$h_2 = u_3x_2 - (u_2 - u_1)x_1 = 0$ — DA is parallel to CB.

Also we have two line equations for x_3 and x_4

$h_3 = x_1x_4 - (x_2 - u_1)x_3 - u_1x_1 = 0$ — E is on BD

$h_4 = u_3x_4 - u_2x_3 = 0$ — E is on AC.

Solving the first two equations, we have $x_1 = u_3$, $x_2 = u_2 - u_1$. Solving the last two equations and using the solutions for x_1 and x_2, we have $x_3 = u_3/2$, $x_4 = u_2/2$. Now substituting the solutions in g, we have $g = 0$. Thus, we prove the theorem.

The above approach is in essence the same used by P.J. Davis to prove Pappus' theorem on the computer [19]. Actually, a general mechanical proof method for affine geometry was already given by D. Hilbert in his classic book [24] (also see Chapter 3).

One defect in the above solution is that it does not specify the degenerate conditions under which the conclusion of the theorem might be invalid, or does not emphasize explicitly the fact that we only prove geometry statements to be generically or generally true (for the definition see Chapter 3). For example, in order to have the above *normal* solutions for $x_1,...,x_4$, we at least have to assume that $u_1 \neq 0$ and $u_3 \neq 0$, i.e., the *nondegenerate condition* that A, B and C are not collinear.

It was Wu Wen-Tsün who first realized the importance of nondegenerate conditions in mechanical geometry theorem proving and introduced a method which can deal with nondegenerate conditions mechanically, [42], [43].

2. Four Examples

Wu's method was introduced as a mechanical method for proving those statements in elementary geometries for which, in their algebraic forms,[1] the hypotheses and the conclusion can be expressed by *polynomial equations.*[2]

For such a geometry statement, after adopting an appropriate coordinate system, the hypotheses can be expressed as a set of polynomial equations:

$$\begin{aligned} &h_1(u_1,...,u_d,x_1,...,x_t) = 0\\ &h_2(u_1,...,u_d,x_1,...,x_t) = 0\\ (2.0)\qquad &\qquad\cdots\\ &h_n(u_1,...,u_d,x_1,...,x_t) = 0, \end{aligned}$$

and the conclusion is also a polynomial equation $g = g(u_1,...,u_d,x_1,...,x_t) = 0$, where $h_1,...,h_n$ and g are polynomials in $\mathbf{Q}[u_1,...,u_d,x_1,...,x_t]$. ($\mathbf{Q}$ denotes the field of rational numbers.) Variables $u_1,...,u_d$ are parameters or independent variables and variables $x_1,...,x_t$ are algebraically dependent on u's under *normal conditions*, being restricted by (2.0). The selection of the parameters u's comes from the assumption that the figure in the geometry statement should be in a general position, an assumption that is often implicit in geometry statements. The use of two different kinds of variables goes back to the Hilbert mechanization theorem for affine geometry (see theorem 62 in [24] or theorem 1.3 in Chapter 3).

Now let us use four examples to illustrate how Wu's method works. The proofs of the four examples were actually produced by the prover, including the selection

[1] For the relationship between axiomatic geometries and number systems, the reader is strongly urged to read [24]. For a brief introduction, see Chapter 3.

[2] Dealing with inequality is generally beyond the method.

of coordinates.

Example (2.1). The same as example 1.3 in Section 1.

The implicit assumption that the parallelogram is in a general position means that any three points among the four points A, B, C, and D can be arbitrarily chosen. Thus u_1, u_2 and u_3 can be chosen as parameters. We have four equations $h_1 = 0, \ldots, h_4 = 0$ for the hypotheses and one equation $g = 0$ for the conclusion (see Section 1.2).

Now one might think that our algebra formulation of the problem would be

$$(2.1.1) \qquad (h_1 = 0 \wedge h_2 = 0 \wedge h_3 = 0 \wedge h_4 = 0) \Rightarrow g = 0.$$

However, (2.1.1) is not a valid statement. The reason is that the important nondegenerate conditions were overlooked. Wu's method can produce a sufficient number of nondegenerate conditions to make (2.1.1) valid.

The next step is to triangulate h_1, h_2, h_3, h_4 so that each polynomial introduces only one new (dependent) variable x_i. Thus h_1, h_2, h_3, h_4 are not in triangular form because h_3 introduces two new variables, x_3 and x_4, at the same time. We can use a simple elimination procedure to obtain a triangular form: let $f_1 = h_1$, $f_2 = h_2$, $f_3 = prem(h_4, h_3, x_4)$,[3] $f_4 = h_4$. Then we have a triangular form:

$$\begin{aligned}
f_1 &= u_1x_1 - u_1u_3 = 0 \\
f_2 &= u_3x_2 - (u_2 - u_1)x_1 = 0 \\
f_3 &= (u_3x_2 - u_2x_1 - u_1u_3)x_3 + u_1u_3x_1 = 0 \\
f_4 &= u_3x_4 - u_2x_3 = 0
\end{aligned}$$

Now we do successive pseudo divisions:

$R_3 = prem(g, f_4, x_4) = (2u_3^2 + 2u_2^2)x_3 - u_3^3 - u_2^2u_3$

$R_2 = prem(R_3, f_3, x_3) = (-u_3^4 - u_2^2u_3^2)x_2 + ((u_2 - 2u_1)u_3^3 + (u_2^3 - 2u_1u_2^2)u_3)x_1 + u_1u_3^4 + u_1u_2^2u_3^2$

$R_1 = prem(R_2, f_2, x_2) = (-u_1u_3^4 - u_1u_2^2u_3^2)x_1 + u_1u_3^5 + u_1u_2^2u_3^3$

$R_0 = prem(R_1, f_1, x_1) = 0.$

Because the *final remainder* R_0 is zero, the theorem follows from appropriate subsidiary conditions. To see this, let us recall the simple and important *remainder formula* (see Section 4) for successive pseudo divisions of g with respect to a triangular form $f_1, \ldots, f_r$:

$$I_1^{s_1} \cdots I_r^{s_r} g = Q_1 f_1 + \ldots + Q_r f_r + R_0$$

where I_k are the leading coefficients of f_k in x_k.

Since $R_0 = 0$, $g = 0$ under $I_k \neq 0$ $(k = 1, \ldots, r)$. The subsidiary conditions $I_k \neq 0$ are usually connected with nondegeneracy and are also called nondegenerate conditions. For this particular case, (2.1.1) is valid under

$$I_1 = u_1 \neq 0$$

[3] $prem(h_4, h_3, x_4)$ denotes the pseudo remainder obtained by dividing h_4 by h_3 in variable x_4. For the basic algebraic operations underlying Wu's method, see Section 4.

$I_2 = u_3 \neq 0$
$I_3 = u_3x_2 - u_2x_1 - u_1u_3 \neq 0$
$I_4 = u_3 \neq 0.$

The conditions $u_1 \neq 0$ and $u_3 \neq 0$ mean that A, B, and C are not collinear. The condition $u_3x_2 - u_2x_1 - u_1u_3 \neq 0$ means that line AC and line BD should have a normal intersection.

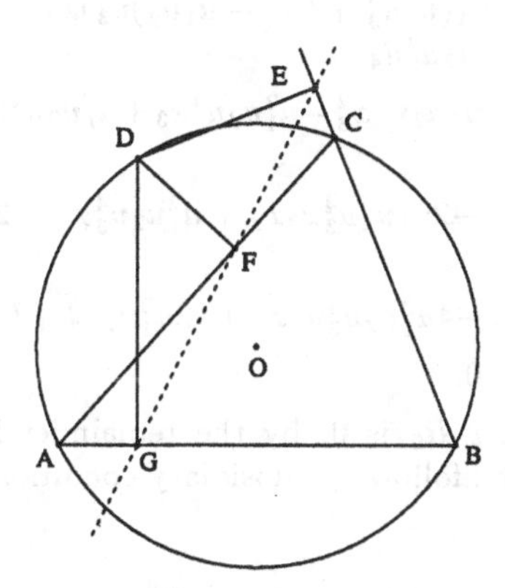

Figure 3

Example (2.2). (Simson's theorem) Let D be a point on the circumscribed circle (O) of triangle ABC. From D three perpendiculars are drawn to the three sides BC, CA and AB of $\triangle ABC$. Let E, F and G be the three feet respectively. Show that E, F and G are collinear (Figure 3).

Let $A = (0,0)$, $B = (u_1, 0)$, $C = (u_2, u_3)$, $O = (x_2, x_1)$, $D = (x_3, u_4)$, $E = (x_5, x_4)$, $F = (x_7, x_6)$, and $G = (x_3, 0)$. Then the hypothesis equations are:

$h_1 = 2u_2x_2 + 2u_3x_1 - u_3^2 - u_2^2 = 0$ — $OA \equiv OC$
$h_2 = 2u_1x_2 - u_1^2 = 0$ — $OA \equiv OB$
$h_3 = -x_3^2 + 2x_2x_3 + 2u_4x_1 - u_4^2 = 0$ — $OA \equiv OD$
$h_4 = u_3x_5 + (-u_2 + u_1)x_4 - u_1u_3 = 0$ — Points E, B and C are collinear
$h_5 = (u_2 - u_1)x_5 + u_3x_4 + (-u_2 + u_1)x_3 - u_3u_4 = 0$ — $DE \perp BC$
$h_6 = u_3x_7 - u_2x_6 = 0$ — Points F, A and C are collinear
$h_7 = u_2x_7 + u_3x_6 - u_2x_3 - u_3u_4 = 0$ — $DF \perp AC$.

The conclusion that points E, F and G are collinear is equivalent to $g = x_4x_7 + (-x_5 + x_3)x_6 - x_3x_4 = 0$. Now we can triangulate $h_1, ..., h_7$ by letting $f_1 = prem(h_1, h_2, x_2)$, $f_2 = h_2$, $f_3 = h_3$, $f_4 = prem(h_4, h_5, x_5)$, $f_5 = h_4$, $f_6 = prem(h_6, h_7, x_7)$, $f_7 = h_7$:

$f_1 = 4u_1u_3x_1 - 2u_1u_3^2 - 2u_1u_2^2 + 2u_1^2u_2 = 0$
$f_2 = 2u_1x_2 - u_1^2 = 0$
$f_3 = -x_3^2 + 2x_2x_3 + 2u_4x_1 - u_4^2 = 0$
$f_4 = (-u_3^2 - u_2^2 + 2u_1u_2 - u_1^2)x_4 + (u_2 - u_1)u_3x_3 + u_3^2u_4 + (-u_1u_2 + u_1^2)u_3 = 0$
$f_5 = u_3x_5 + (-u_2 + u_1)x_4 - u_1u_3 = 0$
$f_6 = (-u_3^2 - u_2^2)x_6 + u_2u_3x_3 + u_3^2u_4 = 0$
$f_7 = u_2x_7 + u_3x_6 - u_2x_3 - u_3u_4 = 0.$

Here we see that $deg(f_3, x_3) = 2$. Thus successive pseudo division is more convenient:

$R_6 = prem(g, f_7, x_7) = (-u_2x_5 - u_3x_4 + u_2x_3)x_6 + u_3u_4x_4$

$R_5 = prem(R_6, f_6, x_6) = (u_2^2u_3x_3 + u_2u_3^2u_4)x_5 + (u_2u_3^2x_3 - u_2^2u_3u_4)x_4 - u_2^2u_3x_3^2 - u_2u_3^2u_4x_3$

$R_4 = prem(R_5, f_5, x_5) = ((u_2u_3^3 + (u_2^3 - u_1u_2^2)u_3)x_3 - u_1u_2u_3^2u_4)x_4 - u_2^2u_3^2x_3^2 + (-u_2u_3^3u_4 + u_1u_2^2u_3^2)x_3 + u_1u_2u_3^3u_4$

$R_3 = prem(R_4, f_4, x_4) = u_1u_2u_3^4x_3^2 - u_1^2u_2u_3^4x_3 + u_1u_2u_3^4u_4^2 + (-u_1u_2u_3^5 + (-u_1u_2^3 + u_1^2u_2^2)u_3^3)u_4$

$R_2 = prem(R_3, f_3, x_3) = -2u_1u_2u_3^4x_3x_2 + u_1^2u_2u_3^4x_3 - 2u_1u_2u_3^4u_4x_1 + (u_1u_2u_3^5 + (u_1u_2^3 - u_1^2u_2^2)u_3^3)u_4$

$R_1 = prem(R_2, f_2, x_2) = -4u_1^2u_2u_3^4u_4x_1 + (2u_1^2u_2u_3^5 + (2u_1^2u_2^3 - 2u_1^3u_2^2)u_3^3)u_4$

$R_0 = prem(R_1, f_1, x_1) = 0.$

Since the final remainder R_0 is 0, by the remainder formula, we have proved Simson's theorem under the following subsidiary conditions:

$$I_1 = 4u_1u_3 \neq 0$$
$$I_2 = 2u_1 \neq 0$$
$$I_4 = -u_3^2 - u_2^2 + 2u_1u_2 - u_1^2 \neq 0$$
$$I_5 = u_3 \neq 0$$
$$I_6 = -u_3^2 - u_2^2 \neq 0$$
$$I_7 = u_2 \neq 0.$$

We will analyze these conditions in Section 6. They are usually connected with nondegeneracy. Now let us look at a less trivial theorem.

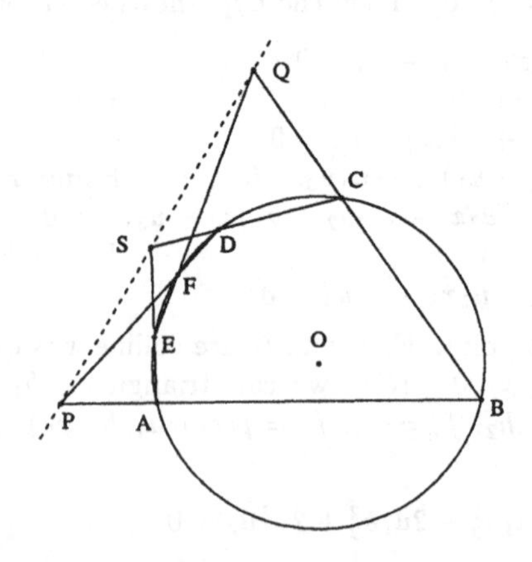

Figure 4

Example (2.3). (Pascal's Theorem). Let A, B, C, D, F and E be six points on a circle (O). Let $P = AB \cap DF$, $Q = BC \cap FE$ and $S = CD \cap EA$. Show that P, Q and S are collinear (Figure 4).

We can let $A = (0,0)$, $O = (u_1, 0)$, $B = (x_1, u_2)$, $C = (x_2, u_3)$, $D = (x_3, u_4)$, $F = (x_4, u_5)$, $E = (x_5, u_6)$, $P = (x_7, x_6)$, $Q = (x_9, x_8)$, and $S = (x_{11}, x_{10})$. Then we have the following equations for the hypotheses.

$h_1 = x_1^2 - 2u_1x_1 + u_2^2 = 0$ $OA \equiv OB$
$h_2 = x_2^2 - 2u_1x_2 + u_3^2 = 0$ $OA \equiv OC$
$h_3 = x_3^2 - 2u_1x_3 + u_4^2 = 0$ $OA \equiv OD$
$h_4 = x_4^2 - 2u_1x_4 + u_5^2 = 0$ $OA \equiv OF$
$h_5 = x_5^2 - 2u_1x_5 + u_6^2 = 0$ $OA \equiv OE$
$h_6 = (u_5 - u_4)x_7 + (-x_4 + x_3)x_6 + u_4x_4 - u_5x_3 = 0$ P is on DF
$h_7 = u_2x_7 - x_1x_6 = 0$ P is on AB
$h_8 = (u_6 - u_5)x_9 + (-x_5 + x_4)x_8 + u_5x_5 - u_6x_4 = 0$ Q is on FE
$h_9 = (u_3 - u_2)x_9 + (-x_2 + x_1)x_8 + u_2x_2 - u_3x_1 = 0$ Q is on BC
$h_{10} = u_6x_{11} - x_5x_{10} = 0$ S is on AE
$h_{11} = (u_4 - u_3)x_{11} + (-x_3 + x_2)x_{10} + u_3x_3 - u_4x_2 = 0$ S is on CD
$g = (x_8 - x_6)x_{11} + (-x_9 + x_7)x_{10} + x_6x_9 - x_7x_8 = 0$ Conclusion: P, Q and S are collinear.

We can let $f_1 = h_1$, $f_2 = h_2$, $f_3 = h_3$, $f_4 = h_4$, $f_5 = h_5$, $f_6 = prem(h_7, h_6, x_7)$, $f_7 = h_7$, $f_8 = prem(h_9, h_8, x_9)$, $f_9 = h_9$, $f_{10} = prem(h_{11}, h_{10}, x_{11})$, $f_{11} = h_{11}$. Then we have a triangular form $f_1, ..., f_{11}$ and we can do successive pseudo divisions $R_{10} = prem(g, f_{11}, x_{11})$, $R_9 = prem(R_{10}, f_{10}, x_{10}), ..., R_0 = prem(R_1, f_1, x_1)$. The computation showed that $R_0 = 0$. Thus we proved Pascal's theorem under the following subsidiary conditions:

$I_6 = u_2x_4 - u_2x_3 + (-u_5 + u_4)x_1 \neq 0$
$I_7 = u_2 \neq 0$
$I_8 = (u_3 - u_2)x_5 + (-u_3 + u_2)x_4 + (-u_6 + u_5)x_2 + (u_6 - u_5)x_1 \neq 0$
$I_9 = u_3 - u_2 \neq 0$
$I_{10} = (u_4 - u_3)x_5 - u_6x_3 + u_6x_2 \neq 0$
$I_{11} = u_6 \neq 0.$

We will analyze these conditions in Section 7. They are usually connected with nondegeneracy. Note that R_5 is a polynomial with 272 terms. Thus by hand calculation, the proof is almost impossible. However, the proof was done on a Symbolics 3600 in less than 2.5 seconds, including the selection of coordinates and the conversion of geometry conditions into polynomial equations.

Example (2.4) (The Butterfly Theorem). A, B, C and D are four points on circle (O). E is the intersection of AC and BD. Through E draw a line perpendicular to OE, meeting AD at F and BC at G. Show that $FE \equiv GE$ (Figure 5).

Let $E = (0,0)$, $O = (u_1, 0)$, $A = (u_2, u_3)$, $B = (x_1, u_4)$, $C = (x_3, x_2)$, $D = (x_5, x_4)$, $F = (0, x_6)$, and $G = (0, x_7)$. Then the hypothesis equations are:

$h_1 = -x_1^2 + 2u_1x_1 - u_4^2 + u_3^2 + u_2^2 - 2u_1u_2 = 0$ $OA \equiv OB$
$h_2 = -x_3^2 + 2u_1x_3 - x_2^2 + u_3^2 + u_2^2 - 2u_1u_2 = 0$ $OA \equiv OC$
$h_3 = -u_3x_3 + u_2x_2 = 0$ Points C, A and E are collinear
$h_4 = -x_5^2 + 2u_1x_5 - x_4^2 + u_3^2 + u_2^2 - 2u_1u_2 = 0$ $OA \equiv OD$
$h_5 = -u_4x_5 + x_1x_4 = 0$ Points D, B and E are collinear

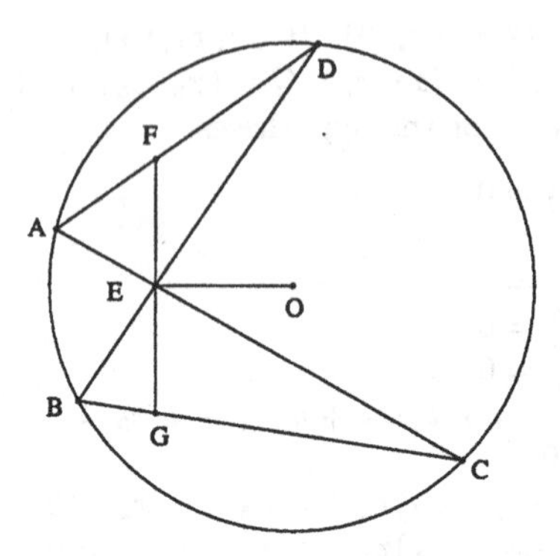

Figure 5

$h_6 = (-x_5 + u_2)x_6 + u_3x_5 - u_2x_4 = 0$ Points F, A and D are collinear
$h_7 = (-x_3 + x_1)x_7 + u_4x_3 - x_1x_2 = 0$ Points G, B and C are collinear.

The conclusion is $g = x_7 + x_6 = 0$.

We can let $f_1 = h_1$, $f_2 = prem(h_2, h_3, x_3)$, $f_3 = h_3$, $f_4 = prem(h_4, h_5, x_5)$, $h_5 = f_5$, $f_6 = h_6$, $f_7 = h_7$ to obtain a triangular form:

$$f_1 = -x_1^2 + 2u_1x_1 - u_4^2 + u_3^2 + u_2^2 - 2u_1u_2 = 0$$
$$f_2 = (-u_3^2 - u_2^2)x_2^2 + 2u_1u_2u_3x_2 + u_3^4 + (u_2^2 - 2u_1u_2)u_3^2 = 0$$
$$f_3 = -u_3x_3 + u_2x_2 = 0$$
$$f_4 = (-x_1^2 - u_4^2)x_4^2 + 2u_1u_4x_1x_4 + (u_3^2 + u_2^2 - 2u_1u_2)u_4^2 = 0$$
$$f_5 = -u_4x_5 + x_1x_4 = 0$$
$$f_6 = (-x_5 + u_2)x_6 + u_3x_5 - u_2x_4 = 0$$
$$f_7 = (-x_3 + x_1)x_7 + u_4x_3 - x_1x_2 = 0.$$

If we do successive pseudo divisions of g with respect to the above triangular form, then the final remainder R_0 is not zero. The reason is that the above triangular form is *reducible* (see Section 3 of Chapter 2). This reducibility comes from a special kind of degeneracy or *ambiguity* when using algebraic equations to encode certain geometric conditions. Equations $h_2 = 0$ and $h_3 = 0$ specify point C: C is on circle (O) and on line AE. However, there are two points satisfying these equations: one is C, which we really want; the other is point A. Because A has been constructed before C, reducibility arises. We might use the complete method of Wu to deal with such reducibility [10]. However, for such special reducibility we can use an elementary method, which is more satisfactory in geometry (see the last footnote of [10], or [11]).

Because $f_2 = 0$ and $f_3 = 0$ have two solutions (one is for $A = (u_2, u_3)$ and the other is for $C = (x_3, x_2)$), $x_2 - u_3$ is a factor of f_2 under the previous geometric conditions: $f_2 = (x_2 - u_3)((-u_3^2 - u_2^2)x_2 - u_3^3 + (-u_2^2 + 2u_1u_2)u_3)$. Thus we can use the division to obtain $f_2' = (-u_3^2 - u_2^2)x_2 - u_3^3 + (-u_2^2 + 2u_1u_2)u_3$ and replace f_2 by f_2'. In the same way we have:

$$f_4 = (x_4 - u_4)f_4' + r$$

where $f_4' = (-x_1^2 - u_4^2)x_4 - u_4x_1^2 + 2u_1u_4x_1 - u_4^3$, and $r = -u_4^2x_1^2 + 2u_1u_4^2x_1 - u_4^4 +$

$(u_3^2 + u_2^2 - 2u_1u_2)u_4^2$.

We have $prem(r, f_1, x_1) = 0$. Hence under the previous conditions $f_i = 0$ and $I_i \neq 0$ $(i = 1, 2, 3)$: $f_4 = (x_4 - u_4)f_4'$. We can replace f_4 by f_4' to obtain the nondegenerate triangular form $f_1, f_2', f_3, f_4', f_5, f_6, f_7$. Now we can do the successive divisions of g with respect to the new triangular form:

$R_6 = prem(R_7, f_7, x_7) = (-x_3 + x_1)x_6 - u_4x_3 + x_1x_2$

$R_5 = prem(R_6, f_6, x_6) = ((u_4 + u_3)x_3 - x_1x_2 - u_3x_1)x_5 + (-u_2x_3 + u_2x_1)x_4 - u_2u_4x_3 + u_2x_1x_2$

$R_4 = prem(R_5, f_5, x_5) = (((-u_4 - u_3)x_1 + u_2u_4)x_3 + x_1^2x_2 + u_3x_1^2 - u_2u_4x_1)x_4 + u_2u_4^2x_3 - u_2u_4x_1x_2$

$R_3 = prem(R_4, f_4', x_4) = ((-u_4^2 - u_3u_4)x_1^3 + (2u_1u_4^2 + 2u_1u_3u_4)x_1^2 + (-u_4^4 - u_3u_4^3 - 2u_1u_2u_4^2)x_1)x_3 + (u_4x_1^4 + ((u_2 - 2u_1)u_4)x_1^3 + u_4^3x_1^2 + u_2u_4^3x_1)x_2 + u_3u_4x_1^4 + (-u_2u_4^2 - 2u_1u_3u_4)x_1^3 + (u_3u_4^3 + 2u_1u_2u_4^2)x_1^2 - u_2u_4^4x_1$

$R_2 = prem(R_3, f_3, x_3) = (-u_3u_4x_1^4 + (u_2u_4^2 + 2u_1u_3u_4)x_1^3 + (-u_3u_4^3 - 2u_1u_2u_4^2 - 2u_1u_2u_3u_4)x_1^2 + (u_2u_4^4 + 2u_1u_2^2u_4^2)x_1)x_2 - u_3^2u_4x_1^4 + (u_2u_3u_4^2 + 2u_1u_3^2u_4)x_1^3 + (-u_3^2u_4^3 - 2u_1u_2u_3u_4^2)x_1^2 + u_2u_3u_4^4x_1$

$R_1 = prem(R_2, f_2', x_2) = 2u_1u_2u_3^2u_4x_1^4 - (2u_1u_2^2u_3u_4^2 + 4u_1^2u_2u_3^2u_4)x_1^3 + (2u_1u_2u_3^2u_4^3 + 4u_1^2u_2^2u_3u_4^2 + (-2u_1u_2u_3^4 + (-2u_1u_2^3 + 4u_1^2u_2^2)u_3^2)u_4)x_1^2 + (-2u_1u_2^2u_3u_4^4 + (2u_1u_2^2u_3^3 + (2u_1u_2^4 - 4u_1^2u_2^3)u_3)u_4^2)x_1$

$R_0 = prem(R_1, f_1, x_1) = 0.$

Since the final remainder R_0 is 0, the theorem follows from the following subsidiary conditions:

$I_2 = u_3^2 + u_2^2 \neq 0$
$I_3 = u_3 \neq 0$
$I_4 = x_1^2 + u_4^2 \neq 0$
$I_5 = u_4 \neq 0$
$I_6 = x_5 - u_2 \neq 0$
$I_7 = x_3 - x_1 \neq 0.$

In addition, the following nondegenerate conditions, which come from reducibility and have been detected by our prover, should be also added: $D \neq B$ (i.e., $x_2 - u_3 \neq 0$); $C \neq A$ (i.e., $x_4 - u_4 \neq 0$).

So far, we have been talking about the elementary use of Wu's method. A number of questions arise.

(1) After successive pseudo divisions, *if the final remainder R_0 is not zero, what conclusion can we infer?*

(2) *In what theory does the method prove theorems?*

These are quite complicated problems. We will discuss them in Chapter 3 and Chapter 4.

(3) The algorithm adds some nondegenerate conditions. *Will the theorem to be proved be weakened by introducing these additional conditions? Especially, will the hypotheses become inconsistent by adding these new conditions?*

The answer to this question can be found in Section 3.2 of Chapter 3.

(4) *Are all those subsidiary conditions necessary? Can those inequations which are necessary for mechanical proofs be translated back into their geometric forms by the program?*

There are no general methods to do so. However, for a large class of geometry statement we have developed a method to solve this problem. See [11] or Section 6 and 7. One of the main aims of [11] is to give a practical solution to the above questions for the quadratic cases in practice.

3. A Summary of Wu's Method

From the above four examples and discussion, we can see that Wu's method consists of the following steps:

Step 1. *Conversion of a geometry statement into the corresponding polynomial equations.*

Step 2. *Triangulation of the hypothesis polynomials using pseudo division.* In the complete method, decomposition or check of the irreducibility is needed in this step (see Chapter 4).

Step 3. *Successive pseudo division to compute the final remainder R_0.* If $R_0 = 0$, then by the remainder formula we can infer the conclusion by adding the subsidiary or nondegenerate conditions available after triangulation. If the final remainder is not zero, then the decomposition or check of irreducibility is needed to make further conclusions. Wu's method is complete for geometric statements involving equality only in metric geometry (Chapter 3), not in Euclidean geometry. However, under certain natural conditions, it is also a decision procedure for Euclidean geometry (see [47], [11], [28], [30] and Chapter 4).

For linear cases (i.e., all $deg(f_i, x_i) = 1$), however, it is complete in *all* geometries. In particular it is also a decision procedure in Euclidean geometry for linear cases. Among 366 geometry problems in the Collection [14], 219 have been expressed as linear cases, including many statements involving the tritangent centers (the incenter and the three excenters) of a triangle. The remaining are quadratic.

Step 4. *Analysis of nondegenerate conditions $I_1 \neq 0, \ldots, I_r \neq 0$.* Some of the conditions $I_i = 0$ can be considered degenerate cases and can be translated back into their geometric forms by the program for at least a large class of geometry statements [11]. The other conditions $I_i \neq 0$ are not necessarily connected with nondegeneracy and have been proved unnecessary for $g = 0$ to be valid if the triangular form $f_1, \ldots, f_r$ is irreducible (see [11], and Sections 6 and 7).

4. Pseudo Division and Successive Pseudo Division

As we see, the basic algebraic operation in the method is pseudo division. Now let us describe this operation.

Let A be a computable commutative ring (e.g., $\mathbf{Q}[y_1, \ldots, y_m]$). Let $f = a_n v^n + \cdots + a_0$, and $h = b_k v^k + \cdots + b_0$ be two polynomials in $A[v]$, where v is a new indeterminate. Suppose k, the leading degree of h in v, is greater than 0. Then the pseudo division proceeds as follows:

Pseudo Division. First let $r = f$. Then repeat the following process until $m = deg(r, v) < k$: $r := b_k r - c_m v^{m-k} h$, where c_m is the leading coefficient of r. It is easy to see that m strictly decreases after each iteration. Thus the process terminates. At the end, we have the *pseudo remainder* $prem(f, h, v) = r = r_0$ and the following formula,

$$b_k^s f = qh + r_0, \quad \text{where } s \leq n - k + 1 \text{ and } deg(r_0, v) < deg(h, v). \tag{4.1}$$

Proof. We fix polynomial h and use induction on $n = deg(f, v)$. If $n < k$, then we have $r_0 = f$ and $f = 0 \cdot h + r_0$. Suppose $n \geq k$ and formula (4.1) is true for those polynomials f, for which $deg(f, v) < n$. After the first iteration, we have $r = b_k f - a_n v^{n-k} h$, where a_n is the leading coefficient of f. Since $deg(r, v) < n$, we have $b_k^t r = q_1 h + r_0$ by the induction hypothesis. Substituting $r = b_k f - a_n v^{n-k} h$ in the last formula, we have (4.1). .QED.

Remark. The most important quantity to measure the complexity of a proof is the maximal number of terms in the polynomials r produced in the above iteration during successive pseudo division. We use *maxt* to denote this number for a given proof. The computer space and time required for a given proof is basically determined by this number, which can be in the thousands for some proofs. In our four examples in section 2, *maxt* = 6 for example (2.1), *maxt* = 8 for example (2.2), *maxt* = 272 for example (2.3), and *maxt* = 18 for example (2.4). Thus, these examples are quite "easy" for the prover.

Suppose we have a triangular form

$$\begin{array}{l} f_1(u_1, \ldots, u_d, x_1) \\ f_2(u_1, \ldots, u_d, x_1, x_2) \\ \cdots \\ f_r(u_1, \ldots, u_d, x_1, \ldots, x_r) \end{array} \tag{4.2}$$

and a polynomial $g = g(u_1, \ldots, u_d, x_1, \ldots, x_r)$. Then we can do successive pseudo divisions as illustrated by the four examples in Section 2: $R_{r-1} = prem(g, f_r, x_r)$, $\ldots$, $R_0 = prem(R_1, f_1, x_1)$. R_0 is called the *final remainder* and is denoted by $prem(g, f_1, \ldots, f_r)$. It is easy to prove the following important proposition:

Proposition (4.3). (The Remainder Formula) Let $f_1, \ldots, f_r$ and R_0 be the same as the above. There are some non-negative integers $s_1, \ldots, s_r$ and polynomials $Q_1, \ldots, Q_r$ such that

(1) $I_1^{s_1} \cdots I_r^{s_r} g = Q_1 f_1 + \ldots + Q_r f_r + R_0$, where the I_i are the leading coefficients (or *initials*) of the f_i.

(2) $deg(R_0, x_i) < deg(f_i, x_i)$, for $i = 1, \ldots, r$.

Proof. We use induction on r. If $r = 1$, then the remainder formula is actually (4.1). Suppose that $r > 1$ and the proposition is true for $r - 1$. Thus we have:

$$I_1^{s_1} \cdots I_{r-1}^{s_{r-1}} R_{r-1} = Q_1 f_1 + \ldots + Q_{r-1} f_{r-1} + R_0,$$

with $deg(R_0, x_i) < deg(f_i, x_i)$, for $i = 1, \ldots, r - 1$. Combining this with $R_{r-1} =$

$I^{s_r} g - Af_r$, we have (1) and (2). .QED.

5. A Simple Triangulation Procedure

There are several ways to triangulate a general polynomial set such as (2.0). From four examples in Section 2, the reader can invent his own triangulation procedures. The following naive procedure was described in [10]. It is very similar to the Gauss elimination procedure and was used subsequently by other authors. For a *complete* triangulation algorithm, proved in detail in [43], [45] and referred to as *Ritt's Principle* by Wu, see Section 4 of Chapter 2.

Suppose we have a set of hypothesis polynomials:

$$
\begin{aligned}
&h_1(u_1, \ldots, u_d, x_1, \ldots, x_r) &= 0\\
&h_2(u_1, \ldots, u_d, x_1, \ldots, x_r) &= 0\\
&\ldots\\
&h_r(u_1, \ldots, u_d, x_1, \ldots, x_r) &= 0.
\end{aligned}
\tag{5.1}
$$

In order to transform it into a triangular form, we first eliminate x_r from $r-1$ polynomials to obtain

$$
\begin{aligned}
&f'_1(u_1, \ldots, u_d, x_1, \ldots, x_{r-1}) = 0\\
&\ldots\\
&f'_{r-1}(u_1, \ldots, u_d, x_1, \ldots, x_{r-1}) = 0\\
&f'_r(u_1, \ldots, u_d, x_1, \ldots, x_{r-1}, x_r) = 0.
\end{aligned}
\tag{5.2}
$$

We then apply the same procedure to x_{r-1} to eliminate it from $r-2$ polynomials, and so on. The program eliminates x_r in the following way.

Case 1. If no x_r appears in any of the f_i, then the program stops and gives the user a warning: something might be wrong in the choice of parameters and dependent variables.

Case 2. If x_r appears in only one f_i, then take this f_i as f_r. We have (5.2).

Case 3. If one of the f_i, say f_r, has degree 1 in x_r, then let $f'_i = prem(f_i, f_r, x_r)$, for $i = 1, ..., r-1$. The f'_i are free of x_r. We have (5.2).

Case 4. Take the two f's which have the minimal positive degrees in x_r, say f_r and f_{r-1}. Suppose that $deg(f_{r-1}, x_r) \le deg(f_r, x_r)$. Let $r_1 = prem(f_r, f_{r-1}, x_r)$, then $deg(r_1, x_r) < deg(f_{r-1}, x_r)$. Usually $deg(r_1, x_r) = deg(f_{r-1}, x_r) - 1$. If $deg(r_1, x_r) > 1$, then let $r_2 = prem(f_{r-1}, r_1, x_r)$, and $r_3 = prem(r_1, r_2, x_r)$, and so on, until for some integer k, we have the following three possibilities:

(1) $deg(r_k, x_r) = 1$. Take r_k and r_{k-1} as new f_r and f_{r-1}, we return to case 3.

(2) $r_k = 0$. This means that f_r and f_{r-1} have a common factor. Since we do not intend to deal with factorization in this simple procedure, we merely stop the program. This situation happened rarely in practice.

(3) $deg(r_{k-1}, x_r) > 1$, but $deg(r_k, x_r) = 0$ (i.e., r_k is free of x_r). Take r_k and r_{k-1} as new f_r and f_{r-1}, then repeat the process beginning from case 2.

In this way, we can transform (5.1) into triangular form (4.2). Since the elementary use of Wu's method is mainly for confirmation of geometry theorems or conjectures, this procedure turned out to be very practical for most geometry theorems [10], [29].

To confirm a statement, we add $I_i \neq 0$, where the I_i are the leading coefficients of the f_i. *Will the theorem to be proved be weakened by introducing these additional conditions? In particular, will the hypotheses become inconsistent by adding these new nondegenerate conditions?* If the following situation happens

$$(5.3) \qquad prem(I_i, f_1, ..., f_{i-1}) = 0,$$

then the hypotheses certainly become inconsistent. Otherwise, with other cautions (check of irreducibility etc), the method proves the statement to be generically or *generally true* (for the definition, see Chapter 3) if the final remainder $R_0 = prem(g, f_1, ..., f_r) = 0$.

If the reader wishes to implement a prover based on the above simple triangulation procedure (it is very efficient) or its variants, (5.3) *should be checked.* For more than four years, the author has experimented with several hundreds of geometry theorems, and for many of those theorems, with several variants. He only found two or three cases of type (5.3) to cause possible inconsistency and found no other inconsistency caused by other reasons. This phenomenon can have the following plausible explanation.

As we mentioned earlier and are emphasizing here, triangulation is essentially to solve dependent variables x's successively. If we assume $I_k \neq 0$, then the x's have normal solutions, i.e., the geometric figure is in a normal position or a nondegenerate position.

This does not mean that the elementary use of the method is absolutely safe or the check of (5.3) is enough. Actually, our present prover [11] takes full precaution at least for quadratic cases (i.e., $deg(f_i, x_i) \leq 2$).

6. Geometry Statements of Constructive Type

The generic or general validity[4] of geometry statements which can be specified at the beginning of Section 2, is inherent to the selection of the parameters u, equations (2.0) and the conclusion $g = 0$, and independent of any particular forms of subsidiary conditions of $I_k \neq 0$. Some $I_k \neq 0$ are not only unnecessary to make $g = 0$ valid, but also are not connected with nondegeneracy. Furthermore, in the case when $I_k \neq 0$ is connected with degeneracy, we do not have a general method to translate it back into its geometric form. However, for a large class of geometry statements, i.e., geometry statements of constructive type defined below, we have a satisfactory solution. Now we define geometry statements of constructive type (for simplicity, we are only talking about plane geometry).

Let Π be a finite set of points. We say line l is constructed *directly* from Π if

l joins two points A and B in Π or
l passes through one point in Π,

[4] For the precise definition, see Chapter 3.

and is parallel to a line joining two other points A and B in Π or
l passes through one point in Π,
and is perpendicular to a line joining two points A and B in Π or
l is perpendicular bisector of AB with A and B in Π.

A line l constructed directly from Π is *well defined* if the two points A and B mentioned in the above definition are distinct.

Likewise, we say a circle is constructed directly from Π if the center of the circle is in Π and its radius r is (AB) where A and B are distinct points in Π.

The lines and circles constructed directly from Π are said to be *in* Π, for brevity.

Definition (6.1). A theorem is of constructive type if the points, lines, and circles in the statement can be constructed in a definite prescribed manner using the following ten constructions, assuming Π to be the set of points already constructed so far:

Construction 1. Taking an arbitrary point.

Construction 2. Drawing an arbitrary line. This can be reduced to taking two arbitrary points.

Construction 3. Drawing an arbitrary circle. This can be also reduced to taking two arbitrary points.

Construction 4. Drawing an arbitrary line through a point in Π. This can be reduced to taking an arbitrary point.

Construction 5. Drawing an arbitrary circle knowing its center in Π. This can be also reduced to taking an arbitrary point.

Construction 6. Taking an arbitrary point on a line in Π.

Construction 7. Taking an arbitrary point on a circle in Π.

Construction 8. Taking the intersection of two lines in Π.

Construction 9. Taking the intersection of a line and a circle in Π.

Construction 10. Taking the intersection of two circles in Π.

The conclusion is a certain (equality) relation among the points thus constructed.[5]

For geometry statements of constructive type, the selection of parameters and dependent variables can be done in a mechanical manner. For example, for construction 1, we can assign two new parameters u_i, u_{i+1} to the coordinates of the new point; for constructions 6 and 7, we can assign a new parameter u_i and a new dependent variable x_k to the two coordinates of the new point; for constructions 8, 9 and 10, we can assign two new dependent variables x_k, x_{k+1} to the two coordinates of the new point.

Remark (6.2). The above assignments of parameters and dependent variables are based on the assumption that the geometric figure can be constructed normally. For example, when we are talking about construction 8, if one of the two lines (say

[5] We emphasize here that we can define wider classes of geometry statements of constructive type than defined here. This definition serves only as an example.

line AB) is degenerate, i.e., $A = B$ is a consequence of the previous constructions, then the geometric figure cannot be constructed normally. This means that $A = B$ is a property of the geometric constructions introduced so far. The reason why this situation is extremely rare in practice is that $A = B$ is a theorem about this geometric configuration, and people (geometers) usually are aware of this theorem before investigating further properties of the configuration.

This does not mean that it is unnecessary for our prover to take extra precautions. Actually, our prover takes these precautions. They involve a check of irreducibility and decomposition. For details, see latter chapters. (Experienced readers might notice that the discussion here is related to the discussion at the end of the last section.)

Also, the geometric meaning of nondegenerate conditions $I_k \neq 0$ can be kept traced because the triangulation procedure now can be under *full control.* Let us look at the following example.

Example (6.3). For Construction 8, we have two line equations for their intersection (x_{k+1}, x_k):

$$l_1 = a_1 x_{k+1} + b_1 x_k + c_1 = 0$$

$$l_2 = a_2 x_{k+1} + b_2 x_k + c_2 = 0.$$

Then we can triangulize them as follows:

$$prem(l_2, l_1, x_{k+1}) = dx_k - d_k = 0$$

$$prem(l_1, l_2, x_k) = dx_{k+1} - d_{k+1} = 0,$$

where $d = a_1b_2 - a_2b_1$, $d_k = a_2c_1 - a_1c_2$, $d_{k+1} = b_1c_2 - b_2c_1$. Now the leading coefficients $I_k = I_{k+1} = d$. From analytic geometry, the condition $d \neq 0$ means that lines l_1 and l_2 should be well defined and have only one common point.

The other constructions can be treated in a similar way. Here we use example (2.2) to illustrate how we can translate all nondegenerate conditions back into their geometric forms.

Example (6.4) (Example (2.2): Simson's theorem revised).

The equality part of the geometric conditions (hypotheses) are:

(6.5)
- $OA = OB$;
- $OA = OC$;
- $OD = OA$;
- $DE \perp BC$;
- B, C and E are collinear;
- $DF \perp AC$;
- A, C and F are collinear;
- $DG \perp AB$;
- A, B and G are collinear.

Note that if $A \neq B$, the condition $OA = OB$ is equivalent to "O is on the perpendicular bisector of AB" *both in geometry and in algebraic equations.* Thus, starting from three arbitrarily chosen points A, B and C, points O, D, E, F, and

G are constructed (in order) as follows: O is on the perpendicular-bisector of A and C and on the perpendicular-bisector of A and B; D is on the circle with the center O and radius OA; E is on the line passing through D and perpendicular to line BC and is on line BC; F is on the line passing through D and perpendicular to line AC and is on line AC; G is on the line passing through D and perpendicular to line AB and is on line AB.

The subsidiary conditions $I_2 \neq 0$, $I_5 \neq 0$ and $I_7 \neq 0$ can be either avoided, using the triangulation in example (6.3), or proved to be unnecessary by a general schema in [11]. The remaining $I_k \neq 0$ can be translated back into their geometric forms:

(1) $I_1 \neq 0$ means the perpendicular bisectors of AB and AC have only one common point. This is equivalent to conditions $A \neq B$, $A \neq C$ and "line AB and AC are not parallel or the same line". Since A is the common point of lines AB and AC, the last condition is equivalent to "A, B and C are not collinear". The inference procedure here is built into the prover.

(2) $I_4 \neq 0$ means that a line passing through D and perpendicular to BC should have only one common point with line BC. This is equivalent to $B \neq C$, and BC is not perpendicular to itself. A line perpendicular to itself is called an *isotropic line.* Thus, $I_4 \neq 0$ is equivalent to "line BC is non-isotropic".

(3) Similarly, $I_6 \neq 0$ is equivalent to "line AC is non-isotropic".

The interesting thing is that here we don't have the corresponding condition for constructing point G. This is implicit in the choice of coordinates $G = (u_4, 0)$. Note, the x-axis should always be non-isotropic. Thus we have a more condition: line AB is non-isotropic.

All of the above analyses have been built into the program in every detail. Thus, in the end, the prover produces the following nondegenerate conditions: Points A, C and B are not collinear; Line BC is non-isotropic; Line AC is non-isotropic; Line AB is non-isotropic.

Note that geometry conditions in nondegenerate conditions do not include degenerate cases. For example, whenever we mention "line AB" we mean it is well defined, i.e., $A \neq B$. However, geometry conditions in (6.5) can include degenerate cases. For example, condition $DF \perp AC$ does not necessarily mean lines DF and AC are well defined. This condition is translated into its algebraic counterpart, which is equivalent to " $D = F$, or $A = C$, or $D \neq F$, $A \neq C$ and line DF is perpendicular to AC." Actually, the theorem is true when $D = F$, as our proof implies.

As we know, isotropic lines do not exist in Euclidean geometry, thus in Euclidean geometry the condition "A, C and B are not collinear" is enough. However, isotropic lines do exist in other models of metric geometry[6] and in Minkowskian geometry (see [13]). Therefore, the "obvious" nondegenerate condition "A, C and B are not collinear" is not enough for those geometries. The full input to our prover is

[6] Euclidean geometry is only one model of the theory of metric geometry introduced by Wu.

```
((cons-sequence A B C O D E F G)
 (eqdistance O A O B)
 (eqdistance O A O C)
 (eqdistance O A O D)
 (collinear G A B)
 (collinear E B C)
 (collinear F A C)
 (perpendicular D E B C)
 (perpendicular D F C A)
 (perpendicular D G A B)
 (collinear E F G)))
```

The first line specifies the order in which new points are introduced (construction sequence). The last line is the conclusion.

The implemented theorem prover also includes constructions, such as taking the midpoint of two points and taking the symmetric reflexion of a line with respect to another line, etc. Actually, we can expand our definition of statements of constructive type to include more constructions. Whenever including a new construction, we have to find the geometric meaning of nondegenerate conditions $I_k \neq 0$ for the new construction if we want to translate them back into their geometric forms and to build in this knowledge in the theorem prover. This has been done in our theorem prover for some constructions other than mentioned above. But for some constructions, we haven't done this yet. This is why for some examples in Part II, polynomial inequations appear in the middle of nondegenerate conditions in geometric forms.

7. Further Discussion of Geometry statements of Constructive Type

For a geometry statement of constructive type, the following four things are intimately related:

(1) The order in which new points are introduced (construction sequence).

(2) The selection of parameters and dependent variables.

(3) The geometry statement is generally true (valid).

(4) Nondegenerate conditions.

As we have seen in the last section, (1) determines (2). Conversely, from (2) we can usually figure out what the construction sequence for the geometry statement should be.

General validity of a geometry statement is relative to a parameter set or to which points can be arbitrarily chosen (in a general position). The choice of parameters (or construction sequences), though not unique and informal, is determined by *the geometric meaning* of the given statement. Different parameter sets (or construction sequences) can lead to confirm the same generic validity but with slightly different nondegenerate conditions. In such cases, we can say the meaning of the geometry statement is unchanged.

Example (7.1). Example (2.3): Pascal's theorem. If we follow the construction as the statement suggests, then the choice of parameters and dependent variables can be that in (2.3). The statement was proved to be generally true. The nondegenerate conditions in geometric forms produced by our prover are: Line AB intersects line DF ($I_6 \neq 0$); Line BC intersects line FE ($I_8 \neq 0$); Line CD intersects line EA ($I_{10} \neq 0$). The reader can obtain these nondegenerate conditions using the same analysis as in example (6.4).[7]

The geometry statement can also been stated as "Given a triangle ABC, and three points D, E and F on the circumscribed circle (O) of ABC, let $P = AB \cap DF$, $Q = BC \cap FE$ and $S = CD \cap EA$, then P, Q and S are collinear." Obviously, this statement does not change the meaning of Pascal's theorem in the usual sense in geometry. According to the construction suggested by this statement, our prover produced an additional nondegenerate condition: A, B and C are not collinear. The choice of variables can be, say, $A = (0,0)$, $B = (u_1, 0)$, $C = (u_2, u_3)$, $O = (x_2, x_1)$, $D = (x_3, u_4)$, $F = (x_4, u_5)$, $E = (x_5, u_6)$, $P = (x_6, 0)$, $Q = (x_8, x_7)$, $S = (x_{10}, x_9)$.

Example (7.2). There are at least eight essentially different construction sequences for the configuration of Simson's theorem that preserve the original meaning of the theorem. They are actually produced by our prover from conditions (6.5) with different point orders in which new points are introduced.

As we have seen, the assignment of parameters and dependent variables is based on the following heuristic rules.

(1) Non-zero coordinates of an arbitrarily chosen point are parameters.

(2) If a point is constructed from one geometric condition, then one of its two coordinates is a parameter and the other is a dependent variable.

(3) The two coordinates of a point constructed from two geometric conditions are dependent variables.

If the point is on the x-axis or on lines parallel or perpendicular to the x-axis, then the assignment can be further simplified as shown in examples (2.2), (2.3) and (2.4). (This simplification is not necessary, but will speed up proofs.)

It should be emphasized that these rules are based on the heuristic rule in algebra: one equation determines one unknown. It is the responsibility of the user to justify whether a point can be arbitrarily chosen or two conditions can really construct a point based on the geometric meaning of a given problem. The following is a very instructive example.

Example (7.3). Let ABC be a triangle, and BE the altitude from B. Show that $AB \equiv CB$ (Figure 6).

We can consider the configuration of the problem to be constructed as follows: Points A, B, and C are arbitrarily chosen; E is on the line passing through B and perpendicular to line AC and is on line AC.

[7] The other three conditions $I_7 \neq 0$, $I_9 \neq 0$ and $I_{11} \neq 0$ are not necessary. This can be done either using the triangulation suggested in example (6.3) or using a general schema in [11].

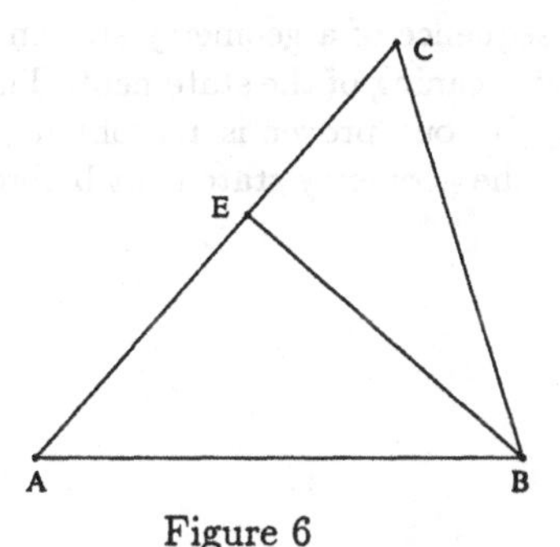

Figure 6

We can assign the coordinate according to the above rules: $A = (0,0)$, $B = (u_1, 0)$, $C = (u_2, u_3)$, $E = (x_1, x_2)$. Then the hypotheses and conclusion are:

$h_1 = u_3x_2 + u_2x_1 - u_1u_2 = 0$ — $EB \perp AC$
$h_2 = u_2x_2 - u_3x_1 = 0$ — E is on AC
$g = u_3^2 + u_2^2 - 2u_1u_2 = 0$ — Conclusion: $AC \equiv BC$.

After triangulation we have

$$f_1 = (u_3^2 + u_2^2)x_1 - u_1u_2^2 = 0$$
$$f_2 = u_2x_2 - u_3x_1 = 0.$$

Now $prem(g, f_1, f_2) \neq 0$. Since $deg(f_1, x_1) = deg(f_2, x_2) = 1$, the statement is generally false in *any geometry.*

However, if we follow the rule that two conditions construct (determine) a point completely and consider the configuration to be constructed such that points A, B, and E are arbitrarily chosen; C is on the line passing through A and perpendicular to line BE and is on line AE. Then, according to our rules, we can assign the coordinates Let $A = (0,0)$, $B = (u_1, 0)$, $E = (u_2, u_3)$, and $C = (x_1, x_2)$. Then we have

$h_1 = u_3x_2 + (u_2 - u_1)x_1 = 0$ — $EB \perp AC$
$h_2 = u_2x_2 - u_3x_1 = 0$ — E is on AC
$g = x_2^2 + x_1^2 - 2u_1x_1 = 0$ — Conclusion: $AC \equiv BC$.

We have a triangular form

$$f_1 = (u_3^2 + u_2^2 - u_1u_2)x_1 = 0$$
$$f_2 = u_2x_2 - u_3x_1 = 0$$

Now $prem(g, f_1, f_2) = 0$. Thus we proved the statement under $u_3^2 + u_2^2 - u_1u_2 \neq 0$ and $u_2 \neq 0$! The real problem is that from the meaning of the statement point E cannot be arbitrarily chosen. The experienced reader might immediately find that the second construction of the configuration is absurd.

Our prover requires the user's responsibility not to introduce "absurd" constructions similar to the above. Though for the above absurd construction and more hidden ones, some heuristics are built into our prover to detect them and give the user warnings, we don't have a general definition of all "absurd" constructions. We require the user to take full responsibility.

To specify construction sequence of a geometry statement of constructive type is actually to specify the exact meaning of the statement. Thus, the above requirement of the user's responsibility by our prover is reasonable, because the user should understand the meaning of the geometry statement before starting to prove it.

Chapter 2. Ritt's Characteristic Set Method

Wu introduced his method in 1977. Later on, he found that the algebraic tools and algorithms he needed were already begun in the work of J. F. Ritt [32], [33]. In [43] and [45], Wu revised Ritt's work for his own need of mechanically proving geometry theorems of certain types. Our presentation here is based on the work of Ritt and Wu. It is essential for understanding the complete method of Wu.

1. The Prerequisite in Algebra

From now on, we will assume the reader has a basic knowledge of algebra. The reader can acquire the knowledge by first reading a few sections of related chapters in standard algebra textbooks. (e.g., [26]).

Thus we assume the reader is familiar with the following notions and results: (commutative) rings, fields, polynomial rings, ideals, prime ideals, radical ideals, quotient rings (or residue-class rings), algebraic and transcendental extensions of fields, the Hilbert basis theorem, Hilbert's Nullstellensatz. For the latter reference, we list the following results, which are not necessarily in their most general forms, but are sufficiently general for our purpose.

Let K be a field, and $A = K[y_1, ..., y_m]$ be a polynomial ring with *indeterminates* $y_1, ..., y_m$. We often use the abbreviation such as $y = y_1, \ldots, y_m$, and $K[y] = K[y_1, \ldots, y_m]$, etc. Unless stated otherwise, all polynomials mentioned in this section are in A.

Theorem (1.1) (the Hilbert Basis Theorem). Every ideal of A is finitely generated. Or equivalently, there are no infinite, strictly increasing sequences of ideals of A.

Thus for every ideal I of A, there are finite polynomials $f_1, \ldots, f_r$ such that $I = (f_1, ..., f_r) = \text{Ideal}(f_1, ..., f_r)$. Here we use the standard notation $(f_1, ..., f_r)$ to denote the ideal of A generated by $f_1, ..., f_r$; i.e., the set $\{Q_1 f_1 + \cdots + Q_r f_r \mid Q_j \in A\}$. We also use Radical($S$) to denote the radical ideal of the ideal generated by polynomial set S, i.e., $\{g \mid g^k \in \text{Ideal}(S) \text{ for some } k > 0\}$.

A field F is said to be an extension field of K (or simply an *extension* of K) if K is a subfield of F. Let F be an extension of K. An element a of F is said to be *algebraic over* K if a is a root of some nonzero polynomial in $K[x]$, where x is an indeterminate. Otherwise, a is said to be *transcendental over* K. Let $a_1, ..., a_r$ be elements in F, then the subfield (resp. subring) generated by K and $a_1, ..., a_r$ is denoted by $K(a_1, ..., a_r)$ (resp. $K[a_1, ..., a_r]$). If $a_1, ..., a_r$ are transcendental over K (i.e., a_1 is transcendental over K, and a_i is transcendental over $K(a_1, ..., a_{i-1})$ for $i = 2, ..., r$), then $K(a_1, ..., a_r) \cong K(x_1, ..., x_r)$, the field of rational functions over K in indeterminates $x_1, ..., x_r$. The following theorem is important for our purpose.

Theorem (1.2). Let F be an extension of the field K and $a \in F$ be algebraic over K, then

(i) $K(a) = K[a]$.

(ii) $K(a) \cong K[x]/(f)$ (here x is an indeterminate) under the mapping $a \mapsto \tilde{x}$, where $f \in K[x]$ is an irreducible polynomial of degree $n \geq 1$, uniquely (up to a nonzero factor in K) determined by the conditions that $f(a) = 0$ and $g(a) = 0$ ($g \in K[x]$) if and only if f divides g; $\tilde{x}$ is the canonical image of x in $K[x]/(f)$.

(iii) Every element in $K(a)$ can be written uniquely in the form $c_{n-1}a^{n-1} + \cdots + c_1 a + c_0$ (with $c_i \in K$).

Example. The ring $\mathbf{R}[x]/(x^2+1)$ is isomorphic to $\mathbf{C}$, the field of complex numbers. Here $\mathbf{R}$ denotes the field of real numbers. In particular, $\tilde{x}^2 = -1$, where $\tilde{x}$ is the canonical image of x in $\mathbf{R}[x]/(x^2+1)$.

We need the elementary theory of factorization and unique factorization domains (UFD), in particular, the following theorem:

Theorem (1.3). Let D be a UFD, then the polynomial ring $D[y_1, ..., y_m]$ is also a UFD. In particular, $K[y_1, ..., y_m]$ is a UFD.

We also need the algorithm aspects of factorization.

Theorem (1.4). Let D be a UFD and K be its quotient field. Let a be an element in any extension of K, algebraic over K; let x be an indeterminate. Suppose there is an algorithm for factorization in D, then

(i) There is an algorithm for factorization in the polynomial rings $D[x]$ and $K[x]$.

(ii) There is an algorithm for factorization in the polynomial ring $K(a)[x]$.

The proof for (1.4), especially for (ii) is by no means trivial, see [41] or [40]. More important to us, we need efficient algorithms for the purpose of our geometry theorem proving. Our factorization algorithm [10] for successive quadratic extensions over fields of rational functions, together with factorization of multivariate polynomials over $\mathbf{Q}$, seems practical for most problems we have found in geometry. (See our solutions of Morley's trisector theorem and V. Thèbault's conjecture in Chapter 4.)

We also need elementary notions in algebraic geometry. The reader can find the following results in [41].

Let E be an extension of the base field K and $A = K[y_1, \ldots, y_m]$ be the polynomial ring in m indeterminates $y_1, \ldots, y_m$ over K.

Definition (1.5). A subset V of E^m is called an *algebraic set* if V is the set of common zeros of all elements of a nonempty polynomial set S, i.e.,

$$V = \{(a_1, ..., a_m) \in E^m \mid f(a_1, ..., a_m) = 0 \text{ for all } f \in S\}.$$

We denote V by $V(S)$.

Remark. $V(S)$ is *dependent* on the extension field E. Unless stated otherwise, throughout this monograph E denotes this field. To emphasize this dependence, we sometimes use the notation E–$V(S)$.

If $I = \text{Ideal}(S)$, i.e., the ideal generated by the polynomial set S, then $V(S) = V(I)$. By the Hilbert basis theorem, I has a finite set of generators $f_1, ..., f_h$. Thus, $V(S) = V(f_1, ..., f_h)$, and we have

Proposition (1.6). Every algebraic set is the set of common zeros of a finite polynomial set.

Let U be a nonempty subset of E^m, then $I =$

$$I(U) = \{f \mid f \in A \text{ and } f(a,..., a_m) = 0 \text{ for all } (a_1,..., a_m) \in U\}$$

is an ideal. We have $S \subset I(V(S))$ and $U \subset V(I(U))$; $V(S) = V(I(V(S)))$ and $I(U) = I(V(I(U)))$.

Proposition (1.7). Let S_1 and S_2 be two polynomial sets, and S_1S_2 be the set of all products of an element of S_1 and an element of S_2. Then:

(i) $V(S_1 \cup S_2) = V(S_1) \cap V(S_2)$.

(ii) $V(S_1S_2) = V(S_1) \cup V(S_2)$.

Thus, the union of a finite set of algebraic sets is an algebraic set; the intersection of any set of algebraic sets is an algebraic set.

Definition (1.8). A nonempty algebraic set is *irreducible* if it cannot be expressed as the union of two proper subsets, each one of which is an algebraic set.

Note that the definition of irreducibility depends on the based field K and on the field E as well.

Example. Let $S = \{y_2^2 + y_1^2,\ y_3^2 - 1 + y_1\}$.

(i) $K = \mathbf{Q}$, $E = \mathbf{C}$, then $V(S)$ is irreducible. See (1.17) and (i) of (3.7).

(ii) $K = E = \mathbf{C}$, then $V(S)$ is reducible.

(iii) $K = \mathbf{Q}$, $E = \mathbf{R}$, then $V(S) = V(S_1) \cup V(S_2)$, where $S_1 = \{y_1,\ y_2,\ y_3 - 1\}$, $S_2 = \{y_1,\ y_2,\ y_3 + 1\}$. Thus, $V(S)$ is reducible.

Proposition (1.9). Let V be a nonempty algebraic set. V is irreducible if and only if $I(V)$ is a prime ideal.

Theorem (1.10). Every algebraic set V in E^m can be uniquely expressed as a union of irreducible algebraic sets, no one containing another. Each irreducible algebraic set is called *a component* of V.

Definition (1.11). The *dimension* of a prime ideal P is the transcendental degree of quotient field of the integral domain $K[y_1,..., y_m]/P$ over the field K. The dimension of an irreducible algebraic set V is the dimension of its prime ideal $I(V)$. The dimension of an algebraic set V is the highest dimension of its components.

Definition (1.12). A *generic zero* of an ideal I is a zero $\mu = (a_1,..., a_m)$ of I in an extension of K, such $f \in I$ if and only if $f(a_1,..., a_m) = 0$.

Theorem (1.13). An ideal I has a generic zero $(a_1,..., a_m)$ in some extension of K if and only if it is a prime ideal not identical to $(1) = A$ (the unit ideal). In that case, $K[a_1,..., a_m]$ is isomorphic to the quotient ring $F = K[y_1,\dots,y_m]/I$ under the mapping $a_i \mapsto \tilde{y}_i$, where the $\tilde{y}_i$ are the canonical images of the y_i in F. Thus, $(\tilde{y}_1,\dots,\tilde{y}_m)$ is a generic zero of I; The degree of I is the transcendental degree of $a_1,..., a_m$ over K, i.e., the number of any maximal algebraically independent (over K) subset of $\{a_1,\dots,a_m\}$.

Definition(1.14). Let V be an irreducible algebraic set, and $P = I(V)$ its prime ideal. Variables $v_1,..., v_s$ among the y are *algebraically independent on* V if P does not contain a polynomial with variables $v_1,..., v_s$ only, or equivalently, there are no polynomials containing $v_1, ..., v_s$ only and vanishing on V. Otherwise, we say $v_1, ..., v_s$ are algebraically dependent on V.

From now on, we will assume E to be **algebraically closed**. There are many forms of Hilbert's Nullstellensatz; the following two are most suitable for our purpose.

Theorem (1.15) (Hilbert's Nullstellensatz). Every polynomial ideal I, not identical to the unit ideal (1) (= A), has a common zero, i.e., $V(I)$ is nonempty.

Theorem (1.16) (Hilbert's Nullstellensatz: another form). For any ideal I of A, $I(V(I)) = \text{Radical}(I)$, i.e., polynomial f vanishes on $V(I)$ (every point in $V(I)$ is a zero of f) if and only if $f^k \in I$ for some integer $k > 0$.

Proposition (1.17). If P is a prime ideal of A not identical to (1), then $V(P)$ is irreducible and $I(V(P)) = P$.

2. ascending chains and Characteristic Sets

Let K be a field, and $y = y_1, y_2, \ldots, y_m$ be indeterminates. Unless stated otherwise, all polynomials mentioned in this section are in $A = K[y_1, \ldots, y_m] = K[y]$. We fix the order of the indeterminates as $y_1 < y_2 < \cdots < y_m$, which is essential for the subsequent discussion. Unless stated otherwise, we assume this order among variables $y_1, ..., y_m$. When talking about an algorithm about A, we always assume that the field K is computable.

Let f be a polynomial. Denote the degree of f in the variable y_i by $deg(f, y_i)$. The *class* of f is the smallest integer c such that f is in $K[y_1, \ldots, y_c]$. We denote it by $class(f)$. If f is in K we define $class(f) = 0$. Let $c = class(f)$ be non-zero and $lv(f)$ denote the *leading variable* y_c of f. Considering f as a polynomial in y_c, let $lc(f)$ denote the *leading coefficient* of f which is also called the *initial* of f. Let $ld(f)$ denote the *leading degree* of f, i.e., $deg(f, y_c)$. A polynomial g is *reduced with respect to* f if $deg(g, y_c) < deg(f, y_c)$, where $c = class(f) > 0$.

Let f and g be in $K[y]$. Suppose that y_i really occurs in f. Considering f and g as polynomials in y_i, we can do pseudo division as defined in Section 4 of Chapter 1. Let $c = class(f) > 0$, then $prem(g, f, y_c)$ is reduced with respect to f; we denote $prem(g, f, y_c)$ simply by $prem(g, f)$.

Definition (2.1). Let $C = f_1, f_2, \ldots, f_r$ be a sequence of polynomials in $K[y]$. We call it a *quasi ascending chain* or a triangular form if either $r = 1$ and $f_1 \neq 0$, or $r > 1$ and $0 < class(f_1)$, $class(f_i) < class(f_j)$ for $i < j$.

Let $f_1, \ldots, f_r$ be a quasi ascending chain with $class(f_1) > 0$. We define $prem(g, f_1, \ldots, f_r)$ inductively to be $prem((prem(g, f_2, \ldots, f_r), f_1)$.

(i) A quasi ascending chain is called an *ascending chain* if f_j is reduced with respect to f_i for $i < j$.

(ii) A quasi ascending chain is called an *ascending chain in Wu's sense* if the initials I_j of the f_j are reduced with respect to f_i for $i < j$.

(iii) A quasi ascending chain is called an *ascending chain in weak sense* if $prem(I_i, f_1, ..., f_r) \neq 0$, for $i = 1, ..., r$.

Obviously, an ascending chain is an ascending chain in Wu's sense; an ascending chain in Wu's sense is an ascending chain in weak sense.

We define a partial order $<$ in $K[y]$: $f < g$ (g is of *higher rank* or *higher* than f) if $class(f) < class(g)$ or $class(f) = class(g) > 0$ and $ld(f) < ld(g)$. If neither $f < g$ nor $g < f$, then we say f and g are of the same rank.

Proposition (2.2). The partial order $<$ in A is *well-founded*, i.e., there are no infinite, strictly decreasing sequences of polynomials $p_1 > p_2 > \cdots > p_n > \cdots$.

Proof. As we know, the well-founded condition is equivalent to that every nonempty set S of polynomials has a minimal element in that partial order $<$; i.e., an element which is not higher than any other element in S. If S contains an element in K, any such element is minimal. Otherwise, a polynomial in S with the least leading degree among the subset of all elements with the least class in S is minimal. In the case of finite S, this provides an algorithm to obtain a minimal element in S. .QED.

Definition (2.3). Let $C = f_1, ... f_r$ and $C_1 = g_1, ..., g_m$ be two ascending chains. We define $C < C_1$ if there is an s such that $s \leq \min(r, m)$ and f_i and g_i are of the same rank for $i < s$ and that $f_s < g_s$, or $m < r$ and f_i and g_i are of the same rank for $i \leq m$.

Proposition (2.4). The partial order $<$ among the set of all ascending chains is well-founded. In other words, every nonempty set Σ of ascending chains has a minimal element.

Proof. Let Σ_1 be all elements of Σ whose first polynomials are minimal among the first polynomials of all elements in Σ. If all elements (ascending chains) in Σ_1 consist of one polynomial, then any element in Σ_1 is minimal in Σ. Suppose that there are elements in Σ_1 which consist of more than one polynomial. We can let Σ_2 be the subset of all elements of Σ_1, whose second polynomials are minimal. If the elements in Σ_2 all have just two polynomials, any of those elements serves our purpose. If not, we continue, reaching a minimal ascending chain in no more than m steps. .QED.

Proposition (2.4) can be rephrased: there are no infinite, strictly decreasing sequences of ascending chains $C_1 > C_2 > \cdots > C_k > \cdots$. This is important for the termination of Ritt's algorithms (see sections 4 and 5).

Definition (2.5). Let S be a nonempty polynomial set. A minimal ascending chain in the set of all chains formed from polynomials in S is called a *characteristic set* of S.

Unless stated otherwise, whenever we talk about a finite polynomial set S, we assume S does not contain zero. By (2.4), every nonempty polynomial set S has a characteristic set.

Proposition (2.6). Let $C = f_1, \ldots, f_r$ be a characteristic set of polynomial set S with $0 < class(f_1)$. Let g be nonzero and reduced with respect to C. Then the set $S_1 = S \cup \{g\}$ has characteristic sets lower than C.

Proof. If $class(g) \leq class(f_1)$, then g alone is an ascending chain lower than C. Suppose $class(g) > class(f_1)$, let $j = max\{i \mid class(f_i) < class(g)\}$, then $f_1, ..., f_j, g$

form an ascending chain lower then C. .QED.

Proposition (2.7). Let $C = f_1, \ldots, f_r$ be an ascending chain in the polynomial set S with $0 < class(f_1)$. C is a characteristic set of S if and only if S contains no nonzero polynomial reduced with respect to C.

Proof. If there is a polynomial g in S reduced with respect to C, then from the proof of (2.6) we have an ascending chain in S lower than C. Thus C is not a characteristic set of S. If C is not a characteristic set of S, then there is an ascending chain $C_1 = g_1, ..., g_m$ in S lower than C. From the definition (2.3), we can find a g_i reduced with respect to C. .QED.

Theorem (2.8). Every nonempty polynomial set S has a characteristic set. In the case of finite set S, there is an ***algorithm*** to construct a characteristic set of S.

Proof. Let f_1 be a polynomial with minimal rank in S. If f_1 is of class zero, then it is a characteristic set of S. Now let f_1 be of positive class. Let S_1 be the set of all polynomials in S, reduced with respect to f_1. If S_1 is empty, then f_1 is a characteristic set of S. Now suppose S_1 is nonempty, then every polynomial in S_1 is of higher class than f_1. Let f_2 be a polynomial with minimal rank in S_1. Let S_2 be the set of all polynomials of in S_1, reduced with respect to f_1, f_2. If S_2 is empty, then f_1, f_2 is a characteristic set. Otherwise continuing, we arrive at a characteristic set of S in no more than m steps. .QED.

Proposition (2.9). Let $f_1, \ldots, f_r$ be a characteristic set of the ideal I.

(i) If $g \in I$, then $prem(g, f_1, \ldots, f_r) = 0$.

(ii) If I is prime, then $prem(g, f_1, \ldots, f_r) = 0 \Rightarrow g \in I$.

proof. Let g be in I. Since $R = prem(g, f_1, \ldots, f_r)$ is in I and reduced with respect to $f_1, \ldots, f_r$, it has to be zero by (2.7). Now conversely suppose that $prem(g, f_1, \ldots, f_r) = 0$, then by the remainder formula we have

$$I_1^{s_1} \cdots I_r^{s_r} g = Q_1 f_1 + \cdots + Q_r f_r,$$

where the I_k are the initials of the f_k. Since the I_k are nonzero and reduced with respect to $f_1, \ldots, f_r$, the I_k are not in the ideal I by (2.6). Thus, if I is a prime ideal, we can conclude that $g \in I$. .QED.

3. Irreducible Ascending Chains

Let $C = f_1, \ldots, f_r$ be an ascending chain, not consisting of a constant. After a **suitable renaming** of the y_j, we may assume that $class(f_1) = d + 1$ and $m = d + r = class(f_r)$, where $d \geq 0$. We distinguish the y_i for $i \leq d$ by calling them u_i and use x_i to denote $lv(f_i)$.

Thus C has the following "triangular" form:

$$
\begin{aligned}
& f_1(u_1, \ldots, u_d, x_1) \\
& f_2(u_1, \ldots, u_d, x_1, x_2) \\
(3.1) \qquad & \cdots \\
& f_r(u_1, \ldots, u_d, x_1, \ldots, x_r).
\end{aligned}
$$

Definition (3.2). An ascending chain $f_1, ..., f_r$ of the form (3.1) is called *irreducible* if each f_i is irreducible in the polynomial ring $K(u)[x_1, \ldots, x_i]/(f_1, ..., f_{i-1})$. Thus the sequence $F_0 = K(u)$, $F_1 = F_0[x_1]/(f_1)$, ..., $F_r = F_{r-1}[x_r]/(f_r) = F_0[x]/(f_1, ..., f_r)$ is a tower of field extensions.[1]

Example (3.3). Let C be the ascending chain $f_1 = x_1^2 - u_1$, $f_2 = x_2^2 - 2x_1x_2 + u_1$. f_1 is irreducible over $F_0 = \mathbf{Q}[u_1]$; but f_2 is reducible over $F_1 = F_0[x_1]/(f_1)$ because $f_2 = (x_2 - x_1)^2$ under $x_1^2 - u_1 = 0$. Thus C is reducible.

Theorem (3.4). Let ascending chain $f_1, ..., f_r$ of the form (3.1) be irreducible,[2] g be a polynomial in $K[u, x]$ and F_r be the same as in (3.2). Let $prem(g, f_1, \ldots, f_r)$ be the remainder obtained from successive pseudo division of g by f_1, ..., f_r. Then the following conditions are equivalent:

(i) $prem(g, f_1, \ldots, f_r) = 0$.

(ii) Let E be an extension field of K. If $\mu = (\tilde{u}_1, \ldots, \tilde{u}_d, \tilde{x}_1, \ldots, \tilde{x}_r)$ in E^{d+r} is a common zero of $f_1, ..., f_r$ with $\tilde{u}_1, \ldots, \tilde{u}_d$ transcendental over K, then μ is also a zero of g, i.e., $g(\mu) = 0$.

(iii) The polynomial g, considered as an element in F_r, is 0; i.e., the image of g in F_r under the canonical mapping is 0.

(iv) There are finite non-zero polynomials $c_1, \ldots, c_s$ in $K[u_1, ..., u_d]$ such that $c_1 \cdots c_s g \in \text{Ideal}(f_1, ..., f_r)$ (in $K[u, x]$).

Proof.

For any polynomial h, let $\tilde{h}$ be the polynomial obtained from h by substituting $u_1, \ldots, u_d, x_1, \ldots x_r$ for $\tilde{u}_1, \ldots, \tilde{u}_d, \tilde{x}_1, \ldots \tilde{x}_r$.

We use induction on k to prove the following assertions (for $0 < k \le r$):

(U) For any polynomial $p = a_s x_k^s + \cdots + a_0$ ($0 < k \le r$, $1 \le s$, $a_i \in K[u, x_1, ..., x_{k-1}]$, $a_s \ne 0$) reduced with respect to $f_1, ..., f_r$, if μ is a zero of p, then $p = 0$.

If $k = 1$, then $p = a_s x_1^s + \cdots + a_0$, with all $a_j \in K[u]$. μ is a zero of p means $\tilde{p} = \tilde{a}_s \tilde{x}_1^s + \cdots + \tilde{a}_0 = 0$. Since p is reduced with respect to f_1, $s < deg(f_1, x_1)$. By the uniqueness of representation in algebraic extension (iii) of (1.2), we have all $\tilde{a}_j = 0$. Since $\tilde{u}_1, \ldots, \tilde{u}_d$ are transcendental over K, all $a_j = 0$. Hence $p = 0$.

Now we want to prove (U) is true for k assuming it is true for $k - 1$. Since μ is a zero of p, $\tilde{p} = \tilde{a}_s \tilde{x}_k^s + \cdots + \tilde{a}_0 = 0$. Since $s < deg(f_k, x_k)$, by (1.2) again, all $\tilde{a}_j = 0$. Thus μ is also a zero of all a_j. Since all a_j are also reduced with respect to $f_1, ..., f_r$, $a_j = 0$ by the induction hypothesis. Hence $p = 0$.

(ii) $\Rightarrow$ (i). Suppose μ is a zero of g. Let $R = prem(g, f_1, ..., f_r)$. We have the remainder formula

$$I_1^{s_1} \cdots I_r^{s_r} g = Q_1 f_1 + \cdots + Q_r f_r + R.$$

Hence μ is a zero of R. Since R is reduced with respect to $f_1, ..., f_r$, $R = 0$.

[1] In contrast to the previous notation, here $(f_1, ..., f_r)$ etc. denotes the polynomial ideal of $K(u)[x]$ (not of $A = K[u, x]$), generated by $f_1, ..., f_r$.

[2] Here the ascending chain can be in weak sense.

(i) ⇒ (ii). Suppose $prem(g, f_1, ..., f_r) = 0$. Then by the remainder formula, we have

$$I_1^{s_1} \cdots I_r^{s_r} g = Q_1 f_1 + \cdots + Q_r f_r, \tag{3.4.1}$$

where the I_k the are initials of the f_k. Since $prem(I_k, f_1, \ldots, f_r) \neq 0$, μ is not a zero of I_k (by (ii) ⇒ (i)). Hence μ is a zero of g.

(iii) is an instance of (ii) when $E = F_r$. Hence (i) and (iii) are equivalent.

(iv) ⇒ (i). Suppose (iv) holds. Let μ be the same as in (ii). Then $c_i(\mu) \neq 0$ because the c_i are non-zeroes in K$[u]$ and $\tilde{u}_1, \ldots, \tilde{u}_d$ are transcendental. Hence $g(\mu) = 0$ and (i) follows.

(i) ⇒ (iv). Now suppose (i) holds. Then we have (3.4.1). Since $p = I_1^{s_1} \cdots I_r^{s_r}$ is nonzero in the field F_r, it has an inverse in F_r. That is, there is a q in $K(u)[x]$ such that $qp - 1$ is in the ideal of $K(u)[x]$ generated by $f_1, ..., f_r$. Clearing the denominator, we have $q_1 p - c$ is in the ideal $(f_1, ..., f_r)$ of A, where c is a non-zero polynomial in $K[u]$. Multiplying (3.4.1) by q_1, we have cg is in the ideal $(f_1, ..., f_r)$ of A. Thus (iv) follows. .QED.

We call $\mu = (\tilde{u}_1, \ldots, \tilde{u}_d, \tilde{x}_1, \ldots, \tilde{x}_r)$ in (ii) a *generic point* of that irreducible ascending chain in field E.

The theorem is no longer true if $f_1, ..., f_r$ is reducible. We can find such an example by letting f_1, f_2 be the same as in example (3.2) and $g = x_2 - x_1$.

Proposition (3.5). Let $f_1, ..., f_r$ be irreducible, g any polynomial. If $prem(g, f_1, ..., f_r) \neq 0$, then there are polynomial q and r such that $qg - r \in$ the ideal $(f_1, ..., f_r)$ of A and $r \in K[u]$.

Proof. This can be seen from the proof of (i) ⇒ (iv) of the above theorem. .QED.

Theorem (3.6). Let $f_1, \ldots, f_r$ be an ascending chain. Suppose that $f_1, \ldots, f_{k-1}$ $(0 < k \leq r)$ is irreducible, but $f_1, \ldots, f_k$ is reducible. Then there are polynomials g and h in $K[u, x]$ reduced with respect to $f_1, \ldots, f_r$ such that $class(g) = class(h) = class(f_k)$ and $gh \in$ the ideal $(f_1, ..., f_k)$ of A.

Proof. By assumption, f_k is reducible in the field $L = F_{k-1}[x_k]$. Thus there are h'' and g'' in L such that $f_k - g''h'' = 0$ with $0 < deg(g'', x_k) < deg(f_k, x_k)$ and $0 < deg(h'', x_k) < deg(f_k, x_k)$.[3] Clearing denominators, we have a polynomial $p = Qf_k - g'h'$, where g' and h' are in $K[u, x_1, \ldots, x_k]$ obtained from g'' and h'' by clearing denominators and $Q \in K[u]$. Polynomial $p = 0$ in L, thus by (iii) of (3.4) $prem(p, f_1, ..., f_{k-1}) = 0$. Thus by the remainder formula, there are integers $s_i \geq 0$ such that $q = I_1^{s_1} \cdots I_{k-1}^{s_{k-1}} g'h'$ is in the ideal $(f_1, ..., f_k)$ of A, where the I_i are the initials of the f_i. Let $g = prem(I_1^{s_1} \cdots I_{k-1}^{s_{k-1}} g', f_1, ..., f_{k-1})$ and $h = prem(h', f_1, ..., f_{k-1})$, then g and h have the specifications in the theorem. .QED.

Theorem (3.7). Let $f_1, .., f_r$ be an irreducible ascending chain and $P =$

$$\{g \mid g \in K[u, x] \text{ and } prem(g, f_1, ..., f_r) = 0\},$$

then

[3] This is one of the key steps in Ritt's decomposition algorithm (5.1). Factorization over algebraic extensions is needed.

(i) P is a prime ideal, with $f_1, ..., f_r$ as its characteristic set.

(ii) A generic point of $f_1, \ldots, f_r$ is a generic zero of prime ideal P.

(iii) If E is algebraically closed, then $dim(V(P)) = d$. A polynomial g vanishes on $V(P)$ if and only $prem(g, f_1, \ldots, f_r) = 0$.

(iv) For any field E, if $prem(g, f_1, \ldots, f_r) = 0$, then polynomial g vanishes on $V(P)$.

Proof. Let μ be a generic point of $f_1, ..., f_r$. By (3.4), $P = \{g \mid g(\mu) = 0\}$. From (1.13), it is now obvious that P is a prime ideal with the generic zero μ. Since there are no nonzero polynomials in P reduced with respect to $f_1, \ldots, f_r$, $f_1, \ldots, f_r$ is a characteristic set of P. If E is algebraically closed, then $I(V(P)) = P$. Therefore, a polynomial g vanishes on $V(P)$ if and only $g \in P$, i.e., $prem(g, f_1, \ldots, f_r) = 0$. The dimension of $V(P)$ is that of P, i.e., d. Since every field E has an algebraically closed extension, (iv) follows from (iii). .QED.

Remark (3.8). If E is not algebraically closed, then $V(P)$ can be irreducible or reducible. However, if $V(P)$ is of dimension d, then $V(P)$ is irreducible. Indeed, let V_1 be one of its components with dimension d. Then $P_1 = I(V_1)$ is a prime ideal containing P and having dimension d. Thus $P = P_1$, $V(P) = V(P_1) = V_1$. $V(P)$ is irreducible. In that case, therefore, a polynomial g vanishes on $V(P)$ if and only if $prem(g, f_1, ..., f_r) = 0$.

Theorem (3.9). Let P be a nontrivial prime ideal of $K[u, x]$, and $f_1, ..., f_r$ be a characteristic set of P. Then $f_1, ..., f_r$ is irreducible.

Proof. As we know from (2.9), $P = \{g \mid prem(g, f_1, .., f_r) = 0\}$. Suppose $f_1, ..., f_r$ is reducible. Then there is a $k > 0$ such that $f_1, ..., f_{k-1}$ is irreducible, but $f_1, \ldots, f_k$ is reducible. By (3.6) we can find two polynomials g and h reduced with respect to $f_1, ..., f_r$ such that $gh \in (f_1, ..., f_k) \subset P$, $0 < deg(f, x_k)$ and $0 < deg(g, x_k)$. Since g and h are reduced with respect to $f_1, ..., f_r$, they are not in P. Therefore, P is not a prime ideal. It contradicts the hypothesis. .QED.

4. A Complete Triangulation Procedure: Ritt's Principle

Now we shall give a complete triangulation procedure, which was implicitly in Ritt's work ([32], [33]) and was rewritten by Wu in detail ([45], [43]). This triangulation procedure is called ***Ritt's Principle*** by Wu. Wu considers it the basis of his method.

Theorem (4.1). (Ritt's Principle). Let $S = \{h_1, ...h_n\}$ be a finite nonempty polynomial set in $A = K[y_1, \ldots, y_m]$, and I be the ideal $(h_1, ..., h_n)$ of A. There is an algorithm to obtain an ascending chain C such that either

(4.2). C consists of a polynomial in $K \cap I$, or

(4.3). $C = f_1, \ldots, f_r$ with $class(f_1) > 0$ and such that $f_i \in I$ and $prem(h_j, f_1, ..., f_r) = 0$ for all $i = 1, \ldots, r$ and $j = 1, \ldots, n$.

Ascending chain C is called ***an extended characteristic set*** of S.

Proof. By (2.8), we can construct a characteristic set C_1 of $S_1 = S$. If C_1 consists of only one nonzero constant, then we have (4.2). Otherwise, we can expand S_1 to S_2 by adding all nonzero remainders of elements of S_1 with respect to C_1. If $S_2 = S_1$, then we have (4.3). Otherwise, we can construct a characteristic set C_2 of S_2. If C_2 does not consist of one nonzero constant, then $C_1 > C_2$ by (2.6) and we can expand S_2 to S_3 using the same procedure. Thus we have a strictly increasing sequence of polynomial sets:

$$S_1 \subset S_2 \subset \cdots,$$

with the corresponding strictly decreasing sequence of characteristic sets

$$C_1 > C_2 > \cdots$$

By (2.4), this decreasing sequence can be only finite. Thus, there is an integer $k \geq 1$ such that either C_k consists of a nonzero constant or $S_k = S_{k+1}$; then we have either (4.2) or (4.3), respectively. .QED.

Corollary (4.4). Let S be the same as in (4.1). Suppose that the case (4.3) happens and $f_1, ..., f_r$ is an extended characteristic set of S. Let I_k be the initials of the f_k. Let $S_k = S \cup \{I_k\}$ and $P = \{g \mid g \in A \text{ and } prem(g, f_1, .., f_r) = 0\}$. For any extension E of the field K, we have

(i) $V(f_1, ..., f_r) - (V(I_1) \cup V(I_2) \cup \cdots \cup V(I_r)) \subset V(P) \subset V(S) \subset V(f_1, ..., f_r)$.

(ii) $V(S) = V(P) \cup V(S_1) \cup \cdots \cup V(S_r)$.

Proof. (i) is an immediate consequence of (4.3) and the remainder formula. Since $V(S) = [V(S) - (V(I_1) \cup \cdots \cup V(I_r))] \cup [V(S) \cap (V(I_1) \cup \cdots \cup V(I_r))]$, (ii) follows from (i) and the standard set manipulations. .QED.

5. Ritt's Decomposition Algorithm

Let E be any extension of the based field K. As before, when we mention the algebraic set $V(S)$, we mean the algebraic set $V(S)$ in E^m.

Theorem (5.1). (Ritt's Decomposition Algorithm). For any finite nonempty polynomial set S of A, there is an algorithm to decide whether Ideal(S) = (1), or (in the opposite case) to decompose

$$V(S) = V(P_1) \cup V(P_2) \cup \cdots \cup V(P_s),$$

where the P_i are prime ideals given by their irreducible characteristic sets.

Proof. Let D be a temporary variable, representing a set of characteristic sets and initialized to be empty. From the polynomial set S, using Ritt's Principle, we can construct an extended characteristic set C of S and the corresponding extended set S' of the set S (i.e., the set S_k in the proof of (4.1)). There are three cases:

Case 1. If C consists of a constant, then $V(S)$ is empty. We stop processing S.

Case 2. The ascending chain $C = f_1, ..., f_r$ is irreducible. Let I_k be the initials of the f_k and $S_k = S' \cup \{I_k\}$, by (ii) of (4.4), we have

$$V(S) = V(S') = V(P_1) \cup V(S_1) \cup \cdots \cup V(S_r),$$

where P_1 is the prime ideal having C as its characteristic set. By (2.6), each S_k has a characteristic set strictly lower than C. We add C to the set D. We can take each S_k as a new S, and repeat the process.

Case 3. The ascending chain $C = f_1, ..., f_r$ is reducible. Then there is an integer $k > 0$ such that $f_1, ..., f_{k-1}$ is irreducible, but $f_1, ..., f_k$ is reducible. By (3.6), we can find[4] two polynomials g and h with $class(f_k) = class(g) = class(h)$, reduced with respect to $f_1, ..., f_r$ and $gh \in (f_1, ..., f_k)$. We have decomposition: $V(S) = V(S') = V(S_1) \cup V(S_2)$, where $S_1 = S' \cup \{g\}$ and $S_2 = S' \cup \{h\}$ have characteristic sets strictly lower than C by (2.6). We can take each of S_1 and S_2 as new S, and repeat the process.

The above process will eventually terminate, because otherwise we would have an infinite, strictly decreasing sequence of ascending chains, which contradicts (2.4). Upon termination, we have two cases:

(1) D is empty. This means that S does not have common zeros in any extensions E of K, in particular in any algebraically closed extensions. By Hilbert's Nullstellensatz, Ideal$(S) = (1) = A$.

(2) $D = \{C_1, ..., C_s\}$, then we have

$$V(S) = V(P_1) \cup V(P_2) \cdots \cup V(P_s),$$

where the P_k are the prime ideals having C_k as their characteristic sets. .QED.

Remark (5.2). If E is *algebraically closed*, then by Hilbert's Nullstellensatz, each $V(P_k)$ is irreducible. In particular, the dimension of $V(P_k)$ in $E^m = m-r$, where r is the number of elements of C_k. The above decomposition gives a decomposition of an algebraic set into the union of irreducible algebraic sets (not necessarily irredundant). Hence, Radical$(S) = P_1 \cap \cdots \cap P_s$.

Remark (5.3). If E is not an algebraically closed field, then it is possible that $V(P_k)$ is reducible or empty. In particular, $V(S)$ can be empty even if $1 \notin$ Ideal(S). We still call $V(P_i)$ a *component* of of $V(S)$.

Remark (5.4). In case $1 \notin$ Idea(S), we can write the decomposition in a slightly different form:

$$V(S) = V(C_1/J_1) \cup \cdots \cup V(C_s/J_s),$$

where the J_i are products of the initials of all elements in C_i and $V(C_i/J_i) = V(C_i) - V(J_i)$. This formula provides a complete solution of a system of polynomial equations up to solving systems of polynomial equations in the form of irreducible ascending chains, assuming the corresponding initials are nonzero. For details see [48].

In an actual implementation of Ritt's algorithms, many details should be taken care of to save the computer time and space.

[4] Here we need the factorization of polynomials over algebraic extensions. Thus we assume field K satisfies conditions in theorem (1.4).

Chapter 3. Algebra and Geometry

Now we have the algebraic background necessary to discuss the complete method of Wu. There are few doubts that geometry statements are confirmed by the method. However, if the method fails to confirm a geometry statement, does it disprove the statement? If so, in what sense? This is quite a complicated problem. There are two major issues behind this problem.

(1) There is a variety of geometries. For example, we have heard about at least affine geometry and Euclidean geometry. Proving or disproving geometry statements in these two geometries is different since the field associated with Euclidean geometry is the field of real numbers, whereas almost any fields can be the fields associated with affine geometry; particularly, they can be algebraically closed fields. To understand this issue fully, we need to discuss the relationship between geometry and algebra (Section 1).

(2) In proving a geometry statement involving equalities only (with certain unspecified additional nondegenerate conditions), Wu's method can produce these additional conditions in inequation forms. From the point of view of logic, this complicates our understanding of proving and disproving a geometry statement. An ideal geometry machine only needs to answer "true" or "false" to a fully specified geometry statement (including nondegenerate conditions). As we will see, that approach, though perfect from the point of view of logic, is not satisfactory from a geometric point of view, especially from the practical point of view of mechanical theorem proving in geometry. This is the topic of Section 2 and Section 3.

1. Axiomatic Geometries and Number Systems

In this section we shall discuss briefly the relationship between geometry and algebra in order to show that the Cartesian product of the field of real numbers is not the *only* realistic (plane) geometry.

We begin with a passage from E. Artin's book "*Geometric Algebra*", [4]:

> We are all familiar with analytic geometry where a point in a plane is described by a pair (x, y) of *real* numbers, a straight line by a linear, a conic by a quadratic equation. Analytic geometry enables us to reduce any elementary geometric problem to a mere algebraic one. The intersection of a straight line and a circle suggests, however, enlarging the system by introducing a new plane whose points are pairs of *complex* numbers. An obvious generalization of this procedure is the following. Let k be a given field; construct a plane whose "points" are the pairs (x, y) of elements of k and define lines by linear equations. ...
>
> A much more fascinating problem is, however, the converse. Given a plane geometry whose objects are the elements of two sets, the set of points and the set of lines; assume that certain axioms of geometric nature are true. Is it possible to find a field k that the points of our geometry can be described by coordinates from k and the lines by linear equations?

This passage suggests that there are two approaches to defining geometry.

(A) The Algebraic Approach. Starting from some number systems E (usually fields), we can define geometry objects and relations between those objects in the Cartesian product E^n (or E^n/E^* in projective geometry). In modern geometry, especially in algebraic geometry, this approach indisputably prevails. If we take this approach, then there are only a few differences between algebra and geometry; geometry can be regarded as a part of algebra.

However, the second approach suggested by Artin is more attractive from the point of view of axiomatic geometry and the point of view of traditional proofs of geometry theorems in Euclidean style.

(B) The Axiomatic Geometry Approach. By this approach we mean the one that was approximately used by Hilbert in his classic book [24]. We shall restrict ourselves to first order logic and assume the reader has a basic knowledge of first order logic; e.g., the contents of the first 23 pages of [35]. Depending on the context, the term "a geometry" has two meanings. When we mention *the theory of a geometry* we meaning *a first order theory* defined in [35], i.e., a formal system in first-order logic. Here "a geometry" means a class of geometry, e.g., affine geometry. On the other hand, when ordinary geometry textbooks mention "a geometry", it usually means *a model* of a certain theory of geometry. Thus Euclidean geometry (i.e., $\mathbf{R}^2$) is an affine geometry; i.e., Euclidean geometry is a model of the theory of affine geometry. Of course, this model has *much richer structures* than an ordinary affine geometry.

Besides the simplicity and beauty of the axiomatic system, a very attractive feature of this approach is that algebra can be introduced in an elegant way (see [24]). For each model of a theory of geometry, we can prove the existence of a number system (usually a field) inherent to that model. This field is called the *field associated with* that model. That model then can be represented as the Cartesian product of its associated field.

1.1. Affine Geometry

Our first example of geometry is the simplest one: affine geometry. This geometry can be illustrated as an example of how to introduce algebra or number systems to geometry. We will use ordinary language to present the theory of affine geometry. The experienced reader can easily translate our presentation into a first order language. The only basic relation in this geometry is that of incidence, i.e., a point A is on a line l, or equivalently, a line l passes through (contains) a point A. Here we use two kinds of objects: points and lines. This can be done in first order language by introducing two unary predicates: Isapoint(x), Isaline(x). In our presentation, there are four groups of axioms for this geometry.

(A1) Axioms of Incidence.

(A1.1) For every two points A and B there exists a line l that contains each of the points A and B.

(A1.2) For every two points A and B there exists no more than one line that contains each of the two points A and B.

(A1.3) There exist at least two points on a line. There exist at least three points that do not lie on a line.

(A2) Axiom of Parallels in the Strong Form. Let l be any line and A a point not on l. Then there exists one and only one line that passes through A and does not intersect l.

As an exercise, it can be proved that the geometry (model) satisfying the above two groups of axioms with the least possible number of points is $\mathbf{Z}_2 \times \mathbf{Z}_2$, where $\mathbf{Z}_2$ is the field with two elements. (Figure 7; note that AC and BD do not intersect.) To avoid unexpected degeneracy in the later reasonings, we need sufficient points on a line. For example, we can introduce the following axiom (A3) to exclude those models whose associated fields have nonzero characteristics. (This restriction is not necessary for general affine geometry.)

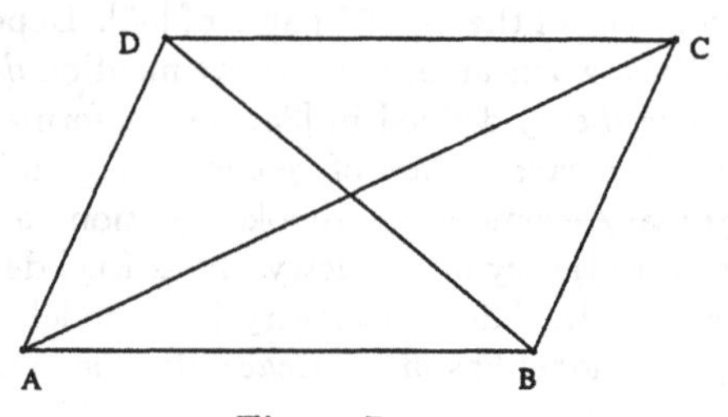

Figure 7

A *parallelogram* is four ordered points A, B, C and D such that no three of them are on the same line and $AB \parallel CD$, $BC \parallel AD$. Using axiom (A2), it is easy to prove that given any three non-collinear points A, B and C, there exists one and only one point D such that $ABCD$ is a parallelogram (Figure 7).

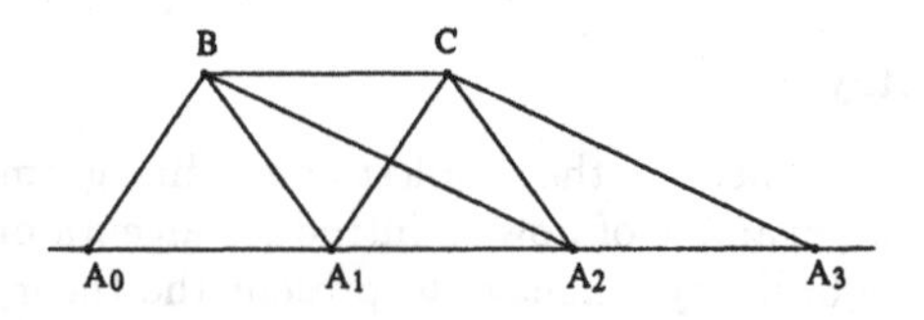

Figure 8

(A3) Axiom of Infinity. Given any two distinct points A_0 and A_1 on a line l and any point B not on l, we can construct parallelograms successively: A_0A_1BC, A_1A_2BC, $A_2A_3BC, \ldots$. Then $A_0, A_1, A_2, A_3, \ldots$ are all distinct points on the line l (Figure 8).

Note that there are an infinite number of axioms in (A3).

(A4) Pappus' Axiom.[1] Let l and l_1 be two distinct lines, and A, B, C and A_1, B_1, C_1 be distinct points on l and l_1, respectively. If $BC_1 \parallel B_1C$ and $AB_1 \parallel A_1B$, then $AC_1 \parallel A_1C$ (Figure 9).

[1] This axiom was called Pascalian axiom in [24].

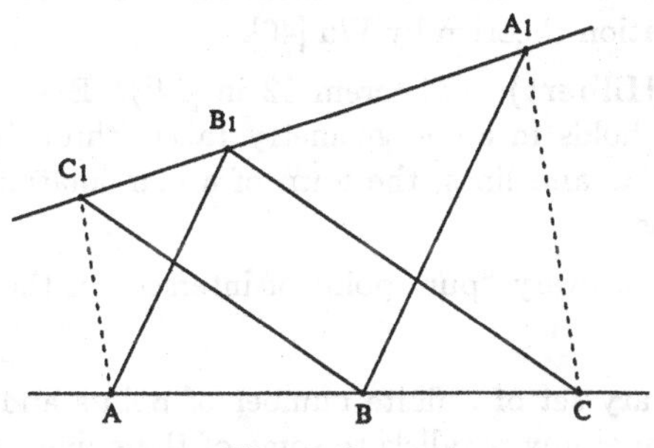

Figure 9

The development based on the above axioms in geometry textbooks is at the model level [24], [4], [45]. This development leads to the classification of models of the theory of affine geometry. Now we will briefly present this result.

Let E be a field with characteristic 0. From E we can construct a *structure* Ω in the sense of [35] as follows. Let

$$\tilde{L} = \{(a, b, c) \mid a, b, c \in \text{E}, \ a \neq 0 \vee b \neq 0\}.$$

We define a relation $\sim$ in $\tilde{L}$ as: $(a, b, c) \sim (a', b', c')$ if and only if there is a $k \in E$ such that $k \neq 0$ and $(a, b, c) = (ka', kb', kc')$. It is easy to see that $\sim$ is an equivalence relation. Let $\tilde{L}/\sim$ (the set of all equivalence classes of $\tilde{L}$) be denoted by L. Define $|\Omega|$ to be $E^2 \cup L$. An element p in $|\Omega|$ is a point if and only if $p \in E^2$ (i.e., $p = (x, y), x, \ y \in E$); an element l in $|\Omega|$ is a line if and only if $l \in L$. A point $p = (x, y)$ is on a line $l = (a, b, c)$ if and only if $ax + by + c = 0$.

It is easy to check the following theorem.

Theorem (1.1). The structure Ω defined above is a model of the theory (A1)–(A4) of affine geometry.

Proof. It is can be easily checked that (A1), (A2) and (A4) are valid in Ω. Especially (A4) can be proved using Wu's method. Since E has characteristic 0, (A3) holds. .QED.

The converse of the above theorem is a much deep result.

Theorem (1.2). Every model (geometry) G of the theory (A1)–(A4) is isomorphic to a structure Ω with some field E of characteristic zero.

The key step of the proof is to introduce the *segment arithmetic* and hence to introduce the field E inherent to G. The field E, uniquely determined by geometry G up to isomorphism, is called the *field associated with* geometry G. Pappus' axiom makes it possible to introduce such a field. If instead of Pappus' axiom we introduce a weaker axiom called Desargues' axiom,[2] then the associated number system is a division ring; i.e., the commutative law of multiplication does not necessarily hold. Each algebraic rule of operation (e.g., associativity of addition) corresponds to a geometry theorem. Pappus' Axiom is intimately connected with commutativity of

[2] See [45], [4].

multiplication. In this connection we will mention a theorem due to Hilbert, called the Hilbert mechanization theorem by Wu [46].

Theorem (1.3) (Hilbert). (Theorem 62 in [24]). Every pure point of intersection theorem that holds in affine geometry takes, through the construction of suitable auxiliary points and lines, the form of a combination of finite number of Pappus' configurations.

According to Hilbert, every "pure point of intersection theorem" can be put in the following form:

> Choose an arbitrary set of a finite number of points and lines. Then draw in a prescribed manner any parallels to some of these lines. Choose any points on some of the lines and draw any lines through some of these points. Then, if connecting lines, points of intersection and parallels are constructed through the points existing already in the prescribed manner, a definite set of finitely many lines is eventually reached, about which the theorem asserts that they either pass through the same point or are parallel.[3]

Following is Hilbert's proof of his theorem (see [24]):

> Consider now the coordinates of the completely arbitrarily chosen points and lines as parameters $u_1, ..., u_n$. Some coordinates of the less arbitrarily chosen points and lines can be considered as additional parameters $u_{n+1}, ..., u_d$. These elements are thus represented by the parameters $u_1, ..., u_d$. The coordinates of all connecting lines, points of intersection and parallels which can be constructed now, become then expressions $A(u_1, ..., u_d)$ which are rationally dependent on these parameters and the assertion of the proposed point of intersection theorem is represented by the statement that some expressions of the type will yield equal values for equal values of the parameters, i.e., the point of intersection theorem will state that certain expressions $R(u_1, ..., u_d)$ depending rationally on certain parameters $u_1, ..., u_d$ vanish as soon as some elements from the segment arithmetic introduced into proposed geometry, are substituted for these parameters. Since the domain of these elements is infinite one concludes by a well-known theorem of algebra that the expressions $R(u_1, ..., u_d)$ must vanish **identically on the basis of the Rules of Operation for fields.** However, in order to prove that the expressions $R(u_1, ..., u_d)$ vanish identically in the segment arithmetic it is sufficient now, as has already been shown for the application of the rules of operation, to apply Pappus' Theorem (Axiom), and the theorem is thus seen.

There are two issues requiring further discussion in the above proof.

(1) In the proof, Hilbert never mentioned nondegenerate conditions necessary to make a geometry statement valid. As we have already seen from many examples, without appropriate nondegenerate conditions, a geometry statement is usually invalid. The algebraic proof of a pure point of intersection theorem is valid if we

[3] From this definition we can see that the class of "pure point of intersection theorems" is a subclass of the class of constructive statements defined earlier in Section 1.6.

assume all denominators appearing in the computation are nonzero. Usually people are not explicitly concerned with those nondegenerate conditions.[3] In other words, "a geometry statement holds" in Hilbert's proof means a geometry statement is *generally true* (see Section 3.1). This issue was clarified in detail by Wu in [46].

(2) Hilbert's main idea was to translate a geometry problem into its algebraic form. Since there are only finite applications of algebraic rules in an algebraic proof of a geometry statement and each algebraic rule takes, through the construction of suitable auxiliary points and lines, the form of a combination of a finite number of Pappus' configurations, the whole algebraic proof also takes, through the construction of suitable auxiliary points and lines, the form of a combination of a finite number of Pappus' configurations. This almost convinces us that an algebraic proof of a geometry statement in affine geometry is (or can be converted into) a formal geometric proof of the statement in the theory. Since we introduce algebra into affine geometry at the model level, we need further clarification to justify that algebraic proofs can be converted into formal proofs in the theory. Especially in theories of more complicated geometries (e.g., metric geometry below), the justification is unclear.

However, algebraic proofs of geometry theorems can be at least regarded as proofs at the model level. Thus, by the Gödel completeness theorem in first order logic [35], an algebraic proof of a geometry statement proves the existence of a formal proof of the geometry statement in the theory.

1.2. Metric Geometry

The next example is metric geometry, introduced by Wu [45]. Metric geometry is a class of affine geometry in which there are two new basic relations "*perpendicularity*" and "(*segment*) *Congruence*". These new relations satisfy two groups of new axioms (*Axiom of Perpendicularity* and *Axiom of Congruence*, see [45], or [11]). These two groups of axioms, together with the axioms of affine geometry, makeup *the theory of metric geometry*. Here we list the theorems on classification of models of the theory of metric geometry.

Let the structure Ω be defined in affine geometry with the following two new features. Two lines $l = (a, b, c)$ and $l' = (a', b', c')$ are perpendicular if and only if $aa' + bb' = 0$. Let $A = (a_1, a_2)$, $B = (b_1, b_2)$, $X = (x_1, x_2)$, $Y = (y_1, y_2)$, $A \neq B$ and $X \neq Y$. Segments AB and XY are congruent (written as $AB \equiv XY$) if and only if $(a_1 - b_1)^2 + (a_2 - b_2)^2 = (x_1 - y_1)^2 + (x_2 - y_2)^2$. We have the following theorems [11]:

Theorem (1.4). If starting field E is a Hilbert field with characteristic 0, then Ω is a model of the theory of metric geometry (a *Hilbert field* is a field in which the sum of the squares of two elements has a square root).

Theorem (1.5). Every model G of the theory of metric geometry is isomorphic to a structure Ω with some Hilbert field E of characteristic zero.

For simplicity, from now on we will write Ω simply as E^2.

---

[3] Also see the proof of Pappus' theorem using a computer by P.J. Davis [19].

Let us discuss the congruence of angles in metric geometry ([43], [45]). In unordered metric geometry, we cannot talk about rays. Thus we cannot define angles in the ordinary way. We can define an angle simply as an unordered pair of lines l_1 and l_2 which have at least one common point, and denote it by $[l_1, l_2]$. Two angles $\alpha = [l_1, l_2]$ and $\beta = [s_1, s_2]$ are congruent (denoted by $\alpha \equiv \beta$) if there is a series of reflexions to map l_1 to s_1 and l_2 to s_2. As we know, in metric geometry reflexions can be defined from more basic relations, "midpoint" and "perpendicular".[4] Let k_1 and k_2 be the slopes of l_1 and l_2 in E^2, respectively. We define function $T(k_1, k_2) =$

$(k_2 + k_1)/(1 - k_1 k_2)$ if $k_1 \neq \infty$, $k_2 \neq \infty$ and $1 - k_1 k_2 \neq 0$
∞ if $1 - k_1 k_2 = 0$
$-1/k_2$ if $k_1 = \infty$ and $k_2 \neq \infty$
$-1/k_1$ if $k_1 \neq \infty$ and $k_2 = \infty$
0 if $k_1 = k_2 = \infty$.

Then we can define the function $tan(l_1, l_2)$ to be $T(-k_1, k_2)$. It can be proved that for non-isotropic lines l_1, l_2, s_1 and s_2, $[l_1, l_2] \equiv [s_1, s_2]$ if and only if $tan(l_1, l_2) = \pm tan(s_1, s_2)$. We define an oriented angle as *an ordered pair* of two lines l_1 and l_2 which have at least one common point, denoting it by $\angle(l_1, l_2)$, and define the congruence of two oriented angles to be $tan(l_1, l_2) = tan(s_1, s_2)$.

From now on, we will only talk about oriented angles and their congruence. Let E be a point distinct from points A and B, then we use $\angle AEB$ to denote the oriented angle $\angle(AE, BE)$. In Figure 1 of Chapter 1 (in the real plane), $\angle AEB$ *is* $\angle AED$ or $\angle CED$, but $\angle AEB \not\equiv \angle DEA$ and $\angle AEB \not\equiv \angle DEC$.

Now we can define the addition of two angles. $\angle(l_1, l_2) \equiv \angle(s_1, s_2) + \angle(t_1, t_2)$ if there is a line l such that $\angle(l_1, l) \equiv \angle(s_1, s_2)$ and $\angle(l, l_2) \equiv \angle(t_1, t_2)$. It can be proved that $\angle(l_1, l_2) \equiv \angle(s_1, s_2) + \angle(t_1, t_2)$ if and only if $tan(l_1, l_2) = T(tan(s_1, s_2), tan(t_1, t_2))$. The congruence of angles is an equivalence relation. The set of all the equivalence classes with the addition of angles form an additive (Abelian) group. In particular, $\angle(l, l) = 0$ and $\angle(l_1, l_2) = -\angle(l_2, l_1)$.

A bisector of $\angle AOB$ can be defined to be a line OC such that $2\angle AOC$ ($= \angle AOC + \angle AOC$) $= \angle AOB$; this is equivalent to "$\angle AOC \equiv \angle COB$" or to "$tan(AOC) = tan(COB)$". There are two bisectors for a given angle.

A trisector of $\angle AOB$ is defined to be a line OC such that $3\angle AOC = \angle AOB$, or equivalently, $tan(3\angle AOC) = tan(AOB)$. Let $tan(AOB) = s$, $tan(AOC) = k$. We have the equation:

$$\frac{3k - k^3}{1 - 3k^2} = s, \text{ or } k^3 - 3sk^2 - 3k + s = 0.$$

If there is a trisector OC of angle $\angle AOB$, then there are three trisectors of $\angle AOB$. In Euclidean geometry, they correspond to $\angle AOC$, $\angle AOC + \pi/3$, $\angle AOC + 2\pi/3$. Being irreducible for the general value s, the above equation cannot distinguish them. A trisector of an angle always exists if the associated field of the geometry is a real closed field or an algebraically closed field.

[4] The discussion here is at the model level.

1.3. Hilbert Geometry

The theory of Hilbert geometry consists of the first four groups of axioms in [24]. As we know, each model in this theory can be represented as E^2, where E is an *ordered* Hilbert field.

Thus we can say Hilbert geometry is *ordered metric geometry*. The reader might ask why we did not include group 5 of axioms in [24] in Hilbert geometry. The axioms in group 5 are not in first-order logic unless we introduce other objects in the formal theory. For example, the first axiom in this group is Archimedean axiom, which can be expressed in first order logic only by introducing natural numbers into the formal theory. The surprising fact is that most developments in Hilbert's book [24] are independent of Archimedean axiom. As people pointed out (see e.g., [45]), Hilbert geometry is in essence Non-Archimedean.

Now suppose Hilbert geometry (a model) E^2 satisfies Archimedean axiom, then we can prove that E is isomorphic to a subfield of $\mathbf{R}$. (Sketch of the proof: by Archimedean axiom we can prove that $\mathbf{Q}$, the field of all rationals, is dense in E. Then we can use the completeness property of $\mathbf{R}$ to prove E is a subfield of $\mathbf{R}$.) Thus, such a model can be regarded as a part of the real plane.

1.4. Tarski Geometry

The theory of Tarski geometry was given fully in [39]. Each model of the theory of Tarski geometry can be represented as R^2, where R is a real closed field [39]. Due to a fundamental contribution by Tarski [38], the theory of Tarski geometry is decidable. Note that the decision procedure is carried out in number systems (real closed fields), i.e., at the model level. This geometry is what Tarski called elementary geometry. As we know, there are Non-Archimedean Tarski geometries. The meaning of the term "elementary geometry" depends on who uses it. Following Wu, we prefer elementary geometry to include all geometries in which no differentiation is involved. Elementary geometry in this sense includes not only all geometries discussed so far, but also those "advanced" geometries, such as Bolyai-Lobachevskian geometry and Riemannian geometry.

The above four geometries are compatible with Euclidean geometry in the sense that Euclidean geometry is a model of the theories of the four geometries. Among geometries incompatible with Euclidean geometry, we mention projective geometry, Bolyai-Lobachevskian geometry, Riemannian geometry, etc. For those geometries, algebra and fields can be also introduced [22], [24], [36], [37], [21], [6].

The inclusion relationship among the models of the four geometries discussed above can be illustrated by the following diagram.

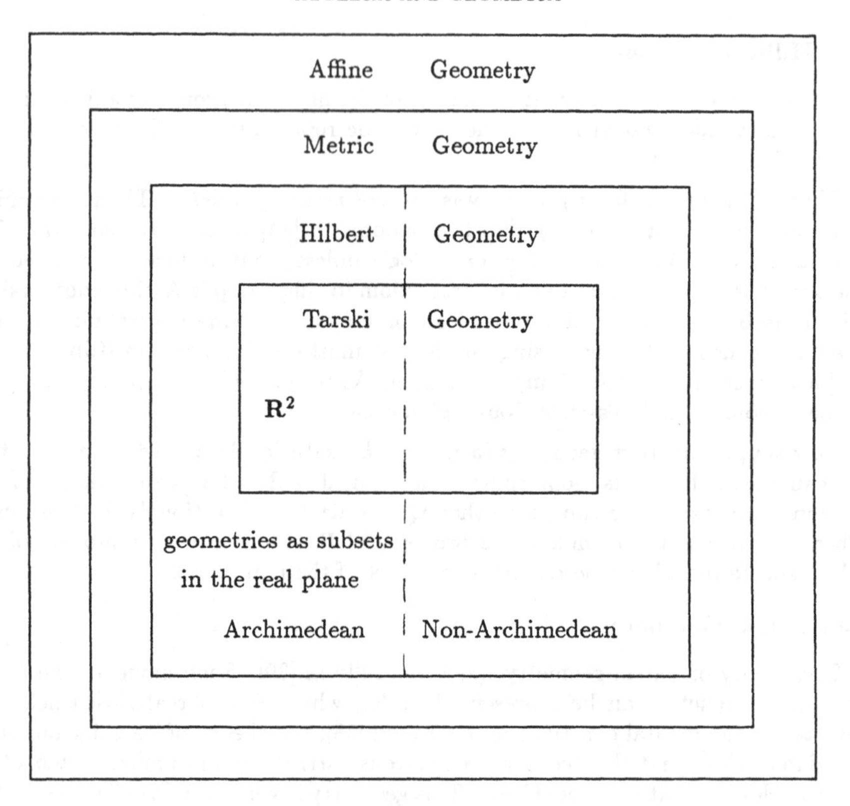

2. On the Algebraic formulation of Geometry Statements

Now we can discuss the problem of algebraic formulation of geometry statements involving equalities only. We shall list several possible formulations and discuss related issues.

For simplicity, we first confine our discussion to a metric geometry whose associated field E is algebraically closed. Furthermore, we only discuss those geometric statements whose hypotheses and conclusion can be expressed by the basic relations in metric geometry – incidence, parallelism, perpendicularity, and congruence with *universal quantifier*; i.e., those geometry statements for which, after adopting appropriate coordinate systems, the hypotheses can be expressed as a set of polynomial equations:

$$h_1(y_1, y_2, ..., y_m) = 0$$

$$\dots$$

$$h_n(y_1, y_2, ..., y_m) = 0,$$

and the conclusion can also be expressed as an equation $g(y_1, ..., y_m) = 0$. Note that the h_i and g are in $\mathbf{Q}[y_1, ..., y_m]$. This is a consequence of the connection between geometry and algebra. In metric geometry, the Axiom of Infinity in section 1.1 implies that the field E associated with any model of metric geometry has characteristic 0. Thus we can assume that E contains $\mathbf{Q}$, the field of rational numbers.

As we know from many examples before, the formula

$$\forall y_i \in E(h_1 = 0 \wedge ... \wedge h_n = 0 \Rightarrow g = 0) \tag{2.0}$$

is not valid for most geometry theorems because some nondegenerate conditions are missing. Note that all nondegenerate conditions are in inequation forms $s_1 \neq 0, ..., s_k \neq 0$, where the s_i are polynomials in $\mathbf{Q}[y_1, ..., y_m]$.

2.1. Formulation F1

Together with $h_1 = 0, ..., h_n = 0$, all nondegenerate conditions necessary for the validity of the statement are specified in the polynomial equation forms $s_1 \neq 0, ..., s_k \neq 0$. Then the algebraic formulation is:

$$\forall y_i \in E[(h_1 = 0 \wedge ... \wedge h_n = 0 \wedge s_1 \neq 0 \wedge ... \wedge s_k \neq 0) \Rightarrow g = 0]. \tag{2.1}$$

Theoretically, this formulation seems perfect and standard. In our opinion, however, finding all nondegenerate conditions essential for the validity of the statements is not easy.

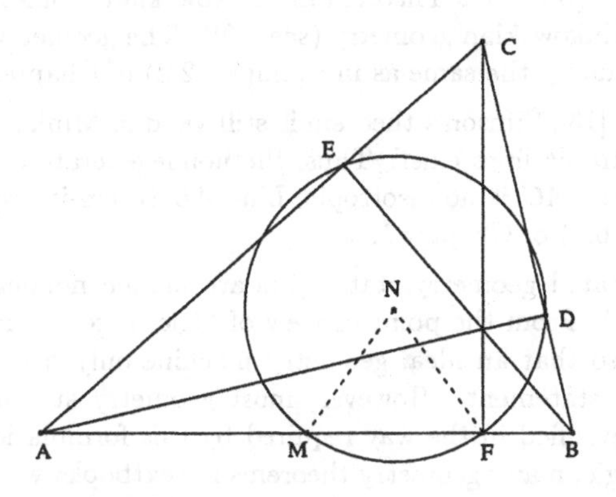

Figure 10

Example (2.2). (One form of the nine-point circle theorem) Let points D, E and F be the three feet of the altitudes of triangle ABC. Let N be such that $NF \equiv NE$ and $NF \equiv ND$, and M be the midpoint of AB. Then $NM \equiv NF$ (Figure 10).

In the above precise specification of the equality part of the statement, besides the condition that A, B and C are not collinear, we need at least the nondegenerate condition that triangle ABC is not a right triangle, which is not easy to detect.

Example (2.3). (M. Paterson) Three similar isosceles triangles, A_1BC, AB_1C, and ABC_1, are erected (either all externally or all internally[5]) on the three respective

[5] In specifying "internally" or "externally" we have to discuss the statement in ordered metric geometry. However, with a proper specification, this is a statement in unordered metric geometry. See example (5.7) of Chapter 4.

sides, BC, CA, AB, of a triangle ABC, then AA_1, BB_1 and CC_1 are concurrent.

Here besides other nondegenerate conditions, we need the condition that the triangle ABC is not an isosceles triangle.

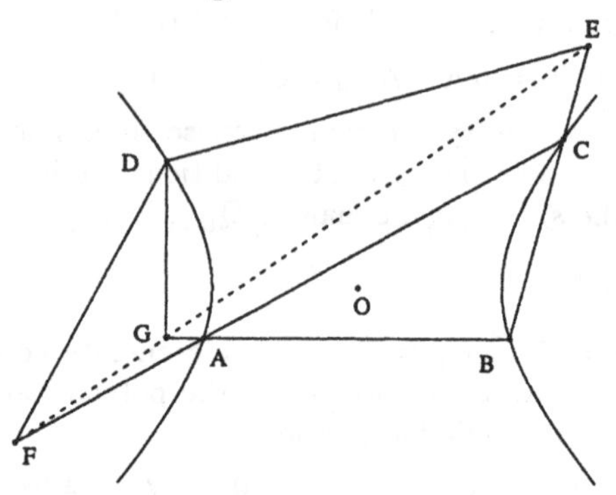

Figure 11

Example (2.4). (Simson's Theorem in Minkowskian geometry, Figure 11). Let us now consider Minkowskian geometry (see [49]). The geometry statement of Simson's theorem is exactly the same as in example (2.2) of Chapter 1.

As we proved in [13], Simson's theorem is still valid in Minkowskian geometry. In this geometry, isotropic lines exist. Thus, the nondegenerate conditions "Line BC is non-isotropic; Line AC is non-isotropic; Line AB is non-isotropic" are necessary. Also see example (6.4) of Chapter 1.

In more complicated geometry statements are hidden nondegenerate conditions even harder to find. From the point of view of logic, a geometry statement should be specified fully so that an ideal geometry machine only needs to answer "true" or "false" to that statement. However, most geometry statements in textbooks have never been specified in the way required by this formulation. Thus, from the point of view of logic, many geometry theorems in textbooks would be false because of missing nondegenerate conditions. One would think that geometry theorems in textbooks should be revised, adding a sufficient number of nondegenerate conditions. As we have already seen, this is a very hard task. Besides, the term "sufficient" is hard to specify.

In this formulation there are no differences between the following two cases: (a) a geometry statement is false because a condition that "a triangle (in the statement) is a right (or isosceles) triangle" is missing; (b) a geometry statement is false because a condition that "a triangle (in the statement) is *not* a right (or isosceles) triangle" is missing. From the point of view of logic, geometry statements in the above cases are both considered false regardless how much they are different from true statements. However, from the point of view of mathematics or geometry, the two cases are essentially different. The geometry statement in (a) should be considered *generally false*, whereas the geometry statement in (b) should be considered *generally true*. (See Section 3.1).

Thus, this theoretically perfect formulation seems to be unsatisfactory from the practical point of view of mechanical theorem proving. The issues we have raised here are important for implementing a practical prover based on this formulation.

In [27], there is a proof method for this formulation and it was claimed: "subsidiary conditions, which are an integral part of a geometry statement, are handled in a natural way in proposed approach." This approach was independently experimented with by the author [16]. In our opinion, subsidiary (nondegenerate) conditions are not handled in a natural way in this approach. For example, in proving Simson's theorem (example (2.2) of Chapter 1) in Euclidean geometry, some "artificial" subsidiary conditions $u_3^2 + u_2^2 - 2u_1u_2 + u_1^2 \neq 0$, $u_3^2 + u_2^2 \neq 0$ *have to be added by the user* to make the proof successful (see examples and discussions in Section 3 of Chapter 5).

One would possibly prefer those formulations and methods which can mechanically produce nondegenerate conditions. Since the term "nondegenerate conditions" hasn't been formalized, one should take extremely precautions. To emphasize this important issue, we first give a formulation appearing in current literature [31], [27], which (with its variants) can cause serious problems.

2.2. Formulation F2

In this formulation, to prove a geometry statement is to find a polynomial s in $\mathbf{Q}[y_1, ..., y_m]$, such that

(2.5) $\quad \forall y_i \in E[(h_1 = 0 \wedge ... \wedge h_n = 0 \wedge s \neq 0) \Rightarrow g = 0]$ and

(2.6) $\quad \neg(\forall y_i \in E[(h_1 = 0 \wedge ... \wedge h_n = 0) \Rightarrow s = 0])$.

Thus any method based on this formulation can deduce a subsidiary condition $s \neq 0$, under which the original formula (2.0) is valid. The algebraic problem in this formulation is well defined. However, the condition $s \neq 0$ often has nothing to do with nondegeneracy in geometry. To make things worse, this formulation can catastrophically change the meaning of a geometry statement to be proved by adding inappropriate subsidiary conditions, thus causing certain kind of *unsoundness*. This is illustrated in the following examples.

Example (2.7). Let ABC be a triangle, and BE the altitude from B. Show that $AB \equiv CB$ (Figure 12).

Let $A = (0,0)$, $B = (y_1, 0)$, $C = (y_4, y_5)$, and $E = (y_2, y_3)$, then we have

$h_1 = y_3y_5 + (y_2 - y_1)y_4 = 0$ $\qquad BE \perp AC$

$h_2 = y_2y_5 - y_3y_4 = 0$ $\qquad E$ is on AC

$g = y_5^2 + y_4^2 - 2y_1y_4 = 0$ $\qquad$ Conclusion: $AB \equiv CB$.

Let $s_1 = y_3^2 + y_2^2 - y_1y_2$. Then s and g satisfy both (2.5) and (2.6). Thus under this formulation, we prove that the triangle ABC in the geometric configuration is isosceles by adding $s \neq 0$. Note that $AB \equiv CB$ is generally false for the given geometric configuration.

Example (2.8). Let $ABCD$ be a parallelogram, O be the intersection of diagonals AC and BD. Show A, B and O are collinear (Figure 13).

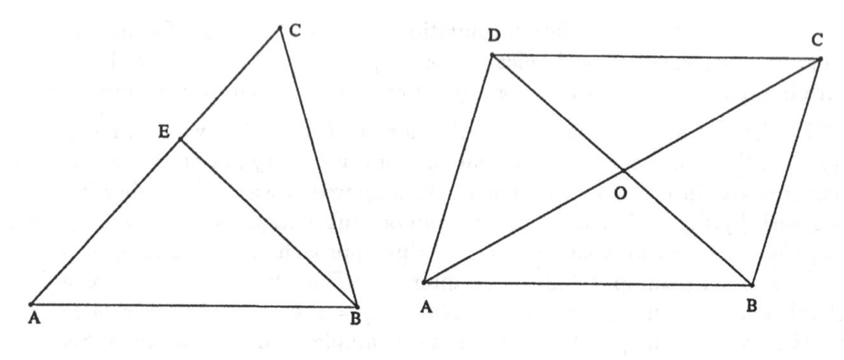

Figure 12 Figure 13

Let $A = (0, 0)$, $B = (u_1, 0)$, $C = (u_2, u_3)$, $D = (x_2, x_1)$, and $O = (x_4, x_3)$, then we have

$h_1 = u_1x_1 - u_1u_3 = 0$	AB is parallel to DC
$h_2 = u_3x_2 - (u_2 - u_1)x_1 = 0$	DA is parallel to CB
$h_3 = x_1x_4 - (x_2 - u_1)x_3 - u_1x_1 = 0$	O is on BD
$h_4 = u_3x_4 - u_2x_3 = 0$	O is on AC
$g = u_1x_3 = 0$	Conclusion: A, B and O are collinear.

Let $s = u_3x_2 - (u_2 - u_1)u_3$. Then s and g satisfy both (2.5) and (2.6). Thus we have proved a statement which is generally false in geometry by adding $s \neq 0$. It is interesting to note that for the above s, "$h_1 = 0, ..., h_4 = 0, s \neq 0 \Rightarrow g_1 = 0$" is not valid, where $g_1 = 2u_2x_4 + 2u_3x_3 - u_3^2 - u_2^2 = 0$ is the condition that $AO \equiv OC$, which is *generally true* in geometry!

3. Formulation F3

This formulation is based on the basic notions in algebraic geometry. First we observe that the equality part of the hypotheses of a geometry statement specified at the beginning of Section 2 defines a *geometric configuration*. In its algebraic form, this configuration can be given by the set of equations $h_1(y_1, ..., y_m) = 0, \ldots, h_n(y_1, ...y_m) = 0$. Let $V = V(h_1, ..., h_n)$ be the common zeroes in E^m, i.e., the algebraic set $V = \{(a_1, ..., a_m) \in E^m \mid h_i(a_1, ..., a_m) = 0 \text{ for all } i\}$.

Each element in V can be regarded as an instance (or a special case) of the given configuration. Note that V includes the degenerate cases of the given configuration. Whether an instance of the configuration should be considered degenerate *is inherent to the configuration and independent of any assertion (conclusion) about the configuration.* To distinguish the degenerate cases from nondegenerate cases, we can use parameters. This was actually done in Hilbert's theorem (1.3). We use the variables $u_1, ..., u_d$ (among the y) to denote the nonzero coordinates of those points which can be arbitrarily chosen and use the variables $x_1, ..., x_r$ (the rest y) to denote the nonzero coordinates of those points which satisfy certain geometric constraints. The variables $u = u_1, ..., u_d$ are called *parameters* or *independent variables* and the variables $x = x_1, ..., x_r$ are called *dependent variables*. We have $m = d + r$. The determination of arbitrarily chosen points of the given configuration, though not unique and informal, is from *the geometric meaning* of the given statement.

Using Ritt's decomposition algorithm (5.1) of Chapter 2,[6] V can be decomposed into irreducible components, represented by their irreducible characteristic sets:

$$(3.1)\quad V = V(P_1^*) \cup ... \cup V(P_c^*) \cup V(P_1^\dagger) \cup ... \cup V(P_e^\dagger) \cup V(P_1) \cup ... \cup V(P_l),$$

where P_i^*, $P_i^\dagger$ and P_i are prime ideals, $V(P_i^*) = V_i^*$ and $V(P_i^\dagger) = V_i^\dagger$ are all those components on which $u_1, \ldots, u_d$ are algebraically independent, and $dim(V_i^*) = d$, $dim(V_i^\dagger) > d$. We call the V_i^* the ***components general for u*** or a ***general component.*** The condition $c > 0$ serves as a necessary criterion to check whether the parameters have been selected correctly.

Remark 1. Even if we use the notion of general positions, it is still possible that some components general for u correspond to what people usually consider degenerate cases. This happens rarely and is due to the ambiguity in specifying points using algebraic equations, similar to that mentioned in example (2.4) of Chapter 1 (the Butterfly theorem). However, for practical problems it is easy to detect such kinds of ambiguities and to exclude them mechanically. Thus we can assume that equations $h_1 = 0, ..., h_n = 0$ have already been ***properly specified*** so that there are no such components general for u or ambiguities.[7]

Remark 2. It is extremely rare that $e > 0$. In more than 600 non-trivial geometry theorems proved by our prover, none fit this case. If $e > 0$, then there might be some hidden property (theorem) of the geometric figure, and the problem is possibly beyond the original intention of the statement (see example (1.6) of the next chapter). In our current implementation we discard such cases, considering them inherently degenerate. However, the program also outputs the related information to let the user take a closer look at the problem. Thus *in what follows we always assume* $e = 0$. This assumption is related to Dimensionality Restriction (1.4) in the next chapter.

Remark 3. For a given geometric configuration, now it is clear that the $V(P_i^*)$ correspond to nondegenerate cases or generic cases, and the $V(P_j)$ correspond to degenerate cases because on those components the u_i are algebraically dependent. Thus the $V(P_i^*)$ are also called *nondegenerate components.*

3.1. The Generic Validity of a Geometry Statement

Now we are in a position to define what is meant by "generally true" or "generally false."

Formulation F3. Let a geometry statement (S), in its algebraic form, be given by $u_1, ..., u_d$, $x_1, ..., x_r$, $h_1, ..., h_n$ and g, where the u, x, h and g are the same as before. (S) is called *generally true* (or *generally valid*) if g vanishes on all nondegenerate

[6] Actually, the about decomposition is independent of any algorithm, see theorem 1.10 of chapter 2.

[7] In our experiments with more than 600 theorems, such an approach was totally successful. See [14] or Part II.

components $V(P_i^*)$ of $V = V(h_1, ..., h_n)$. "(S) is *not generally true*" is the negation of "(S) is generally true". If g vanishes on none of the nondegenerate components $V(P_i^*)$, then (S) is called *generally false*.

Remark. For any polynomial c, if c vanishes on all nondegenerate components $V(P_i^*)$, we also say that the assertion $c = 0$ is *generally true* for the given geometric configuration. If c vanishes on none of the nondegenerate components $V(P_i^*)$, then we say that the assertion $c = 0$ is *generally false* for the given configuration.

For more than 95 percent of the problems we encountered in practice, $c = 1$. In such case, "not generally true" is equivalent to "generally false" and we have a clear-cut answer: either a geometry statement is generally true, or it is *generally false*. However, we also found a few problems belonging to the following case:

The polynomial g vanishes on some but not all nondegenerate components.

The answer to such problems is not clear cut. Some people may say that the geometry statement is not generally true because it is not true in some nondegenerate cases. However, others could counter that under some subsidiary conditions, which exclude those *nondegenerate* cases, the geometry statement is true. In our opinion, these problems should be investigated further. They often become more interesting. If the original problem comes from Euclidean geometry, then it is usually connected with order relation. We should use the order relation to obtain a complete answer. We will give examples in the next chapter (examples (5.3), (5.4), (5.5) and (5.6) of the next chapter). Here we only give an example to illustrate formulation F3.

Example (3.2). Example of parallelogram. The same as example (2.8). Under these geometric conditions, we want to decide whether (i) $AO \equiv CO$; (ii) point A, O and B are collinear.

The choice of coordinates are the same as in example (2.8). Since points A, B and C can be arbitrarily chosen, u_1, u_2 and u_3 can be chosen as parameters. Once these points are normally fixed, the remaining points D and O can be determined. Thus their coordinates x_1, x_2, x_3 and x_4 are algebraically dependent on u_1, u_2 and u_3 under normal (nondegenerate) conditions, being restricted by equations $h_1 = 0, ..., h_4 = 0$. We have two conjectures: $g_1 = 2u_2x_4 + 2u_3x_3 - u_3^2 - u_2^2 = 0$ ($AO \equiv CO$); $g_2 = u_1x_3 = 0$ (A, O and B are collinear). According to Ritt's decomposition algorithm, we can decompose $V(h_1, ..., h_4)$ into four irreducible components:[8]

$$V = V(P_1^*) \cup V(P_1) \cup V(P_2) \cup V(P_3),$$

where P_1^*, P_1, P_2, and P_3 have

$C_1^* = \{f_1, f_2, f_3, f_4\}$, ($f_1, ..., f_4$ are the same as in example (2.1) in Chapter 1)

$C_1 = \{u_3, x_1, x_4\}$,

$C_2 = \{u_1, u_3x_2 - u_2x_1, u_2x_4 - u_3x_3\}$, and

$C_3 = \{u_2, u_3, x_1, x_2 - u_1\}$

for their characteristic sets, respectively. Thus we have a unique nondegenerate

---

[8] Actually, we don't have to get all irreducible components to decide the general validity of a geometry theorem. See Section 2 of the next Chapter.

component $V(P_1^*)$. To check whether $g_1 = 0$ or $g_2 = 0$ are generally true, we only need to check whether g_1 or g_2 vanishes on $V(P_1^*)$. According to (3.7) of Chapter 2, it is equivalent to checking whether $prem(g_1, C_1^*) = 0$ or $prem(g_2, C_1^*) = 0$. Now $prem(g_1, C_1^*) = 0$ but $prem(g_2, C_1^*) \neq 0$. Thus, (i) is generally true, but (ii) is generally false.

Note that $dim(V(P_1)) = dim(V(P_2)) = 4 > dim(V(P_1^*)) = 3$. *Dimensions of degenerate components can be even higher than that of general components.* Thus, we cannot distinguish degenerate cases from general cases based only on dimension consideration.

3.2. Identifying Nondegenerate Conditions

Formulation F3 actually specifies what nondegenerate conditions should be for a given geometric configuration. We have already seen a practical method for giving sufficient subsidiary conditions so that the conclusion $g = 0$ is valid under $h_1 = 0, \ldots, h_n = 0$ and the conditions $I_1 \neq 0, \ldots, I_r \neq 0$ (where the I_i are initial of the f_i obtained from triangulation procedure; see examples in Chapter 1).

Example (3.3). Example (3.2) revised. Thus, the nondegenerate conditions are $I_1 = u_1 \neq 0$, $I_2 = u_3 \neq 0$, $I_3 = u_3x_2 - u_2x_1 - u_1u_3 \neq 0$, $I_4 = u_3 \neq 0$. (Also see Example (2.1) of Chapter 1.)

A polynomial in $\mathbf{Q}[u, x]$ is said to be a *u-polynomial* if it is nonzero and in $\mathbf{Q}[u]$. In this example (or in examples (2.2), (2.3), and (2.4) in Chapter 1), the ascending chain (in weak sense) $f_1, \ldots, f_r$ is irreducible. By (3.5) of Chapter 2, for each initial I_i, there are a polynomial q_i and a u-polynomial c_i such that $q_iI_i - c_i \in (f_1, \ldots, f_r)$. Under $f_1 = 0, \ldots, f_r = 0$, $c_i \neq 0 \Rightarrow I_i \neq 0$. Thus we have u-polynomials $c_1, \ldots, c_r$, such that $c_i \neq 0$, $f_i = 0 \Rightarrow g = 0$. In general, we have the following theorem:

Theorem (3.4). We use the same notations in Section 3.1. For the given geometric configuration defined by $h_1 = 0, \ldots, h_n = 0$, there is a u-polynomial s such that *any* assertion (conclusion) $g = 0$ is generally true for the given configuration if and only if $s \cdot g$ is in Radical$(h_1, \ldots, h_n)$, i.e., the radical ideal generated by $h_1, \ldots, h_n$.

Proof. let

(3.5) $$V(h_1, .., h_n) = V_1^* \cup \ldots \cup V_c^* \cup V_1 \cup \ldots \cup V_l,$$

where the V_i^* are all components general for u, and the u are algebraically dependent on the V_i. Since u are algebraically dependent on each V_j, there is a u-polynomial s_j such that s_j vanishes on V_j. Let $s = s_1 \cdots s_l$ (if $l = 0$, then let $s = 1$).

(i) Suppose $g = 0$ is generally true, i.e., g vanishes on each V_i^*. Since

$$V(h_1, \ldots, h_n) - V(s) \subset V_1^* \cup \ldots \cup V_c^*$$

and g vanishes on $V_1^* \cup \ldots \cup V_c^*$, sg vanishes on $V(h_1, \ldots, h_n)$, i.e., $sg \in$ Radical$(h_1, \ldots, h_n)$ by Hilbert's Nullstellensatz.

(ii) Conversely, if $sg \in$ Radical$(h_1, \ldots, h_n)$ for the above u-polynomial s, then sg vanishes on $V(h_1, \ldots, h_n)$. Since s contains the u only, s does not vanishes on each V_k^*. Since each V_k^* is irreducible, g vanishes on each V_k^*. .QED.

Note that s depends only on the configuration (the set of hypotheses) and is independent of a particular assertion (conclusion) $g = 0$ about the configuration.

Remark (3.6). Actually, the polynomial s in (3.4) is not necessarily a u–polynomial. The assertion $g = 0$ is generally true for the given configuration if and only if there is a polynomial s such that $s = 0$ is generally false for the given configuration and $sg \in \text{Radical}(h_1, ..., h_n)$. The proof is the same as the proof of (3.4).

Now we address the questions raised in section 2 of Chapter 1. The first question is

(1) *Will the theorem to be proved be weakened by introducing these additional conditions? Especially, will the hypotheses become inconsistent by adding these new nondegenerate conditions?*

First, we point out that the method proves the statement to be generally true, independent of the number of subsidiary conditions.

Theorem (3.7). Let the notations be the same as in theorem (3.4). Let s be the polynomial in the proof of theorem (3.4). Let t be any u–polynomial. Then

$$T = I(V(h_1, ..., h_n) - V(s)) = I(V(h_1, ..., h_n) - (V(s) \cup V(t)),$$

where the notation $I(X)$ denotes the set of all polynomials vanishing on the set X in E^m.

Proof. This is an immediate consequence of theorem (3.4). .QED.

Obviously, the set T is an ideal of $K[u, x]$, which represents all assertions generally true for the given configuration.

The following fact can be used to explain to what extent a statement is weakened by adding subsidiary conditions.

Proposition (3.8). Suppose we have decomposition (3.5), let $c_1, ..., c_f$ be a finite set of u–polynomials. Then for each V_i^*, $V_i^* - V(c_1) \cup ... \cup V(c_f)$ is nonempty.

Proof. Let $c = c_1 \cdots c_f$, then $V(c) = V(c_1) \cup ... \cup V(c_f)$. Since c contains the u only and the u are algebraically independent on V_i^*, c does not vanish on V_i^*. The proposition follows. .QED.

Remark (3.9). From Proposition (3.8), we can see that each nondegenerate component V_i^* contains at least an infinite number of cases of the form $V(c_i)$. Thus deleting such finite cases (or equivalently, adding finite subsidiary conditions) will not weaken the statement "severely". Especially, it will not cause inconsistency.

The second question is

(2) *Are all those subsidiary conditions necessary? Can those inequations which are necessary for mechanical proofs be translated back into their geometric forms by the program?*

As we have already seen, some subsidiary conditions produced by the method are not necessarily connected with nondegeneracy and are not necessary for the validity of the statement. The remaining have geometric meanings and are connected with nondegeneracy. For statements of constructive type, the program can do this analysis, and translate those which can be considered nondegenerate conditions in the ordinary sense, back into their geometric forms. However, so far there are no mechanical methods to do this in general cases. We strongly encourage people who wish to implement a geometry theorem prover based on similar methods to translate algebraic inequations into their geometric forms as mechanically as possible.

3.3. The Generic Validity of a Geometry Statement in An Arbitrary Field

The above formulation is for geometries with associated fields algebraically closed. Now we extend the formulation to geometries with any associated fields. Our starting point is the decomposition (3.1). Since the associated field E now is not necessarily algebraically closed, this decomposition is generally not an irreducible decomposition. However, the u are algebraically dependent on any components of the algebraic sets $V(P_i)$, and thus we can regard $V(P_i)$ as degenerate cases. We will still call any $V(P_i^*)$ a component general for u or a nondegenerate component (note that $V(P_i^*)$ can be reducible or even empty).

Formulation F3 for Any Fields. If the polynomial g vanishes on all $V(P_i^*)$, then we say the geometry statement (or $g = 0$) is *generally true*. If the polynomial g vanishes on none of the $V(P_i^*)$, then the statement (or $g = 0$) is *generally false*.

Theorem (3.10). If a geometry statement (S) is generally true in a model of metric geometry whose associated field is algebraically closed, it is generally true in any other model G of metric geometry.

Proof. Let E be the field associated with G. Theorem (3.4) implies that if (S) is generally true in an algebraically closed field, then it is generally true in any other model whose associated field is algebraically closed. In particular, it is generally true in a model whose associated field $\overline{E}$ is algebraically closed and contains E. Therefore (S) is also generally true in the model G. .QED.

Definition (3.11). A statement is said to be *a generic statement* in E if all nondegenerate components $V(P_i^*)$ are of dimension d.

Theorem (3.12). For a generic statement (S) in E, (S) is generally true (in E) if *and only if* it is true in an algebraically closed field.

Proof. Suppose (S) is generically true in E. Let $g = 0$ be the conclusion of (S). For each i, since g vanishes on $V(P_i^*)$ and $V(P_i^*)$ has dimension d, $g \in P_i^*$ by (3.8) of Chapter 2. Thus g vanishes on F–$V(P_i^*)$ for any algebraically closed field F. (S) is generally true in any algebraically closed fields. .QED.

Remark (3.13). Suppose $dim(V(P_i^*)) = d$ for some i. According to Remark (3.8) of Chapter 2, proposition (3.8) is still true for this i. Thus remark (3.9) can also apply to this case. Thus for a generic statement (in some field), adding finite subsidiary conditions $c_1 \neq 0, \ldots, c_f \neq 0$ (here the c_i are u–polynomials) will not weaken the statement.

Chapter 4. The Complete Method of Wu

Now we shall present the complete method of Wu in detail. Generally, this method works for any field but is complete only for those geometries whose associated fields are algebraically closed.

1. Ritt's Principle Revised

For a geometry statement (S), suppose its hypotheses (with nondegenerate conditions unspecified) can be algebraically expressed by a set of polynomial equations

$$h_1(u_1, ..., u_d, x_1, ..., x_t) = 0$$
$$\cdots$$
$$h_n(u_1, ..., u_d, x_1, ..., x_t) = 0,$$

and the conclusion is also a polynomial equation $g = g(u_1, ..., u_d, x_1, ..., x_t) = 0$, where $h_1, ..., h_n$ and g are polynomials in $\mathbf{Q}[u_1, ..., u_d, x_1, ..., x_t]$. Here $u_1, ..., u_d$ are parameters and $x_1, ..., x_t$ are dependent variables.

Now we present Ritt's Principle in the context of geometry theorem proving: we can let $K = \mathbf{Q}[u_1, ..., u_d]$, $x_i = y_i$, $t = m$, and $I = (h_1, ..., h_n)$. Note that Ritt's principle also works when K is the ring $\mathbf{Q}[u]$. Thus, applying Ritt's principle to $h_1, ..., h_n$, we can obtain an ascending chain C such that either

Case (1.1). C consists of a polynomial in $K \cap I$, or

Case (1.2). $C = f_1, \ldots, f_r$ with $class(f_1) > 0$ and such that $f_i \in I$ and $prem(h_j, f_1, ..., f_r) = 0$ for all $i = 1, \ldots, r$ and $j = 1, \ldots, n$.

In the case of (1.1), we have a polynomial in the parameters u only. This cannot happen if the parameters are properly chosen. Whenever this happens the program stops and outputs the related information. The user should reformulate the problem or choose the parameters in an appropriate way. If the statement is of constructive type and the choice of parameters is done by the program, this means that there possibly exist certain unexpected properties for the geometric figure in the statement.

Example (1.3). In example (2.2) of Chapter 1, if we wish to discuss properties of the circumscribed circle of "triangle" EFG and let its center P be (x_9, x_8), then this situation happens. The reason is that EFG is a line, not a triangle.

In the case of (1.2), we usually need the following restriction:

Dimensionality Restriction (1.4). Every dependent variable x_i appears as a leading variable of an f_i. This means that $t = r$, and $x_i = lv(f_i)$. Thus the ascending chain $f_1, ..., f_r$ has the following form:

$$\begin{array}{ll} & f_1(u_1, \ldots, u_d, x_1) \\ & f_2(u_1, \ldots, u_d, x_1, x_2) \\ (1.5) & \cdots \\ & f_r(u_1, \ldots, u_d, x_1, \ldots, x_r) \end{array}$$

Suppose some x_i does not appear as a leading variable in any of the f_k, then we can rename x_i to be a new parameter as we did in Section 3 of Chapter 2. This can be done by programs. But in view of problems similar to example (7.3) of Chapter 1, we prefer the user to take the responsibility of renaming variables if the parameters were specified by the user. If the statement is of constructive type and the choice of parameters was done by the program, then this means that there possibly exist certain unexpected properties for the geometric configuration in the statement. When discussing the formulation problem, we have already assumed the dimensionality restriction (see remark 2 in Section 3.1 of Chapter 3).

Example (1.6). Let AD, BE and CF be the three medians of triangle ABC. Let G be the intersection of AD and BE, H be the intersection of CF and AD, and let P be a point on "line" GH ...

Let $A = (0,0)$, $B = (u_1, 0)$, $C = (u_2, u_3)$, $D = (x_1, x_2)$, $E = (x_3, x_4)$, $F = (x_5, 0)$, $G = (x_7, x_6)$, $H = (x_9, x_8)$, and $P = (x_{10}, u_4)$. If we set corresponding equations $h_1, ..., h_{10}$, then x_{10} will not appear in the ascending chain $f_1, ..., f_9$ obtained from $h_1, ..., h_{10}$ using Ritt's principle. Here the "unexpected property" is that $G = H$ is generally true.

2. Ritt's Decomposition Algorithm Revised

Let $u_1, \ldots, u_d, x_1, \ldots x_r$ and $H = \{h_1, ..., h_n\}$ be the same as in the previous section. First we present a simplified Ritt's decomposition algorithm in the context of geometry theorem proving to obtain all general (nondegenerate) components. The presentation here is our clarification of Wu's work [43], [45].

Algorithm (2.1). (General Component Decomposition). First set S to be H.

Main Step. Apply Ritt's principle as revised in the last section to the polynomial set S to obtain an ascending chain C.

Case 1. C consists of a polynomial $p \in \text{Ideal}(S) \cap \mathbf{Q}[u]$. If S is the initial polynomial set H, then we stop the program and output the related information. Thus we assume this case does not happen. If If S is not the initial polynomial set H, then we record this polynomial p.

Case 2. C is reducible. Then we have case 3 in Ritt's decomposition algorithm (Section 5 of Chapter 2). Thus, we have at least two polynomial sets S_1, S_2 such that $V(S) = V(S_1) \cup V(S_2)$. Take S_1 and S_2 as new S and for each new S, repeat the main step.

Case 3. C is irreducible, but does not satisfy the dimensionality restriction (1.4). This means the geometric configuration possibly has richer properties than we expected. Whenever this happens, the problem should be investigated further by the user. The program stops and outputs related information. Thus we assume this case does not happen (see Remark 2 of Section 3.1 of Chapter 3).

Case 4. C is irreducible and satisfies the dimensionality restriction (1.4). Then record C.

As in Ritt's original decomposition algorithm, each new S obtained from case 2 (if any) has an extended characteristic set strictly lower than that of its predecessor. Thus the procedure will terminate. Upon termination, we have a set of polynomial sets $S_1, ..., S_c$ with extended irreducible characteristic sets C_1, ...,C_c (from case 4) in the form of (1.5) and a set of polynomials p_1,...,p_s (from case 1). It is easy to see that $V(H) - V(p_1 \cdots p_s) \subset V(S_1) \cup ... \cup V(S_c)$. Let P_i be the prime ideal with characteristic set C_i. Then

Theorem (2.2). $V(P_1)$,...,$V(P_c)$ are all components of $V(H)$, general for u.

Proof. Obviously, $V(P_1)$,...,$V(P_c)$ are components general for u. There are no any other components general for u. The reason is as follows. Further decomposition in Ritt's algorithm will start from case 1 or case 4. In case 1, since there is a u–polynomial $p \in$ Ideal(S), the u are algebraically dependent on all components of $V(S)$. In case 4, the further decomposition is to deal with polynomial sets $S \cup \{I_i\}$, where the I_i are the initials of the ascending chain C. By proposition (3.5) of Chapter 2, Ideal($S \cup \{I_i\}$) contain a u–polynomial. Thus, the u are algebraically dependent on all components of $V(S \cup \{I_i\})$. .QED.

The complete method of Wu is based on this version of decomposition algorithm of Ritt. The main difference between this version and the original version is in the efficiency. 219 of 366 theorems in the *Collection* [14] are linear, i.e., $f_1, ..., f_r$ obtained from $h_1, ..., h_n$ are linear in its leading variables. Thus one main step (Ritt's principle) is sufficient (the irreducibility is trivial). It is very fast, usually within 1-2 seconds. However, if we use the original algorithm to obtain all components, then it will take much longer.

3. Complete Method of Wu – Irreducible Cases

Let all notations, (S), $h_1, ..., h_n$, g and $f_1, ..., f_r$, be the same as in Section 1. As we have seen from Chapter 1, if $prem(g, f_1, ..., f_r) = 0$, then (S) is valid under $I_k \neq 0$, where the I_k are the initials (leading coefficients) of the f_k. This is elementary use of the method to confirm a geometry statement. But an implicit problem exists (see discussion in Section 5 of Chapter 1).

Example (3.1). Let $h_1 = x_1^2, h_2 = x_1x_2$, $g = x_2 + x_1$. Then $f_1 = h_1, f_2 = h_2$ and $prem(g, f_1, f_2) = 0$. Thus, $(f_1 = 0, f_2 = 0, x_1 \neq 0) \Rightarrow g = 0$. However, by adding $x_1 \neq 0$, the hypotheses become inconsistent.

Now we suppose the ascending chain $f_1, ..., f_r$ obtained from $h_1 ..., h_n$ by Ritt's principle as revised in the section 1 or its variants[1] is irreducible and satisfies Dimensionality Restriction (1.4). Such a geometry statement (S) is said to be an *irreducible statement* or an *irreducible problem.* This definition depends on the choice of parameters.

Theorem (3.2). Let (S), $h_1, ..., h_n$, g, and $f_1, ..., f_r$ be the same as above. Suppose $f_1, ..., f_r$ is irreducible. If $prem(g, f_1, ..., f_r) = 0$, then

(i) (S) is generally true in all geometries (fields), and

(ii) For all fields, $(h_1 = 0, ..., h_n = 0,\ I_1 \neq 0, ..., I_r \neq 0) \Rightarrow g = 0$, where the I_k

---

[1] It can be an ascending chain in weak sense.

are the of the initials f_k.

Proof. (i) Let P be the prime ideal with the characteristic set $f_1, ..., f_r$. Then by theorem (2.2), $V(P)$ is the only component general for u. By (3.7) of Chapter 2, $prem(g, f_1, ..., f_r) = 0$ implies g vanishes on $V(P)$. Thus (S) is generally true (see Section 3.1 and 3.3 of Chapter 3). (ii) follows from the remainder formula. .QED.

Theorem (3.3). Let (S), $h_1, ..., h_n$, g, and $f_1, ..., f_r$ be the same as above. Suppose ascending chain $f_1, ..., f_r$ is irreducible. Let P be the prime ideal with the characteristic set $f_1, ..., f_r$. Let E be the field associated with the geometry G. $prem(g, f_1, ..., f_r) = 0$ is a necessary condition for (S) to be generally true if one of the following conditions is satisfied:

(i) For E, $V(P)$ is of degree d.

(ii) E is algebraically closed field.

(iii) $f_1, ..., f_r$ has a generic point in E.

Proof. (i) Suppose $dim(V(P)) = d$. By remark (3.8) in chapter 2, $V(P)$ is irreducible and $I(V(P)) = P$. Thus g vanishes on $V(P)$ if and only if $g \in P$, i.e., $prem(g, f_1, ..., f_r) = 0$. (ii) and (iii) follow from (i) (see (3.7) of chapter 2). .QED.

Theorem (3.4). Let (S) be an irreducible geometry statement. (S) is a generally true statement in metric geometry, i.e., in *all* models of the theory of metric geometry, only if $prem(g, f_1, ..., f_r) = 0$.

Proof. Since $\mathbf{C}^2$ is a metric geometry, where $\mathbf{C}$ is the field of complex numbers, the theorem follows from (ii) of (3.3). .QED.

Theorem (3.5). Let (S) be an irreducible statement in Euclidean geometry $\mathbf{R}^2$, where $\mathbf{R}$ is the field of real numbers. Suppose $f_1, ..., f_r$ has a generic point in $\mathbf{R}$. (S) is generally true in Euclidean geometry only if $prem(g, f_1, ..., f_r) = 0$.

Proof. This is an immediately consequence of (iii) of (3.3). .QED.

By (3.5), to disprove a statement in Euclidean geometry requires proving the existence of generic points in $\mathbf{R}$. The notion of generic points seems "non-elementary". However, the following theorem establishes an elementary understanding of generic points in $\mathbf{R}$.

Theorem (3.6). Let $f_1, \ldots, f_r$ be an irreducible ascending chain. The following two conditions are equivalent:

(i) There are intervals $U_1, \ldots, U_d$ in $\mathbf{R}$ such that for all $u_i \in U_i$, $f_1 = 0, \ldots, f_r = 0$ has a solution in $\mathbf{R}$.

(ii) There exists a real generic point $(e_1, \ldots, e_d, \tilde{x}_1, \ldots, \tilde{x}_r)$ of $f_1, \ldots, f_r$.

Proof.

(i) $\Rightarrow$ (ii). Since each set U_i is uncountable, we can take $e_i \in U_i$ with $e_1, \ldots, e_d$ algebraically independent over $\mathbf{Q}$. By (i), $f_1, \ldots, f_r$ has a zero $(e_1, \ldots, e_d, \tilde{x}_1, \ldots, \tilde{x}_r)$ in $\mathbf{R}$. So $(e_1, \ldots, e_d, \tilde{x}_1, \ldots, \tilde{x}_r)$ is a generic point of $f_1, \ldots, f_r$.

(ii) $\Rightarrow$ (i). Suppose $(e_1, \ldots, e_d, \tilde{x}_1, \ldots, \tilde{x}_r)$ is a real generic point for $f_1, ..., f_r$. By Tarski's theorem, for the system of equations $f_1 = 0, \ldots, f_r = 0$, there are finite sets Φ_i $(i = 1, \ldots, m)$ of polynomial equations, inequations and inequalities in parameters $u_1, \ldots, u_d$ alone, such that $f_1, \ldots, f_r$ has a solution for the x in $\mathbf{R}$ for $u_i = c_i$ $(1 \le i \le d)$ iff the c_i satisfy all the conditions of one of the sets Φ_i. So $e_1, \ldots, e_d$ satisfy one of the sets Φ_i, say Φ_k. Φ_k cannot contain nonzero polynomial equations because $e_1, \ldots, e_d$ are algebraically independent. Thus Φ_k contains inequalities and inequations only. By continuity of polynomials, there exist intervals $U_1, \ldots, U_d$ containing $e_1, \ldots, e_d$ respectively, such that Φ_k is satisfied for all $c_i \in U_i$ $(1 \le i \le d)$. Thus for all $u_i = c_i \in U_i$ $(1 \le i \le d)$, $f_1, \ldots, f_r$ has solutions. QED.

Example (3.7). The same as example (2.1) of Chapter 1. After triangulation, we have an ascending chain f_1, f_2, f_3, f_4 with $deg(x_i, f_i) = 1$ for all i. Thus the ascending chain is *trivially* irreducible and its real generic points *trivially* exist. Suppose we have a conjecture that A, E and B are collinear. The corresponding equation is $g_1 = u_1 x_2 = 0$. Then $prem(g_1, f_1, f_2, f_3, f_4) \ne 0$. Thus we *disproved the conjecture in any geometry, in particular, in Euclidean geometry.*

As we mentioned earlier, 219 among the 366 theorems in the *Collection* [14] can be expressed as linear cases (i.e., $deg(f_i, x_i) = 1$ for all f_i), including many statements involving the tritangent centers (the incenter and the three excenters) of a triangle. Thus we can prove or disprove any conjectures about the configurations of such problems mechanically in all geometries, including Euclidean geometry.

Example (3.8). The same as example (2.2) (Simson's theorem) of Chapter 1. After triangulation, we have an ascending chain $f_1, \ldots, f_7$ with all f_i linear but f_3 which is quadratic (i.e., $deg(f_3, x_3) = 2$). Our prover can quickly check the irreducibility of the triangular form. By theorem (3.4), our prover can prove or disprove any conjectures about the Simson configuration in metric geometry.

To disprove conjectures about the Simson configuration in Euclidean geometry, however, requires proving the existence of real solutions. In order to verify the condition (i) in (3.6), we can draw the diagram on the plane (Figure 3). It is evident that u_1, u_2 and u_3 can be arbitrarily chosen at least in some intervals in $\mathbf{R}$ since they are the coordinates of the vertices of an arbitrarily chosen triangle. The variable u_4 can also be arbitrarily chosen in some small interval in $\mathbf{R}$ because $D = (x_3, u_4)$ is an arbitrary point in the circumscribed circle of the triangle. In that small interval, solutions for x_3 exist. The condition (i) in theorem (3.6) is satisfied for the Simson configuration. Therefore, we can prove and *disprove* any conjectures about the Simson configuration in Euclidean geometry.

The above reasoning for the existence of real solutions is not mechanical, but it is plausible. For geometric configurations in ordinary theorems of Euclidean geometry, we can always draw the corresponding diagrams on the plane. This is a plausible reason to justify the conditions (i) in (3.6) and to disprove geometry statements in Euclidean geometry using our method. This is also the *real reason* behind our success in using Wu's method to prove so many theorems in Euclidean geometry. We give the following negative example to show the incompleteness of the method for Euclidean geometry.

Example (3.9). Let A and B be two distinct points, let M be the midpoint of

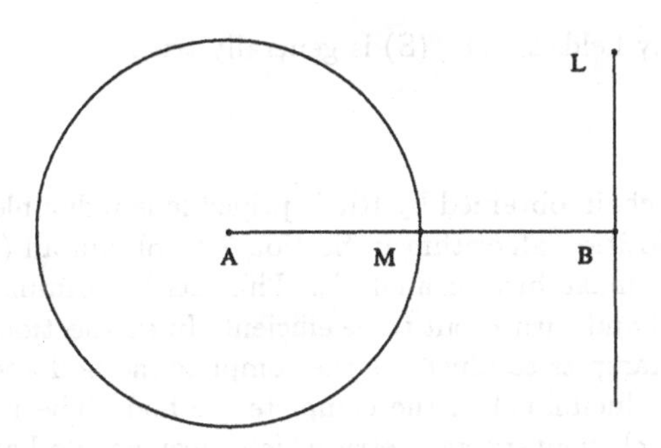

Figure 14

A and B. Let C be one of the intersections of the circle $(A, (AM))$ with the line L passing through B and perpendicular to AB, then $AC \equiv AB$ (Figure 14).

This statement is not valid in $\mathbf{C}^2$, but is generally valid in Euclidean geometry. However, the method cannot prove this statement in Euclidean geometry.

4. Complete Method of Wu – General Cases

Now it is easy to see how to handle general cases. First we use the General Component Decomposition Algorithm (2.1) to obtain all general components $V(P_1),...,V(P_c)$ of $V(h_1,...,h_n)$, where each prime ideal P_i has a characteristic set C_i of the form (1.5). Irreducible problems correspond to cases when $c = 1$.

Theorem (4.1). If $prem(g, C_i) = 0$ $(i = 1, ..., c)$, then (S) is generally true in all geometries.

Proof. This is immediate from the Formulation F3 of Section 3.1 or Section 3.3 of Chapter 3 and (3.7) of Chapter 2. .QED.

Theorem (4.2). Let E be the field associated with the geometry considered. $prem(g, C_i) = 0$ for all i are necessary conditions for (S) to be generally true if one of the following conditions is satisfied:

(i) For E, all $V(P_i)$ is of degree d $(i = 1, ..., c)$.

(ii) E is an algebraically closed field.

(iii) Each ascending chain C_i has a generic point in E.

In each of cases (i)–(iii), if $prem(g, C_i) \neq 0$ for all i, then (S) is generally false.

Proof. It is similar to the proof of (3.3). .QED.

If $prem(g, C_i) \neq 0$ for some i but not for all i, then the statement should be under further investigation. We will give several examples in the next section (examples (5.3), (5.4), (5.5), and (5.6)).

Theorem (4.3). If (S) is generally true for an algebraically closed field E, then it is generally true for all fields E.

Proof. Suppose (S) is generally true for an algebraically closed field E. Then by (4.2), $prem(g, C_i) = 0$ for all $i = 1, ..., c$. By (iv) of (3.7) of Chapter 2, g vanishes on all $V(P_i)$ for any fields E, i.e., (S) is generally true. .QED.

5. Examples

If the ascending chain obtained by Ritt's principle is reducible, generally we need to use the decomposition algorithm in Section 2 to obtain all (irreducible) general components and to make further analysis. This has been implemented at least for quadratic cases [11] and turned out to be efficient. In this section we will give several more interesting examples to illustrate the complete method and the implication of irreducibility (or reducibility) in the complete method. The implication reveals a remarkable fact in elementary geometry which most people have been unaware of until recently (see (5.A)–(5.C), and examples (5.1)–(5.5)).

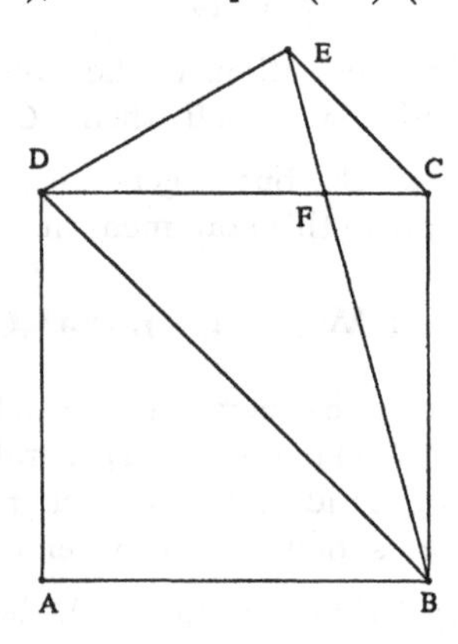

Figure 15

Example (5.1). Let $ABCD$ be a square. CG is parallel to the diagonal BD. Point E is on CG such that $BE \equiv BD$. F is the intersection of BE and DC. Show that $DF \equiv DE$ (Figure 15).

Let $A = (0,0)$, $B = (u_1,0)$, $C = (u_1,u_1)$, $D = (0,u_1)$, $E = (x_1,x_2)$ and $F = (x_3,u_1)$. Then the hypotheses can be expressed by the following polynomial equations:

$$h_1 = x_2^2 + x_1^2 - 2u_1x_1 - u_1^2 = 0 \qquad BE \equiv BD$$
$$h_2 = -u_1x_2 - u_1x_1 + 2u_1^2 = 0 \qquad CE \parallel BD$$
$$h_3 = -x_2x_3 + u_1x_2 + u_1x_1 - u_1^2 = 0 \qquad F \text{ is on } BE.$$

The conclusion ($DF \equiv DE$) can be expressed by $g = x_3^2 - x_2^2 + 2u_1x_2 - x_1^2 - u_1^2 = 0$. The triangulation procedure gives the following triangular form:

$$f_1 = 2u_1^2x_1^2 - 6u_1^3x_1 + 3u_1^4 = 0$$
$$f_2 = -u_1x_2 - u_1x_1 + 2u_1^2 = 0$$
$$f_3 = -x_2x_3 + u_1x_2 + u_1x_1 - u_1^2 = 0$$

It is easy to check *even by hand* that the ascending chain f_1, f_2, f_3 is *irreducible.* The fact that $deg(f_1, x_1) = 2$ can be interpreted as follows. Given a real general value $\tilde{u}_1$ (i.e., $\tilde{u}_1$ is transcendental over $\mathbf{Q}$) of u_1, there are two generic points of

f_1, f_2, f_3 in $\mathbf{R}$,[2] i.e., two subcases of the problem: one is in Figure 15 and the other in Figure 16. Thus, if we take the high school proof of the subcase in Figure 15 for granted, then $prem(g, f_1, f_2, f_3) = 0$ by theorem (3.3). Hence by theorem (3.2), the subcase in Figure 16 is also generally valid.

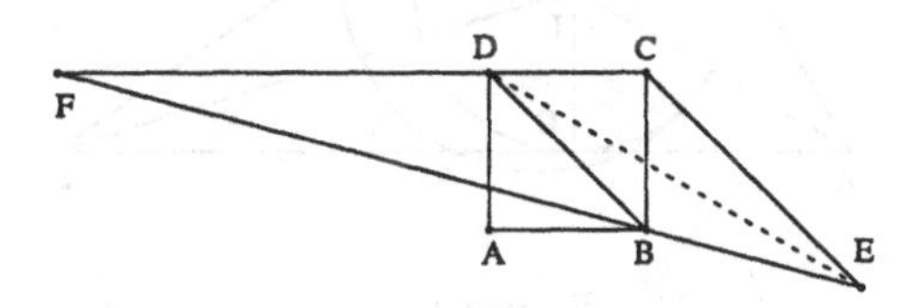

Figure 16

Of course, the two subcases can be easily checked by the computer: $prem(g, f_1, f_2, f_3) = 0$ with $maxt = 8$. Because this is an irreducible problem, our equations $f_1 = 0$, $f_2 = 0$ and $f_3 = 0$ even *cannot* distinguish the two subcases.

Let us sum up the implications in this example:

For a given geometry statement (S) involving equality only, let $h_1 = 0, \ldots, h_n = 0$ be hypothesis polynomials, and $f_1, \ldots, f_r$ be the ascending chain obtained from the h_i using Ritt's Principle as revised in Section 1 and satisfying Dimensionality Restriction (1.4). Assuming that the product of the leading degree of the f_i is greater than one (so there are more than one subcase), then

(5.A). If $f_1, \ldots, f_r$ is irreducible and (S) is valid in one subcase, then $prem(g, f_1, \ldots, f_r) = 0$. Hence (S) is valid in all other subcases.

(5.B). If $f_1, \ldots, f_r$ is irreducible and (S) is not valid in one subcase then (S) is not valid in all other subcases.

(5.C). If (S) is valid in one subcase and $prem(g, f_1, \ldots, f_r) \neq 0$, then $f_1, \ldots, f_r$ is *reducible*.

Following are more examples to illustrate (5.A)–(5.C).

Example (5.2) (Feuerbach's theorem). *The nine-point circle* (N) *of a triangle is tangent to the incircle and to each of the excircles of the triangle* (Figure 17).[3]

Superficially, the fact that (N) touches the incircle and *at the same time* also touches the three excircles seems a coincidence. However, from the algebraic formulation of geometry problems, this fact is natural once the irreducibility has been checked.

---

[2] If the reader is still not happy with the informal proof of the existence of real solutions, he can find one by hand, or consider the geometric configuration to be in complex geometry $\mathbf{C}^2$. For the remaining examples in this section, we always assume the corresponding real generic solutions exist, as ordinary geometric textbooks implicitly assume.

[3] This statement is for Euclidean geometry. In metric geometry, we cannot even distinguish the incenter from the excenters.

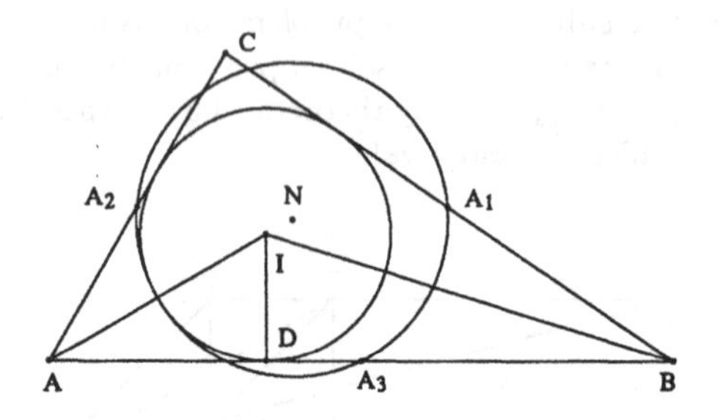

Figure 17

Suppose that we try to prove that circle (N) touches only the incircle (I). We can use the bisectors of angles to determine incenter I. However, if we use a polynomial equation to specify the bisector of an angle, we always have an ambiguity: we have an irreducible polynomial of degree two, which means that there are *two* bisectors of an angle. One is internal; the other is external.[4] Because of the irreducibility, we cannot distinguish one from the other. To make the discussion concrete, we use the following geometric conditions to specify the hypotheses of the problem:

$$
\begin{array}{ll}
 & \tan(\angle CBI) = \tan(\angle IBA) \ \text{(i.e., } I \text{ is on the bisectors of } \angle ABC) \\
 & \tan(\angle CAI) = \tan(\angle IAB) \ \text{(i.e., } I \text{ is on the bisectors of } \angle BAC) \\
 & A, B \text{ and } D \text{ are collinear} \\
 & ID \perp AB \\
(5.2.1) & A_1 \text{ is the midpoint of } B \text{ and } C \\
 & A_2 \text{ is the midpoint of } A \text{ and } C \\
 & A_3 \text{ is the midpoint of } A \text{ and } B \\
 & NA_3 \equiv NA_1 \\
 & NA_3 \equiv NA_2.
\end{array}
$$

The conclusion can be expressed as: the circle (N) with center N and radius NA_3 is tangent to the circle (I) with center I and radius ID.

There are many ways to choose the coordinates. One way is to let $A = (0,0)$, $B = (u_1, 0)$, $C = (u_2, u_3)$, $I = (x_1, x_2)$, $D = (x_1, 0)$, $A_1 = (x_3, x_4)$, $A_2 = (x_5, x_6)$, $A_3 = (x_7, 0)$, and $N = (x_8, x_9)$.[5] Then the hypothesis polynomials corresponding to geometric conditions (5.2.1) are:

$$
\begin{array}{ll}
h_1 = u_1u_3x_2^2 + ((2u_1u_2 - 2u_1^2)x_1 - 2u_1^2u_2 + 2u_1^3)x_2 - u_1u_3x_1^2 + 2u_1^2u_3x_1 - u_1^3u_3 = 0 & \\
 & \tan(CBI) = \tan(IBA) \\
h_2 = u_1u_3x_2^2 + 2u_1u_2x_1x_2 - u_1u_3x_1^2 = 0 & \tan(CAI) = \tan(IAB) \\
h_3 = 2x_3 - u_2 - u_1 = 0 & \\
h_4 = 2x_4 - u_3 = 0 & A_1 \text{ is the midpoint of } B \text{ and } C \\
h_5 = 2x_5 - u_2 = 0 &
\end{array}
$$

[4] This is in ordered geometry. In unordered geometry we cannot talk about "internal" or "external".

[5] The assignment of the coordinates was done by the program with the input of conditions (5.2.1) and a *construction sequence* A, B, C, I, D, A_1, A_2, A_3, N.

$h_6 = 2x_6 - u_3 = 0$ — A_2 is the midpoint of A and C
$h_7 = 2x_7 - u_1 = 0$ — A_3 is the midpoint of A and B
$h_8 = 2x_6x_9 + (-2x_7 + 2x_5)x_8 + x_7^2 - x_6^2 - x_5^2 = 0$ — $NA_3 \equiv NA_2$
$h_9 = 2x_4x_9 + (-2x_7 + 2x_3)x_8 + x_7^2 - x_4^2 - x_3^2 = 0$ — $NA_3 \equiv NA_1$.

The conclusion that circle (N) is tangent to circle (I) is equivalent to $g = ((8x_2x_7 - 8x_1x_2)x_8 - 4x_2x_7^2 + 4x_1^2x_2)x_9 + (-4x_7^2 + 8x_1x_7 + 4x_2^2 - 4x_1^2)x_8^2 + (4x_7^3 - 4x_1x_7^2 + (-8x_2^2 - 4x_1^2)x_7 + 4x_1^3)x_8 - x_7^4 + (4x_2^2 + 2x_1^2)x_7^2 - x_1^4 = 0$. The triangulation procedure gives the following triangular form:[6]

$f_1 = 4x_1^4 - 8u_1x_1^3 + (-4u_3^2 - 4u_2^2 + 4u_1u_2 + 4u_1^2)x_1^2 + (4u_1u_3^2 + 4u_1u_2^2 - 4u_1^2u_2)x_1 - u_1^2u_3^2 = 0$
$f_2 = (2x_1 + 2u_2 - 2u_1)x_2 - 2u_3x_1 + u_1u_3 = 0$
$f_3 = 2x_3 - u_2 - u_1 = 0$
$f_4 = 2x_4 - u_3 = 0$
$f_5 = 2x_5 - u_2 = 0$
$f_6 = 2x_6 - u_3 = 0$
$f_7 = 2x_7 - u_1 = 0$
$f_8 = 4x_8 - 2u_2 - u_1 = 0$
$f_9 = 4u_3x_9 - u_3^2 + u_2^2 - u_1u_2 = 0$

The fact that $deg(f_1, x_1) = 4$ means there are four solutions for I: one is the incenter and the other three are the excenters. Now f_1 is irreducible in x_1,[7] which means that we cannot distinguish the four centers by the above equations. If one subcase (e.g., the case of incenter) is true, then all four subcases are true by (5.A). Thus the following assertion that

(5.2.2) *If the nine point circle touches the incircle it also touches the other three excircles*

is a consequence of our theory. Of course, Feuerbach's theorem can be easily proved by our prover (with $maxt = 42$ and 2.7 seconds of CPU time on a Symbolics 3600).

Many theorems involving the inscribed circle of a triangle have a phenomenon similar to that of (5.2.2). Here we mention two of them: the Gergonne point theorem and Euler's formula ($d^2 = R^2 - 2Rr$). The reader can find many such examples in the *Collection* [14] and Part II. This is really a remarkable fact in elementary geometry. Conscious of this fact, the author was aware of a more general solution to (5.5) (Thèbault's conjecture, see below) when he tried to solve the problem mechanically.

Remark. There are many *construction sequences* to satisfy the *same* geometric condition (5.2.1), and the construction sequences are intimately connected with the choices of parameters u and dependent variables x [11]. A very instructive one is to let $D = (0, 0)$, $A = (u_1, 0)$, $I = (0, u_2)$, $B = (u_3, 0)$, $C = (x_1, x_2)$, $A_1 = (x_3, x_4)$, $A_2 = (x_5, x_6)$, $A_3 = (x_7, 0)$, and $N = (x_8, x_9)$. This choice of coordinates was produced by the program from the same geometric conditions (5.2.1), but from a

[6] This is the simplified characteristic set (see [17] or section 5.2 of Chapter 5). The factors containing integers, variables u_1 and u_3 were taken out of the polynomials.

[7] Our prover can factor multivariate polynomials over **Q** as well as quadratic polynomials of successive quadratic extensions of fields of rational functions.

different construction sequence. Then the hypothesis polynomials corresponding to (5.2.1) are

$h_1 = ((u_2^2-u_1^2)u_3 - u_1u_2^2+u_1^3)x_2+(-2u_1u_2u_3+2u_1^2u_2)x_1+2u_1^2u_2u_3-2u_1^3u_2 = 0$ $\tan(CBI) = \tan(IBA)$

$h_2 = (u_3^3-u_1u_3^2-u_2^2u_3+u_1u_2^2)x_2+(2u_2u_3^2-2u_1u_2u_3)x_1-2u_2u_3^3+2u_1u_2u_3^2 = 0$ $\tan(CAI) = \tan(IAB)$

$h_3 = 2x_3 - x_1 - u_3 = 0$

$h_4 = 2x_4 - x_2 = 0$ A_1 is the midpoint of B and C

$h_5 = 2x_5 - x_1 - u_1 = 0$

$h_6 = 2x_6 - x_2 = 0$ A_2 is the midpoint of A and C

$h_7 = 2x_7 - u_3 - u_1 = 0$ A_3 is the midpoint of A and B

$h_8 = 2x_6x_9 + (-2x_7+2x_5)x_8 + x_7^2 - x_6^2 - x_5^2 = 0$ $NA_3 \equiv NA_2$

$h_9 = 2x_4x_9 + (-2x_7+2x_3)x_8 + x_7^2 - x_4^2 - x_3^2 = 0$ $NA_3 \equiv NA_1$.

The simplified characteristic set is

$f_1 = (u_1u_3+u_2^2)x_1 - u_2^2u_3 - u_1u_2^2 = 0$
$f_2 = (u_1u_3+u_2^2)x_2 - 2u_1u_2u_3 = 0$
$f_3 = (2u_1u_3+2u_2^2)x_3 - u_1u_3^2 - 2u_2^2u_3 - u_1u_2^2 = 0$
$f_4 = (u_1u_3+u_2^2)x_4 - u_1u_2u_3 = 0$
$f_5 = (2u_1u_3+2u_2^2)x_5 + (-u_2^2-u_1^2)u_3 - 2u_1u_2^2 = 0$
$f_6 = (u_1u_3+u_2^2)x_6 - u_1u_2u_3 = 0$
$f_7 = 2x_7 - u_3 - u_1 = 0$
$f_8 = (4u_1u_3+4u_2^2)x_8 - u_1u_3^2 + (-3u_2^2-u_1^2)u_3 - 3u_1u_2^2 = 0$
$f_9 = (8u_1u_2u_3+8u_2^3)x_9 + (-u_2^2+u_1^2)u_3^2 - 4u_1u_2^2u_3 + u_2^4 - u_1^2u_2^2 = 0$

Now $deg(f_i, x_i) = 1$. So ascending chain $f_1, \ldots, f_9$ is trivially irreducible, and we have only one case instead of four as in the previous choice of the coordinates. Where are the other 3 cases? If we look at the constructions corresponding to the present choice of u's and x's closely, we can find that the problem has been formulated in a slightly different way:

(5.2.3) *The nine point circle of the triangle whose three sides touch a given circle touches this given circle.*

If we first fix a triangle, we know that there are four such circles touching the three sides of the given triangle. (5.2.3) implicitly says that any of the four circles is tangent to the nine-point circle. *We will use a similar interpretation in example* (5.5) *without repeating the argument.*

Example (5.3). On the two sides AC and BC of triangle ABC, two square $ACDE$ and $BCFG$ are drawn. M is the midpoint of AB. Show $DF \equiv 2CM$ (Figure 18).

Let $A = (u_1, 0)$, $B = (u_2, u_3)$, $C = (0,0)$, $D = (0, u_1)$, $F = (x_1, x_2)$, and $M = (x_3, x_4)$, then

$h_1 = x_2^2 + x_1^2 - u_3^2 - u_2^2 = 0$ $CF = BC$

$h_2 = u_3x_2 + u_2x_1 = 0$ $CF \perp BC$

$h_3 = 2x_3 - u_2 - u_1 = 0$

$h_4 = 2x_4 - u_3 = 0$ M is the midpoint of A and B

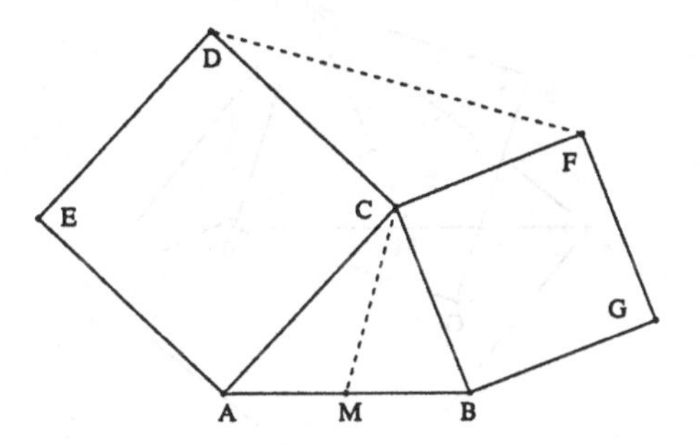

Figure 18

$g = 4x_4^2 + 4x_3^2 - x_2^2 + 2u_1x_2 - x_1^2 - u_1^2 = 0$ Conclusion: $DF \equiv 2CM$.

The triangular form obtained from $h_1, ..., h_4$ is

$f_1 = (u_3^2 + u_2^2)x_1^2 - u_3^4 - u_2^2u_3^2 = 0$
$f_2 = u_3x_2 + u_2x_1 = 0$
$f_3 = 2x_3 - u_2 - u_1 = 0$
$f_4 = 2x_4 - u_3 = 0.$

However, $prem(g, f_1, ..., f_4) \neq 0$. Thus, if the original geometry statement is true, then $f_1, ..., f_4$ *must be reducible*. Actually, $f_1 = (u_3^2 + u_2^2)(x_1 - u_3)(x_1 + u_3)$. Thus, Ritt's decomposition algorithm gives two general components of the configuration: $C' = f_1', f_2, f_3, f_4$ and $C'' = f_1'', f_2, f_3, f_4$, where $f_1' = x_1 - u_3$, $f'' = x_1 + u_3$. Computer results showed that $prem(g, C') = 0$, but $prem(g, C'') \neq 0$. Thus the conclusion is true for one subcase of the configuration, but not for the other.

Our prover can go a step further. Let $\delta = -u_1^2u_3^2/(u_1^2u_2x_2 - u_1^2u_3x_1)$ and Δ be $[(D$ and B on either sides of $AC) \wedge (A$ and F on either sides of $BC)] \vee [(D$ and B on the same side of $AC) \wedge (A$ and F on the same side of $BC)]$. Then $\delta > 0$ iff Δ holds; $\delta < 0$ iff Δ does not hold.

Computation (for the method see [12]) showed that $\delta = u_3^2/(u_3^2+u_2^2)$ for component C' and $\delta = -u_3^2/(u_3^2 + u_2^2)$. Thus we can "see" $\delta > 0$ for component C' if we assume A, B and C are not collinear (i.e., $u_1u_3 \neq 0$). Thus the statement is valid if and *only if* Δ holds, which corresponds to cases in Figures 18 and 19. However, "see" is still not a mechanical procedure.

As we have seen from this example, decomposition or factorization is, in essence, a case analysis approach. For example, if we want to decide whether "$AB \equiv DF$" (i.e., $g_1 = x_2^2 - 2u_1x_2 + x_1^2 - u_3^2 - u_2^2 + 2u_1u_2 = 0$) is valid for this geometric configuration, we will find $prem(g_1, C') \neq 0$, but $prem(g_1, C'') = 0$. Thus, "$AB \equiv DF$" is valid for the other subcase.

Now we come to a harder problem. One of the most surprising theorems in elementary geometry was discovered about 1899 by F. Morley. He mentioned it to his friends, who spread it over the world in the form of mathematical gossip. At last, after ten years, a trigonometrical proof by Satyanarayana and an elementary proof

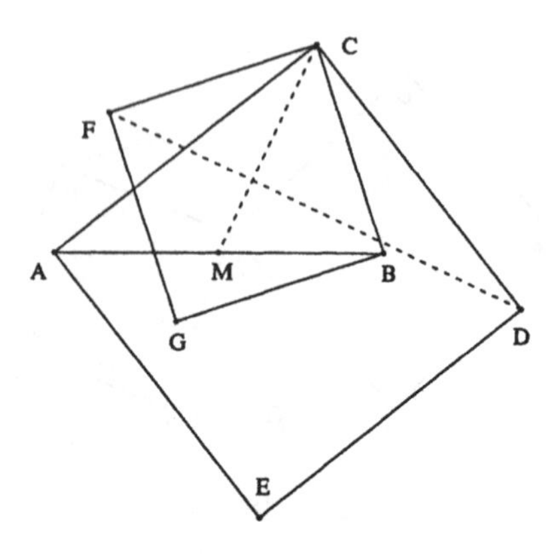

Figure 19

by M. T. Naraniengar were published.[8] This theorem intrigued mathematicians for the past three quarters of a century. It is now simply known as *Morley's trisector theorem.*

Example (5.4). (Morley's trisector theorem). The three points of intersection of the adjacent trisectors of the angles of any triangle form an equilateral triangle (Figure 20).

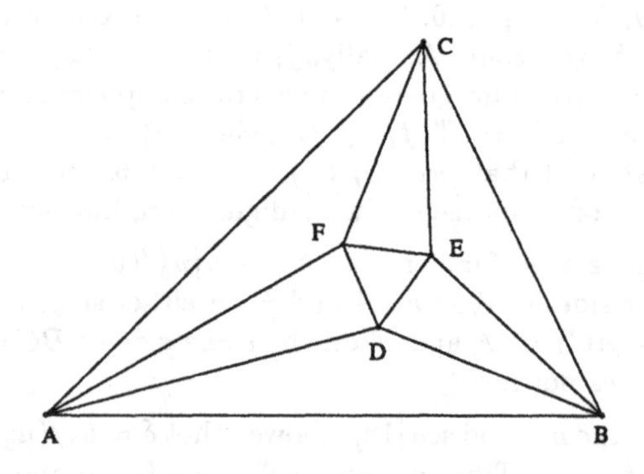

Figure 20

Let ABC be a triangle. As we know, each angle of triangle ABC has three trisectors. (See section 1.2 of Chapter 3). Thus, we have $3^3 = 27$ *Morley triangles* formed by the trisectors of a triangle. First, let us fix one of the three trisectors of $\angle CAB$ and one of the three trisectors of $\angle CBA$, and let D be their intersection. As we know, there are 9 such points D. Once D is fixed, C is uniquely determined to be the intersection of two lines defined from points A, B and D. Thus we can specify the problem as follows.

Let $A = (0,0)$, $B = (u_1, 0)$, $D = (u_2, u_3)$, $C = (x_2, x_1)$, $F = (x_4, x_3)$, and $E = (x_6, x_5)$. We have the following equations for the hypotheses:

[8] This passage is from Professor H.S.M. Coxeter's book [15].

$h_1 = (u_3^3+(-3u_2^2+6u_1u_2-3u_1^2)u_3)x_2+((-3u_2+3u_1)u_3^2+u_2^3-3u_1u_2^2+3u_1^2u_2-u_1^3)x_1-u_1u_3^3+(3u_1u_2^2-6u_1^2u_2+3u_1^3)u_3 = 0$ $\quad tan(ABC) = tan(3\angle ABD)$

$h_2 = (u_3^3-3u_2^2u_3)x_2+(-3u_2u_3^2+u_2^3)x_1 = 0$ $\quad tan(BAC) = tan(3\angle BAD)$

$h_3 = (u_1u_3x_2-u_1u_2x_1)x_4+(u_1u_2x_2+u_1u_3x_1)x_3 = 0$ $\quad tan(BAD) = tan(FAC)$

$h_4 = (3x_1x_2^2-2u_1x_1x_2-x_1^3)x_4^3+((-3x_2^3+3u_1x_2^2+9x_1^2x_2-3u_1x_1^2)x_3-6x_1x_2^3+3u_1x_1x_2^2-6x_1^3x_2+3u_1x_1^3)x_4^2+((-9x_1x_2^2+6u_1x_1x_2+3x_1^3)x_3^2+(6x_2^4-6u_1x_2^3-6u_1x_1^2x_2-6x_1^4)x_3+3x_1x_2^4+6x_1^3x_2^2+3x_1^5)x_4+(x_2^3-u_1x_2^2-3x_1^2x_2+u_1x_1^2)x_3^3+(6x_1x_2^3-3u_1x_1x_2^2+6x_1^3x_2-3u_1x_1^3)x_3^2+(-3x_2^5+3u_1x_2^4-6x_1^2x_2^3+6u_1x_1^2x_2^2-3x_1^4x_2+3u_1x_1^4)x_3-u_1x_1x_2^4-2u_1x_1^3x_2^2-u_1x_1^5 = 0$ $\quad tan(ACB) = tan(3\angle ACF)$

$h_5 = (u_1u_3x_2+(-u_1u_2+u_1^2)x_1-u_1^2u_3)x_6+((u_1u_2-u_1^2)x_2+u_1u_3x_1-u_1^2u_2+u_1^3)x_5-u_1^2u_3x_2+(u_1^2u_2-u_1^3)x_1+u_1^3u_3 = 0$ $\quad tan(ABD) = tan(EBC)$

$h_6 = ((2x_1x_2-u_1x_1)x_4+(-x_2^2+u_1x_2+x_1^2)x_3-x_1x_2^2-x_1^3)x_6+((-x_2^2+u_1x_2+x_1^2)x_4+(-2x_1x_2+u_1x_1)x_3+x_2^3-u_1x_2^2+x_1^2x_2-u_1x_1^2)x_5+(-x_1x_2^2-x_1^3)x_4+(x_2^3-u_1x_2^2+x_1^2x_2-u_1x_1^2)x_3+u_1x_1x_2^2+u_1x_1^3 = 0$ $\quad tan(ACF) = tan(ECB).$

We have two conclusions: $g_1 = x_6^2-2u_2x_6+x_5^2-2u_3x_5-x_4^2+2u_2x_4-x_3^2+2u_3x_3 = 0$ $(DE \equiv DF)$, and $g_2 = x_6^2-2x_4x_6+x_5^2-2x_3x_5+2u_2x_4+2u_3x_3-u_3^2-u_2^2 = 0$ $(ED \equiv FE)$.

triangulizing $h_1, ..., h_6$, we have an ascending chain:

$$f_1 = I_1x_1 + ...$$
$$f_2 = I_2x_2 + ...$$
$$f_3 = I_3x_3^3 + ...$$
$$f_4 = I_4x_4 + ...$$
$$f_4 = I_5x_5 + ...$$
$$f_6 = I_6x_6 + ...$$

The ascending chain $f_1, ..., f_6$ is reducible. Thus we have to decompose it. Since there is only one polynomial f_3 whose leading degree is greater than 1, no factorization over algebraic extensions is involved. However, it took about 10 minutes to obtain the two general components with the following simplified characteristic sets:

$$C' = p_1, p_2, p_3', p_4, p_5, p_6 \text{ and}$$
$$C'' = p_1, p_2, p_3'', p_4, p_5, p_6,$$

where

$p_1 = (3u_3^4+(6u_2^2-6u_1u_2-u_1^2)u_3^2+3u_2^4-6u_1u_2^3+3u_1^2u_2^2)x_1-u_3^5+(6u_2^2-6u_1u_2+3u_1^2)u_3^3+(-9u_2^4+18u_1u_2^3-9u_1^2u_2^2)u_3$

$p_2 = (3u_3^4+(6u_2^2-6u_1u_2-u_1^2)u_3^2+3u_2^4-6u_1u_2^3+3u_1^2u_2^2)x_2-3u_2u_3^4+(10u_2^3-18u_1u_2^2+9u_1^2u_2)u_3^2-3u_2^5+6u_1u_2^4-3u_1^2u_2^3$

$p_3' = (3u_3^4+(6u_2^2-6u_1u_2-u_1^2)u_3^2+3u_2^4-6u_1u_2^3+3u_1^2u_2^2)x_3^2+((-12u_2^2+16u_1u_2)u_3^3+(-12u_2^4+24u_1u_2^3-12u_1^2u_2^2)u_3)x_3-4u_2^2u_3^4+(12u_2^4-24u_1u_2^3+12u_1^2u_2^2)u_3^2 = 0$

$p_3'' = (3u_3^4+(6u_2^2-6u_1u_2-u_1^2)u_3^2+3u_2^4-6u_1u_2^3+3u_1^2u_2^2)x_3+2u_1u_2u_3^3+(-6u_1u_2^3+12u_1^2u_2^2-6u_1^3u_2)u_3$

$p_4 = 2u_2u_3x_4+(u_3^2-u_2^2)x_3$

$p_5 = (((-6u_2+3u_1)u_3^4+(-12u_2^3+18u_1u_2^2-4u_1^2u_2-u_1^3)u_3^2-6u_2^5+15u_1u_2^4-12u_1^2u_2^3+3u_1^3u_2^2)x_3-2u_2u_3^5+(4u_2^3-12u_1u_2^2+6u_1^2u_2)u_3^3+(6u_2^5-12u_1u_2^4+6u_1^2u_2^3)u_3)x_5+$

$$((-2u_2+2u_1)u_3^5+(4u_2^3-6u_1^2u_2+2u_1^3)u_3^3+(6u_2^5-18u_1u_2^4+18u_1^2u_2^3-6u_1^3u_2^2)u_3)x_3=0$$
$$p_6=(((-6u_2+3u_1)u_3^4+(-12u_2^3+18u_1u_2^2-4u_1^2u_2-u_1^3)u_3^2-6u_2^5+15u_1u_2^4-12u_1^2u_2^3+3u_1^3u_2^2)x_3-2u_2u_3^5+(4u_2^3-12u_1u_2^2+6u_1^2u_2)u_3^3+(6u_2^5-12u_1u_2^4+6u_1^2u_2^3)u_3)x_6+(u_3^6+(-3u_2^2+6u_1u_2-3u_1^2)u_3^4+(-u_2^4+16u_1u_2^3-24u_1^2u_2^2+8u_1^3u_2)u_3^2+3u_2^6-6u_1u_2^5+3u_1^2u_2^4)x_3+2u_1u_2u_3^5+(-4u_1u_2^3+12u_1^2u_2^2-6u_1^3u_2)u_3^3+(-6u_1u_2^5+12u_1^2u_2^4-6u_1^3u_2^3)u_3$$

Now the computer results showed that $prem(g_1, C') = prem(g_2, C') = 0$. Note that $deg(p_3', x_3) = 2$. This means that for each of the nine points D, there are two points F such that triangles FED are equilateral. Thus we have proved the most general form of Morley's trisector theorem:

Among 27 triangles formed by the trisectors of a triangle, 18 are equilateral.

Wu Wen-Tsün first gave a mechanical proof of this general form of the theorem [43]. By an ingenious technique, he succeeded in specifying the problem to be irreducible. His solution is better than ours in terms of computer time and space. However, the solution here is more instructive for understanding the complete method of Wu.

At last, we come to a problem to which we have original contributions. This was a conjecture proposed by V. Thèbault of France in 1938 and proved by K.B. Taylor of England in 1983. We state the problem as follows (see [3]):

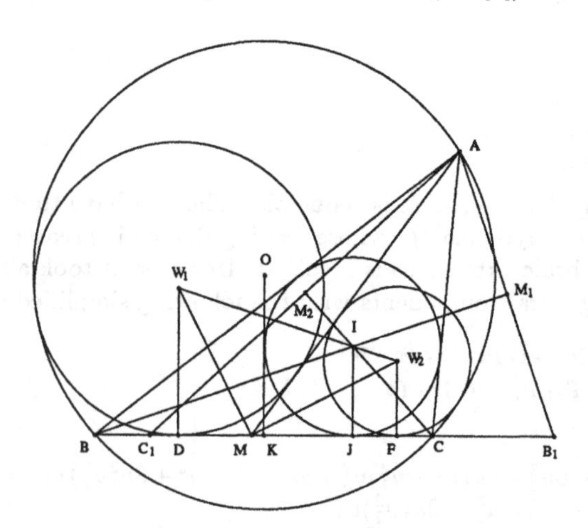

Figure 21

Example (5.5). Through the vertex A of a triangle ABC a straight line AM is drawn, cutting the side BC in M. Let O and I be the centers of the circumscribed circle (O) and the *inscribed circle* (I) of ABC. The circle (w_1) and (w_2) with centers w_1 and w_2 are each tangent to (O) and the first is tangent also to two sides of angle AMB, while the second is tangent to the two sides of angle AMC. Prove that the straight line joining w_1 and w_2 passes through I (Figure 21).

There are many ways to specify the problem in terms of construction (or of choice of u's and x's). To understand the nature of the problem, we first let $B = (0,0)$, $M = (u_1, 0)$, $A = (u_2, u_3)$, $w_1 = (x_2, x_3)$, $O = (1, x_1)$, $D = (x_2, 0)$, $K = (1,0)$, $C = (x_4, 0)$, $I = (x_5, x_6)$, $w_2 = (x_7, x_8)$, $F = (x_7, 0)$, where D, F and K are projections of w_1, w_2 and O on line BC respectively.

$h_1 = -2u_3x_1 + u_3^2 + u_2^2 - 2u_2 = 0$ $\qquad OB \equiv OA.$

$h_2 = u_1u_3x_3^2 + ((2u_1u_2 - 2u_1^2)x_2 - 2u_1^2u_2 + 2u_1^3)x_3 - u_1u_3x_2^2 + 2u_1^2u_3x_2 - u_1^3u_3 = 0$
w_1 is on the bisectors of $\angle AMB$

$h_3 = 16x_3^2 + (16x_1x_2^2 - 32x_1x_2)x_3 - 4x_2^4 + 16x_2^3 - 16x_2^2 = 0$
circle (w_1) is tangent to circle (O)

$h_4 = x_4 - 2 = 0$ $\qquad K$ is the midpoint of B and C

$h_5 = u_3x_4x_6^2 + ((-2x_4^2 + 2u_2x_4)x_5 + 2x_4^3 - 2u_2x_4^2)x_6 - u_3x_4x_5^2 + 2u_3x_4^2x_5 - u_3x_4^3 = 0$
I is on the bisectors of $\angle ACB$

$h_6 = -u_3x_4x_6^2 - 2u_2x_4x_5x_6 + u_3x_4x_5^2 = 0$ $\qquad I$ is on the bisectors of $\angle ABC$

$h_7 = (-x_5 + x_2)x_8 + (x_6 - x_3)x_7 - x_2x_6 + x_3x_5 = 0$ $\qquad w_2$ is on w_1I

$h_8 = x_3x_8 + (x_2 - u_1)x_7 - u_1x_2 + u_1^2 = 0$ $\qquad w_2M \perp w_1M.$

The conclusion that circle (w_2) is tangent to circle (O) is equivalent to $g = 4x_8^2 + (4x_1x_7^2 - 8x_1x_7)x_8 - x_7^4 + 4x_7^3 - 4x_7^2 = 0$. Now the simplified characteristic set is

$f_1 = -2u_3x_1 + u_3^2 + u_2^2 - 2u_2 = 0$

$f_2 = u_3^2x_2^8 + \ldots$

$f_3 = ((2u_3^2 + 2u_2^2 - 4u_2)x_2^2 + (-4u_3^2 - 4u_2^2 + 8u_1)x_2 + 8u_1u_2 - 8u_1^2)x_3 - u_3x_2^4 + 4u_3x_2^3 - 8u_1u_3x_2 + 4u_1^2u_3 = 0$

$f_4 = x_4 - 2 = 0$

$f_5 = x_5^4 - 4x_5^3 + (-u_3^2 - u_2^2 + 2u_2 + 4)x_5^2 + (2u_3^2 + 2u_2^2 - 4u_2)x_5 - u_3^2 = 0$

$f_6 = (x_5 + u_2 - 2)x_6 - u_3x_5 + u_3 = 0$

$f_7 = (x_3x_6 + (x_2 - u_1)x_5 - x_3^2 - x_2^2 + u_1x_2)x_7 - x_2x_3x_6 + (x_3^2 - u_1x_2 + u_1^2)x_5 + u_1x_2^2 - u_1^2x_2 = 0$

$f_8 = x_3x_8 + (x_2 - u_1)x_7 - u_1x_2 + u_1^2 = 0.$

The fact that $deg(f_2, x_2) = 8$ means that there are eight solutions for w_1. In fact, those solutions also include w_2: f_2 is an irreducible polynomial in x_2. However, we have distinguished w_1 from w_2 by the condition $w_1M \perp w_2M$. If we fix *one of the two bisectors* of $\angle BMA$, then there are four w_1 on it, and the other four w_2 are on the other bisector. Thus, the above algebraic reformulation of the problem states: for any one of the four w_1 and for any one of the four I, there exist a w_2 such that w_1, w_2 and I are collinear.

If $f_1, \ldots, f_8$ is irreducible and if the original geometry statement is true, then there should be 16 such lines instead of one as the problem originally proposed. However, computation showed that $prem(g, f_1, \ldots, f_r) \neq 0$. By (5.C), this means $f_1, \ldots, f_8$ is *reducible* if the original conjecture is true.

In order to make the problem within the reach of the available facilities of our prover[9] and the computation within the limit of space and time permitted by our machine (Symbolics 3600), we have to reformulate the problem. Let $D = (0,0)$, $B = (1,0)$, $M = (u_2, 0)$, $w_1 = (0, u_3)$, $A = (u_4, x_1)$, $O = (x_2, x_3)$, $K = (x_2, 0)$,

[9] Generally, our prover can only deal with quadratic cases completely.

$C = (x_4,0)$, $C_1 = (x_5,0)$, $B_1 = (x_6,0)$, $M_1 = (x_7,x_8)$, $M_2 = (x_9,x_{10})$, $I = (x_{11},x_{12})$, $w_2 = (x_{13},x_{14})$, and $F = (x_{13},0)$. We specify geometric conditions and the corresponding equations as follows (see Figure 21).

$h_1 = (-u_3^2 + u_2^2)x_1 + 2u_2u_3u_4 - 2u_2^2u_3 = 0$ w_1 is on the bisectors of $\angle AMB$
$h_2 = (8u_3x_2 - 4u_3)x_3 + (4u_3^2 - 4)x_2^2 + (-8u_3^2 + 4)x_2 + 4u_3^2 - 1 = 0$
circle (O) is tangent to circle (w_1)
$h_3 = -2x_1x_3 + (-2u_4 + 2)x_2 + x_1^2 + u_4^2 - 1 = 0$ $OB = OA$
$h_4 = x_4 - 2x_2 + 1 = 0$ K is the midpoint of BC
$h_5 = -x_5^2 + 2x_4x_5 - 2u_4x_4 + x_1^2 + u_4^2 = 0$ $CC_1 \equiv CA$
$h_6 = -x_6^2 + 2x_6 + x_1^2 + u_4^2 - 2u_4 = 0$ $BB_1 \equiv BA$
$h_7 = -2x_7 + x_6 + u_4 = 0$
$h_8 = -2x_8 + x_1 = 0$ M_1 is the midpoint of AB_1
$h_9 = -2x_9 + x_5 + u_4 = 0$
$h_{10} = -2x_{10} + x_1 = 0$ M_2 is the midpoint of AC_1
$h_{11} = (-x_9 + x_4)x_{12} + x_{10}x_{11} - x_4x_{10} = 0$ I is on CM_2
$h_{12} = (-x_7 + 1)x_{12} + x_8x_{11} - x_8 = 0$ I is on BM_1
$h_{13} = -x_{11}x_{14} + (x_{12} - u_3)x_{13} + u_3x_{11} = 0$ w_2 is on w_1I
$h_{14} = u_3x_{14} - u_2x_{13} + u_2^2 = 0$ $w_1M \perp w_2M$.

The conclusion that (w_2) is tangent to (O) is equivalent to $g = (4x_2^2 - 8x_2 + 4)x_{14}^2 + (4x_3x_{13}^2 - 8x_2x_3x_{13} + (8x_2 - 4)x_3)x_{14} - x_{13}^4 + 4x_2x_{13}^3 + (-4x_2^2 - 4x_2 + 2)x_{13}^2 + (8x_2^2 - 4x_2)x_{13} - 4x_2^2 + 4x_2 - 1 = 0$.

Now $h_1, \ldots, h_{14}$ can be decomposed[10] into the following four general components:

$C_1 = f_1, \ldots, f_4,\ A_5x_5 + B_5,\ A_6x_6 + B_6,\ f_7, \ldots, f_{14}$
$C_2 = f_1, \ldots, f_4,\ A_5x_5 - B_5,\ A_6x_6 + B_6,\ f_7, \ldots, f_{14}$
$C_3 = f_1, \ldots, f_4,\ A_5x_5 + B_5,\ A_6x_6 - B_6,\ f_7, \ldots, f_{14}$
$C_4 = f_1, \ldots, f_4,\ A_5x_5 - B_5,\ A_6x_6 - B_6,\ f_7, \ldots, f_{14}$,

where

$f_1 = (-u_3^2 + u_2^2)x_1 + 2u_2u_3u_4 - 2u_2^2u_3$
$f_2 = ((4u_3^2 - 4)x_1 - 8u_3u_4 + 8u_3)x_2^2 + (4u_3x_1^2 + (-8u_3^2 + 4)x_1 + 4u_3u_4^2 + 4u_3u_4 - 8u_3)x_2 - 2u_3x_1^2 + (4u_3^2 - 1)x_1 - 2u_3u_4^2 + 2u_3$
$f_3 = -2x_1x_3 + (-2u_4 + 2)x_2 + x_1^2 + u_4^2 - 1$
$f_4 = x_4 - 2x_2 + 1$
$f_7 = -2x_7 + x_6 + u_4$
$f_8 = -2x_8 + x_1$
$f_9 = -2x_9 + x_5 + u_4$
$f_{10} = -2x_{10} + x_1$
$f_{11} = (2x_1x_9 - 2x_1x_7 - 2x_1x_4 + 2x_1)x_{11} - 2x_1x_9 + 2x_1x_4x_7$
$f_{12} = (2x_7 - 2)x_{12} - x_1x_{11} + x_1$
$f_{13} = (u_3x_{12} - u_2x_{11} - u_3^2)x_{13} + (u_3^2 + u_2^2)x_{11}$
$f_{14} = u_3x_{14} - u_2x_{13} + u_2^2$
$A_5 = ((u_2-1)u_3^7 + (u_2^2-u_2)u_3^5 + (-u_2^5+u_2^4)u_3^3 + (-u_2^6+u_2^5)u_3)u_4^2 + ((-3u_2^2+2u_2+$

[10] This is the first major application of our factoring algorithm for successive quadratic extension fields (see [10]).

$1)u_3^7+(2u_2^4-4u_2^3+2u_2^2)u_3^5+(u_2^6+2u_2^5-3u_2^4)u_3^3)u_4+(2u_2^3-2u_2)u_3^7+(-2u_2^5+2u_2^3)u_3^5$

$B_5=(((-4u_2^2+4u_2)u_3^7+(4u_2^4-12u_2^3+8u_2^2)u_3^5+(8u_2^5-12u_2^4+4u_2^3)u_3^3+(4u_2^6-4u_2^5)u_3)u_4^2+((8u_2^3-8u_2)u_3^7+(-8u_2^5+8u_2^4+8u_2^3-8u_2^2)u_3^5+(-8u_2^6+8u_2^4)u_3^3)u_4+(-4u_2^4-4u_2^3+4u_2^2+4u_2)u_3^7+(4u_2^6+4u_2^5-4u_2^4-4u_2^3)u_3^5)x_2+((-u_2+1)u_3^7+(-2u_2^3+u_2^2+u_2)u_3^5+(-u_2^5-u_2^4+2u_2^3)u_3^3+(-u_2^6+u_2^5)u_3)u_4^3+((5u_2^2-4u_2-1)u_3^7+(6u_2^4+2u_2^3-8u_2^2)u_3^5+(u_2^6+2u_2^5-u_2^4-2u_2^3)u_3^3+(-2u_2^6+2u_2^5)u_3)u_4^2+((-8u_2^3+2u_2^2+6u_2)u_3^7+(-4u_2^5-4u_2^4+4u_2^3+4u_2^2)u_3^5+(2u_2^6+2u_2^5-4u_2^4)u_3^3)u_4+(4u_2^4+2u_2^3-4u_2^2-2u_2)u_3^7+(-2u_2^5+2u_2^3)u_3^5$

$A_6=(u_3^5-u_2^4u_3)u_4-2u_2u_3^5+2u_2^3u_3^3$

$B_6=(((4u_2-4)u_3^5+(-4u_2^3+8u_2^2-4u_2)u_3^3+(-4u_2^4+4u_2^3)u_3)u_4+(-4u_2^2+4)u_3^5+(4u_2^4-4u_2^2)u_3^3)x_2+(u_3^5+2u_2^2u_3^3+u_2^4u_3)u_4^2+(-4u_2u_3^5+(-4u_2^3-2u_2^2+2u_2)u_3^3+(2u_2^4-2u_2^3)u_3)u_4+(4u_2^2+2u_2-2)u_3^5+(-2u_2^3+2u_2^2)u_3^3.$

If the original proposed conjecture is true (as proved by Taylor in 1983), then there exists one component C_i such that $prem(g, C_i) = 0$. Thus, by combining the proof of Taylor with our decomposition we can infer the following conclusion:

For each of the four points w_1 there is an I and a w_2 such that w_1, I and w_2 are collinear.

By *symmetric consideration*, I should be different for different w_1. Hence we finally come to a *generalization of Thébault-Taylor's theorem*:

For each of the four centers I (the incenter and the three excenters) there exist a w_1 and a w_2 such that w_1, I and w_2 are collinear.

In order to confirm our results mechanically and independently of the proof by Taylor, we have to find such a component among C_1, C_2, C_3 and C_4. The computation needs an exceedingly large among of memory. Our mechanical proof succeeded just at the end of 1985. This is component C_1. The total CPU time (on a Symbolics 3600) for computing $prem(g, C_1) = 0$ was about 44 hours, 99% of which was spent on the manipulation of polynomials of more than 100,000 terms. The largest remainder produced has 674,927 (= *maxt*) terms.

A geometric problem which can be specified as a reducible problem, sometimes can be also specified as an irreducible problem.

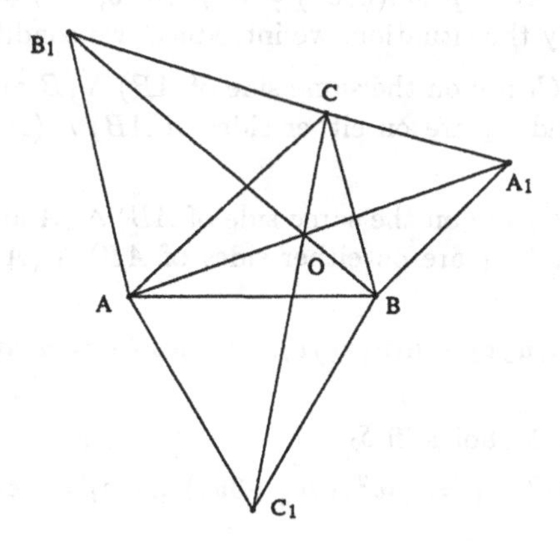

Figure 22

Example (5.6). Three equilateral triangles A_1BC, AB_1C and ABC_1 are erected on the three respective sides BC, CA and AB of a triangle ABC, then lines AA_1, BB_1 and CC_1 are concurrent (Figure 22).

Let $A = (0,0)$, $B = (u_1,0)$, $C = (u_2,u_3)$, $B_1 = (x_1,x_2)$, $C_1 = (x_3,x_4)$, $A_1 = (x_5,x_6)$ and $O = (x_7,x_8)$. We have the following equations for the hypotheses and conclusion:

$h_1 = -x_2^2 - x_1^2 + u_3^2 + u_2^2 = 0$ $\quad AC \equiv AB_1$
$h_2 = 2u_3x_2 + 2u_2x_1 - u_3^2 - u_2^2 = 0$ $\quad B_1A \equiv B_1C$
$h_3 = 2u_1x_3 - u_1^2 = 0$ $\quad C_1A \equiv C_1B$
$h_4 = -x_4^2 - x_3^2 + u_1^2 = 0$ $\quad AB \equiv AC_1$
$h_5 = -x_6^2 - x_5^2 + 2u_1x_5 + u_3^2 + u_2^2 - 2u_1u_2 = 0$ $\quad BC \equiv A_1B$
$h_6 = 2u_3x_6 + (2u_2 - 2u_1)x_5 - u_3^2 - u_2^2 + u_1^2 = 0$ $\quad A_1B \equiv A_1C$
$h_7 = (-x_3 + u_2)x_8 + (x_4 - u_3)x_7 - u_2x_4 + u_3x_3 = 0$ C, C_1 and O are collinear
$h_8 = (-x_1 + u_1)x_8 + x_2x_7 - u_1x_2 = 0$ $\quad B$, B_1 and O are collinear.
$g = x_5x_8 - x_6x_7 = 0$ $\quad$ The conclusion: A, O and A_1 are collinear.

Polynomial set $h_1, ..., h_8$ can be decomposed into four components general for u:
$C' = f_1, f_2, f_3, f_4', f_5', f_6, f_7, f_8$, $\quad C' = f_1, f_2, f_3, f_4'', f_5', f_6, f_7, f_8$,
$C''' = f_1, f_2, f_3, f_4', f_5'', f_6, f_7, f_8$, $\quad C'''' = f_1, f_2, f_3, f_4'', f_5'', f_6, f_7, f_8$, where

$f_1 = (-4u_3^2 - 4u_2^2)x_1^2 + (4u_2u_3^2 + 4u_2^3)x_1 + 3u_3^4 + 2u_2^2u_3^2 - u_2^4$
$f_2 = 2u_3x_2 + 2u_2x_1 - u_3^2 - u_2^2$
$f_3 = 2u_1x_3 - u_1^2$
$f_4' = (2u_3^3 + 2u_2^2u_3)x_4 + (2u_1u_3^2 + 2u_1u_2^2)x_1 - u_1u_2u_3^2 - u_1u_2^3$
$f_4'' = (2u_3^3 + 2u_2^2u_3)x_4 + (-2u_1u_3^2 - 2u_1u_2^2)x_1 + u_1u_2u_3^2 + u_1u_2^3$
$f_5' = (2u_3^2 + 2u_2^2)x_5 + (2u_3^2 + 2u_2^2)x_1 + (-2u_2 - u_1)u_3^2 - 2u_2^3 - u_1u_2^2$
$f_5'' = (2u_3^2 + 2u_2^2)x_5 + (-2u_3^2 - 2u_2^2)x_1 - u_1u_3^2 - u_1u_2^2$
$f_6 = 2u_3x_6 + (2u_2 - 2u_1)x_5 - u_3^2 - u_2^2 + u_1^2$
$f_7 = ((-x_1+u_1)x_4+x_2x_3-u_2x_2+u_3x_1-u_1u_3)x_7+(u_2x_1-u_1u_2)x_4+(-u_1x_2-u_3x_1 + u_1u_3)x_3 + u_1u_2x_2$
$f_8 = (-x_1 + u_1)x_8 + x_2x_7 - u_1x_2.$

$prem(g, C'') = 0$. However, $prem(g, C') \neq 0$, $prem(g, C''') \neq 0$, and $prem(g, C'''')$ $\neq 0$. In order to clarify the situation, we introduce two conditions.

Let Δ_1 be [(C and C_1 are on the same side of AB) $\wedge$ (B and B_1 are on the same side of AC)] $\vee$ [(C and C_1 are on either sides of AB) $\wedge$ (B and B_1 are on either sides of AC)].

Let Δ_2 be [(C and C_1 are on the same side of AB) $\wedge$ (A and A_1 are on the same side of BC)] $\vee$ [(C and C_1 are on either sides of AB) $\wedge$ (A and A_1 are on either sides of BC)].

Let $\delta_1 = -u_1^2u_3^2/(u_1u_2x_2 - u_1u_3x_1)x_4$, and let $\delta_2 = u_1^2u_3^2/((u_1u_2 - u_1^2)x_4x_6 - u_1u_3x_4x_5 + u_1^2u_3x_4)$.

Δ_1 holds iff $\delta_1 > 0$; Δ_2 holds iff $\delta_2 > 0$.

Now for component C', $\delta_1 = -4u_3^2/(3u_3^2+3u_2^2)$ and $\delta_2 = -4u_3^2/3(u_3^2+u_2^2-2u_1u_2+u_1^2)$.

For component C'', $\delta_1 = 4u_3^2/(3u_3^2+3u_2^2)$ and $\delta_2 = 4u_3^2/3(u_3^2+u_2^2-2u_1u_2+u_1^2)$.

For component C''', $\delta_1 = -4u_3^2/(3u_3^2+3u_2^2)$ and $\delta_2 = 4u_3^2/3(u_3^2+u_2^2-2u_1u_2+u_1^2)$.

For component C'''', $\delta_1 = 4u_3^2/(3u_3^2+3u_2^2)$ and $\delta_2 = -4u_3^2/3(u_3^2+u_2^2-2u_1u_2+u_1^2)$.

Thus, the conclusion is true when the three triangles are all outside ΔABC, or all "inside" ΔABC.

Remark. Traditional proofs of geometry theorems depend heavily on diagrams. Figure 22 suggests only one case; i.e., the three triangles are all outside ΔABC. Thus traditional proofs is usually valid only for that case. Most people are even unaware of whether there is the other case (i.e., the three triangles are all "inside" ΔABC, Figure 23) or are not sure whether or not the other case is also valid. From our theory, however, it is quite natural that if one case is valid then the other case is also valid.

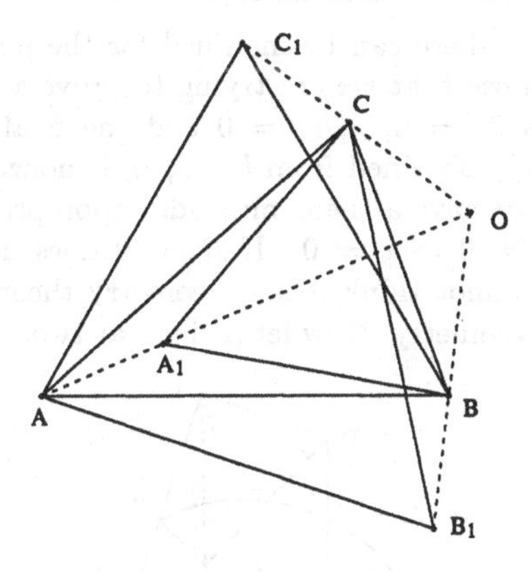

Figure 23

This problem, specified in a reducible way, can be also specified in an irreducible way. We will give a more general version of the above theorem:

Example (5.7) (M. Paterson). Three similar isosceles triangles, A_1BC, AB_1C, and ABC_1, are erected on the three respective sides, BC, CA, AB, of a triangle ABC, then AA_1, BB_1 and CC_1 are concurrent (Figure 22).

Let $A=(0,0)$, $B=(u_1,0)$, $C=(u_2,u_3)$, $C_1=(x_2,x_1)$, $A_1=(x_4,x_3)$, $B_1=(x_6,x_5)$, and $O=(x_8,x_7)$. We can specify the statement in the following way.

$h_1 = 2u_1x_1 - u_1^2 = 0$ $\quad C_1A \equiv C_1B$

$h_2 = (u_1u_3x_1 + (u_1u_2 - u_1^2)u_4)x_3 + ((-u_1u_2 + u_1^2)x_1 + u_1u_3u_4)x_2 - u_1^2u_3x_1 + (-u_1^2u_2 + u_1^3)u_4 = 0$ $\quad \tan(BAC_1) = \tan(CBA_1)$

$h_3 = (2u_2 - 2u_1)x_3 + 2u_3x_2 - u_3^2 - u_2^2 + u_1^2 = 0$ $\quad A_1B \equiv A_1C$

$h_4 = (u_1u_3x_1 + u_1u_2u_4)x_5 + (-u_1u_2x_1 + u_1u_3u_4)x_4 + (-u_1u_3^2 - u_1u_2^2)u_4 = 0$ $\quad \tan(BAC_1) = \tan(ACB_1)$

$h_5 = 2u_2x_5 + 2u_3x_4 - u_3^2 - u_2^2 = 0$ $\quad B_1A \equiv B_1C$

$h_6 = x_4x_7 + (-x_5 + u_1)x_6 - u_1x_4 = 0$ $\quad O$ is on line BB_1

$h_7 = x_2x_7 - x_3x_6 = 0$ $\quad O$ is on line AA_1

$g = (u_4 - u_3)x_7 + (-x_1 + u_2)x_6 + u_3x_1 - u_2u_4 = 0$

The conclusion: Points O, C and C_1 are collinear.

Though the geometry statement specified in this way does not exactly belong to the class of geometry statement of constructive type defined in section 6 of Chapter 1, it is a statement of constructive type: Points A, B and C are arbitrarily chosen; Point C_1 is on the perpendicular-bisector of A and B; Point A_1 is on the perpendicular-bisector of B and C and on the line defined by $tan(BAC_1) = tan(CBA_1)$; point B_1 is on the perpendicular-bisector of A and C and on the line defined by $tan(BAC_1) = tan(ACB_1)$.

Thus, the problem became not only irreducible but also linear. Our prover proved the theorem in 1.0 seconds with $maxt = 80$. The novice should study this problem carefully to understand the real trick behind this specification and find all nondegenerate conditions (in geometric forms) for the statement to be valid (i.e., all geometric conditions to make the constructions normal).

The method developed here can be modified for the purpose of finding new geometry theorems. Suppose that we are trying to prove a theorem with conclusion $g = 0$ and hypotheses $h_1 = 0, ..., h_n = 0$ and the final remainder R_0 with the triangular form $f_1, \ldots, f_r$ obtained from $h_1, ..., h_n$ is nonzero. If we add a new hypothesis $R_0 = 0$, then we have a theorem: under appropriate subsidiary conditions $R_0 = 0, h_1 = 0, ..., h_n = 0 \Rightarrow g = 0$. If $R_0 = 0$ does not have a clear meaning in geometry, then we cannot think it is a geometry theorem, at least it is not an interesting theorem in geometry. Now let us look at two examples.

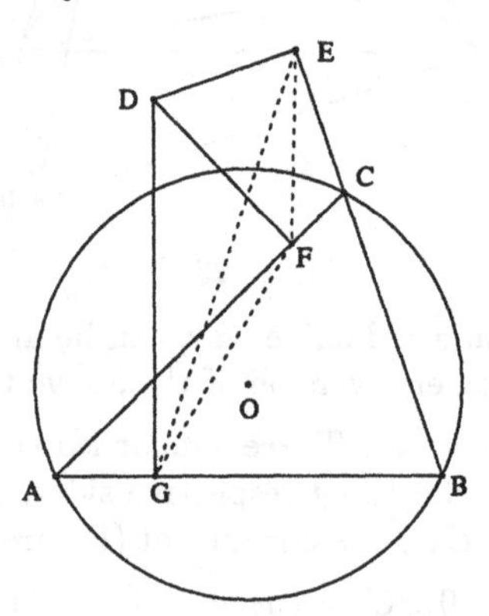

Figure 24

Example (5.8). Let ABC be a triangle and D be a moving point. From D three perpendicular lines are drawn to the three sides of the triangle ABC. Let the feet be E, F, and G. If the area of the directed triangle EFG keeps constant, what is the locus of point D? (See Figure 24.)

Let the area be $a/2$ and $A = (0,0)$, $B = (u_1, 0)$, $C = (u_2, u_3)$, $D = (x, y)$, $E = (x_2, x_1)$, $F = (x_4, x_3)$ and $G = (x, 0)$. Then all geometric conditions can be expressed by the following equations:

$$h_1 = u_3 x_2 + (-u_2 + u_1)x_1 - u_1 u_3 = 0 \qquad E \text{ is on line } BC$$

$h_2 = (u_2 - u_1)x_2 + u_3x_1 - u_3y + (-u_2 + u_1)x = 0$ — $ED \perp BC$
$h_3 = u_3x_4 - u_2x_3 = 0$ — F is on line AC
$h_4 = u_2x_4 + u_3x_3 - u_3y - u_2x = 0$ — $FD \perp AC$
$g = x_1x_4 + (-x_2 + x)x_3 - xx_1 + a = 0$
The area of oriented triangle EFG is $a/2$.

considering $u_1, ..., u_4, a, x, y$ as parameters, we can first triangulize $h_1, ..., h_4$:

$$f_1 = (u_3^2 + u_2^2 - 2u_1u_2 + u_1^2)x_1 - u_3^2y + ((-u_2 + u_1)u_3)x + (u_1u_2 - u_1^2)u_3 = 0$$
$$f_2 = u_3x_2 + (-u_2 + u_1)x_1 - u_1u_3 = 0$$
$$f_3 = (u_3^2 + u_2^2)x_3 - u_3^2y - u_2u_3x = 0$$
$$f_4 = u_2x_4 + u_3x_3 - u_3y - u_2x = 0.$$

Then R_0, the final remainder of g with respect to $f_1, ..., f_4$, is:[11]

$R_0 = u_1u_3^3x^2 + u_1u_3^3y^2 - u_1^2u_3^3x + (-u_1u_3^4 + (-u_1u_2^2 + u_1^2u_2)u_3^2)y + (u_3^4 + (2u_2^2 - 2u_1u_2 + u_1^2)u_3^2 + u_2^4 - 2u_1u_2^3 + u_1^2u_2^2)a.$

The equation $R_0 = 0$ means point D is on a circle (O). If $a = 0$, then $R_0 = u_1u_3^3x^2 + u_1u_3^3y^2 - u_1^2u_3^3x + (-u_1u_3^4 + (-u_1u_2^2 + u_1^2u_2)u_3^2)y = 0$ is the equation for the circle passing through the three vertices $A = (0,0)$, $B = (u_1, 0)$, $C = (u_2, u_3)$, with the same center as that of the circle (O). Thus we come to

A generalization of Simson's theorem. The locus of points D that keep the area of the oriented triangle EFG constant is a circle with the circumcenter of triangle ABC as its center.

The author found this theorem using the method described here in the spring of 1983. Later, he learned from P. Le Chenadec that this was a known theorem due to J. D. Gergonne of last century. The following is another interesting problem of this kind.

Example (5.9). (M. Paterson's Problem). Let us recall example (5.7). Now, what is the locus of the points O of concurrency as the areas of the three similar triangles are varied between 0 and infinity?

Let $A = (0,0)$, $B = (u_1, 0)$, $C = (u_2, u_3)$, $O = (x, y)$, $C_1 = (x_2, x_1)$, $B_1 = (x_4, x_3)$.

$h_1 = (y - u_3)x_2 + (-x + u_2)x_1 - u_2y + u_3x = 0$ — C_1 is on line OC
$h_2 = 2u_1x_2 - u_1^2 = 0$ — $C_1A \equiv C_1B$
$h_3 = yx_4 + (-x + u_1)x_3 - u_1y = 0$ — B_1 is on line OB
$h_4 = 2u_2x_4 + 2u_3x_3 - u_3^2 - u_2^2 = 0$ — $B_1A \equiv B_1C$
$g = (u_1u_3x_2 + u_1u_2x_1)x_4 + (-u_1u_2x_2 + u_1u_3x_1)x_3 + (-u_1u_3^2 - u_1u_2^2)x_1 = 0$
$\tan(BAC_1) = \tan(ACB_1)$.

As we have done in the last example, we can first triangulize $h_1, ..., h_4$:

$$f_1 = (2u_1x - 2u_1u_2)x_1 + (2u_1u_2 - u_1^2)y - 2u_1u_3x + u_1^2u_3 = 0$$
$$f_2 = 2u_1x_2 - u_1^2 = 0$$
$$f_3 = (2u_3y + 2u_2x - 2u_1u_2)x_3 + (-u_3^2 - u_2^2 + 2u_1u_2)y = 0$$
$$f_4 = 2u_2x_4 + 2u_3x_3 - u_3^2 - u_2^2 = 0$$

Then R_0, the final remainder of g with respect to $f_1, ..., f_4$, is (see the last foot-

[11] Some expected factors were already removed during successive divisions [11].

note):

$$R_0 = u_1^2(u_3^2 + u_2^2)(ay^2 + 2bxy + cu_3x^2 + (u_1u_3^2 - u_1u_2^2 - u_1^2u_2)y + (2u_1u_2 - u_1^2)u_3x),$$

where $a = (2u_2 - u_1)u_3$, $b = (-u_3^2 + u_2^2 - u_1u_2 + u_1^2)$, $c = -a$.

From analytic geometry we know that this is a hyperbola since $ac-b^2 = -a^2-b^2 < 0$ is generally true. Independently, M. Paterson also got the same result.[12] For a complete description of the method, see [11].

[12] Here was Professor Paterson's response: "Your locus equation shows that we get a rectangular hyperbola. It was the rectangularity which shocked me first when I plotted some trials by computer. I subsequently derived a very messy proof by exhausting coordinate geometry (by hand!)"

Chapter 5.
Geometry Theorem Proving Using The Gröbner Basis Method

The Gröbner basis method was introduced by B. Buchberger in 1965 [7]. It is a powerful algorithm tool for solving many important problems in polynomial ideal theory. The success of Wu's method stimulated researchers to apply the Gröbner basis method to the same class of geometry theorems which can be addressed by Wu's method. Recently, applications of the Gröbner basis method to geometry theorem proving have been reported [16], [27], [31]. In a recent experiment, we proved 358 among the 366 theorems in the *Collection* [14] using our approach based on the Gröbner basis method [16] (method (2.4) below). In this chapter we will present our approach. We will also compare Wu's method with the Gröbner basis method in the context of proving geometry theorems. In section 1, we give a brief review of the basic notions in the Gröbner basis method necessary for our purpose. Buchberger's paper [8] is an excellent and comprehensive review.

1. A Review of the Gröbner Basis Method

As in chapter 2, let K be a (computable) field and $A = K[y_1, ..., y_n]$ be the polynomial ring over K with indeterminates $y_1, \ldots, y_n$.[1]

Definition (1.1). A total order $<$ among the set M of all monomials (i.e., the power products of the y; 0 is not considered a monomial) in $K[y]$ is called a *compatible ordering* (with multiplication), if

(i) $1 < m$ for all $m \neq 1$.

(ii) $m_1 < m_2 \Rightarrow sm_1 < sm_2$, for all s, m_1, $m_2 \in M$.

Example (1.1.1). The following two compatible orderings are mostly in common use in the Gröbner basis method.

(i) **The purely lexicographic ordering.** $y_1^{s_1} \cdots y_n^{s_n} < y_1^{t_1} \cdots y_n^{t_n}$ if there is $k \leq n$ such that $s_i = t_i$ for $i < k$ and $s_k < t_k$.

(ii) **The total degree ordering.** $y_1^{s_1} \cdots y_n^{s_n} < y_1^{t_1} \cdots y_n^{t_n}$ if $s = s_1 + \cdots + s_n < t = t_1 + \cdots + t_n$, or $s = t$ and there is $k \leq n$ such that $s_i = t_i$ for $i < k$, and $s_k < t_k$.

Theorem (1.2). Any compatible ordering is well-ordered.

Proof. Suppose $m_1 > m_2 > \cdots > m_k > \cdots$ is an infinite, strictly decreasing sequence of monomials. We assert that such a sequence does not exist. One can use induction on n to prove the assertion. Here we give an indirect proof using the Hilbert basis theorem. Let $J_k = (m_1, ..., m_k)$ be the polynomial ideals in $K[y_1, ..., y_n]$, generated by monomials $m_1, ..., m_k$. Then $J_1 \subset J_2 \subset \cdots$ is an ascending sequence of ideals. By the Hilbert basis theorem, there is an integer t such that

[1] We are only interested in Gröbner bases in polynomial rings over fields. There are many generalizations of fields to certain commutative rings.

$J_t = J_{t+k}$ for all $k > 0$. Thus $m_{t+1} = p_1m_1 + \cdots + p_tm_t$. Hence $m_{t+1} \geq m_s$ for some $s < t+1$ by (ii) of (1.1). This contradicts our assumption. .QED.

In what follows, we fix a compatible ordering $<$; all discussions are relative to this ordering. Thus a nonzero polynomial p can be expressed in the form:

$$p = c_1m_1 + \cdots + c_km_k,$$

where the c_i are nonzeros in K and the m_i are decreasing monomials: $m_1 > \cdots > m_k$. This representation is unique. The monomial m_1 is called the *leading monomial* of p, c_1m_1 the *leading term* of p. We also say that $m_1,...,m_k$ are *monomials of p*.

Let S be a polynomial set and g be a polynomial. Polynomial g is *b-reducible* modulo S if there is an $f \in S$ such that the leading monomial m_1 of f is a factor of a monomial m of g. In that case, letting c_1m_1 be the leading term of f, $f = c_1m_1+f_1$ and $m = sm_1$, we can replace m in g by $-sf_1/c_1$ to obtain a polynomial g_1. We say g is reduced to g_1 modulo S via one step. Obviously, $g - g_1 \in \text{Ideal}(S)$. If g_1 is b-reducible, then we can repeat the process to obtain a new polynomial g_2. We call this process *reduction*. From theorem (1.2), it is not hard to prove that this process terminates. Thus, g is eventually reduced to a polynomial g_k, b-irreducible modulo S. This polynomial is called a *normal form* of g modulo S. Note that a normal form of polynomial g modulo S is not necessarily unique. Different reductions of g modulo S can lead to different normal forms.

Definition (1.3). A set of generators G of an ideal $I \subset K[y]$ is called a *Gröbner basis* of I if any reduction of any polynomial g of $K[y]$ modulo G leads to the same normal form of g.

Theorem (1.4). If G is a Gröbner basis of ideal I, then $g \in I$ if and only if the normal form of g modulo G is zero.

Proof. See [5].

This theorem means that once we have a finite Gröbner basis of an ideal I, then we can decide the ideal membership easily. Now we present Buchberger's algorithm to construct a finite Gröbner basis of an ideal I.

Definition (1.5). Let $p_1 = c_1m_1 + f_1$ and $p_2 = c_2m_2 + f_2$ be two nonconstant polynomials with leading terms c_1m_1 and c_2m_2, respectively. Let m be the (monic) least common multiplier of m_1 and m_2, s_1 and s_2 be monomials such that $m = s_1m_1 = s_2m_2$. Then polynomial $p = s_1f_1 - (c_1/c_2)s_2f_2$ is called the *S-polynomial* of p_1 and p_2.

Note that if $\text{GCD}(m_1, m_2) = 1$, then the S-polynomial p of p_1 and p_2 can be reduced to zero modulo $\{p_1, p_2\}$.

Algorithm (1.6). (Buchberger's algorithm for constructing Gröbner bases). Given a finite set S of generators of an ideal I, we can use the following procedure to construct a Gröbner basis of I.

Form all S-polynomials of all pairs of polynomials in $S_0 = S$. Reduce these S-polynomials to their normal forms modulo S_0 and add any nonzero normal forms to S to get a new polynomial set S_1. Repeat the process for S_1 to obtain a new polynomial set S_2, and so on.

Theorem (1.7). (Buchberger). The procedure in (1.6) always terminates, i.e., for the sequence $S_0 \subset S_1 \subset S_2 \subset \cdots$, there is k such that $S_k = S_{k+1}$ and S_k is a Gröbner basis of I.

Proof. Considering the leading monomials of polynomials of the set S_i in the order in which new polynomials are added, we can use the similar reasoning as in the proof of (1.2) to prove the termination of the procedure; i.e., there is k such that $S_k = S_{k+1}$. For the proof of the fact that S_k is a Gröbner basis of I, see [5].

Our actual implementation is that, whenever a new (reduced) S-polynomial is added to the polynomial set, we reduce each polynomial modulo others in certain order until every polynomial is b-reduced modulo the others; then we form new S-polynomials again. In this way we can obtain a *reduced Gröbner basis* of I, i.e., a Gröbner basis of I, each polynomial of which is monic and reduced modulo the remaining polynomials. For reduced Gröbner bases, we have the following uniqueness theorem:

Theorem (1.8) (Buchberger). For a given compatible order, the reduced Gröbner basis of an ideal I is unique regardless of the starting set S of generators $p_1, ..., p_m$ of I.

Proof. See [9].

Unless stated otherwise, whenever we mention a Gröbner basis of an ideal $(p_1, ..., p_m)$, we mean this unique reduced Gröbner basis and denote it by $\mathrm{GB}(p_1, ..., p_m)$.

2. Proof Methods for Formulation F3

Basically, the Gröbner basis method can be used to prove the same class of geometry theorems that can be addressed by Wu's method. Our approach to proving geometry theorems using the Gröbner basis method is based on the following theorems.

Let (S), $u = u_1, ..., u_d$, $x = x_1, ..., x_r$, $h_1, ..., h_n$, and g be the same as in Chapter 4. We assume that the field E associated with the geometry in consideration is algebraically closed. As we already know (theorem (4.3) of Chapter 4), if we can confirm (S) to be generally true in E, then (S) is generally true in any field. Let I be the ideal generated by $h_1, ..., h_n$ in $\mathbf{Q}[u, x]$, J be the ideal generated by $h_1, ..., h_n$ in $\mathbf{Q}(u)[x]$. Obviously, $I \subset J$ and $g \in J$ if and only if there is a u–polynomial p such that $pg \in I$. Let $L = \mathrm{Radical}(h_1, ..., h_n)$ in $\mathbf{Q}(u)[x]$; i.e., Radical(J).

Theorem (2.1). Let all notations be the same as in the previous paragraph. The geometry statement (S) is generally true (in E) if and only if $g \in L$.

Proof. Suppose (S) is generally true, then by theorem (3.4) of Chapter 3, there is a u-polynomial s such that $sg \in \mathrm{Radical}(I)$ of $\mathbf{Q}[u, x]$, i.e, there is an integer $t > 0$ such that $(sg)^t \in I$. Hence $g^t \in J$, i.e., $g \in L$.

Conversely, suppose that $g \in L$, then there is an integer $t > 0$, $g^t \in J$. Thus there is a u–polynomial s such that $sg^t \in I$. By (3.4) of Chapter 3, g is generally true. .QED.

Theorem (2.2). If $g \in J$, then (S) is generally true.

proof. It is obvious from theorem (2.1) since $g \in J \subset L$. .QED.

Theorem (2.3). Let all notations be the same as in (2.1) and z be a new variable other than the u and the x. $\mathrm{GB}(h_1, ..., h_n, zg - 1) = \{1\}$ (in $\mathbf{Q}(u)[x, z]$) if and only if $g \in L$. Thus, (S) is generally true if and only if $\mathrm{GB}(h_1, ..., h_n, zg - 1) = \{1\}$.

Proof. If $g \in L$, then $h_1, ..., h_n, zg - 1$, considered as polynomials of $\mathbf{Q}(u)[x, z]$, do not have common zeros in any extension of $\mathbf{Q}(u)$, in particular in algebraically closed extensions of $\mathbf{Q}(u)$. Thus, by Hilbert's Nullstellensatz, the ideal $(h_1, ..., h_n, zg-1)$ of $\mathbf{Q}(u)[x, z]$ is the unit ideal (1), i.e., $\mathrm{GB}(h_1, ..., h_n, zg-1) = \{1\}$. If $\mathrm{GB}(h_1, ..., h_n, zg-1) = \{1\}$, then there are q_i in $\mathbf{Q}(u)[x, z]$ such that

$$q_1 h_1 + \cdots + q_n h_n + q_{n+1}(zg - 1) = 1.$$

Let $z = 1/g$, and clearing denominators, we have

$$a_1 h_1 + \cdots + a_n h_n = g^t,$$

for some $t \geq 0$ and polynomials $a_i \in \mathbf{Q}(u)[x]$. Thus $g \in L$. .QED.

The main method we use in our geometry theorem prover is based on theorems (2.1), (2.2) and (2.3):

Method (2.4). We assume $x_1 < \cdots < x_r$ and the purely lexicographic ordering among monomials of $\mathbf{Q}(u)[x]$ (see example (1.1.1) of the last section). Note that variables $u_1, ..., u_d$ are now in the based field $\mathbf{Q}(u)$; there are no orderings among the u.

Step 1. Compute the Gröbner basis of $h_1, ..., h_n$ in $\mathbf{Q}(u)[x]$; then use it to reduce g to see whether $g \in J$. If it is, then (S) is generally true. If not, then

Step 2. Compute the Gröbner basis GB of $h_1, ..., h_n, zg - 1$ in $\mathbf{Q}(u)[x]$ to see whether it is $\{1\}$. (S) is generally true if and only if $\mathrm{GB} = \{1\}$. (Note that E is algebraically closed.)

We can also find subsidiary conditions by collecting all denominators (removing repeated factors) during the computation of the Gröbner basis.

Step 1 can be considered as a first approximation. However, for all theorems we have found in practice, $J = L$. Thus, step 1 is usually sufficient.

Example (2.5). Example (2.1) of Parallelogram in Chapter 1.

In $\mathbf{Q}(u_1, u_2, u_3)[x_1, x_2, x_3, x_4]$, $\mathrm{GB}(h_1, ..., h_4, zg - 1) = \{1\}$. Thus the statement has been proved to be generally true. However, more directly, we can compute $\mathrm{GB}(h_1, ..., h_4) =$

$x_1 - u_3$
$x_2 - (u_2 - u_1)$
$x_3 - u_3/2$
$x_4 - u_2/2.$

The GB reduces the conclusion polynomial $g = 2u_2x_4 + 2u_3x_3 - u_3^2 - u_2^2$ to zero. Thus the statement is generally true. The subsidiary conditions (obtained by collecting all different prime factors in denominators in the computation of the GB) are: $u_3 \neq 0$, $u_1 \neq 0$.

Example (2.6). Example (2.2) of Simson's theorem in Chapter 1.

GB = GB($h_1, ..., h_7$) =

$$x_1 - (u_3^2 + u_2^2 - u_1u_2)/2u_3$$
$$x_2 - u_1/2$$
$$x_3^2 - (u_1u_3x_3 - u_3u_4^2 + (u_3^2 + u_2^2 - u_1u_2)u_4)/u_3$$
$$x_4 - ((u_2 - u_1)u_3x_3 + u_3^2u_4 + (-u_1u_2 + u_1^2)u_3)/(u_3^2 + u_2^2 - 2u_1u_2 + u_1^2)$$
$$x_5 - ((u_2^2 - 2u_1u_2 + u_1^2)x_3 + (u_2 - u_1)u_3u_4 + u_1u_3^2)/(u_3^2 + u_2^2 - 2u_1u_2 + u_1^2)$$
$$x_6 - (u_2u_3x_3 + u_3^2u_4)/(u_3^2 + u_2^2)$$
$$x_7 - (u_2^2x_3 + u_2u_3u_4)/(u_3^2 + u_2^2).$$

The Gröbner basis reduces the conclusion polynomial g to zero. Thus we proved Simson's theorem to be generally true. The subsidiary conditions needed for the computation are: $u_2 \neq 0$, $u_1 \neq 0$, $u_3 \neq 0$, $u_3^2 + u_2^2 - 2u_1u_2 + u_1^2 \neq 0$, $u_3^2 + u_2^2 \neq 0$.

Example (2.7). The Butterfly theorem: example (2.4) of Chapter 1.

We choose slightly different coordinates: $E = (0,0)$, $O = (u_1, 0)$, $A = (u_2, u_3)$, $B = (u_4, x_1)$, $C = (x_2, x_3)$, $D = (x_4, x_5)$, $F = (0, x_6)$, and $G = (0, x_7)$. Then the hypotheses and the conclusion are:

$$h_1 = x_1^2 + u_4^2 - 2u_1u_4 - u_3^2 - u_2^2 + 2u_1u_2 = 0$$
$$h_2 = x_3^2 + x_2^2 - 2u_1x_2 - u_3^2 - u_2^2 + 2u_1u_2 = 0$$
$$h_3 = u_2x_3 - u_3x_2 = 0$$
$$h_4 = x_5^2 + x_4^2 - 2u_1x_4 - u_3^2 - u_2^2 + 2u_1u_2 = 0$$
$$h_5 = u_4x_5 - x_1x_4 = 0$$
$$h_6 = (x_4 - u_2)x_6 + u_2x_5 - u_3x_4 = 0$$
$$h_7 = (x_2 - u_4)x_7 + u_4x_3 - x_1x_2 = 0$$
$$g = x_7 + x_6 = 0$$

However, neither GB($h_1, ...h_7, zg - 1$) = {1}, nor GB($h_1, ..., h_7$) reduces g to zero. The reason is the same as we explained in example (2.4) of Chapter 1 and in Remark 1 in Section 3.1 of Chapter 3: in specifying of point C and B we have ambiguities. These ambiguities can be easily detected by the program and, in the case of using the Gröbner basis method, can be removed by adding two new equations $s_1 = (x_2 - u_2)v_1 - 1 = 0$ ($A \neq C$), $s_2 = (x_4 - u_4)v_2 - 1 = 0$ ($B \neq D$). Thus GB($s_1, s_2, h_1, ..., h_7$) =

$$v_1 + (2u_1u_4 + u_3^2 + u_2^2 - 2u_1u_2)/(2u_1u_4^2 + (2u_3^2 + 2u_2^2 - 4u_1u_2)u_4)$$
$$v_2 + (u_3^2 + u_2^2)/(2u_2u_3^2 + 2u_2^3 - 2u_1u_2^2)$$
$$x_1^2 + u_4^2 - 2u_1u_4 - u_3^2 - u_2^2 + 2u_1u_2$$
$$x_2 + (u_2u_3^2 + u_2^3 - 2u_1u_2^2)/(u_3^2 + u_2^2)$$
$$x_3 + (u_3^3 + (u_2^2 - 2u_1u_2)u_3)/(u_3^2 + u_2^2)$$
$$x_4 + ((u_3^2 + u_2^2 - 2u_1u_2)u_4)/(2u_1u_4 + u_3^2 + u_2^2 - 2u_1u_2)$$
$$x_5 + ((u_3^2 + u_2^2 - 2u_1u_2)x_1)/(2u_1u_4 + u_3^2 + u_2^2 - 2u_1u_2)$$
$$x_6 + ((u_2u_3^2 + u_2^3 - 2u_1u_2^2)x_1 + (-u_3^3 + (-u_2^2 + 2u_1u_2)u_3)u_4)/((u_3^2 + u_2^2)u_4 + u_2u_3^2 + u_2^3 - 2u_1u_2^2)$$
$$x_7 - ((u_2u_3^2 + u_2^3 - 2u_1u_2^2)x_1 + (-u_3^3 + (-u_2^2 + 2u_1u_2)u_3)u_4)/((u_3^2 + u_2^2)u_4 + u_2u_3^2 +$$

$u_2^3 - 2u_1u_2^2)$.

The Gröbner basis reduces g to zero. Thus, we proved the Butterfly theorem. In the computation we need the following subsidiary conditions: $(u_3^2+u_2^2)u_4+u_2u_3^2+u_2^3-2u_1u_2^2 \neq 0$, $u_4-u_2 \neq 0$, $2u_1u_4+u_3^2+u_2^2-2u_1u_2 \neq 0$, $u_4 \neq 0$, $u_1u_4+u_3^2+u_2^2-2u_1u_2 \neq 0$, $u_3^2+u_2^2 \neq 0$, $u_2 \neq 0$, $u_3^2+u_2^2-u_1u_2 \neq 0$.

Note that the computation of the Gröbner basis is very sensitive to the variable ordering. If we choose the same coordinates that are in example (2.4) of chapter 1, then the computation would take much longer. For the present choice of coordinates, the computation took about 24 seconds.

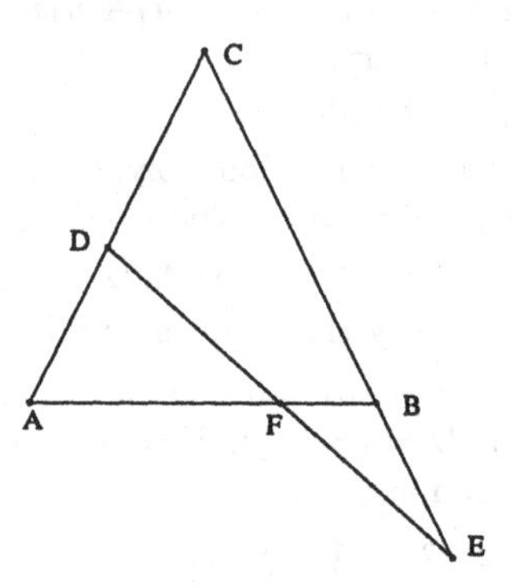

Figure 25

Example (2.8). Let ABC be a triangle such that $AC \equiv BC$. D is a point on AC; E is a point on BC such that $AD \equiv BE$. F is the intersection of DE and AB. Show $DF \equiv EF$ (Figure 25).

This theorem could not be confirmed by the elementary use of Wu's method (without decomposition), as described in chapter 1 or in the first two sections in [10]. Let $A = (0,0)$, $B = (u_1,0)$, $C = (x_1,u_2)$, $D = (u_3,x_2)$, $E = (x_3,x_4)$, and $F = (x_5,0)$. The hypotheses and the conclusion are

$$h_1 = 2u_1x_1 - u_1^2 = 0 \qquad AC \equiv BC$$
$$h_2 = -x_1x_2 + u_2u_3 = 0 \qquad D \text{ is on } AC$$
$$h_3 = x_4^2 + x_3^2 - 2u_1x_3 - x_2^2 - u_3^2 + u_1^2 = 0 \qquad AD \equiv BE$$
$$h_4 = (-x_1+u_1)x_4 + u_2x_3 - u_1u_2 = 0 \qquad E \text{ is on } BC$$
$$h_5 = (-x_4+x_2)x_5 + u_3x_4 - x_2x_3 = 0 \qquad F \text{ is on } ED$$
$$g = (2x_3 - 2u_3)x_5 - x_4^2 - x_3^2 + x_2^2 + u_3^2 = 0 \qquad \text{Conclusion: } DF \equiv EF.$$

The Gröbner Basis is

$$x_5 - (2u_3+u_1)/2$$
$$x_4 + 2u_2u_3/u_1$$
$$x_3 - (u_3+u_1)$$
$$x_2 - 2u_2u_3/u_1$$
$$x_1 - u_1/2,$$

which reduces g to zero. Therefore the statement has been proved to be generally true. The subsidiary conditions needed for the proof are: $u_3 \neq 0$, $2u_3 - u_1 \neq 0$,

$u_2 \neq 0$, $4u_2^2 + u_1^2 \neq 0$, $u_1 \neq 0$.

The reader can find many other examples using this method in [16] and in Part II.

The following theorem gives another method for finding subsidiary conditions.

Theorem (2.9). Let all notations be the same as in (2.1) and z be a new variable other than the u and the x. (S) is generally true if and only if the Gröbner basis of $I_z = (h_1, ..., h_n, gz - 1)$ in $\mathbf{Q}[u, x, z]$ (in purely lexicographic orderings $u < x$ and $u < z$) contains at least a u–polynomial.

Proof. Suppose (S) is generally true. Let J_z be the ideal generated by $h_1, ..., h_n$, $gz - 1$ in $\mathbf{Q}(u)[x, z]$. Then by (2.3), $J_z = (1)$. A polynomial $a \in J_z$ if and only if there is a u–polynomial s such that $sa \in I_z$. Since $1 \in J_z$, there is a u–polynomial s such that $s = s \cdot 1 \in I_z$. Since we are using purely lexicographic orderings with $u < x$ and $u < z$, $\mathrm{GB}(h_1, ..., h_n, gz - 1)$ of I_z contains at least a u–polynomial. Conversely, suppose $\mathrm{GB}(h_1, ..., h_n, gz - 1)$ of I_z contains a u–polynomial s. Then there are polynomials q_i in $\mathbf{Q}[u, x, z]$ such that

$$q_1 h_1 + \cdots + q_n h_n + q_{n+1}(zg - 1) = s.$$

Let $z = 1/g$, and clearing denominators, we have

$$a_1 h_1 + \cdots + a_n h_n = sg^t,$$

for some $t \geq 0$ and $a_i \in \mathbf{Q}[u, x]$. Thus (S) is generally true. .QED.

Method (2.10). Compute the Gröbner basis GB of ideal $(h_1, ..., h_n, gz - 1)$ in $\mathbf{Q}[u, x, z]$ in a purely lexicographic ordering $u < x$, $u < z$ (where z is a new variable) to see whether it contains a u–polynomial. (S) is generally true if and only if GB contains a u–polynomial. In that case, let $s_1, \ldots, s_k$ be all u–polynomials in GB, then

$$\forall y_i \in E[(h_1 = 0 \wedge \cdots \wedge h_n = 0) \wedge (s_1 \neq 0 \vee \cdots \vee s_k \neq 0) \Rightarrow g = 0].$$

The idea of this method is due to D. Kapur [27]. However, Kapur's original method is for Formulation F2 and can cause serious problems, as shown in examples (2.7) and (2.8) of Chapter 3. This method can produce subsidiary conditions weaker than method (2.4) or Wu's method. But it is extremely time-consuming. Therefore, we are using method (2.4) in our prover.

3. A Proof Method for Formulation F1

Formulation F1 in Section 2 of Chapter 3 requires that nondegenerate conditions of a geometry statement be fully specified by the user. Thus the equivalent algebraic formula of a geometry statement is

(3.1) $\forall y_i(h_1 = 0 \wedge \cdots \wedge h_n = 0 \wedge s_1 \neq 0 \wedge \cdots \wedge s_k \neq 0 \Rightarrow g = 0)$,

where $h_1, ..., h_n$, $s_1, ..., s_k$, g are polynomials in $\mathbf{Q}[y_1, ..., y_m]$. Obviously, (3.1) is equivalent to

(3.2) $\forall y_i[\exists z_j(h_1 = 0 \wedge \cdots \wedge h_n = 0 \wedge s_1 z_1 - 1 = 0 \wedge \cdots \wedge s_k z_k - 1 = 0) \Rightarrow g = 0]$, where $z_1, ..., z_k$ are new variables. Since $z_1, ..., z_k$ do not occur in g, (3.2) is equivalent to

(3.3) $\forall y_i \forall z_j[(h_1 = 0 \wedge \cdots \wedge h_n = 0 \wedge s_1 z_1 - 1 = 0 \wedge \cdots \wedge s_k z_k - 1 = 0) \Rightarrow g = 0]$.

Our method [16] is based on the following theorem.

Theorem (3.4). For an algebraically closed field E, (3.3) is valid if and only if the ideal $I = (h_1, ..., h_n, s_1 z_1 - 1, ..., s_k z_k - 1, gv - 1)$ of $\mathbf{Q}[y_1, ..., y_m]$ is the unit ideal, where v is a new variable.

Proof. Suppose $I = (1)$. Similar to the proof of theorem (2.3), we can infer that there is $t > 0$ such that $g^t \in (h_1, ..., h_n, s_1 z_1 - 1, ..., s_k z_k - 1)$; thus (3.3) is valid. If (3.3) is valid, then polynomials $h_1, ..., h_n, s_1 z_1 - 1, ..., s_k z_k - 1, gv - 1$ do not have common zeros in E. Because E is algebraically closed, $I = (1)$ by Hilbert's Nullstellensatz. .QED.

Method (3.5). (A proof method for Formulation F1). Check whether GB($h_1, ...,$ $h_n, s_1 z_1 - 1, ..., s_k z_k - 1, gv - 1$) of $\mathbf{Q}[y_1, ..., y_m]$ is $\{1\}$. If so, then the statement is confirmed in any field. If not, then the statement is not valid in any algebraically closed field E, but not necessarily in other fields E.

Example (3.6). Example (2.5): Parallelogram again. It seems plausible that "A, B and C are non-collinear" (i.e., $u_1 u_3 \neq 0$) is the only nondegenerate condition. Thus we can ask whether or not

$$(h_1 = 0, h_2 = 0, h_3 = 0, h_4 = 0, u_1 u_3 \neq 0) \Rightarrow g = 0.$$

In an algebraically closed field, this is equivalent to checking whether GB($h_1, ..., h_4,$ $u_1 u_3 z_1 - 1, gz - 1$) $= \{1\}$. Our prover showed this is the case. Thus we proved the theorem.

As a first approximation, we can compute GB $=$ GB($h_1, ..., h_4$, $u_1 u_3 z_1 - 1$) in $\mathbf{Q}[y]$, then check whether GB reduces g to zero. GB($h_1, ..., h_4, u_1 u_3 z_1 - 1$) $=$

$$\begin{array}{l} x_1 - u_3 \\ x_2 - (u_2 - u_1) \\ x_3 - u_3/2 \\ x_4 - u_2/2 \\ z_1 u_1 u_3 - 1, \end{array}$$

which reduces g to zero. Compare this with example (2.5).

Remark. The reader might think the nondegenerate condition "line AC and line BD intersect" (have only one common point) is redundant because this is always the case for a nondegenerate parallelogram. However, it is not easy to find a traditional proof of this trivial fact. (Please try it!) Note that if AC and BD do not intersect normally, then the problem is meaningless (degenerate). Again, this extremely simple example reveals the fact that a nondegenerate condition, obvious to one person, might not be obvious to another and difficult to a third. In the real plane, the success of a traditional proof depends heavily on many "trivial facts." People accept these "trivial facts" simply by looking at the corresponding diagrams.

Example (3.7). Example (2.6): Simson's theorem.

By looking at diagrams (Figure 3), we are sure that the "obvious" nondegenerate condition "A, B and C are non-collinear", i.e., $u_1u_3 \neq 0$, is enough. We can think the algebraic equivalence of Simson's theorem (fully specified) is: $[(h_1 = 0, ..., h_7 = 0, u_1u_3 \neq 0) \Rightarrow g = 0]$.

However, $\mathrm{GB}(h_1, ..., h_7, u_1u_3z_1 - 1, gz - 1) \neq \{1\}$. Thus the theorem cannot be confirmed. The real situation is that if the Gröbner basis method confirms a statement, then the statement is valid in all fields. We have to add other nondegenerate conditions $c_1 = u_3^2 + u_2^2 - 2u_1u_2 + u_1^2 \neq 0$, $c_2 = u_3^2 + u_2^2 \neq 0$.

Our prover confirmed that $\mathrm{GB}(h_1, ..., h_7, u_1u_3z_1 - 1, (u_3^2 + u_2^2 - 2u_1u_2 + u_1^2)z_2 - 1, (u_3^2 + u_2^2)z_3 - 1, gz - 1) = \{1\}$. Therefore the theorem has been confirmed.

We can also compute $\mathrm{GB}(h_1, ..., h_7, u_1u_3z_1 - 1, (u_3^2 + u_2^2 - 2u_1u_2 + u_1^2)z_2 - 1, (u_3^2 + u_2^2)z_3 - 1)$ first. Then check whether the GB rewrites g to zero. Our prover also confirmed the theorem in this way.

Remark (3.8). As we have seen from this example, one has to add some seemingly "unnatural" nondegenerate conditions ($c_1 \neq 0$, $c_2 \neq 0$) in order to make the proof successful. Ordinary users will find it difficult to understand these conditions, and it is almost impossible for them to find such conditions. This is the drawback of this method and any method based on formulation F1 in general (see the discussion in Section 2 of Chapter 3). In comparison with the method (2.4), this method is slow, at least in our current implementation. For example, it took about 30 seconds to prove Simson's theorem using this method, whereas it only took 2.5 seconds using the method (2.4) in the previous section. For this method, the time for computing Gröbner bases is more sensitive to variable ordering. For example, it would take much longer if we choose z_1, z_2, and z_3 to be the highest variables.

4. Connections Between Characteristic Sets and Gröbner Bases[2]

Let $u_1, ..., u_d$, $x_1, ..., x_r$, $h_1, ..., h_n$ be the same as in Chapter 4 with $\mathbf{Q}$ replaced by any (computable) filed K. Furthermore, suppose that $\{h_1, \ldots, h_n\}$ satisfies the *dimensionality restriction* i.e., case 3 of of algorithm (2.1) of Chapter 4 will not happen (also see Remark 2 of Section 3.1 of Chapter 3). Let $f_1, ..., f_r$ be obtained from $h_1, ..., h_n$ using Ritt's principle as revised in Section 1 of Chapter 4. let I and J be the ideals of $K[u, x]$ and $K(u)[x]$ respectively, and generated by $h_1, \ldots, h_n$. Let $P = K[u, x] \cap J$. Obviously, $I \subset P$, and $g \in P$ iff there is a $c \in K[u]$ such that $cg \in I$. In this section I, J and P *always* denote the above ideals. In this section, a Gröbner basis of an ideal always means the unique reduced Gröbner basis, unless stated otherwise. First we have:

Proposition (4.1). Let the notation be the same as above. Suppose $\{h_1, \ldots, h_n\}$ satisfies the dimensionality restriction. For the following three conditions

(i) $f_1, \ldots, f_r$ is an irreducible ascending chain,
(ii) J is maximal,
(iii) $V = V(h_1, \ldots, h_n)$ has only one component general for u,

---
[2] The reader can skip this Section for the first reading.

we have (i) $\Longrightarrow$ (ii) $\Longrightarrow$ (iii).

Proof. If (i) holds then the sequence of ring extensions

$$K(u)[x_1]/(f_1) \subset K(u)[x_1, x_2]/(f_1, f_2) \subset \ldots \subset K(u)[x]/(f)$$

is actually a sequence of field extensions. Thus the (f) generate a maximal ideal, but it is contained in J. Suppose (ii) holds and suppose there were two components of V general for u. Then these would correspond to two maximal ideals of $K(u)[x]$ both containing J. This is a contradiction. .QED.

Clearly (iii) does not imply (ii) but only that Radical(J) be maximal. Also if (ii) holds we do not necessarily have an irreducible ascending chain. This is shown by the actual geometry example (2.8), or by the more trivial example $\{x_1(x_1 - u_1), x_2x_1 - u_2\}$. In $K(u)[x]$ u_2 and hence the x_1 is invertible, so that we just get $x_1 - u_1$ as the generator of J.

Theorem (4.2). With the same notations as in (4.1), suppose $\{h_1, \ldots, h_n\}$ satisfies the dimensionality restriction. Let $G = \{b_1, \ldots, b_j\}$ be the Gröbner Basis of J. If J is a prime ideal (or more strongly, $f_1, ..., f_r$ is irreducible) then

(i) $j = r$.

(ii) b_i is a monic polynomial in x_i with coefficients in $K(u)[x_1, \ldots, x_{i-1}]$.

(iii) Let $b_i' = c_ib_i \in I$ where $c_i \in K[u, x]$ and $\notin P$, then $b_1', \ldots, b_r'$ is an irreducible characteristic set of prime ideal P.

Proof. (i) and (ii). Since J is prime, we know by the dimensionality restriction that in fact J is maximal in $K(u)[x]$. Let $F = K(u)[x]/J$. We induct on r. The induction step is based on the fact that if $E \subset E[t]/A$ is a field extension where t is an indeterminate, then the ideal A is generated by a single monic polynomial in $E[t]$. Thus if we let F_i be the subfields of F generated by the first i of the $\bar{x}_k$ where $\bar{x}_k$ is canonical image of x_k in F, and let A_i be their defining ideal then A_i is generated by adding a monic polynomial d_i in $F_{i-1}[x_i]/A_{i-1}$ to A_{i-1}. Thus we can choose inverse images $b_1, \ldots, b_r$ of $d_1, \ldots, d_r$ in $K(u)[x]$ such that b_i are monic and $deg(b_j, x_i) < deg(b_i, x_i)$ for $i < j$. It is easy to see that $b_1, \ldots, b_r$ are generators of J. But such a set of *reduced* generators has no overlaps among the leading monomials of the generators, hence is the Gröbner basis by the uniqueness of reduced Gröbner bases.

(iii) Let $C = b_1', \ldots, b_r'$. Because we have field extensions, the b_i' are irreducible in $K(u)[x_1, \ldots, x_i]/(b_1, \ldots, b_{i-1}) = K(u)[x_1, \ldots, x_i]/(b_1', \ldots, b_{i-1}')$. Thus C is an irreducible ascending chain. Let b be a polynomial in P reduced with respect to C. Since the canonical image $\bar{b}$ of b in the field $F = K(u)[x]/(b_1, \ldots, b_r) = K(u)[x]/(b_1', \ldots, b_r')$ is zero, it follows that b is zero. Thus C is a characteristic set of P. Since $P = K[u, x] \cap J$ and J is prime, P is prime. .QED.

In practice, for 95% theorems from geometry, $f_1, \ldots, f_r$ is irreducble; thus computing GB$(h_1, ..., h_n)$ in $K(u)[x]$ in method (2.4) is actually computing a special kind of characteristic sets of P. It usually takes more time than using Ritt's principle (or its variants) to compute $f_1, ..., f_r$.

In Ritt's method, a prime ideal is given by its characteristic set, which is generally not a set of generators of that ideal. The following theorem gives an algorithm[2] for constructing the Gröbner basis of a prime ideal P from its (irreducible) characteristic set.

Theorem (4.3). Let $f_1(u, x_1), ..., f_r(u, x_1, ..., x_r)$ be an irreducible characteristic set of a prime ideal P of $K[u, x]$, $I_1, ..., I_r$ be the initials of $f_1, ..., f_r$, and z be a new variable. Let G_z be a Gröbner basis of the ideal $(f_1, ..., f_r, I_1 \cdots I_r z - 1)$ in the ring $K[u, x, z]$. The polynomial set $G = G_z \cap K[u, x]$ is a Gröbner basis of the prime ideal P. Here the Gröbner bases are in any (fixed) purely lexicographic ordering $u < z$ and $x < z$.

Proof. Let $I = I_1 \cdots I_r$ and $g_z = Iz - 1$. Since $u < z$ and $x < z$, G is clearly a Gröbner basis of the ideal $Q = (f_1, ..., f_r, g_z) \cap K[u, x]$. Thus it is enough to prove that $Q = P$.

Let $g \in Q$, then we have

$$g = A_1 f_1 + \cdots + A_r f_r + A_{r+1}(I_1 \cdots I_r z - 1),$$

where the A_i are in $K[u, x, z]$. Let μ be a generic point of $f_1, ..., f_r$, then $g(\mu) = A_{r+1}(\mu, z)(I_1(\mu) \cdots I_r(\mu) z - 1)$. Since the $I_i(\mu) \neq 0$ and $g(\mu)$ does not contain the variable z, $A_{r+1}(\mu, z)$ must be zero. Hence $g(\mu) = 0$. Thus, by (3.4) of Chapter 2, $g \in P$.

Now let $g \in P$. Then by the remainder formula we have

$$I_1^{s_1} \cdots I_r^{s_r} g = A_1 f_1 + \cdots + A_r f_r.$$

Multiplying the above equality by $B_1^{s_1} \cdots B_r^{s_r}$, where

$$B_i = z \cdot \prod_{j \neq i} I_j,$$

we have $g = B_1^{s_1} \cdots B_r^{s_r}(A_1 f_1 + \cdots + A_r f_r) - g((Iz)^s - 1)$, where $s = s_1 + \cdots + s_r$. Hence $g \in (f_1, ..., f_r, g_z)$. Since g does not contain the variable z, $g \in Q$. .QED.

The above algorithm gives a method for finding new theorems in geometry. Let a geometric configuration be given by a set of polynomial equations $h_1(u, x) = 0, ..., h_n(u, x) = 0$. Let $f_1, ..., f_r$ be obtained from $h_1, ..., h_n$ using triangulation procedure (Section 1 of Chapter 4) be irreducible. Then we have the following decomposition (see (3.1) of Chapter 3 and theorem (2.2) of Chapter 4):

$$V(h_1, ..., h_n) = V(P) \cup V_{\text{deg}},$$

where P is the prime ideal having $f_1, ..., f_r$ for its characteristic set and V_{deg} consists of all degenerate components. An assertion $g = 0$ about the given configuration is generally true if and only if g vanishes on $V(P)$, or equivalently, $g \in P$. Thus the *reduced* Gröbner basis of P (in the purely lexicographic ordering $u < x$) can be regarded as *a minimal set of all the simplest assertions which are generally true for the given configuration.* This set is dependent on the variable orderings. We give the following example to illustrate the method.

[2] W.F. Schelter and J.G. Yang told the author this algorithm.

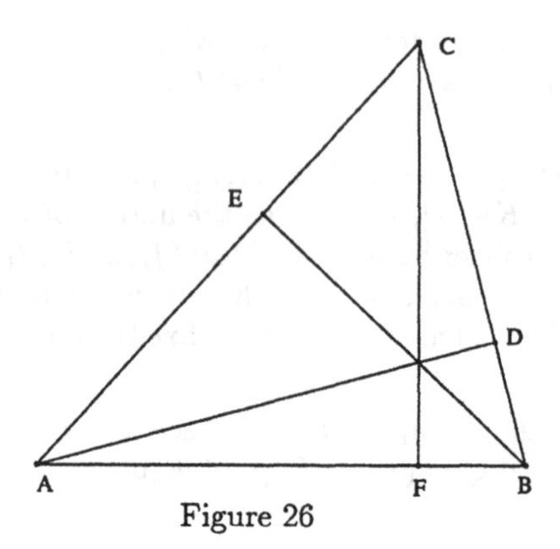

Figure 26

Example (4.4). The configuration of a triangle ABC with the three altitudes AD, BE and CF (Figure 26).

Let $A = (u_1, 0)$, $B = (u_2, 0)$, $C = (0, u_3)$, $F = (0, 0)$, $D = (x_1, x_2)$, $E = (x_3, x_4)$. Then the configuration can be expressed by the following set of hypothesis equations.

$$\begin{aligned}
h_1 &= u_3x_2 - u_2x_1 + u_1u_2 = 0\\
h_2 &= u_2x_2 + u_3x_1 - u_2u_3 = 0\\
h_3 &= u_3x_4 - u_1x_3 + u_1u_2 = 0\\
h_4 &= u_1x_4 + u_3x_3 - u_1u_3 = 0.
\end{aligned}$$

The irreducible characteristic set obtained from h_1, h_2, h_3, h_4 is:

$$\begin{aligned}
f_1 &= (u_3^2 + u_2^2)x_1 - u_2u_3^2 - u_1u_2^2 = 0\\
f_2 &= u_2x_2 + u_3x_1 - u_2u_3 = 0\\
f_3 &= (u_3^2 + u_1^2)x_3 - u_1u_3^2 - u_1^2u_2 = 0\\
f_4 &= u_1x_4 + u_3x_3 - u_1u_3 = 0.
\end{aligned}$$

GB = GB($f_1, ..., f_4, u_1u_2(u_3^2 + u_1^2)(u_3^2 + u_2^2)z - 1$) in the purely lexicographic ordering $u < x < z$ contains 28 polynomials, of which 12 are in $\mathbf{Q}[u, x]$. They are:

$$\begin{aligned}
g_1 &= x_4^2 + x_3^2 - (u_2 + u_1)x_3 + u_1u_2\\
g_2 &= x_2x_4 - (x_1x_3 - u_1u_2)\\
g_3 &= x_1x_4 + x_2x_3\\
g_4 &= u_3x_4 - (u_1x_3 - u_1u_2)\\
g_5 &= u_1x_4 + u_3x_3 - u_1u_3\\
g_6 &= u_1x_2x_3 - (u_3x_1x_3 - u_1u_3x_1)\\
g_7 &= u_2x_1x_3 + (u_1x_1 - u_1u_2)x_3 - u_1u_2x_1\\
g_8 &= u_3^2x_3 + u_1^2x_3 - u_1u_3^2 - u_1^2u_2\\
g_9 &= x_2^2 + x_1^2 - (u_2 + u_1)x_1 + u_1u_2\\
g_{10} &= u_3x_2 - (u_2x_1 - u_1u_2)\\
g_{11} &= u_2x_2 + u_3x_1 - u_2u_3\\
g_{12} &= u_3^2x_1 + u_2^2x_1 - u_2u_3^2 - u_1u_2^2,
\end{aligned}$$

which form the reduced Gröbner basis G of the prime ideal P having $f_1, ..., f_4$ as its characteristic set. Let us look at the simplest one among them: $g_3 = x_1x_4 + x_2x_3 = 0$. The interpretation of the above equation is as follows:

Assertion 1. Let the geometry configuration be the same as in example (4.4). From D and E two perpendiculars are drawn to the side AB. Let the feet be D_1 and E_1, respectively. Then $DD_1 \cdot FE_1 = EE_1 \cdot FD_1$ (Figure 27).

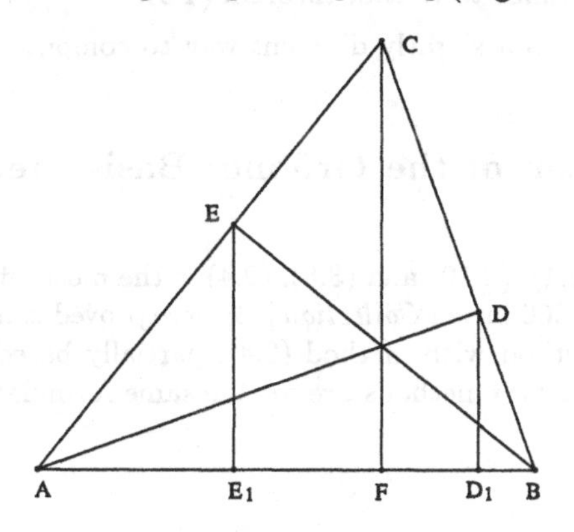

Figure 27

This is a theorem whose traditional proof is nontrivial. (Please try it.) Usually, polynomials of degree two in the Gröbner basis G may have nice interpretations in geometry. For example, we can interpret $g_1 = 0$ and $g_2 = 0$ as follows:

Assertions 2 and 3. With the same notations as in Assertion 1, assume that triangle ABC is acute. Let FB and FC be the positive directions of the x-axis and the y-axis, respectively. (Then we have $FE_1 = -x_3$, $FA = -u_1$, $FD_1 = x_1$, $FB = u_2$, etc.) The following two assertions are generally true:

(i) $EF^2 + FE_1 \cdot (FB - FA) = FB \cdot FA$;

(ii) $DD_1 \cdot EE_1 + FE_1 \cdot FD_1 = FB \cdot FA$.

Remark (4.5). The other 16 polynomials of GB which contain the variables z are quite complicated. Even from the final 28 polynomials, we can see that the expense of the computation is still very high, taking 591 seconds on a Symbolics 3600. If we choose the variable order $z < u < x$, then the reduced Gröbner basis $(f_1, ..., f_4, u_1u_2(u_3^2 + u_1^2)(u_3^2 + u_2^2)z - 1)$ contains only 7 polynomials. It only took 20 seconds to complete the computation. However, the Gröbner basis in this ordering has nothing to do with the Gröbner basis of the prime ideal P. From this example, we can see that computing Gröbner bases of the *same* ideal in different purely lexicographic orderings might mean computing totally different things in another sense. This is one reason to justify the experimental observation in [8] that "the complexity of the algorithm for constructing Gröbner bases is extremely sensitive to a permutation of variables when purely lexicographical ordering is used." Also see Remark (3.8).

Remark (4.6). For irreducible configuration and with the same notations as before, let $z_1, ..., z_r$ be new variables. Then we have

(i) $(f_1, ..., f_r, I_1 \cdots I_r z - 1) = (h_1, ..., h_n, I_1 \cdots I_r z - 1)$.

(ii) $(f_1, ..., f_r, I_1 z_1 - 1, ..., I_r z_r - 1) \cap K[u, x] = P$.

Proof. (i) By (4.3) of Chapter 2, $(f_1, ..., f_r, I_1 \cdots I_r z-1) \subset (h_1, ..., h_n, I_1 \cdots I_r z-1)$ and each $h_i \in P$. Thus the two ideals are actually identical.

(ii) The proof is similar to that of theorem (4.3). .QED.

Property (ii) provides a slightly different way to compute Gröbner bases of the prime ideal P.

5. A Comparison of the Gröbner Basis Method with Wu's Method

Among methods (2.4), (2.10) and (3.5), (2.4) is the most efficient. In a recent experiment, 359 among 366 in the *Collection* [14] were proved using method (2.4). Now we compare Wu's method with method (2.4), partially based on that experiment. First we note that the two methods are for the same formulation, i.e., Formulation F3.

5.1. The Scope

Basically the two methods have the same scope: they can address those geometry statements, for which the hypotheses and the conclusion can be expressed as polynomial equations with certain unspecified nondegenerate (subsidiary) conditions in inequation forms. The two methods can confirm whether a geometry statement is generally valid and, in the case of general validity, produce subsidiary conditions in a natural way.

Wu's method was originally aimed at proving geometry theorems. Thus, it can address more delicate problems in geometry. For example, at least for geometry statements of constructive type, it can delete unnecessary subsidiary conditions and translate the remaining back into their geometric forms [11]. Moreover, it can deal with reducible problems, such as Morley's trisector theorem specified in Chapter 4 and V. Thèbault's conjecture (examples (5.4) and (5.5) of Chapter 4). The Gröbner basis method alone cannot solve such problems unless factorization (decomposition) is introduced. As we have seen in Section 5 of Chapter 4, factorization (decomposition) is in essence a case analysis approach. This is a useful approach, as shown in example (5.3), (5.4), (5.5) and (5.6) of Chapter 4. On the other hand, the Gröbner basis method can confirm some theorems which need factorization using Wu's method. (See example (2.8).)

5.2. The Efficiency

So far, there are no theoretical studies comparing the efficiency of these two methods. In our current implementation, the Gröbner basis method is generally slower than Wu's method for most problems. For timing comparisons in experiments see [16], [14] and Part II of this monograph. Our observation is based on at least 600 nontrivial geometry theorems.

Here we would emphasize that comparisons are often implementation dependent. For example, we proved Pascal's theorem in 1,905 seconds on a Symbolics 3600 using method (2.4); the same example was used and proved in [31] in 13,335 seconds on an IBM 4341 using a method similar to (2.4).[3] Also in [31], there are several examples which could not be proved within 2.4 M bytes of working space using their implementation of Wu's method. Some of those examples were from Wu's work. One is the Pascal Conic Theorem discovered by Wu [44]. For the input specified in [44] and the proof produced by our prover, $maxt = 1412$. Only about 500 K bytes were needed on a Symbolics 3600. Note that programs in LISP need more space than programs in other languages. The *same proof* was produced by Wu himself on a small computer HP9835A using about 100 K bytes of working space [44].

One reason that the characteristic set method is generally faster than the Gröbner basis method is that computing a GB usually takes more time than computing a CS. Once we have the CS or the GB, we can use them to reduce the conclusion polynomial g to see if it is zero. In this step, the situation is not clear-cut. There are many problems which favor GB. However, there are also many other problems which favor the CS version.

One reason that some problems favor GB is that there is a cancellation of common factors containing the u's only, when working over $\mathbf{Q}(u)$. Since at least for irreducible problems, GB is a special kind of CS. This tells us we should probably do a similar cancellation for CS (removing common factors of coefficients of the f) for those problems for which the reduction (successive division) produce large polynomials. We call such a CS a *simplified characteristic set.*

In this monograph we only discuss the basic algorithms, without going into implementation details in speeding up the algorithms. There are many ways to speed up the characteristic set method in the context of geometry theorem proving:

(5.1) Separation Principle ([43]). Let the ascending chain $f_1, ..., f_r$ be irreducible and $g = g_n t^n + \ldots + g_0$, where the g_i are polynomials in the u and the x, and t is a new variable. Then $prem(g, f_1, ..., f_r) = 0$ if and only if $prem(g_i, f_1, .., f_r) = 0$ for each i. Thus, in intermediate steps, instead of working on a large polynomial, we can work on several smaller ones. It saves not only space, but also time.

(5.2) Redundant Factors. During successive divisions, in the intermediate remainders often appear certain redundant factors, some of which can be predicted (see [11]). Removing those redundant factors is another way to speed up the proof of a geometry theorem.

(5.3) Using Simplified or partially simplified ascending chain. We will briefly discuss these strategies in Part II.

[3] The method in [31] includes only step 1 of our method (2.4); thus it is incomplete. It is unable to prove, e.g., "$x^2 = 0 \Rightarrow x = 0$".

References

[1] N. Altshiller-Court, *College Geometry,* Barnes & Noble, Inc., 1952.

[2] C. F. Adler, *Modern Geometry,* McGraw–Hill Book Company Inc., 1958.

[3] K.B. Taylor, "Three Circles with Collinear Centers", *American Mathematical Monthly* **90** (1983), 486-487.

[4] E. Artin, *Geometric Algebra,* Interscience Publishing, Inc., 1957.

[5] L. Bachmair and B. Buchberger, "A Simplifed Proof of the Characterization Theorem for Gröbner Bases", *ACM SIGSAM Bull* **14(4)** (1980), 29–34.

[6] F. Bachmann and K. Reidemeister, "Die Metrishe Form in der Absoluten und der Elliptischen Geometrie", *Math. Annalen* **29** (1937), 748–765.

[7] B. Buchberger, "An Algorithm Criterion for the Solvability of a System of Algebraic Equations" (German), *Aequationes Math.* **4** (1970), 374-383.

[8] B. Buchberger, "Gröbner Bases: An Algorithmic Method in Polynomial Ideal Theory", Chapter 6 in *Recent Trends in Multidimensional Systems Theory,* N.K. Bose (ed.) D. Reidel Publ. Comp. 1985.

[9] B. Buchberger, "Some Properties of Gröbner bases for Polynomial ideals", *JACM SIGSAM Bull.* **10(4)** (1976), 19-24.

[10] S.C. Chou, "Proving Elementary Geometry Theorems Using Wu's Algorithm", in *Automated Theorem Proving: After 25 years,* Ed. By W.W. Bledsoe and D. Loveland, AMS Contemporary Mathematics Series **29** (1984), 243-286.

[11] S.C. Chou, "Proving and Discovering Theorems in Elementary Geometries Using Wu's Method", PhD Thesis, Department of Mathematics, University of Texas, Austin (1985).

[12] S.C. Chou, "A Method for Mechanical Derivation of Formulas in Elementary Geometry", February 1986, submitted to and accepted by *Journal of Automated Reasoning.*

[13] S.C. Chou and H.P. Ko, "On Mechanical Theorem Proving in Minkowskian Plane Geometry", *Proc. of Symp. of Logic in Computer Science,* pp187-192, 1986.

[14] S.C. Chou, " Proving Geometry Theorems Using Wu's Method: A Collection of Geometry Theorems Proved Mechanically", Technical Report 50, Institute for Computing Science, University of Texas at Austin, July 1986.

[15] H.S.M. Coxeter, *Introduction to Geometry,* John Wiley & Sons, Inc, 1969.

[16] S.C. Chou and W.F. Schelter, "Proving Geometry Theorems with Rewrite Rules", *Journal of Automated Reasoning,* **2(4)** (1986), 253-273.

[17] S.C. Chou, W. Schelter and G.J. Yang, "Characteristic Sets and Gröbner Bases in Proving Geometry Theorems", Preprint, March 1986.

[18] G.E. Collins, "Quantifier Elimination for Real Closed Fields by Cylindrical Algebraic Decomposition", *Lecture Notes In Computer Science,* **33** (1975),

Springer-Verlag, Berlin, 134-183.

[19] P. J. Davis and E. Cerutti, "FORMAC meets Pappus", *The American Mathematical Monthly,* **76** (1969), 895 - 905.

[20] H. Gelernter, J.R. Hanson and D.W. Loveland, "Empirical explorations of the geometry-theorem proving machine", *Proc. West. Joint Computer Conf.*, 143-147, 1960.

[21] M. J. Greenberg, *Euclidean and Non-Euclidean Geometries, Development and History,* Freemann and Company, 1973.

[22] R. Hartshorne, *Foundations of Projective Geometry,* Harvard University, 1967.

[23] J. Hadamard, *Lecons de Geometrie Elementaire,* I, Paris, 1931.

[24] D. Hilbert, *Foundations of Geometry,* Open Court Publishing Company, La Salla, Illinois, 1971.

[25] Howard Eves, *A Survey of Geometry, Volume I,* Allyn and Bacon, Inc., Boston, 1963.

[26] T. W. Hungerford, *Algebra,* Springer-Verlag, 1978.

[27] D. Kapur, "Geometry Theorem Proving Using Hilbert's Nullstellensatz", in Proceedings of the 1986 Symposium on Symbolic and Algebraic Computation, 202-208.

[28] H.P. Ko and M.A. Hussain, "A Study of Wu's Method - a Method to Prove Certain Theorems in Elementary Geometry", to appear in the *Proceedings of 1985 International Conference on Combinatorics, Graph Theory, and Computing.*

[29] H.P. Ko and Moayyed A. Hussain, "ALGE-Prover – an Algebraic Geometry Theorem Proving Software", Technical Report, 85CRD139, General Electric Company, 1985.

[30] H.P. Ko and S.C. Chou, " A Decision Method for Certain Algebraic Goemetry Problems" Preprint, Submitted to *Rocky Mountain Journal of Mathematics.*

[31] B. Kutzler and S. Stifter, "Automated Goemetry Theorem Proving Using Buchberger's Algrithm", in Proceedings of the 1986 Symposium on Symbolic and Algebraic Computation, 209-214.

[32] R. F. Ritt, *Differential Equation from Algebraic Standpoint,* AMS Colloquium Publications Volume 14, New York, 1938.

[33] R. F. Ritt, *Differential Algebra,* AMS Colloquium Publications, New York, 1950.

[34] A. Seidenberg, "A New Decision Method for Elementary Algebra", *Annals of Math.*, **60** (1954), 365–371.

[35] J. R. Shoenfield, *Mathematical Logic,* Addison-Wesley Publishing Company, (1967).

[36] P. Szasz, "Direct Introduction of Weierstrass Homogeneous Coordinates in the Hyperbolic Plane, on the Basis of the End-Calculus of Hilbert", in *The*

Axiomatic Method, Ed. by Tarski et al., 97-113, 1959.

[37] G.W. Szmielew, "Some Metamathematical Problems Concerning Elementary Hyperbolic Geometry", in *The Axiomatic Method*, Ed. by Tarski et al., 30–52, 1959.

[38] A. Tarski, *A Decision Method for Elementary Algebra and Geometry,* Second Edition, Berkeley and Los Angeles, 1951.

[39] A. Tarski, "What is Elementary Goemetry", in *The Axiomatic Method*, Ed. by Tarski et al., 1959.

[40] B. H. Träger, "Algebraic Factoring and Rational Function Integration", Proc. of 176 ACM Symp. On Symbolic and Algebraic Computation, 1976, 219–226.

[41] B. L. Van Der Waerden, *Modern Algebra,* the English Edition, Frederick Ungar Publishing Co., 1948.

[42] Wu Wen-tsün, "On the Decision Problem and the Mechanization of Theorem Proving in Elementary Geometry", *Scientia Sinica* **21** (1978), 157-179.

[43] Wu Wen-tsün, "Basic Principles of Mechanical Theorem Proving in Geometries", *J. of Sys. Sci. and Math. Sci.* **4(3)**, 1984, 207-235, republished in *Journal of Automated Reasoning* **2**(**4**) (1986), 221-252.

[44] Wu Wen-tsün, "Some Recent Advances in Mechanical Theorem-Proving of Geometries", in *Automated Theorem Proving: After 25 years*, American Mathematical Society, Contemporary Mathematics **29** (1984), 235-242.

[45] Wu Wen-tsün, *Basic Principles of Mechanical Theorem Proving in Geometries,* (in Chinese) Peking 1984.

[46] Wu Wen-tsün, "Toward Mechanization of Geometry – Some Comments on Hilbert's 'Grundlagen der Geometrie'", Acta Mathematica Scientia **2** (1982), 125-138.

[47] Wu Wen-tsün, "Mechanical Theorem Proving in Elementary Geometry and Differential Geometry", *Proc. 1980 Beijing Symposium on Differential Geometry and Differential Equations* **2** (1982), 125-138, Science Press.

[48] Wu Wen-tsün, "On Zeros of Algebraic Equations –An Application of Ritt Principle, *Kexue Tongbao* **31(1)** (1986), 1-5.

[49] I.M. Yaglom, *A Simple Non-Euclidean Geometry and its Physical Basis,* Springer-Verlag, New York, 1979.

[50] I.M. Yaglom, *Geometry Transformations* III, Random House, Inc., New York, 1973.

Part II

512 Theorems Mechanically Proved

Explanations

1. General Remarks

This is a collection of 512 geometry theorems proved by the computer program described in my thesis [11], which was based on the pioneering work of Wu Wen-Tsün. The method and underlying theory have been explained in detail in Part I of this monograph. These 512 were selected from 366 examples in *Collection* [14] and 244 examples in another technical report. All 512 theorems can be confirmed by Wu's method without using factorization.

In our current implementation of the characteristic set method, we use two strategies:

(1) Using characteristic sets in weak sense with partial simplification. Characteristic sets (CS) produced in this strategy are usually not reduced. If a polynomial f in a CS is monic, then we use it to reduce the other polynomials. If the coefficients of f has a common factors, then we remove it from f. This strategy turns our to be most effective in practice. In this collections, only one problems could not be confirmed within reasonable space using this strategy (its $maxt > 50,000$). Our timing and *maxt* in this collections is exactly for this strategy expect one, A-303, which was confirmed by this strategy, but without any simplifications.

(2) Using (reduced) characteristic sets with full simplification. In Ritt's original method, common factors of a polynomial were never taken care of. We will remove these common factors whenever possible. We call such a characteristic set a *simplified characteristic set* (SCS). It takes more time to produce an SCS. However, for many cases, an SCS reduces the conclusion polynomial g to zero faster than using strategy (1). For example, the only problem in this collection with which strategy (1) had difficulty, was solved easily in 112 seconds using this strategy. When the SCS strategy yields better results than strategy (1), we also listed the timing for the SCS strategy.

We also experimented with the Gröbner basis methods using the same set of 512 theorems and succeeded in proving 477. The Gröbner basis method had difficulty proving the remaining 35 theorems (the proofs of which need more than 4 hours). The SCS strategy has difficulty in proving 17 theorems in this collection. For the reference, we put the timing for the Gröbner basis method (method (2.4) in Section 5.2) in Appendix 1.

Most of this document was generated mechanically. A typical entry in this collection includes an identification number, a diagram, an exact statement of the theorem in the form of a construction, and a note on the "nondegenerate" conditions of the theorem. The latter are unstated constraints on the original construction that are necessary to insure that the statement is a theorem, e.g., that certain points are not collinear. Some entries include a conventional (informal) statement of the theorem in English. The diagrams and exact constructions are generated mechanically from the input to my theorem prover. The nondegenerate conditions are generated from the theorem-prover's output. I personally wrote only the informal statements of the theorems and the occasional remarks. All such text is printed in *italics*.

How to Read the Examples. Various examples are mainly taken from six sources. Examples with identification numbers beginning with the letters 'A","C", "Y", "E", and "H" are from [1], [15], [50], [25]and [23], respectively. Examples with identification numbers beginning with the letter "M" are from *A Collection of Problems in High School Mathematics* (in Chinese, Peking, 1981).

In order to explain how to read the problems in this collection, let us consider "**Example 144** (A65-4: 883.08s, 16537; 620.28s, 13017)". The identification number A65-4 means the theorem is found in book [1], page 65, problem number 4. "883.08s, 16537" means that the proof took 883.08 seconds on a Symbolics 3600 with $maxt = 16537$ using strategy (1). "620.28s, 13017" denotes the timing and the $maxt$ for strategy (2). (We include them because the SCS yielded better results; otherwise we omitted these two items.) For theorems with identification numbers in the form A-n, n is the relevant article in [1]. Thus, problem A-282 (the same as example (2.2) in the previous section) is found in article 282 of [1].

The exact statement for problem A-282, above, was generated by a computer program from the input to my prover. The actual input was:

```
((CONS-SEQUENCE A B C O D E F G)
(PERP-BISECT O A C)
(EQDISTANCE O A O D)
(COLLINEAR E B C)
(PERPENDICULAR D E B C)
(COLLINEAR F A C)
(PERPENDICULAR D F A C)
(COLLINEAR G A B)
(PERP-BISECT O A B)
(PERPENDICULAR D G A B)
(COLLINEAR E F G))
```

The diagram for A-282 was generated from the input above. The user (in this case, me) was asked to select locations for points A, B, C and D. The actual location selected is irrelevant to the proof obtained by the prover.

The italicized text was typed by me.

The geometry conditions in the exact statements are transformed to their corresponding algebraic equations. However, the algebraic equations include the degenerate cases. For example, the condition $DE \perp BC$ includes the degenerate cases when $C = B$ or $E = D$. (See section 2.)

The nondegenerate conditions for A-282 were generated by the prover in the form of polynomial inequations. One of the contributions in my thesis is to exclude some unnecessary inequations and translate others back into their corresponding geometric forms mechanically. Since the prover is for the *metric geometry*, all four conditions are necessary: A, C and B are not collinear; Line BC is non-isotropic; Line AC is non-isotropic; Line AB is non-isotropic. An isotropic line is a line which is perpendicular to itself. In Euclidean geometry (or in ordered geometries) isotropic lines do not exist. Thus the first condition is enough for the validity of Simson's theorem in Euclidean geometry.

Geometry conditions in nondegenerate conditions do not include degenerate cases. For example, when we mention line AB we mean it is well defined, i.e., $A \neq B$; when we mention line AB and line CD intersect, we mean that they have exactly one common point. As a global convention, we always assume that the points with coordinates $(0,0)$ and $(u_1,0)$ form the x-axis of the coordinate system, thus $u_1 \neq 0$ is a implicit condition for all problems.

Notation Convention. We generally use the standard notations in elementary geometry. Depending on the context, AB can denote the line passing through the points A and B or the segment determined by A and B. $\overline{AB}$ denotes the directed segment from A to B. $\nabla(ABC)$ denotes twice the signed area of directed triangle ABC. $\Diamond ABCD$ denotes twice the signed area of directed quadrilateral $ABCD$ (not necessarily convex, see C3.11).

2. Algebraic Representations of Geometric Conditions

In this section we shall list all algebraic equations or polynomials for the geometric conditions or quantities appearing in the exact geometry statements (with unspecified nondegenerate conditions) of the 512 examples.

Let points $A = (a_1,a_2)$, $B = (b_1,b_2)$, $C = (c_1,c_2)$ $D = (d_1,d_2)$, $E = (e_1,e_2)$, $F = (f_1,f_2)$, $G = (g_1,g_2)$ and $H = (h_1,h_2)$.

Here we would emphasize that the methods described in Part I generally can only deal with unordered geometry or the geometry statements not involving order in ordered geometry. Thus the methods generally cannot deal with the length of a segment. But the square of the length of a segment is actually a notion in unordered geometry (see [45]). We denote this quantity by $\overline{AB}^2$ for the segment AB. In a coordinate system, $\overline{AB}^2 = (a_1-b_1)^2+(a_2-b_2)^2$.

1. "*Points A, B and C are collinear.*" (Or "*A is on line BC*"; or "*B is on line CA*"). Its equivalent polynomial equation is $(a_1-b_1)(b_2-c_2)-(a_2-b_2)(b_1-c_1)=0$.

2. "*AB* $\parallel$ *CD*": $(a_1-b_1)(c_2-d_2)-(a_2-b_2)(c_1-d_1)=0$. Note that this equation includes degenerate cases, i.e., cases when $A=B$ or $C=D$. More exactly, its geometric equivalence is:

$$(A=B)\vee(C=D)\vee[(A\neq B)\wedge(C\neq D)\wedge\ (AB \text{ is parallel to } CD)].$$

3. "*AB* $\perp$ *CD*": $(a_1-b_1)(c_1-d_1)+(a_2-b_2)(c_2-d_2)=0$. Note again that this equation includes degenerate cases. In what follows, we will not list those degenerate cases. The reader should have no difficulty listing them.

4. "*A is on the circle with diagonal BC*". This is equivalent to "$BA \perp CA$". The reason that the program produces the former instead of the latter is to make the geometry statement fit the constructive type defined in Section 6 of Chapter 1.

5. "$AB \equiv CD$": $\overline{AB}^2 - \overline{CD}^2 = 0$.

6. "*B is the midpoint of A and C*": $2b_1 - a_1 - c_1 = 0$, and $2b_2 - a_2 - c_2 = 0$. If A, B and C are collinear, then one of the two equations is enough.

7. "$AB \cdot CD = EF \cdot GH$": $\overline{AB}^2 \cdot \overline{CD}^2 - \overline{EF}^2 \cdot \overline{GH}^2 = 0$.

8. "$3AB = 7CD$": $9\overline{AB}^2 - 49\overline{CD}^2 = 0$.

9. "*Points A, B, C and D are on the same circle*". Its polynomial equation is equivalent to that of $tan(\angle ACD) - tan(\angle BCD) = 0$.

10. "*Point A is on the radical axis of circle (B, CD) and circle (E, FG)*". The radical axis of the two circles is a line whose points X satisfy $\overline{XB}^2 - \overline{XE}^2 - (\overline{CD}^2 - \overline{FG}^2) = 0$. If two circles have two common points (this is always the case if the field associated with the geometry is algebraically closed), then the radical axis is the line joining the two common points.

11. "*Algebraic sum of segments AB, CD, EF = 0*". As we know, the addition (or subtraction) of two segments needs order and is generally beyond unordered geometry. But in unordered geometry we still can deal with some weak version. Let $x^2 = \overline{AB}^2$, $y^2 = \overline{CD}^2$ and $z^2 = \overline{EF}^2$. The above condition is equivalent to $(x - y - z)(x - y + z)(x + y - z)(x + y + z) = 0$.

12. "*Circle (A, BC) is tangent to circle (D, EF)*". This is equivalent to "algebraic sum of segments AD, BC, $EF = 0$." In unordered geometry there is no notion that the two circles are tangent internally or externally.

As we mentioned before, the addition (or subtraction) of segments needs order. However, if these segments are parallel or on the same line, then we can deal with the addition and subtraction completely, using directed segments. We use $\overline{AB}$ to denote the directed segment from A to B. The following quantities and conditions (13–17) are related to directed segments. Note that these quantities and conditions are meaningful **only if** all segments involved are on the same line or on parallels.

13. The quantity "$\overline{AB}/\overline{CD}$" or "$\frac{\overline{AB}}{\overline{CD}}$" denotes the ratio of the directed segments of $\overline{AB}$ over $\overline{CD}$, which is $(b_1 - a_1)/(d_1 - c_1) = (b_2 - a_2)/(d_2 - c_2)$. Of course, the last two quantities are meaningful only if the line is not parallel to the y–axis or x–axis. We have discussed the problem that a special subsidiary condition will not weaken the theorem to be proved in detail in section 3 of Chapter 3.

14. "*C and D are inverse wrpt circle (A, AB)*". Its polynomial form is equivalent to that of $\frac{\overline{AC}}{\overline{AB}} - \frac{\overline{AB}}{\overline{AD}} = 0$.

15. "*Points A, B, C and D form a harmonic set*". Its polynomial form is equivalent to that of $\frac{\overline{AC}}{\overline{CB}} - \frac{\overline{DA}}{\overline{DB}} = 0$.

16. The quantity "*crossratio*($ABCD$)" is equal to $\frac{\overline{CA}}{\overline{CB}} \cdot \frac{\overline{DB}}{\overline{DA}}$.

17. The condition "$\overline{AB} - \overline{CD} - \overline{EF} + \overline{EA} = 0$" is converted to $(b_1 - a_1) - (d_1 - c_1) - (f_1 - e_1) + (a_1 - e_1) = 0$ or $(b_2 - a_2) - (d_2 - c_2) - (f_2 - e_2) + (a_2 - e_2) = 0$. (We emphasize once again that all points must be on the same line and the line is not parallel to the y–axis or x–axis.)

In Hilbert's classic book "*Foundations of geometry*", the definition of the angle and the axioms of angle congruence need order. However, we can use a different definition of the (oriented) angle in unordered geometry. This has been discussed in detail in section 1.2 of Chapter 3. According to this definition, the addition and subtraction of angles can be fully dealt with in unordered geometry. From the algebraic point of view, this definition is more natural. The generic nature of geometry statements about the angle congruence and addition is more clear in this definition. (See examples in Chapter 4 and more examples in this collection.) I don't claim that Wu's presentation ([45]) in section 1.2 of Chapter 3 is better. If we need to measure an angle (e.g., in Euclidean geometry), this definition is not suitable.

18. "tan(ABC) = tan(DEF)". Its polynomial form is equivalent to that of $tan(\angle ABC) - tan(\angle DEF) = 0$. Its corresponding geometric condition is $\angle ABC \equiv \angle DEF$.

19. "tan($\angle ABC + \angle A_1B_1C_1 + \angle A_2B_2C_2$) − tan($\angle DEF$) = 0." Its corresponding geometric condition is $\angle ABC + \angle A_1B_1C_1 + \angle A_2B_2C_2 \equiv \angle DEF$.

20. The quantity "$\nabla(ABC)$" is twice the signed area of directed triangle ABC and equal to $(a_1 - b_1)(b_2 - c_2) - (a_2 - b_2)(b_1 - c_1)$.

21. The quantity "$\lozenge(ABCD)$" is twice the signed area of directed quadrilateral $ABCD$ and equal to $\nabla(ABC) + \nabla(CDA)$.

Theorems Proved Mechanically by Wu's Method

Example 1 (Pappus: 1.52s, 28). *(Pappus' Theorem). Let ABC and $A_1B_1C_1$ be two lines, and $P = AB_1 \cap A_1B$, $Q = AC_1 \cap A_1C$, $S = BC_1 \cap B_1C$. Then P, Q and S are collinear.*

Points A, B, A_1, B_1 are arbitrarily chosen. Points C, C_1, P, Q, S are constructed (in order) as follows: C is on line AB; C_1 is on line A_1B_1; P is on line A_1B; P is on line AB_1; Q is on line AC_1; Q is on line A_1C; S is on line B_1C; S is on line BC_1. The **conclusion**: Points P, Q and S are collinear.

$A = (0,0)$, $B = (u_1,0)$, $A_1 = (u_2,u_3)$, $B_1 = (u_4,u_5)$, $C = (u_6,0)$, $C_1 = (x_1,u_7)$, $P = (x_3,x_2)$, $Q = (x_5,x_4)$, $S = (x_7,x_6)$.

The **nondegenerate** conditions: $A \neq B$; $A_1 \neq B_1$; Line AB_1 intersects line A_1B; Line A_1C intersects line AC_1; Line BC_1 intersects line B_1C.

Remark. Note that there are six Pappus lines for one Pappus configuration. We denote the Pappus line in the statement by [123]. *The other five are* [312], [231], [213], [321], *and* [132].

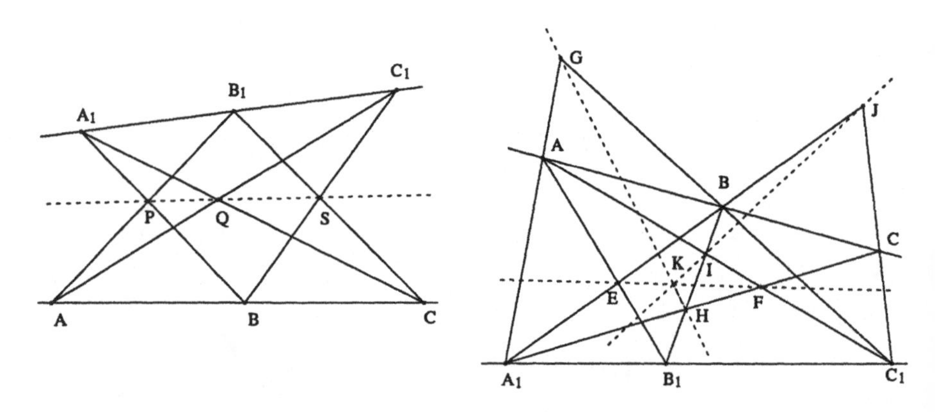

Pappus — Pappus-point

Example 2 (Pappus-point: 4.87s, 201). *Pappus Point Theorem. The three Pappus lines* [123], [312] *and* [231] *are concurrent. So are the Pappus lines* [213], [321] *and* [132].

Points A, B, A_1, C_1 are arbitrarily chosen. Points B_1, C, E, F, G, H, I, J, K are constructed (in order) as follows: B_1 is on line A_1C_1; C is on line AB; E is on line A_1B; E is on line AB_1; F is on line A_1C; F is on line AC_1; G is on line BC_1; G is on line AA_1; H is on line BB_1; H is on line CA_1; I is on line BB_1; I is on line AC_1; J is on line BA_1; J is on line CC_1; K is on line GH; K is on line EF. The **conclusion**: Points I, J and K are collinear.

$A = (0,0)$, $B = (u_1,0)$, $A_1 = (u_2,u_3)$, $C_1 = (0,u_4)$, $B_1 = (x_1,u_5)$, $C = (u_6,0)$,

$E = (x_3, x_2)$, $F = (0, x_4)$, $G = (x_6, x_5)$, $H = (x_8, x_7)$, $I = (0, x_9)$, $J = (x_{11}, x_{10})$, $K = (x_{13}, x_{12})$.

The **nondegenerate** conditions: $A_1 \neq C_1$; $A \neq B$; Line AB_1 intersects line A_1B; Line AC_1 intersects line A_1C; Line AA_1 intersects line BC_1; Line CA_1 intersects line BB_1; Line AC_1 intersects line BB_1; Line CC_1 intersects line BA_1; Line EF intersects line GH.

Remark. This is a possibly new theorem found by Wu. But it also can be regarded as a degenerate case of Steiner's theorem.

Example 3 (Ams-1: 4.05s, 45). *Starting from five points A, B, C, D and E with A, B, C collinear, new lines and points of intersection are formed. ED, IG, LK and JH are concurrent.*

Points A, B, D, E are arbitrarily chosen. Points C, F, I, J, L, G, K, H, O are constructed (in order) as follows: C is on line AB; F is on line ED; F is on line AB; I is on line DB; I is on line AE; J is on line CD; J is on line AE; L is on line CD; L is on line BE; G is on line BE; G is on line AD; K is on line CE; K is on line DB; H is on line EC; H is on line AD; O is on line IG; O is on line ED. The **conclusions**: (1) Points O, L and K are collinear; (2) Points O, J and H are collinear.

$A = (0,0)$, $B = (u_1, 0)$, $D = (u_2, u_3)$, $E = (0, u_4)$, $C = (u_5, 0)$, $F = (x_1, 0)$, $I = (0, x_2)$, $J = (0, x_3)$, $L = (x_5, x_4)$, $G = (x_7, x_6)$, $K = (x_9, x_8)$, $H = (x_{11}, x_{10})$, $O = (x_{13}, x_{12})$.

The **nondegenerate** conditions: $A \neq B$; Line AB intersects line ED; Line AE intersects line DB; Line AE intersects line CD; Line BE intersects line CD; Line AD intersects line BE; Line DB intersects line CE; Line AD intersects line EC; Line ED intersects line IG.

Remark. This is a possibly new theorem found by our numerical example search.

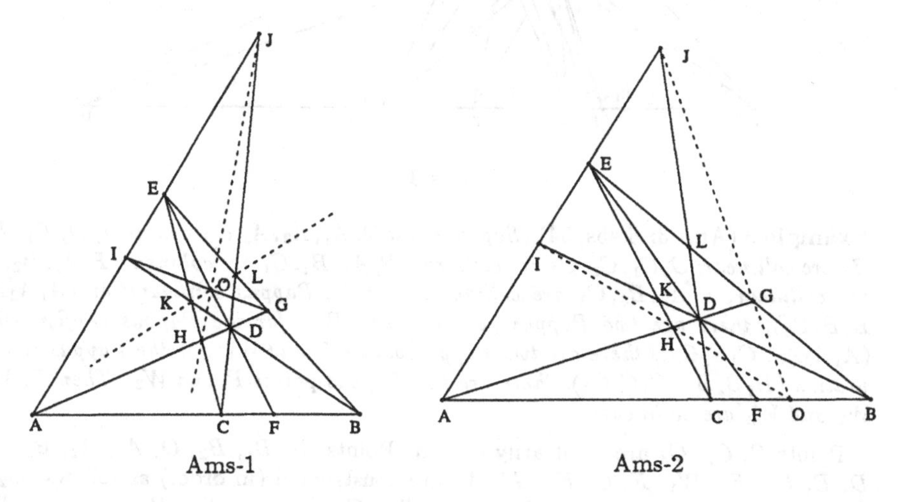

Ams-1 Ams-2

Example 4 (Ams-2: 1.65s, 5). *For the same configuration as in AMS-1, (1) AB,*

GJ and HI are collinear; (2) AB, GK and IL are collinear; (3) AB, HL and JK are collinear.

Points A, B, D, E are arbitrarily chosen. Points C, F, I, J, L, G, K, H, O are constructed (in order) as follows: C is on line AB; F is on line ED; F is on line AB; I is on line DB; I is on line AE; J is on line CD; J is on line AE; L is on line CD; L is on line BE; G is on line BE; G is on line AD; K is on line CE; K is on line DB; H is on line EC; H is on line AD; O is on line GJ; O is on line AB. The **conclusion**: Points O, H and I are collinear.

$A = (0,0)$, $B = (u_1, 0)$, $D = (u_2, u_3)$, $E = (u_4, u_5)$, $C = (u_6, 0)$, $F = (x_1, 0)$, $I = (x_3, x_2)$, $J = (x_5, x_4)$, $L = (x_7, x_6)$, $G = (x_9, x_8)$, $K = (x_{11}, x_{10})$, $H = (x_{13}, x_{12})$, $O = (x_{14}, 0)$.

The **nondegenerate** conditions: $A \neq B$; Line AB intersects line ED; Line AE intersects line DB; Line AE intersects line CD; Line BE intersects line CD; Line AD intersects line BE; Line DB intersects line CE; Line AD intersects line EC; Line AB intersects line GJ.

Remark. This is a possibly new theorem found by our numerical example search.

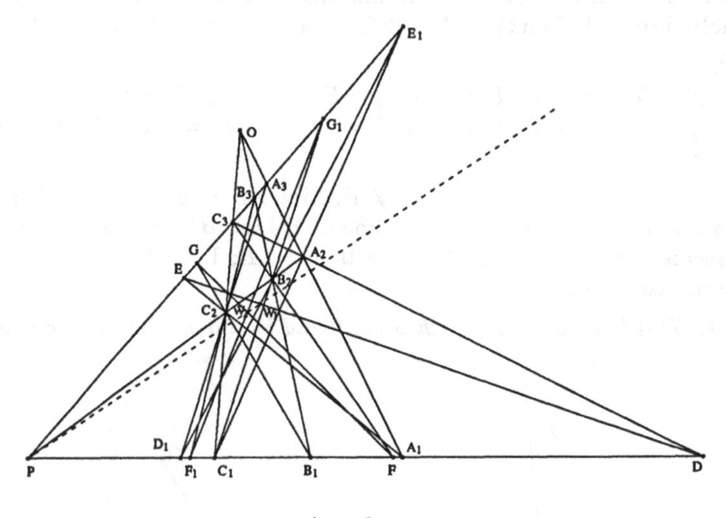

Ams-3

Example 5 (Ams-3: 6.45s, 94). *Suppose that O, A_1, A_2, A_3 are collinear; O, B_1, B_2, B_3 are collinear; O, C_1, C_2, C_3 are collinear; P, A_1, B_1, C_1 are collinear; P, A_2, B_2, C_2 are collinear; P, A_3, B_3, C_3 are collinear. For the Pappus configuration ($A_1A_2A_3$, $B_1B_2B_3$), there are two Pappus points P and W_3; for the Pappus configuration ($A_1A_2A_3$, $C_1C_2C_3$), there are two Pappus points P and W_1; for the Pappus configuration ($B_1B_2B_3$, $C_1C_2C_3$), there are two Pappus points P and W_2. Then P, W_1, W_2 and W_3 are collinear.*

Points P, C_1, C_3 are arbitrarily chosen. Points C_2, B_1, B_2, O, A_1, A_2, B_3, A_3, D, E, D_1, E_1, W_1, F, G, F_1, G_1, W_2 are constructed (in order) as follows: C_2 is on line C_1C_3; B_1 is on line C_1P; B_2 is on line C_2P; O is on line B_1B_2; O is on line

C_1C_3; A_1 is on line C_1P; A_2 is on line C_2P; A_2 is on line OA_1; B_3 is on line C_3P; B_3 is on line OB_1; A_3 is on line PC_3; A_3 is on line OA_1; D is on line C_3A_2; D is on line C_1A_1; E is on line C_2A_1; E is on line C_3A_3; D_1 is on line C_2A_3; D_1 is on line C_1A_1; E_1 is on line C_3A_3; E_1 is on line C_1A_2; W_1 is on line D_1E_1; W_1 is on line DE; F is on line C_3B_2; F is on line C_1B_1; G is on line C_2B_1; G is on line C_3B_3; F_1 is on line C_2B_3; F_1 is on line C_1B_1; G_1 is on line C_3B_3; G_1 is on line C_1B_2; W_2 is on line F_1G_1; W_2 is on line FG. The **conclusion**: Points P, W_1 and W_2 are collinear.

$P = (0,0)$, $C_1 = (u_1,0)$, $C_3 = (0,u_2)$, $C_2 = (x_1,u_3)$, $B_1 = (u_4,0)$, $B_2 = (x_2,u_5)$, $O = (x_4,x_3)$, $A_1 = (u_6,0)$, $A_2 = (x_6,x_5)$, $B_3 = (0,x_7)$, $A_3 = (0,x_8)$, $D = (x_9,0)$, $E = (0,x_{10})$, $D_1 = (x_{11},0)$, $E_1 = (0,x_{12})$, $W_1 = (x_{14},x_{13})$, $F = (x_{15},0)$, $G = (0,x_{16})$, $F_1 = (x_{17},0)$, $G_1 = (0,x_{18})$, $W_2 = (x_{20},x_{19})$.

The **nondegenerate** conditions: $C_1 \neq C_3$; $C_1 \neq P$; $C_2 \neq P$; Line C_1C_3 intersects line B_1B_2; $C_1 \neq P$; Line OA_1 intersects line C_2P; Line OB_1 intersects line C_3P; Line OA_1 intersects line PC_3; Line C_1A_1 intersects line C_3A_2; Line C_3A_3 intersects line C_2A_1; Line C_1A_1 intersects line C_2A_3; Line C_1A_2 intersects line C_3A_3; Line DE intersects line D_1E_1; Line C_1B_1 intersects line C_3B_2; Line C_3B_3 intersects line C_2B_1; Line C_1B_1 intersects line C_2B_3; Line C_1B_2 intersects line C_3B_3; Line FG intersects line F_1G_1.

Remark. This is a possibly new theorem found by us.

Example 6 (Pas-2: 29.6s, 1368). *Pascal's theorem for general conics. Let A, B, C, D, F and E be six points on a conic. Let $P = AB \cap DE$, $Q = BC \cap EF$ and $S = CD \cap FA$. P, Q and S are collinear.*

Point A is on the conic with parameters u_1, u_2, u_3, u_4, u_5; Point B is on the conic with parameters u_1, u_2, u_3, u_4, u_5; Point C is on the conic with parameters u_1, u_2, u_3, u_4, u_5; Point D is on the conic with parameters u_1, u_2, u_3, u_4, u_5; Point E is on the conic with parameters u_1, u_2, u_3, u_4, u_5; Point F is on the conic with parameters u_1, u_2, u_3, u_4, u_5; P is on line AB; P is on line DE; Q is on line BC; Q is on line EF; S is on line CD; S is on line FA. The **conclusion**: Points P, Q and S are collinear.

$A = (x_1,u_6)$, $B = (x_2,u_7)$, $C = (x_3,u_8)$, $D = (x_4,u_9)$, $E = (x_5,u_{10})$, $F = (x_6,u_{11})$, $P = (x_7,x_8)$, $Q = (x_9,x_{10})$, $S = (x_{11},x_{12})$.

The **nondegenerate** conditions: Line DE intersects line AB; Line EF intersects line BC; Line FA intersects line CD.

Remark. Let $G = (x_2,x_1)$. "Point G is on the conic with parameters u_1, u_2, u_3, u_4, u_5" means $x_2^2 + (u_1x_1 + u_3)x_2 + u_2x_1^2 + u_4x_1 + u_5 = 0$. It is not necessary to check this general version if we prove Pascal's theorem for circles. Since maxt = 1368, 200K byte memory is enough for the proof.

Example 7 (Pas-3: 77.9s, 1398). *The converse of Pascal's theorem. If $P = AB \cap DE$, $Q = BC \cap EF$ and $S = CD \cap FA$ are collinear, then A, B, C, D, E and F are on the same conic.*

Point A is on the conic with parameters u_1, u_2, u_3, u_4, u_5; Point B is on the conic with parameters u_1, u_2, u_3, u_4, u_5; Point C is on the conic with parameters u_1, u_2,

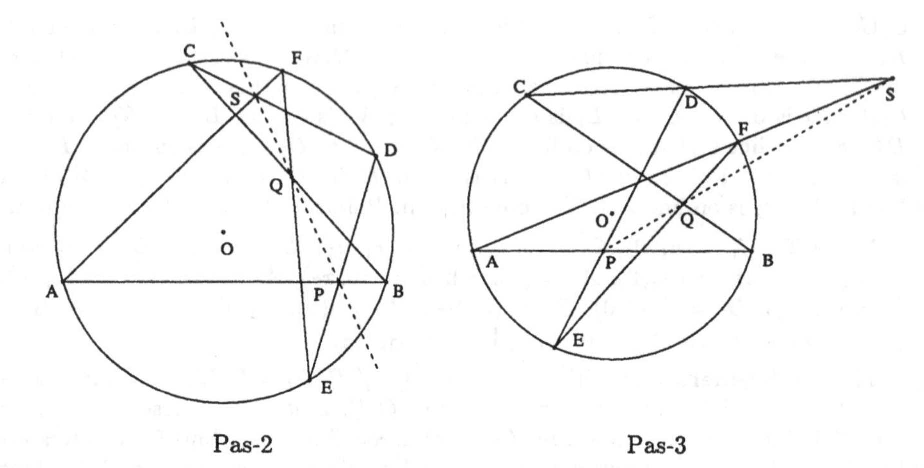

Pas-2 Pas-3

u_3, u_4, u_5; Point D is on the conic with parameters u_1, u_2, u_3, u_4, u_5; Point E is on the conic with parameters u_1, u_2, u_3, u_4, u_5; Q is on line BC; P is on line AB; P is on line DE; S is on line CD; S is on line PQ; F is on line QE; F is on line SA. The **conclusion**: Point F is on the conic with parameters u_1, u_2, u_3, u_4, u_5.

$A = (x_1, 0)$, $B = (x_2, 0)$, $C = (x_3, u_8)$, $D = (0, x_4)$, $E = (0, x_5)$, $Q = (x_6, u_{11})$, $P = (x_7, x_8)$, $S = (x_9, x_{10})$, $F = (x_{11}, x_{12})$.

The **nondegenerate** conditions: $u_2 \neq 0$; $u_2 \neq 0$; $B \neq C$; Line DE intersects line AB; Line PQ intersects line CD; Line SA intersects line QE.

Example 8 (Pas-4: 6.93s, 344). *Let A, B, C, D, F and E be six points with $P = AB \cap DE$, $Q = BC \cap EF$ and $S = CD \cap FA$ collinear. Then $P_1 = AC \cap DE$, $Q_1 = BE \cap CF$ and $S_1 = AB \cap FD$ are collinear.*

Points B, A, D, E, S are arbitrarily chosen. Points C, P, Q, F, S_1, P_1, Q_1 are constructed (in order) as follows: C is on line SD; P is on line DE; P is on line AB; Q is on line PS; Q is on line BC; F is on line SA; F is on line QE; S_1 is on line BA; S_1 is on line FD; P_1 is on line DE; P_1 is on line AC; Q_1 is on line EB; Q_1 is on line CF. The **conclusion**: Points P_1, Q_1 and S_1 are collinear.

$B = (0, 0)$, $A = (u_1, 0)$, $D = (u_2, u_3)$, $E = (u_4, u_5)$, $S = (u_6, u_7)$, $C = (x_1, u_8)$, $P = (x_2, 0)$, $Q = (x_4, x_3)$, $F = (x_6, x_5)$, $S_1 = (x_7, 0)$, $P_1 = (x_9, x_8)$, $Q_1 = (x_{11}, x_{10})$.

The **nondegenerate** conditions: $S \neq D$; Line AB intersects line DE; Line BC intersects line PS; Line QE intersects line SA; Line FD intersects line BA; Line AC intersects line DE; Line CF intersects line EB.

Remark. We can use the condition in Pas-3 as the definition of "six points are on the same conic". This theorem is to check Pascal's theorem using new definition.

Example 9 (Pas-conic-1: 14.43s, 400). *Given five points A_0, A_1, A_2, A_3 and A_4, then points $A_0A_1 \cap A_2A_3$, $A_0A_1 \cap A_2A_4$, $A_0A_2 \cap A_1A_3$, $A_0A_2 \cap A_1A_4$, $A_0A_3 \cap A_1A_2$, $A_0A_4 \cap A_1A_2$ are on the same conic.*

Points A_0, A_1, A_2, A_3, A_4 are arbitrarily chosen. Points P_0, P_1, P_2, P_3, P_4, P_5,

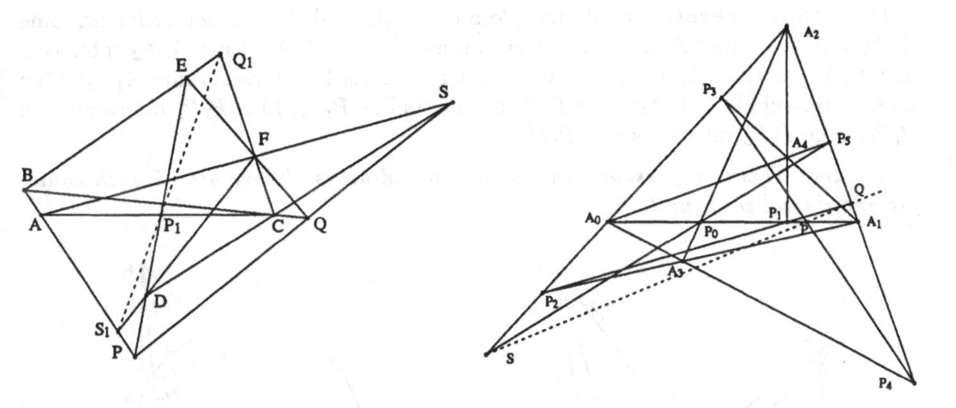

Pas-4 Pas-conic-1

P, Q, S are constructed (in order) as follows: P_0 is on line A_2A_3; P_0 is on line A_0A_1; P_1 is on line A_2A_4; P_1 is on line A_0A_1; P_2 is on line A_1A_3; P_2 is on line A_0A_2; P_3 is on line A_1A_4; P_3 is on line A_0A_2; P_4 is on line A_1A_2; P_4 is on line A_0A_3; P_5 is on line A_1A_2; P_5 is on line A_0A_4; P is on line P_3P_4; P is on line P_0P_1; Q is on line P_4P_5; Q is on line P_1P_2; S is on line P_5P_0; S is on line P_2P_3. The **conclusion**: Points P, Q and S are collinear.

$A_0 = (0,0)$, $A_1 = (u_1,0)$, $A_2 = (u_2,u_3)$, $A_3 = (u_4,u_5)$, $A_4 = (u_6,u_7)$, $P_0 = (x_1,0)$, $P_1 = (x_2,0)$, $P_2 = (x_4,x_3)$, $P_3 = (x_6,x_5)$, $P_4 = (x_8,x_7)$, $P_5 = (x_{10},x_9)$, $P = (x_{11},0)$, $Q = (x_{13},x_{12})$, $S = (x_{15},x_{14})$.

The **nondegenerate** conditions: Line A_0A_1 intersects line A_2A_3; Line A_0A_1 intersects line A_2A_4; Line A_0A_2 intersects line A_1A_3; Line A_0A_2 intersects line A_1A_4; Line A_0A_3 intersects line A_1A_2; Line A_0A_4 intersects line A_1A_2; Line P_0P_1 intersects line P_3P_4; Line P_1P_2 intersects line P_4P_5; Line P_2P_3 intersects line P_5P_0.

Remark. This is a possibly new theorem found by us. There are 60 such conics for one configuration.

Example 10 (Pas-conic-2: 31.7s, 534). *Given six points A_0, A_1, A_2, A_3, A_4 and A_5 on one conic, then points $A_0A_1 \cap A_2A_3$, $A_0A_1 \cap A_4A_5$, $A_0A_2 \cap A_1A_3$, $A_0A_3 \cap A_1A_2$, $A_0A_4 \cap A_1A_5$, $A_0A_5 \cap A_1A_4$ are on the same conic.*

Points A_0, A_1, A_2 are arbitrarily chosen. Points O, A_3, A_4, A_5, P_0, P_1, P_2, P_3, P_4, P_5, X, Y, Z are constructed (in order) as follows: $OA_0 \equiv OA_2$; $OA_0 \equiv OA_1$; $A_3O \equiv OA_0$; $A_4O \equiv OA_0$; $A_5O \equiv OA_0$; P_0 is on line A_2A_3; P_0 is on line A_0A_1; P_1 is on line A_4A_5; P_1 is on line A_0A_1; P_2 is on line A_1A_3; P_2 is on line A_0A_2; P_3 is on line A_1A_2; P_3 is on line A_0A_3; P_4 is on line A_1A_5; P_4 is on line A_0A_4; P_5 is on line A_1A_4; P_5 is on line A_0A_5; X is on line P_3P_4; X is on line P_0P_1; Y is on line P_4P_5; Y is on line P_1P_2; Z is on line P_5P_0; Z is on line P_2P_3. The **conclusion**: Points X, Y and Z are collinear.

$A_0 = (0,0)$, $A_1 = (u_1,0)$, $A_2 = (u_2,u_3)$, $O = (x_2,x_1)$, $A_3 = (x_3,u_4)$, $A_4 = (x_4,u_5)$, $A_5 = (x_5,u_6)$, $P_0 = (x_6,0)$, $P_1 = (x_7,0)$, $P_2 = (x_9,x_8)$, $P_3 = (x_{11},x_{10})$, $P_4 = (x_{13},x_{12})$, $P_5 = (x_{15},x_{14})$, $X = (x_{16},0)$, $Y = (x_{18},x_{17})$, $Z = (x_{20},x_{19})$.

The **nondegenerate** conditions: Points A_0, A_1 and A_2 are not collinear; Line A_0A_1 intersects line A_2A_3; Line A_0A_1 intersects line A_4A_5; Line A_0A_2 intersects line A_1A_3; Line A_0A_3 intersects line A_1A_2; Line A_0A_4 intersects line A_1A_5; Line A_0A_5 intersects line A_1A_4; Line P_0P_1 intersects line P_3P_4; Line P_1P_2 intersects line P_4P_5; Line P_2P_3 intersects line P_5P_0.

Remark. This is a possibly new theorem found by us. There are 45 such conics for one Pascal configuration.

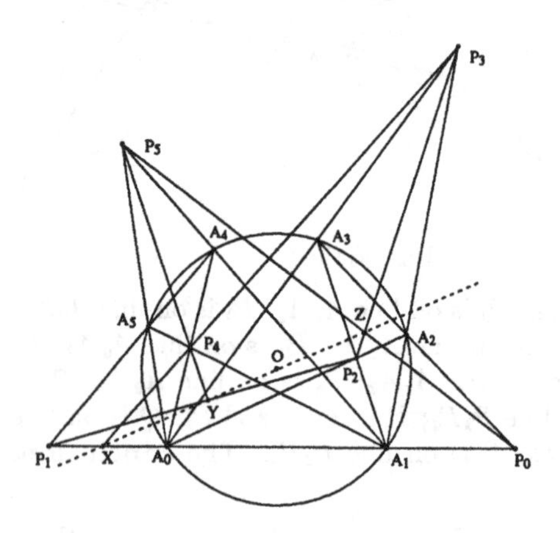

Pas-conic-2

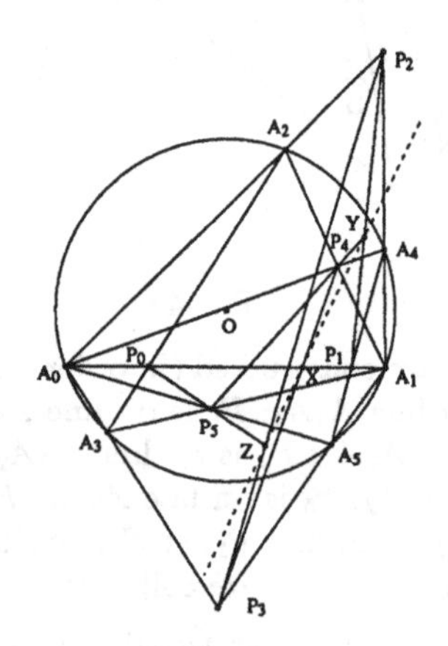

Pas-conic-3

Example 11 (Pas-conic-3: 37.3s, 586). *Given six points A_0, A_1, A_2, A_3, A_4 and A_5 on one conic, then points $A_0A_1 \cap A_2A_3$, $A_0A_1 \cap A_4A_5$, $A_0A_2 \cap A_1A_4$, $A_0A_3 \cap A_1A_5$, $A_0A_4 \cap A_1A_2$, $A_0A_5 \cap A_1A_3$ are on the same conic.*

Points A_0, A_1, A_2 are arbitrarily chosen. Points O, A_3, A_4, A_5, P_0, P_1, P_2, P_3, P_4, P_5, X, Y, Z are constructed (in order) as follows: $OA_0 \equiv OA_2$; $OA_0 \equiv OA_1$; $A_3O \equiv OA_0$; $A_4O \equiv OA_0$; $A_5O \equiv OA_0$; P_0 is on line A_2A_3; P_0 is on line A_0A_1; P_1 is on line A_4A_5; P_1 is on line A_0A_1; P_2 is on line A_1A_4; P_2 is on line A_0A_2; P_3 is on line A_1A_5; P_3 is on line A_0A_3; P_4 is on line A_1A_2; P_4 is on line A_0A_4; P_5 is on line A_1A_3; P_5 is on line A_0A_5; X is on line P_3P_4; X is on line P_0P_1; Y is on line P_4P_5; Y is on line P_1P_2; Z is on line P_5P_0; Z is on line P_2P_3. The **conclusion**: Points X, Y and Z are collinear.

$A_0 = (0,0)$, $A_1 = (u_1,0)$, $A_2 = (u_2,u_3)$, $O = (x_2,x_1)$, $A_3 = (x_3,u_4)$, $A_4 = (x_4,u_5)$, $A_5 = (x_5,u_6)$, $P_0 = (x_6,0)$, $P_1 = (x_7,0)$, $P_2 = (x_9,x_8)$, $P_3 = (x_{11},x_{10})$, $P_4 = (x_{13},x_{12})$, $P_5 = (x_{15},x_{14})$, $X = (x_{16},0)$, $Y = (x_{18},x_{17})$, $Z = (x_{20},x_{19})$.

The **nondegenerate** conditions: Points A_0, A_1 and A_2 are not collinear; Line A_0A_1 intersects line A_2A_3; Line A_0A_1 intersects line A_4A_5; Line A_0A_2 intersects line A_1A_4; Line A_0A_3 intersects line A_1A_5; Line A_0A_4 intersects line A_1A_2; Line A_0A_5 intersects line A_1A_3; Line P_0P_1 intersects line P_3P_4; Line P_1P_2 intersects line P_4P_5; Line P_2P_3 intersects line P_5P_0.

Remark. This is a possibly new theorem found by us. There are 90 such conics for one Pascal configuration.

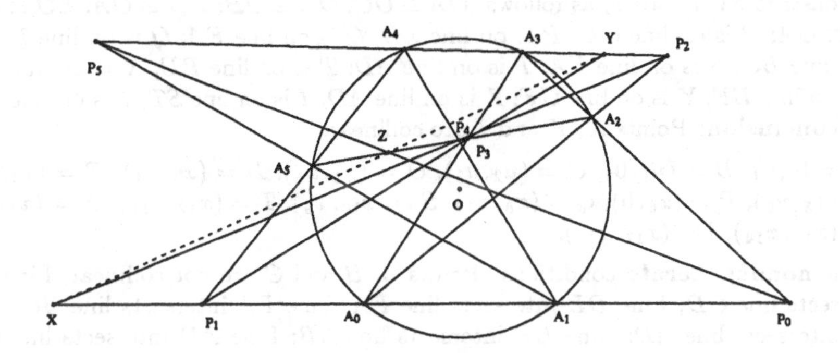

Pas-conic-4

Example 12 (Pas-conic-4: 261.35s, 3670). *Given six points A_0, A_1, A_2, A_3, A_4 and A_5 on one conic, then points $A_0A_1 \cap A_2A_3$, $A_0A_1 \cap A_4A_5$, $A_0A_2 \cap A_3A_4$, $A_0A_3 \cap A_2A_5$, $A_1A_4 \cap A_2A_5$, $A_1A_5 \cap A_3A_4$ are on the same conic.*

Points A_0, A_1 are arbitrarily chosen. Points O, A_2, A_3, A_4, A_5, P_0, P_1, P_2, P_3, P_4, P_5, X, Y, Z are constructed (in order) as follows: $OA_0 \equiv OA_1$; $A_2O \equiv OA_0$; $A_3O \equiv OA_0$; $A_4O \equiv OA_0$; $A_5O \equiv OA_0$; P_0 is on line A_2A_3; P_0 is on line A_0A_1; P_1 is on line A_4A_5; P_1 is on line A_0A_1; P_2 is on line A_3A_4; P_2 is on line A_0A_2; P_3 is on line A_2A_5; P_3 is on line A_0A_3; P_4 is on line A_2A_5; P_4 is on line A_1A_4; P_5 is on line A_3A_4; P_5 is on line A_1A_5; X is on line P_4P_2; X is on line A_0A_1; Y is on line A_3A_4; Y is on line P_1P_3; Z is on line P_5P_0; Z is on line A_2A_5. The **conclusion**: Points X, Y and Z are collinear.

$A_0 = (0,0)$, $A_1 = (u_1,0)$, $O = (x_1,u_2)$, $A_2 = (x_2,u_3)$, $A_3 = (x_3,u_4)$, $A_4 = (x_4,u_5)$, $A_5 = (x_5,u_6)$, $P_0 = (x_6,0)$, $P_1 = (x_7,0)$, $P_2 = (x_9,x_8)$, $P_3 = (x_{11},x_{10})$, $P_4 = (x_{13},x_{12})$, $P_5 = (x_{15},x_{14})$, $X = (x_{16},0)$, $Y = (x_{18},x_{17})$, $Z = (x_{20},x_{19})$.

The **nondegenerate** conditions: $A_0 \neq A_1$; Line A_0A_1 intersects line A_2A_3; Line A_0A_1 intersects line A_4A_5; Line A_0A_2 intersects line A_3A_4; Line A_0A_3 intersects line A_2A_5; Line A_1A_4 intersects line A_2A_5; Line A_1A_5 intersects line A_3A_4; Line A_0A_1 intersects line P_4P_2; Line P_1P_3 intersects line A_3A_4; Line A_2A_5 intersects line P_5P_0.

Remark. This is a possibly new theorem found by Wu in 1980. There are 60 such conics for one configuration. For the above three statements, our prover also confirmed (disproved) that the conclusions are no longer true if the starting six points are not on the same conic. Without loss of generality (using projection), we use a circle instead of a general conic in the statements. We also confirmed the statements in affine geometry and in projective geometry.

Example 13 (Steiner: 10.08s, 342). *Steiner's theorem. Given six points A, B, C, D, E, and F on a circle (or a conic), the three Pascal lines [ABEDCF], [CDAFEB], [EFCBAD] are concurrent.*

Points A, B, C are arbitrarily chosen. Points O, D, E, F, P, Q, S, T, X, Y, I

are constructed (in order) as follows: $OA \equiv OC$; $OA \equiv OB$; $DO \equiv OA$; $EO \equiv OA$; $FO \equiv OA$; P is on line CD; P is on line AB; Q is on line FA; Q is on line DE; S is on line BC; S is on line FA; T is on line AD; T is on line BE; X is on line AB; X is on line EF; Y is on line CF; Y is on line AD; I is on line ST; I is on line PQ. The **conclusion**: Points X, Y and I are collinear.

$A = (0,0)$, $B = (u_1,0)$, $C = (u_2,u_3)$, $O = (x_2,x_1)$, $D = (x_3,u_4)$, $E = (x_4,u_5)$, $F = (x_5,u_6)$, $P = (x_6,0)$, $Q = (x_8,x_7)$, $S = (x_{10},x_9)$, $T = (x_{12},x_{11})$, $X = (x_{13},0)$, $Y = (x_{15},x_{14})$, $I = (x_{17},x_{16})$.

The **nondegenerate** conditions: Points A, B and C are not collinear; Line AB intersects line CD; Line DE intersects line FA; Line FA intersects line BC; Line BE intersects line AD; Line EF intersects line AB; Line AD intersects line CF; Line PQ intersects line ST.

Remark. There are 20 Steiner points for one Pascal configuration.

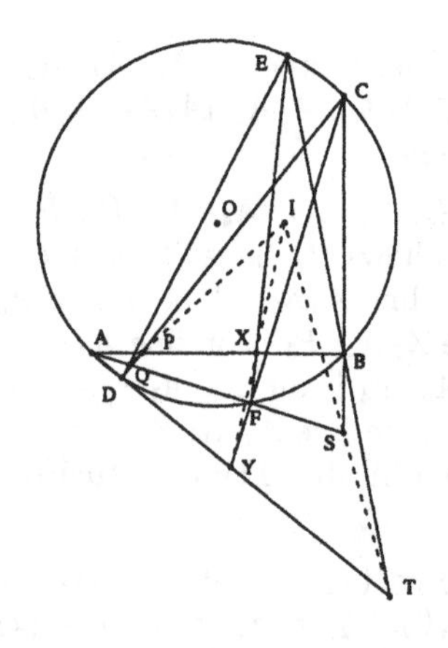

Steiner

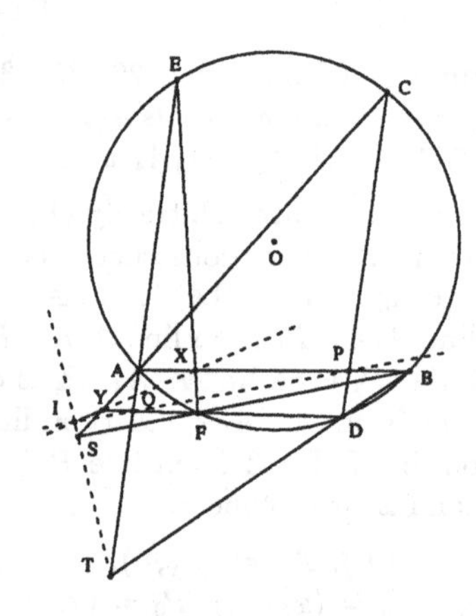

Kirkman

Example 14 (Kirkman: 8.78s, 203). *Kirkman's theorem. Given six points A, B, C, D, E, and F on a circle (or a conic), the three Pascal lines [BAECDF], [CDBFEA], [FECABD] are concurrent.*

Points A, B, C are arbitrarily chosen. Points O, D, E, F, P, Q, S, T, X, Y, I are constructed (in order) as follows: $OA \equiv OC$; $OA \equiv OB$; $DO \equiv OA$; $EO \equiv OA$; $FO \equiv OA$; P is on line CD; P is on line AB; Q is on line DF; Q is on line AE; S is on line AC; S is on line BF; T is on line BD; T is on line AE; X is on line AB; X is on line EF; Y is on line DF; Y is on line AC; I is on line ST; I is on line PQ. The **conclusion**: Points X, Y and I are collinear.

$A = (0,0)$, $B = (u_1,0)$, $C = (u_2,u_3)$, $O = (x_2,x_1)$, $D = (x_3,u_4)$, $E = (x_4,u_5)$, $F = (x_5,u_6)$, $P = (x_6,0)$, $Q = (x_8,x_7)$, $S = (x_{10},x_9)$, $T = (x_{12},x_{11})$, $X = (x_{13},0)$, $Y = (x_{15},x_{14})$, $I = (x_{17},x_{16})$.

The **nondegenerate** conditions: Points A, B and C are not collinear; Line AB intersects line CD; Line AE intersects line DF; Line BF intersects line AC; Line

AE intersects line BD; Line EF intersects line AB; Line AC intersects line DF; Line PQ intersects line ST.

Remark. There are 60 Kirkman points for one Pascal configuration.

Example 15 (Gauss-line: 1.4s, 6). *Gauss' theorem. The midpoints of the three diagonals of a complete quadrilateral are collinear. Let us call this line the Gauss line for the given four lines.*

Points A_0, A_1, A_2, A_3 are arbitrarily chosen. Points X, Y, M_1, M_2, M_3 are constructed (in order) as follows: X is on line A_0A_3; X is on line A_1A_2; Y is on line A_2A_3; Y is on line A_1A_0; M_1 is the midpoint of A_1 and A_3; M_2 is the midpoint of A_0 and A_2; M_3 is the midpoint of X and Y. The **conclusion**: Points M_1, M_2 and M_3 are collinear.

$A_0 = (0,0)$, $A_1 = (u_1,0)$, $A_2 = (u_2,u_3)$, $A_3 = (0,u_4)$, $X = (0,x_1)$, $Y = (x_2,0)$, $M_1 = (x_3,x_4)$, $M_2 = (x_5,x_6)$, $M_3 = (x_7,x_8)$.

The **nondegenerate** conditions: Line A_1A_2 intersects line A_0A_3; Line A_1A_0 intersects line A_2A_3.

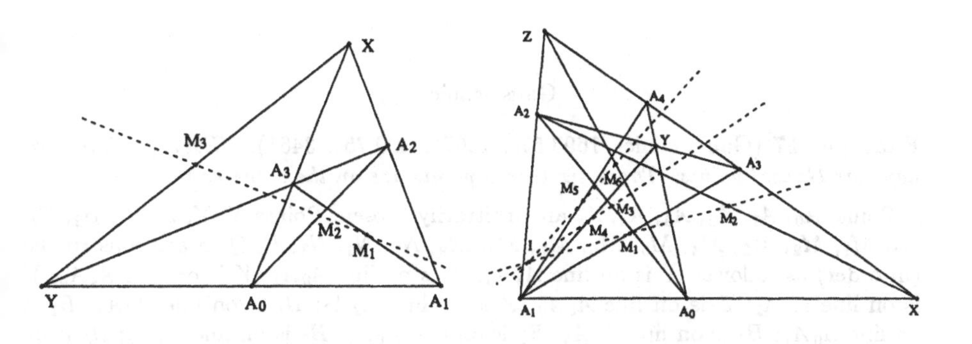

Gauss-line Gauss-point

Example 16 (Gauss-point: 3.27s, 65). *Given five lines, we have five Gauss' lines. These five Gauss' lines are concurrent. Let us call this point of concurrency the Gauss point for the given five points.*

Points A_1, A_0, A_3, A_4, A_2 are arbitrarily chosen. Points X, Y, Z, M_1, M_2, M_3, M_4, M_5, M_6, I are constructed (in order) as follows: X is on line A_3A_4; X is on line A_1A_0; Y is on line A_2A_3; Y is on line A_0A_4; Z is on line A_4A_3; Z is on line A_1A_2; M_1 is the midpoint of A_1 and A_3; M_2 is the midpoint of X and A_2; M_3 is the midpoint of A_0 and A_2; M_4 is the midpoint of Y and A_1; M_5 is the midpoint of A_1 and A_4; M_6 is the midpoint of A_0 and Z; I is on line M_3M_4; I is on line M_1M_2. The **conclusion**: Points I, M_5 and M_6 are collinear.

$A_1 = (0,0)$, $A_0 = (u_1,0)$, $A_3 = (u_2,u_3)$, $A_4 = (u_4,u_5)$, $A_2 = (0,u_6)$, $X = (x_1,0)$, $Y = (x_3,x_2)$, $Z = (0,x_4)$, $M_1 = (x_5,x_6)$, $M_2 = (x_7,x_8)$, $M_3 = (x_9,x_{10})$, $M_4 =$

(x_{11}, x_{12}), $M_5 = (x_{13}, x_{14})$, $M_6 = (x_{15}, x_{16})$, $I = (x_{18}, x_{17})$.

The **nondegenerate** conditions: Line A_1A_0 intersects line A_3A_4; Line A_0A_4 intersects line A_2A_3; Line A_1A_2 intersects line A_4A_3; Line M_1M_2 intersects line M_3M_4.

Remark. This is a possibly new theorem found by Wang Tong-Ming. Let us call the point of concurrency the Gauss Point for the five lines.

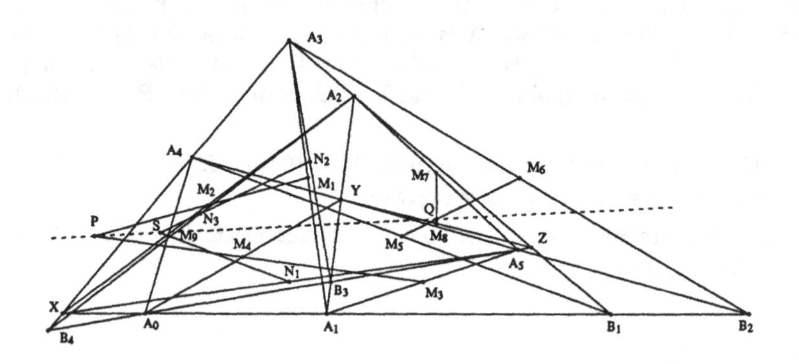

Gauss-conic

Example 17 (Gauss-conic: 1693.57s, 15674; 420.75s, 3461). *Given six lines, we have six Gauss' points. These six Gauss points are on the same conic.*

Points A_0, A_1, A_2, A_3, A_4, A_5 are arbitrarily chosen. Points X, Y, Z, B_1, B_2, B_3, B_4, M_1, M_2, M_3, M_4, M_5, M_6, M_7, M_8, M_9, N_1, N_2, N_3, P, Q, S are constructed (in order) as follows: X is on line A_3A_4; X is on line A_0A_1; Y is on line A_4A_5; Y is on line A_1A_2; Z is on line A_5A_0; Z is on line A_2A_3; B_1 is on line A_2A_3; B_1 is on line A_0A_1; B_2 is on line A_4A_5; B_2 is on line A_0A_1; B_3 is on line A_1A_2; B_3 is on line A_0A_5; B_4 is on line A_3A_4; B_4 is on line A_0A_5; M_1 is the midpoint of A_1 and A_3; M_2 is the midpoint of A_2 and X; M_3 is the midpoint of A_1 and A_5; M_4 is the midpoint of A_0 and Y; M_5 is the midpoint of B_1 and A_4; M_6 is the midpoint of B_2 and A_3; M_7 is the midpoint of A_2 and A_5; M_8 is the midpoint of Z and Y; M_9 is the midpoint of A_0 and A_4; N_1 is the midpoint of A_5 and X; N_2 is the midpoint of A_3 and B_3; N_3 is the midpoint of A_2 and B_4; P is on line M_3M_4; P is on line M_1M_2; Q is on line M_7M_8; Q is on line M_5M_6; S is on line N_2N_3; S is on line M_9N_1. The **conclusion**: Points P, Q and S are collinear.

$A_0 = (0,0)$, $A_1 = (u_1,0)$, $A_2 = (u_2,u_3)$, $A_3 = (u_4,u_5)$, $A_4 = (u_6,u_7)$, $A_5 = (0,u_8)$, $X = (x_1,0)$, $Y = (x_3,x_2)$, $Z = (0,x_4)$, $B_1 = (x_5,0)$, $B_2 = (x_6,0)$, $B_3 = (0,x_7)$, $B_4 = (0,x_8)$, $M_1 = (x_9,x_{10})$, $M_2 = (x_{11},x_{12})$, $M_3 = (x_{13},x_{14})$, $M_4 = (x_{15},x_{16})$, $M_5 = (x_{17},x_{18})$, $M_6 = (x_{19},x_{20})$, $M_7 = (x_{21},x_{22})$, $M_8 = (x_{23},x_{24})$, $M_9 = (x_{25},x_{26})$, $N_1 = (x_{27},x_{28})$, $N_2 = (x_{29},x_{30})$, $N_3 = (x_{31},x_{32})$, $P = (x_{34},x_{33})$, $Q = (x_{36},x_{35})$, $S = (x_{38},x_{37})$.

The **nondegenerate** conditions: Line A_0A_1 intersects line A_3A_4; Line A_1A_2

intersects line A_4A_5; Line A_2A_3 intersects line A_5A_0; Line A_0A_1 intersects line A_2A_3; Line A_0A_1 intersects line A_4A_5; Line A_0A_5 intersects line A_1A_2; Line A_0A_5 intersects line A_3A_4; Line M_1M_2 intersects line M_3M_4; Line M_5M_6 intersects line M_7M_8; Line M_9N_1 intersects line N_2N_3.

Remark. This is a possibly new theorem found by us.

Example 18 (Pappus-dual: 1.45s, 36). *The dual of Pappus' theorem.*

Points O, A, B, C, O_1 are arbitrarily chosen. Points A_1, B_1, C_1, I are constructed (in order) as follows: A_1 is on line O_1B; A_1 is on line OC; B_1 is on line OA; B_1 is on line O_1C; C_1 is on line OB; C_1 is on line O_1A; I is on line BB_1; I is on line AA_1. The **conclusion**: Points C, C_1 and I are collinear.

$O = (0,0)$, $A = (u_1, 0)$, $B = (u_2, u_3)$, $C = (u_4, u_5)$, $O_1 = (u_6, u_7)$, $A_1 = (x_2, x_1)$, $B_1 = (x_3, 0)$, $C_1 = (x_5, x_4)$, $I = (x_7, x_6)$.

The **nondegenerate** conditions: Line OC intersects line O_1B; Line O_1C intersects line OA; Line O_1A intersects line OB; Line AA_1 intersects line BB_1.

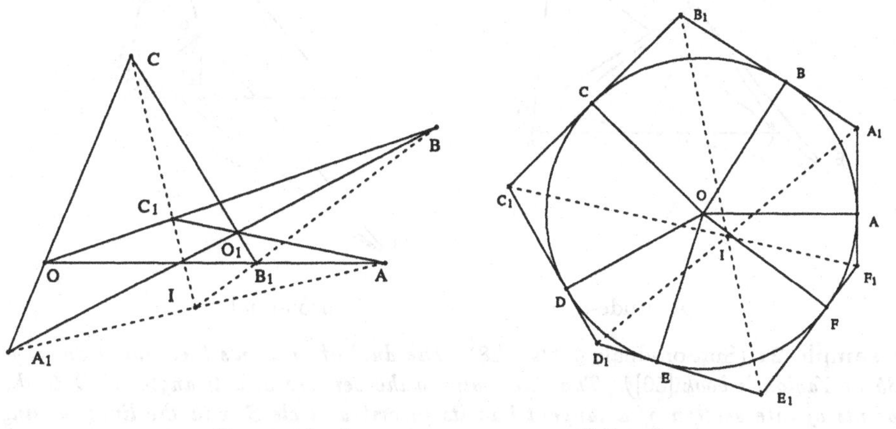

Pappus-dual Brianchon

Example 19 (Brianchon: 12.72s, 456). *Brianchon's theorem: the dual of Pascal theorem. Here we proved the theorem for circles.*

Points O, A are arbitrarily chosen. Points B, C, D, E, F, A_1, B_1, C_1, D_1, E_1, F_1, I are constructed (in order) as follows: $BO \equiv OA$; $CO \equiv OA$; $DO \equiv OA$; $EO \equiv OA$; $FO \equiv OA$; $A_1B \perp BO$; $A_1A \perp AO$; $B_1C \perp CO$; $B_1B \perp BO$; $C_1D \perp DO$; $C_1C \perp CO$; $D_1E \perp EO$; $D_1D \perp DO$; $E_1F \perp FO$; $E_1E \perp EO$; F_1 is on line A_1A; $F_1F \perp FO$; I is on line B_1E_1; I is on line A_1D_1. The **conclusion**: Points C_1, F_1 and I are collinear.

$O = (0,0)$, $A = (u_1, 0)$, $B = (x_1, u_2)$, $C = (x_2, u_3)$, $D = (x_3, u_4)$, $E = (x_4, u_5)$, $F = (x_5, u_6)$, $A_1 = (u_1, x_6)$, $B_1 = (x_8, x_7)$, $C_1 = (x_{10}, x_9)$, $D_1 = (x_{12}, x_{11})$, $E_1 = (x_{14}, x_{13})$, $F_1 = (u_1, x_{15})$, $I = (x_{17}, x_{16})$.

The **nondegenerate** conditions: Points A, O and B are not collinear; Points B, O and C are not collinear; Points C, O and D are not collinear; Points D, O and E

are not collinear; Points E, O and F are not collinear; Line FO is not perpendicular to line A_1A; Line A_1D_1 intersects line B_1E_1.

Example 20 (Altitude-dual: 1.02s, 27). *The dual of the orthocenter theorem.*

Points A, B, C, O are arbitrarily chosen. Points D, E, F are constructed (in order) as follows: D is on line BC; $DO \perp AO$; E is on line CA; $EO \perp BO$; F is on line AB; $FO \perp CO$. The **conclusion**: Points E, F and D are collinear.

$A = (0,0)$, $B = (u_1,0)$, $C = (u_2,u_3)$, $O = (u_4,u_5)$, $D = (x_2,x_1)$, $E = (x_4,x_3)$, $F = (x_5,0)$.

The **nondegenerate** conditions: Line AO is not perpendicular to line BC; Line BO is not perpendicular to line CA; Line CO is not perpendicular to line AB.

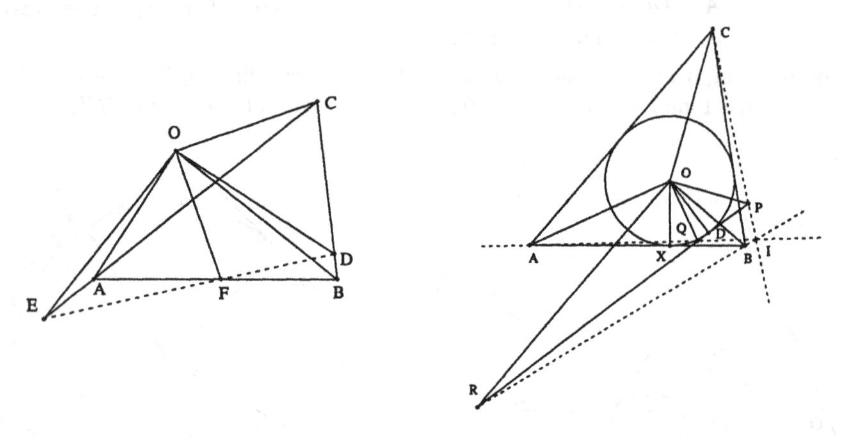

Altitude-dual Simson-dual

Example 21 (Simson-dual: 4.33s, 128). *The dual of Simson's theorem (from page 85 of Yaglom's book [50]). The lines joining the vertices of a triangle ABC to the points of intersection of a tangent l to its inscribed circle S with the lines passing through the center O of S and parallel to the bisectors of the exterior angles at these vertices are concurrent.*

Points A, B, O are arbitrarily chosen. Points C, X, D, P, Q, R, I are constructed (in order) as follows: $\tan(ABO) = \tan(OBC)$; $\tan(BAO) = \tan(OAC)$; X is on line AB; $XO \perp AB$; $DO \equiv OX$; $PD \perp DO$; $PO \perp CO$; $QD \perp DO$; $QO \perp AO$; $RD \perp DO$; $RO \perp BO$; I is on line BR; I is on line AQ. The **conclusion**: Points C, P and I are collinear.

$A = (0,0)$, $B = (u_1,0)$, $O = (u_2,u_3)$, $C = (x_2,x_1)$, $X = (u_2,0)$, $D = (x_3,u_4)$, $P = (x_5,x_4)$, $Q = (x_7,x_6)$, $R = (x_9,x_8)$, $I = (x_{11},x_{10})$.

The **nondegenerate** conditions: Points O, A and B are not collinear; $\angle AOB$ is not right; Line AB is non-isotropic; Points C, O and D are not collinear; Points A, O and D are not collinear; Points B, O and D are not collinear; Line AQ intersects line BR.

Remark. Actually, our proof is for the four tritangent circles.

Example 22 (E1.5-19: 2.82s, 11). *Let ABC be any triangle and $ABDE$, $ACFG$ any parallelograms described on AB and AC. Let DE and FG meet in H and draw BL and CM equal and parallel to HA. Then area($BCML$) = area($ABDE$) + area($ACFG$).*

Points A, B, C, D, F are arbitrarily chosen. Points E, G, H, R, S, L are constructed (in order) as follows: $EA \parallel DB$; $ED \parallel BA$; $GA \parallel FC$; $GF \parallel CA$; H is on line FG; H is on line DE; R is on line BC; R is on line HA; S is on line HA; $\frac{\overline{HA}}{\overline{RS}} - 1 = 0$; $LB \parallel HA$; $LS \parallel BC$. The **conclusion**: $\nabla(BLC) - (\nabla(ACF) + \nabla(ADB)) = 0$.

$A = (0,0)$, $B = (u_1, 0)$, $C = (u_2, u_3)$, $D = (u_4, u_5)$, $F = (u_6, u_7)$, $E = (x_1, u_5)$, $G = (x_3, x_2)$, $H = (x_4, u_5)$, $R = (x_6, x_5)$, $S = (x_8, x_7)$, $L = (x_{10}, x_9)$.

The **nondegenerate** conditions: Points B, A and D are not collinear; Points C, A and F are not collinear; Line DE intersects line FG; Line HA intersects line BC; $x_4 \neq 0$; Line BC intersects line HA.

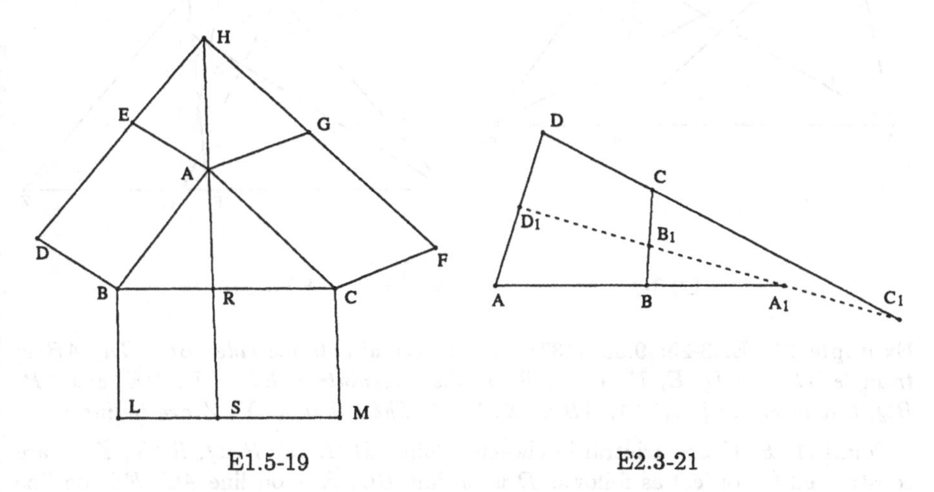

E1.5-19 E2.3-21

Example 23 (E2.3-21: 1.52s, 56). *A generalization of Menelaus' Theorem.*

Points A, B, C, D are arbitrarily chosen. Points A_1, B_1, C_1, D_1 are constructed (in order) as follows: A_1 is on line AB; B_1 is on line BC; C_1 is on line A_1B_1; C_1 is on line CD; D_1 is on line A_1B_1; D_1 is on line DA. The **conclusion**: $\frac{\overline{AA_1}}{\overline{A_1B}} \frac{\overline{BB_1}}{\overline{B_1C}} \frac{\overline{CC_1}}{\overline{C_1D}} \frac{\overline{DD_1}}{\overline{D_1A}} - 1 = 0$.

$A = (0,0)$, $B = (u_1, 0)$, $C = (u_2, u_3)$, $D = (u_4, u_5)$, $A_1 = (u_6, 0)$, $B_1 = (x_1, u_7)$, $C_1 = (x_3, x_2)$, $D_1 = (x_5, x_4)$.

The **nondegenerate** conditions: $A \neq B$; $B \neq C$; Line CD intersects line A_1B_1; Line DA intersects line A_1B_1.

Example 24 (E2.3-22: 6.37s, 160). *A converse of the above.*

Points A, B, C, D are arbitrarily chosen. Points A_1, B_1, C_1, D_1, I, J are constructed (in order) as follows: A_1 is on line AB; B_1 is on line BC; C_1 is on line

CD; $\frac{\overline{AA_1}}{\overline{A_1B}} \frac{\overline{BB_1}}{\overline{B_1C}} \frac{\overline{CC_1}}{\overline{C_1D}} \frac{\overline{DD_1}}{\overline{D_1A}} - 1 = 0$; D_1 is on line DA; I is on line A_1B_1; I is on line AC; J is on line A_1D_1; J is on line BD. The **conclusions**: (1) Points I, C_1 and D_1 are collinear; (2) Points J, B_1 and C_1 are collinear.

$A = (0,0)$, $B = (u_1,0)$, $C = (u_2,u_3)$, $D = (u_4,u_5)$, $A_1 = (u_6,0)$, $B_1 = (x_1,u_7)$, $C_1 = (x_2,u_8)$, $D_1 = (x_4,x_3)$, $I = (x_6,x_5)$, $J = (x_8,x_7)$.

The **nondegenerate** conditions: $A \neq B$; $B \neq C$; $C \neq D$; $(-u_1u_4x_1 + ((-u_2+u_1)u_4)u_6 + u_1u_2u_4)x_2 + ((-u_4^2+u_2u_4)u_6 + u_1u_4^2)x_1 + (u_2u_4^2 - u_1u_2u_4)u_6 - u_1u_2u_4^2 \neq 0$; $-u_5 \neq 0$; Line AC intersects line A_1B_1; Line BD intersects line A_1D_1.

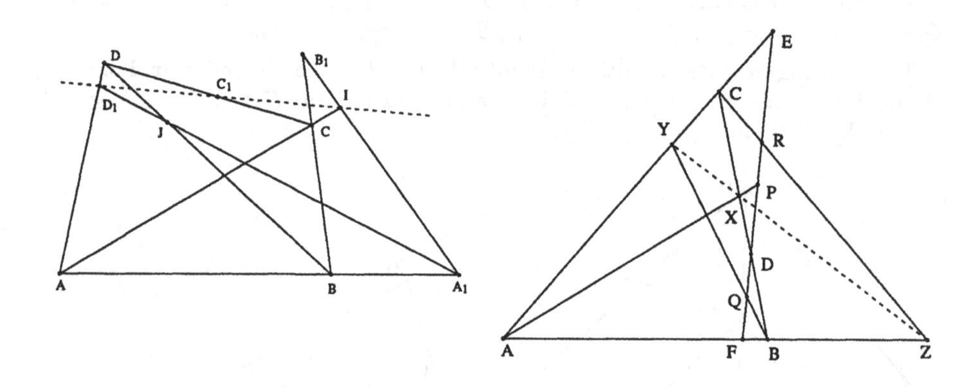

E2.3-22 E2.3-25

Example 25 (E2.3-25: 9.55s, 433). *A transversal cuts the sides BC, CA, AB of triangle ABC in D, E, F. P, Q, R are the midpoints of EF, FD, DE, and AP, BQ, CR intersect BC, CA, AB in X, Y, Z. Show that X, Y, Z are collinear.*

Points A, B, C are arbitrarily chosen. Points D, E, F, P, Q, R, X, Y, Z are constructed (in order) as follows: D is on line BC; E is on line AC; F is on line ED; F is on line AB; $PF \equiv PE$; P is on line FE; $QF \equiv QD$; Q is on line FE; $RE \equiv RD$; R is on line FE; X is on line AP; X is on line BC; Y is on line BQ; Y is on line AC; Z is on line CR; Z is on line AB. The **conclusion**: Points X, Y and Z are collinear.

$A = (0,0)$, $B = (u_1,0)$, $C = (u_2,u_3)$, $D = (x_1,u_4)$, $E = (x_2,u_5)$, $F = (x_3,0)$, $P = (x_5,x_4)$, $Q = (x_7,x_6)$, $R = (x_9,x_8)$, $X = (x_{11},x_{10})$, $Y = (x_{13},x_{12})$, $Z = (x_{14},0)$.

The **nondegenerate** conditions: $B \neq C$; $A \neq C$; Line AB intersects line ED; Line FE is non-isotropic; Line FE is non-isotropic; Line FE is non-isotropic; Line BC intersects line AP; Line AC intersects line BQ; Line AB intersects line CR.

Remark. This theorem has been also confirmed in Bolyai-Lobachevskian geometry by our prover.

Example 26 (E2.3-26: 86.13s, 1934). *Let O and U be two points in the plane of triangle ABC. Let AO, BO, CO intersect the opposite sides BC, CA, AB in P,*

Q, R. Let PU, QU, RU intersect QR, RP, PQ respectively in X, Y, Z. Show that AX, BY, CZ are concurrent.

Points A, B, C, O, U are arbitrarily chosen. Points P, Q, R, X, Y, Z, I are constructed (in order) as follows: P is on line BC; P is on line AO; Q is on line CA; Q is on line BO; R is on line AB; R is on line CO; X is on line QR; X is on line PU; Y is on line RP; Y is on line QU; Z is on line PQ; Z is on line RU; I is on line BY; I is on line AX. The **conclusion**: Points C, Z and I are collinear.

$A = (0,0)$, $B = (u_1,0)$, $C = (u_2,u_3)$, $O = (u_4,u_5)$, $U = (u_6,u_7)$, $P = (x_2,x_1)$, $Q = (x_4,x_3)$, $R = (x_5,0)$, $X = (x_7,x_6)$, $Y = (x_9,x_8)$, $Z = (x_{11},x_{10})$, $I = (x_{13},x_{12})$.

The **nondegenerate** conditions: Line AO intersects line BC; Line BO intersects line CA; Line CO intersects line AB; Line PU intersects line QR; Line QU intersects line RP; Line RU intersects line PQ; Line AX intersects line BY.

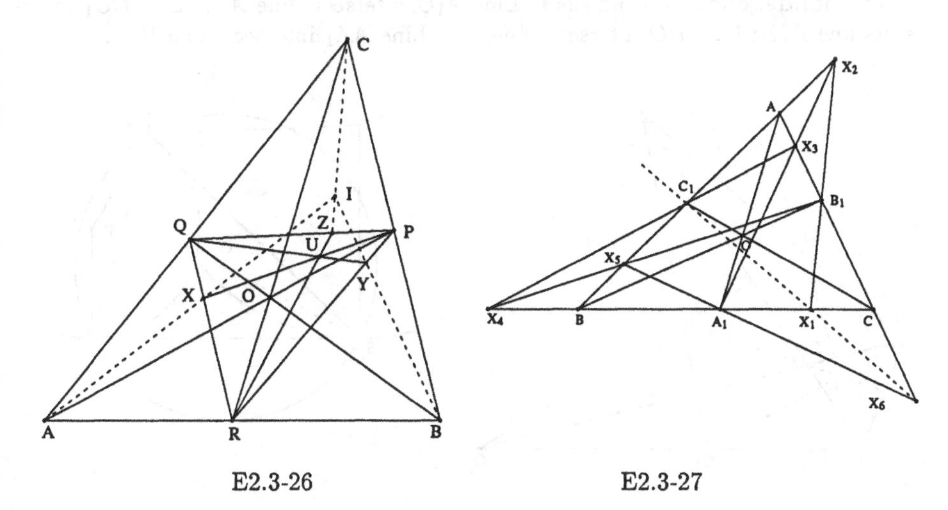

E2.3-26 E2.3-27

Example 27 (E2.3-27: 4.15s, 67). *Nehring's theorem (1942). Let AA_1, BB_1, CC_1 be three concurrent cevian lines for triangle ABC. Let X_1 be a point on BC, $X_2 = X_1B_1 \cap BA$, $X_3 = X_2A_1 \cap AC$, $X_4 = X_3C_1 \cap CB$, $X_5 = X_4B_1 \cap BA$, $X_6 = X_5A_1 \cap AC$, $X_7 = X_6C_1 \cap CB$. Show $X_7 = X_1$.*

Points B, C, A, O are arbitrarily chosen. Points A_1, B_1, C_1, X_1, X_2, X_3, X_4, X_5, X_6 are constructed (in order) as follows: A_1 is on line BC; A_1 is on line AO; B_1 is on line AC; B_1 is on line BO; C_1 is on line AB; C_1 is on line CO; X_1 is on line BC; X_2 is on line BA; X_2 is on line X_1B_1; X_3 is on line AC; X_3 is on line X_2A_1; X_4 is on line CB; X_4 is on line X_3C_1; X_5 is on line BA; X_5 is on line X_4B_1; X_6 is on line AC; X_6 is on line X_5A_1. The **conclusion**: Points X_6, X_1 and C_1 are collinear.

$B = (0,0)$, $C = (u_1,0)$, $A = (u_2,u_3)$, $O = (u_4,u_5)$, $A_1 = (x_1,0)$, $B_1 = (x_3,x_2)$, $C_1 = (x_5,x_4)$, $X_1 = (u_6,0)$, $X_2 = (x_7,x_6)$, $X_3 = (x_9,x_8)$, $X_4 = (x_{10},0)$, $X_5 = (x_{12},x_{11})$, $X_6 = (x_{14},x_{13})$.

The **nondegenerate** conditions: Line AO intersects line BC; Line BO intersects line AC; Line CO intersects line AB; $B \neq C$; Line X_1B_1 intersects line BA; Line

X_2A_1 intersects line AC; Line X_3C_1 intersects line CB; Line X_4B_1 intersects line BA; Line X_5A_1 intersects line AC.

Example 28 (E2.4-16: 1.85s, 30). *In a hexagon $AC_1BA_1CB_1$, BB_1, C_1A, A_1C are concurrent and CC_1, A_1B, B_1A are concurrent. Prove that AA_1, B_1C, C_1B are also concurrent.*

Points A, C_1, B, A_1, C are arbitrarily chosen. Points O, H, B_1, I are constructed (in order) as follows: O is on line AC_1; O is on line A_1C; H is on line A_1B; H is on line CC_1; B_1 is on line AH; B_1 is on line BO; I is on line B_1C; I is on line AA_1. The **conclusion**: Points C_1, B and I are collinear.

$A = (0,0)$, $C_1 = (u_1,0)$, $B = (u_2,u_3)$, $A_1 = (u_4,u_5)$, $C = (u_6,u_7)$, $O = (x_1,0)$, $H = (x_3,x_2)$, $B_1 = (x_5,x_4)$, $I = (x_7,x_6)$.

The **nondegenerate** conditions: Line A_1C intersects line AC_1; Line CC_1 intersects line A_1B; Line BO intersects line AH; Line AA_1 intersects line B_1C.

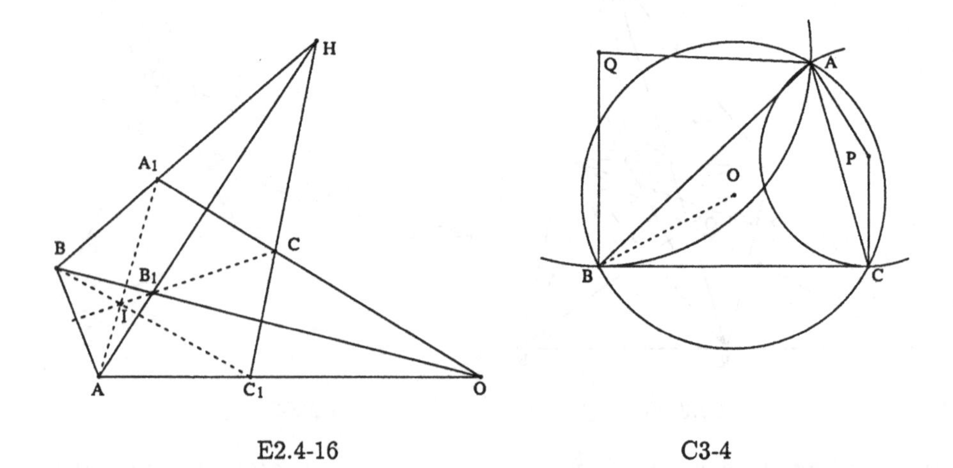

E2.4-16 C3-4

Example 29 (C3-4: 0.82s, 30). *In a triangle ABC, let p and q be the radii of two circles through A, touching side BC at B and C, respectively. Then $pq = R^2$.*

Points B, C, A are arbitrarily chosen. Points O, Q, P are constructed (in order) as follows: $OC \equiv OB$; $OA \equiv OB$; $QB \perp BC$; $QA \equiv QB$; $PC \perp CB$; $PA \equiv PC$. The **conclusion**: $QB \cdot PC = OB \cdot OB$.

$B = (0,0)$, $C = (u_1,0)$, $A = (u_2,u_3)$, $O = (x_2,x_1)$, $Q = (0,x_3)$, $P = (u_1,x_4)$.

The **nondegenerate** conditions: Points A, B and C are not collinear; Points A, B and C are not collinear; Points A, C and B are not collinear.

Example 30 (C18-4: 1.13s, 7). *Let H and O be the orthocenter and circumcenter of triangle ABC. Show $\angle HAO = |B - C|$.*

Points B, C, A are arbitrarily chosen. Points O, D are constructed (in order) as follows: $OB \equiv OC$; $OB \equiv OA$; D is on line BC; $DA \perp BC$. The **conclusion**: $\tan(\angle ABC + \angle ACB) - \tan(\angle OAD) = 0$.

$B = (0,0)$, $C = (u_1,0)$, $A = (u_2,u_3)$, $O = (x_2,x_1)$, $D = (u_2,0)$.

The **nondegenerate** conditions: Points B, A and C are not collinear; Line BC is non-isotropic.

Remark. Note our specification of the problem to deal with addition of angles.

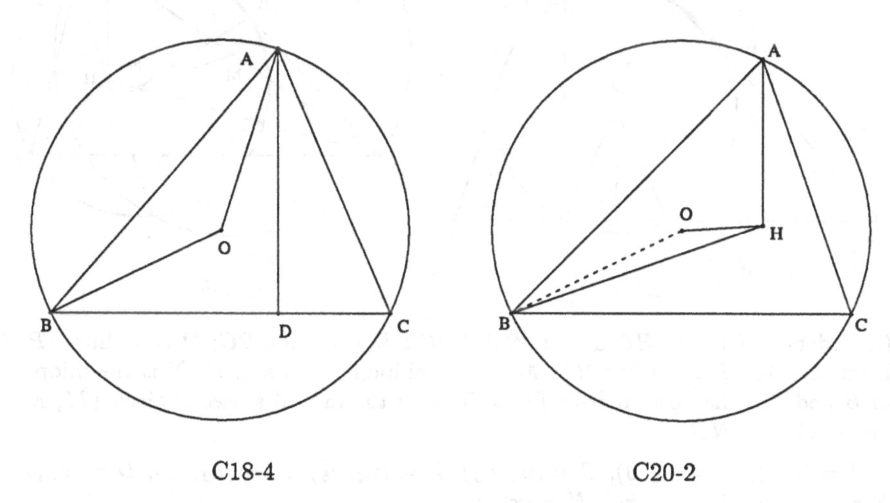

C18-4 C20-2

Example 31 (C20-2: 0.58s, 9). $OH^2 = 9R^2 - a^2 - b^2 - c^2$.

Points B, C, A are arbitrarily chosen. Points O, H are constructed (in order) as follows: $OC \equiv OB$; $OA \equiv OB$; $HB \perp AC$; $HA \perp BC$. The **conclusion**: $\overline{OH}^2 - (9\,\overline{OB}^2 - \overline{AB}^2 - (\overline{AC}^2 + \overline{BC}^2)) = 0$.

$B = (0,0)$, $C = (u_1,0)$, $A = (u_2,u_3)$, $O = (x_2,x_1)$, $H = (u_2,x_3)$.

The **nondegenerate** conditions: Points A, B and C are not collinear; Points B, C and A are not collinear.

Example 32 (C22-5: 1.46s, 27). *The nine-point circle cuts the sides of the triangle at angles* $|B - C|$, $|C - A|$, $|A - B|$.

Points A, B, C are arbitrarily chosen. Points A_1, B_1, C_1, F, N, L are constructed (in order) as follows: A_1 is the midpoint of B and C; B_1 is the midpoint of A and C; C_1 is the midpoint of B and A; F is on line AB; $FC \perp AB$; $NC_1 \equiv NA_1$; $NC_1 \equiv NB_1$; L is the midpoint of F and C_1. The **conclusion**: $\tan(\angle CAB + \angle CBA) - \tan(\angle LNF) = 0$.

$A = (0,0)$, $B = (u_1,0)$, $C = (u_2,u_3)$, $A_1 = (x_1,x_2)$, $B_1 = (x_3,x_4)$, $C_1 = (x_5,0)$, $F = (u_2,0)$, $N = (x_7,x_6)$, $L = (x_8,0)$.

The **nondegenerate** conditions: Line AB is non-isotropic; Points C_1, B_1 and A_1 are not collinear.

Example 33 (C2.45: 1.21s, 12). *If circles are constructed on two cevians as diameters, their radical axis passes through the orthocenter H of the triangle.*

Points A, B, C, P are arbitrarily chosen. Points H, D, E, M, N are constructed

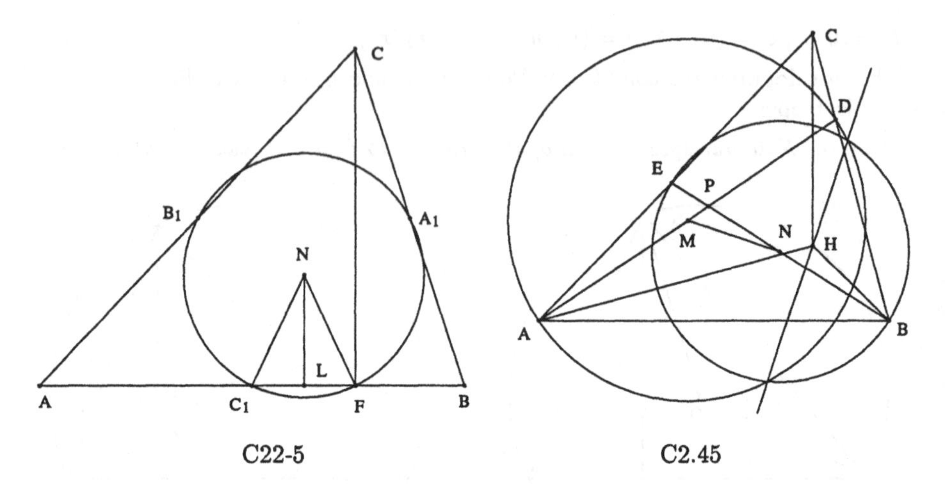

C22-5 C2.45

(in order) as follows: $HC \perp AB$; $HA \perp BC$; D is on line BC; D is on line AP; E is on line AC; E is on line BP; M is the midpoint of A and D; N is the midpoint of B and E. The **conclusion**: Point H is on the radical axises of circle (M, MA) and circle (N, NB).

$A = (0,0)$, $B = (u_1, 0)$, $C = (u_2, u_3)$, $P = (u_4, u_5)$, $H = (u_2, x_1)$, $D = (x_3, x_2)$, $E = (x_5, x_4)$, $M = (x_6, x_7)$, $N = (x_8, x_9)$.

The **nondegenerate** conditions: Points B, C and A are not collinear; Line AP intersects line BC; Line BP intersects line AC.

Example 34 (C25-3: 0.92s, 19). *If lines PB and PD, outside a parallelogram $ABCD$, make equal angles with the sides BC and DC, respectively, then $\angle CPB \equiv \angle DPA$.*

Points A, B, C are arbitrarily chosen. Points D, P are constructed (in order) as follows: $DA \parallel CB$; $DC \parallel BA$; $\tan(CBP) = \tan(PDC)$. The **conclusion**: $\tan(DPA) = \tan(CPB)$.

$A = (0,0)$, $B = (u_1, 0)$, $C = (u_2, u_3)$, $D = (x_1, u_3)$, $P = (x_2, u_4)$.

The **nondegenerate** conditions: Points B, A and C are not collinear; Points D, C and B are not collinear.

Example 35 (C45-2: 1.37s, 12). *Let ABC be an equilateral triangle inscribed in a circle with center O, and let P be any point on the circle. Then the Simson line of P bisects the radius OP.*

Points B, C are arbitrarily chosen. Points A, O, P, D, F, I are constructed (in order) as follows: $AB \equiv BC$; $AB \equiv AC$; $OB \equiv OC$; $OB \equiv OA$; $PO \equiv OB$; D is on line BC; $DP \perp BC$; F is on line AB; $FP \perp AB$; I is on line OP; I is on line DF. The **conclusion**: I is the midpoint of O and P.

$B = (0,0)$, $C = (u_1, 0)$, $A = (x_2, x_1)$, $O = (x_4, x_3)$, $P = (x_5, u_2)$, $D = (x_5, 0)$, $F = (x_7, x_6)$, $I = (x_9, x_8)$.

The **nondegenerate** conditions: Line BC is non-isotropic; Points B, A and C are

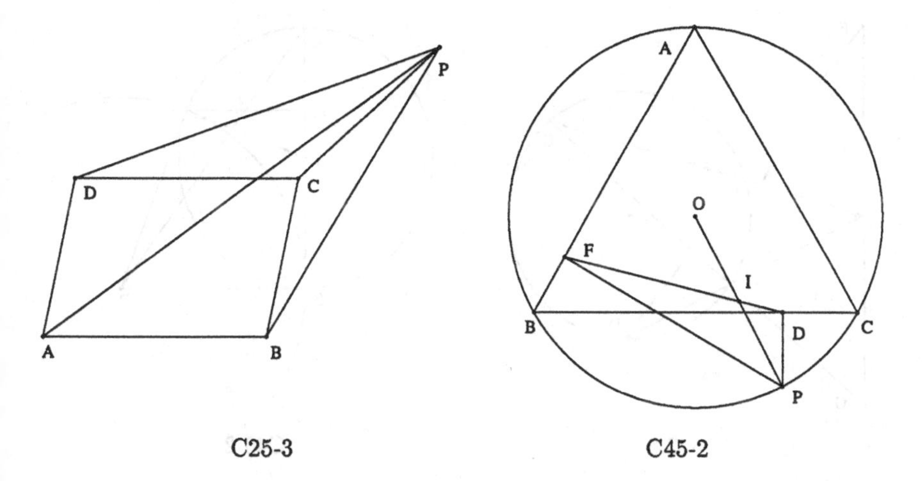

C25-3 C45-2

not collinear; Line BC is non-isotropic; Line AB is non-isotropic; Line DF intersects line OP.

Example 36 (C2.81: 1.12s, 16). *The Butterfly theorem.*

Points E, O, A are arbitrarily chosen. Points B, C, D, F, G are constructed (in order) as follows: $BO \equiv OA$; C is on line AE; $CO \equiv OA$; D is on line BE; $DO \equiv OA$; F is on line AD; $FE \perp EO$; G is on line BC; G is on line EF. The **conclusion**: E is the midpoint of F and G.

$E = (0,0)$, $O = (u_1,0)$, $A = (u_2,u_3)$, $B = (x_1,u_4)$, $C = (x_3,x_2)$, $D = (x_5,x_4)$, $F = (0,x_6)$, $G = (0,x_7)$.

The **nondegenerate** conditions: Line AE is non-isotropic; Line BE is non-isotropic; Line EO is not perpendicular to line AD; Line EF intersects line BC. In addition, the following nondegenerate conditions, which come from reducibility and have been detected by our prover, should be also added: $D \neq B$; $C \neq A$.

Remark. Note that the position of point E can make the diagrams "completely different". Compare the diagram with figure 4 of Part I. In traditional proofs, it often needs two separate proofs for the two different diagrams. However, our single proof is for both of them.

Example 37 (C46-2: 0.58s, 3). *Let PT and PB be two tangents to a circle, AB the diameter through B, and TH the perpendicular from T to AB. Then AP bisects TH.*

Points B, P are arbitrarily chosen. Points O, T, A, H, I are constructed (in order) as follows: $OB \perp BP$; $TO \equiv OB$; $TP \equiv PB$; O is the midpoint of B and A; H is on line AB; $HT \perp AB$; I is on line TH; I is on line AP. The **conclusion**: I is the midpoint of T and H.

$B = (0,0)$, $P = (u_1,0)$, $O = (0,u_2)$, $T = (x_2,x_1)$, $A = (0,x_3)$, $H = (0,x_1)$, $I = (x_4,x_1)$.

The **nondegenerate** conditions: $B \neq P$; Line PO is non-isotropic; Line AB is

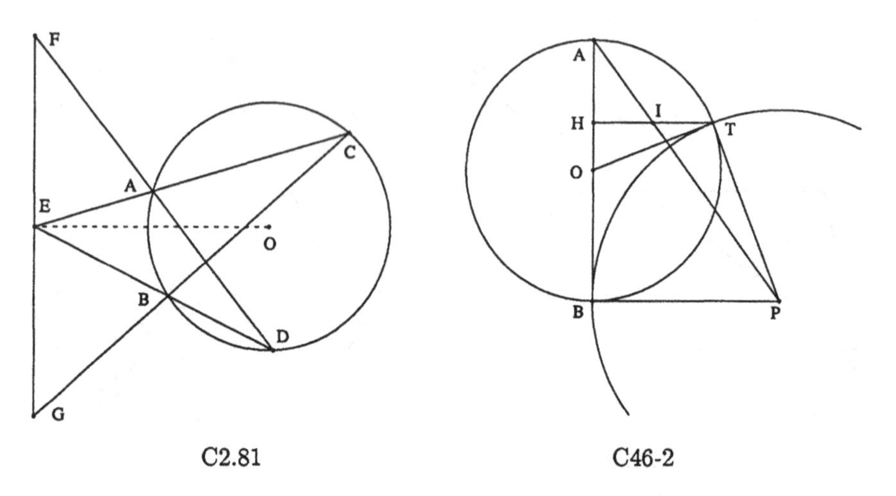

C2.81 C46-2

non-isotropic; Line AP intersects line TH. In addition, the following nondegenerate conditions, which come from reducibility and have been detected by our prover, should be also added: $T \neq B$.

Example 38 (C47-3: 0.9s, 11). *Let incircle (with center I) of $\triangle ABC$ touch the side BC at X and let A_1 be the midpoint of this side. Then the line A_1I (extended) bisectors AX.*

Points B, C, I are arbitrarily chosen. Points A, A_1, X, O are constructed (in order) as follows: $\tan(BCI) = \tan(ICA)$; $\tan(CBI) = \tan(IBA)$; A_1 is the midpoint of B and C; X is on line BC; $XI \perp BC$; O is on line A_1I; O is on line AX. The **conclusion**: O is the midpoint of A and X.

$B = (0,0)$, $C = (u_1,0)$, $I = (u_2,u_3)$, $A = (x_2,x_1)$, $A_1 = (x_3,0)$, $X = (u_2,0)$, $O = (x_5,x_4)$.

The **nondegenerate** conditions: Points I, B and C are not collinear; $\angle BIC$ is not right; Line BC is non-isotropic; Line AX intersects line A_1I.

Remark. Actually, our proof is for the four tritangent circles.

Example 39 (Morley: 15.78s, 567). *The Morley trisector theorem. The points of intersection of the adjacent trisectors of the angles of any triangle are the vertices of an equilateral triangle.*

Points A, B, D are arbitrarily chosen. Points C, F, E are constructed (in order) as follows: $\tan(\angle CBA) - \tan(3\angle DBA) = 0$; $\tan(\angle CAB) - \tan(3\angle DAB) = 0$; x_3 $x_3 - 3 = 0$; $\tan(DAB) = \tan(CAF)$; $\tan(\angle BAD + \angle DBA + \angle ACF) - x_3 = 0$; $\tan(ABD) = \tan(EBC)$; $\tan(ACF) = \tan(ECB)$. The **conclusion**: $\tan(\angle EDF) - x_3 = 0$.

$A = (0,0)$, $B = (u_1,0)$, $D = (u_2,u_3)$, $C = (x_2,x_1)$, $F = (x_5,x_4)$, $E = (x_7,x_6)$.

The **nondegenerate** conditions: $3u_1u_3^5+(6u_1u_2^2-6u_1^2u_2-u_1^3)u_3^3+(3u_1u_2^4-6u_1^2u_2^3+3u_1^3u_2^2)u_3 \neq 0$; $u_3^3 - 3u_2^2u_3 \neq 0$; $(((u_1u_2 - u_1^2)u_3^2 + u_1u_2^3 - u_1^2u_2^2)x_2^2 + ((u_1u_2 - u_1^2)u_3^2 + u_1u_2^3-u_1^2u_2^2)x_1^2)x_3+(u_1u_3^3+u_1u_2^2u_3)x_2^2+(u_1u_3^3+u_1u_2^2u_3)x_1^2 \neq 0$; $u_1u_3x_2-u_1u_2x_1 \neq 0$;

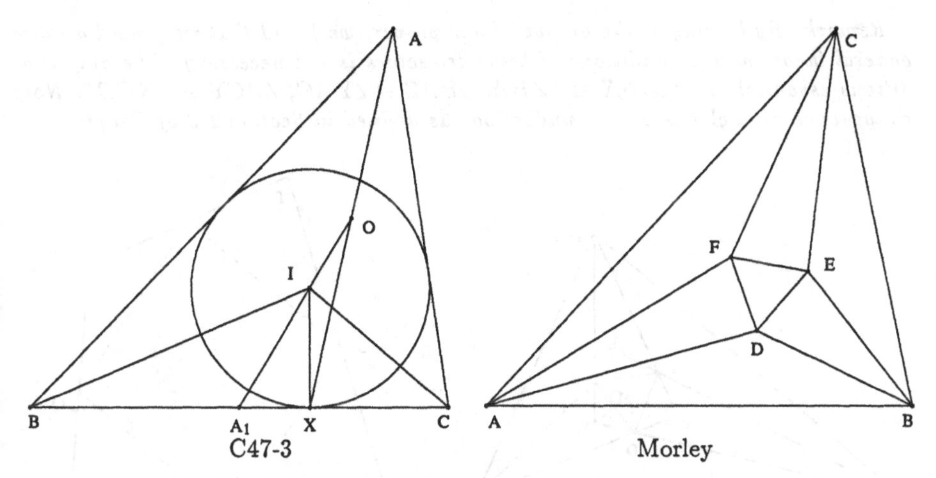

C47-3 Morley

$(-u_1u_3x_2^3+((-u_1u_2+u_1^2)x_1+2u_1^2u_3)x_2^2+(-u_1u_3x_1^2+(2u_1^2u_2-2u_1^3)x_1-u_1^3u_3)x_2+(-u_1u_2+u_1^2)x_1^3+(-u_1^3u_2+u_1^4)x_1)x_5+((u_1u_2-u_1^2)x_2^3+(-u_1u_3x_1-2u_1^2u_2+2u_1^3)x_2^2+((u_1u_2-u_1^2)x_1^2+2u_1^2u_3x_1+u_1^3u_2-u_1^4)x_2-u_1u_3x_1^3-u_1^3u_3x_1)x_4+u_1u_3x_2^4-2u_1^2u_3x_2^3+(2u_1u_3x_1^2+u_1^3u_3)x_2^2-2u_1^2u_3x_1^2x_2+u_1u_3x_1^4+u_1^3u_3x_1^2 \neq 0$.

Remark. The specification here was due to Wu Wen-Tsun. It took Wu some time to come up with this tricky specification. Thus our prover actually proved the most general form of the theorem: among 27 triangles formed by the trisectors, 18 are equilateral. Any further references to this example should mention Wu's work on this example.

Example 40 (C49-1: 218.18s, 4906). *Let the angle trisectors CF and BD (extended) meet at V, AD and CE at W, AF and BE at U. Then the three lines UD, VE, WF are concurrent.*

Points A, B, D are arbitrarily chosen. Points C, F, E, U, V, W, O are constructed (in order) as follows: $\tan(\angle CBA) - \tan(3\angle DBA) = 0$; $\tan(\angle CAB) - \tan(3\angle DAB) = 0$; $x_3\ x_3 - 3 = 0$; $\tan(DAB) = \tan(CAF)$; $\tan(\angle BAD + \angle DBA + \angle ACF) - x_3 = 0$; $\tan(ABD) = \tan(EBC)$; $\tan(ACF) = \tan(ECB)$; U is on line BE; U is on line AF; V is on line BD; V is on line CF; W is on line CE; W is on line AD; O is on line VE; O is on line UD. The **conclusion**: Points F, W and O are collinear.

$A = (0,0)$, $B = (u_1, 0)$, $D = (u_2, u_3)$, $C = (x_2, x_1)$, $F = (x_5, x_4)$, $E = (x_7, x_6)$, $U = (x_9, x_8)$, $V = (x_{11}, x_{10})$, $W = (x_{13}, x_{12})$, $O = (x_{15}, x_{14})$.

The **nondegenerate** conditions: $3u_1u_3^5+(6u_1u_2^2-6u_1^2u_2-u_1^3)u_3^3+(3u_1u_2^4-6u_1^2u_2^3+3u_1^3u_2^2)u_3 \neq 0$; $u_3^3-3u_2^2u_3 \neq 0$; $(((u_1u_2-u_1^2)u_3^2+u_1u_2^3-u_1^2u_2^2)x_2^2+((u_1u_2-u_1^2)u_3^2+u_1u_2^3-u_1^2u_2^2)x_1^2)x_3+(u_1u_3^3+u_1u_2^2u_3)x_2^2+(u_1u_3^3+u_1u_2^2u_3)x_1^2 \neq 0$; $u_1u_3x_2-u_1u_2x_1 \neq 0$; $(-u_1u_3x_2^3+((-u_1u_2+u_1^2)x_1+2u_1^2u_3)x_2^2+(-u_1u_3x_1^2+(2u_1^2u_2-2u_1^3)x_1-u_1^3u_3)x_2+(-u_1u_2+u_1^2)x_1^3+(-u_1^3u_2+u_1^4)x_1)x_5+((u_1u_2-u_1^2)x_2^3+(-u_1u_3x_1-2u_1^2u_2+2u_1^3)x_2^2+((u_1u_2-u_1^2)x_1^2+2u_1^2u_3x_1+u_1^3u_2-u_1^4)x_2-u_1u_3x_1^3-u_1^3u_3x_1)x_4+u_1u_3x_2^4-2u_1^2u_3x_2^3+(2u_1u_3x_1^2+u_1^3u_3)x_2^2-2u_1^2u_3x_1^2x_2+u_1u_3x_1^4+u_1^3u_3x_1^2 \neq 0$; Line AF intersects line BE; Line CF intersects line BD; Line AD intersects line CE; Line UD intersects line VE.

Remark. By looking at the output of our prover, we found that we proved a more general theorem: the conditions of being trisectors is not necessary. The only conditions essential are "$\angle CBX \equiv \angle ZBA$, $\angle BAZ \equiv \angle YAC$, $\angle ACY \equiv \angle XCB$". Note congruence of angles is always understood as defined in Section 1.3 of Chapter 3.

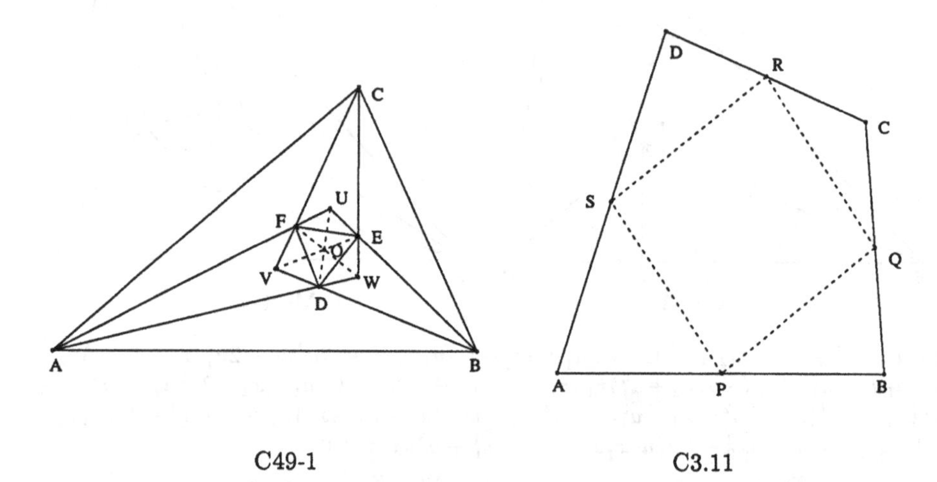

C49-1 C3.11

Example 41 (C3.11: 0.75s, 10). *The figure formed when the midpoints of the sides of a quadrangle are joined in order is a parallelogram, and its area is half that of the quadrangle.*

Points A, B, C, D are arbitrarily chosen. Points Q, R, S, P are constructed (in order) as follows: Q is the midpoint of B and C; R is the midpoint of C and D; S is the midpoint of D and A; P is the midpoint of A and B. The **conclusion**: $\diamond(ABCD) - 2\diamond(QRSP) = 0$.

$A = (0,0)$, $B = (u_1,0)$, $C = (u_2,u_3)$, $D = (u_4,u_5)$, $Q = (x_1,x_2)$, $R = (x_3,x_4)$, $S = (x_5,x_6)$, $P = (x_7,0)$.

The **nondegenerate** conditions: none.

Remark. Here the quadrangle is not necessarily convex and the area is understood as oriented, that is oriented-area(ABCD) = oriented-area(ABC) + oriented-area(CDA). It is easy to check (using our prover or by hand calculation) that this definition is only depends on the orientation of A, B, C, D. Thus oriented-area(ABCD) = oriented-area(DABC), oriented-area(ABCD) = − oriented-area(A DCB), etc.

Example 42 (C3.14: 0.63s, 7). *If a quadrangle $ABCD$ has its opposite sides AD and BC (extended) meeting at W, while X and Y are the midpoints of the diagonals AC and BD, then $(WXY) = 1/4(ABCD)$.*

Points B, C, D, A are arbitrarily chosen. Points X, Y, W are constructed (in order) as follows: X is the midpoint of A and C; Y is the midpoint of B and D; W is on line AD; W is on line BC. The **conclusion**: $\diamond(ABCD) - 4\,\nabla(WXY) = 0$.

$B = (0,0)$, $C = (u_1,0)$, $D = (u_2,u_3)$, $A = (u_4,u_5)$, $X = (x_1,x_2)$, $Y = (x_3,x_4)$,

$W = (x_5, 0)$.

The **nondegenerate** conditions: Line BC intersects line AD.

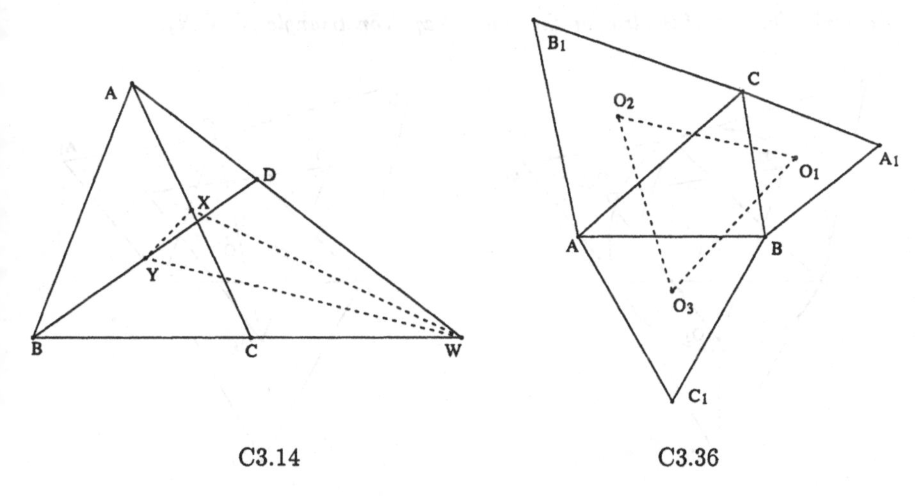

C3.14 C3.36

Example 43 (C3.36: 27.73s, 1003; 3.07s, 20). *(The Napoleon triangle). If equilateral triangles are erected externally (or internally) on the sides of any triangle, their centers form an equilateral triangle.*

Points A, B, C are arbitrarily chosen. Points C_1, A_1, B_1, O_1, O_2, O_3 are constructed (in order) as follows: $C_1A \equiv AB$; $C_1A \equiv C_1B$; $\tan(C_1AB) = \tan(A_1BC)$; $A_1C \equiv A_1B$; $\tan(C_1AB) = \tan(CAB_1)$; $B_1A \equiv B_1C$; $O_1A_1 \equiv O_1B$; $O_1C \equiv O_1B$; $O_2A \equiv O_2B_1$; $O_2A \equiv O_2C$; $O_3C_1 \equiv O_3A$; $O_3A \equiv O_3B$. The **conclusion**: $O_3O_1 \equiv O_3O_2$.

$A = (0,0)$, $B = (u_1, 0)$, $C = (u_2, u_3)$, $C_1 = (x_2, x_1)$, $A_1 = (x_4, x_3)$, $B_1 = (x_6, x_5)$, $O_1 = (x_8, x_7)$, $O_2 = (x_{10}, x_9)$, $O_3 = (x_{12}, x_{11})$.

The **nondegenerate** conditions: Line AB is non-isotropic; $(2u_1u_3^2 + 2u_1u_2^2 - 4u_1^2u_2 + 2u_1^3)x_2 \neq 0$; $(2u_1u_3^2 + 2u_1u_2^2)x_2 \neq 0$; Points C, B and A_1 are not collinear; Points A, C and B_1 are not collinear; Points A, B and C_1 are not collinear.

Remark. Actually, our proof is for both inner and outer Napoleon triangles.

Example 44 (C65-2-a: 3.35s, 70; 2.98s, 20). *In the above notation, the lines A_1O_1, B_1O_2, C_1O_3 all pass through O, the circumcenter of $\triangle ABC$.*

Points A, B, C are arbitrarily chosen. Points O, C_1, A_1, B_1, O_1, O_2, O_3 are constructed (in order) as follows: $OA \equiv OC$; $OA \equiv OB$; $C_1A \equiv AB$; $C_1A \equiv C_1B$; $\tan(C_1AB) = \tan(A_1BC)$; $A_1C \equiv A_1B$; $\tan(C_1AB) = \tan(CAB_1)$; $B_1A \equiv B_1C$; $O_1A_1 \equiv O_1B$; $O_1C \equiv O_1B$; $O_2A \equiv O_2B_1$; $O_2A \equiv O_2C$; $O_3C_1 \equiv O_3A$; $O_3A \equiv O_3B$. The **conclusion**: Points O_1, A_1 and O are collinear.

$A = (0,0)$, $B = (u_1, 0)$, $C = (u_2, u_3)$, $O = (x_2, x_1)$, $C_1 = (x_4, x_3)$, $A_1 = (x_6, x_5)$, $B_1 = (x_8, x_7)$, $O_1 = (x_{10}, x_9)$, $O_2 = (x_{12}, x_{11})$, $O_3 = (x_{14}, x_{13})$.

The **nondegenerate** conditions: Points A, B and C are not collinear; Line AB is non-isotropic; $(2u_1u_3^2+2u_1u_2^2-4u_1^2u_2+2u_1^3)x_4\neq 0$; $(2u_1u_3^2+2u_1u_2^2)x_4\neq 0$; Points C, B and A_1 are not collinear; Points A, C and B_1 are not collinear; Points A, B and C_1 are not collinear.

Remark. Our proof is also for the inner Napoleon triangle $N_1N_2N_3$.

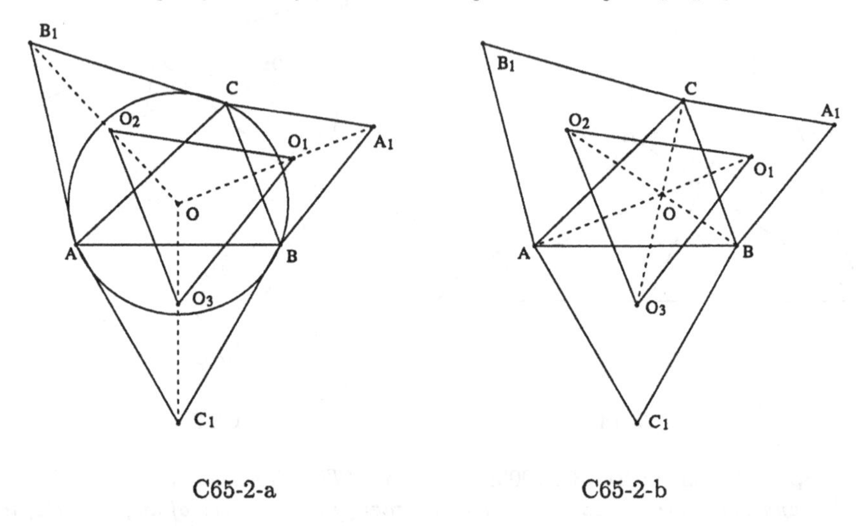

C65-2-a C65-2-b

Example 45 (C65-2-b: 6.37s, 273; 4.28s, 100). *In the above notation, the lines AO_1, BO_2, CO_3 are concurrent.*

Points A, B, C are arbitrarily chosen. Points C_1, A_1, B_1, O_1, O_2, O_3, O are constructed (in order) as follows: $C_1A\equiv AB$; $C_1A\equiv C_1B$; $\tan(C_1AB)=\tan(A_1BC)$; $A_1C\equiv A_1B$; $\tan(C_1AB)=\tan(CAB_1)$; $B_1A\equiv B_1C$; $O_1A_1\equiv O_1B$; $O_1C\equiv O_1B$; $O_2A\equiv O_2B_1$; $O_2A\equiv O_2C$; $O_3C_1\equiv O_3A$; $O_3A\equiv O_3B$; O is on line BO_2; O is on line AO_1. The **conclusion**: Points O, C and O_3 are collinear.

$A=(0,0)$, $B=(u_1,0)$, $C=(u_2,u_3)$, $C_1=(x_2,x_1)$, $A_1=(x_4,x_3)$, $B_1=(x_6,x_5)$, $O_1=(x_8,x_7)$, $O_2=(x_{10},x_9)$, $O_3=(x_{12},x_{11})$, $O=(x_{14},x_{13})$.

The **nondegenerate** conditions: Line AB is non-isotropic; $(2u_1u_3^2+2u_1u_2^2-4u_1^2u_2+2u_1^3)x_2\neq 0$; $(2u_1u_3^2+2u_1u_2^2)x_2\neq 0$; Points C, B and A_1 are not collinear; Points A, C and B_1 are not collinear; Points A, B and C_1 are not collinear; Line AO_1 intersects line BO_2.

Remark. Our proof is also for N_1, N_2 and N_3.

Example 46 (C69-3: 88.13s, 3279; 8.82s, 252). *Let C and F be any points on the respective sides AE and BD of a parallelogram $AEBD$. Let M and N denote the points of intersection of CD and FA and of EF and BC. Let the line MN meet DA at P and EB at Q. Then $AP\equiv QB$.*

Points A, E, B are arbitrarily chosen. Points D, C, F, M, N, P, Q are constructed (in order) as follows: $DA\parallel BE$; $DB\parallel EA$; C is on line AE; F is on line BD; M is on line FA; M is on line CD; N is on line BC; N is on line EF; P is on line DA; P is on line MN; Q is on line EB; Q is on line MN. The **conclusion**: $AP\equiv BQ$.

$A = (0,0)$, $E = (u_1, 0)$, $B = (u_2, u_3)$, $D = (x_1, u_3)$, $C = (u_4, 0)$, $F = (u_5, u_3)$, $M = (x_3, x_2)$, $N = (x_5, x_4)$, $P = (x_7, x_6)$, $Q = (x_9, x_8)$.

The **nondegenerate** conditions: Points E, A and B are not collinear; $A \neq E$; $B \neq D$; Line CD intersects line FA; Line EF intersects line BC; Line MN intersects line DA; Line MN intersects line EB.

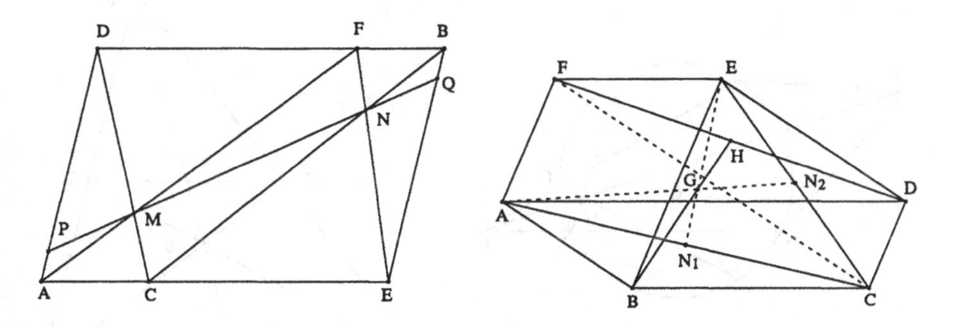

C69-3 C73-1

Example 47 (C73-1: 4.02s, 24). *If a hexagon $ABCDEF$ has two opposite sides BC and EF parallel to the diagonal AD and two opposite sides CD and FA parallel to the diagonal BE, while the remaining sides DE and AB also are parallel, then the third diagonal CF is parallel to AB, and the centroids of $\triangle ACE$ and $\triangle BDF$ coincide.*

Points B, C, E are arbitrarily chosen. Points D, A, F, N_1, N_2, G, H are constructed (in order) as follows: $DC \parallel BE$; $AB \parallel ED$; $AD \parallel CB$; $FA \parallel DC$; $FE \parallel CB$; N_1 is the midpoint of A and C; N_2 is the midpoint of C and E; G is on line AN_2; G is on line EN_1; H is on line DF; H is on line GB. The **conclusions**: (1) $CF \parallel AB$; (2) H is the midpoint of D and F.

$B = (0,0)$, $C = (u_1, 0)$, $E = (u_2, u_3)$, $D = (x_1, u_4)$, $A = (x_2, u_4)$, $F = (x_3, u_3)$, $N_1 = (x_4, x_5)$, $N_2 = (x_6, x_7)$, $G = (x_9, x_8)$, $H = (x_{11}, x_{10})$.

The **nondegenerate** conditions: $B \neq E$; Line CB intersects line ED; Points C, B and D are not collinear; Line EN_1 intersects line AN_2; Line GB intersects line DF.

Example 48 (C76-1: 24.32s, 894). *If five of six vertices of a hexagon lie on a circle, and the three pairs of opposite sides meet at three collinear points, then the sixth vertex lies on the same circle.*

Points A, B, C are arbitrarily chosen. Points O, D, E, P, S, Q, F are constructed (in order) as follows: $OA \equiv OC$; $OA \equiv OB$; $DO \equiv OA$; $EO \equiv OA$; P is on line AB; S is on line EA; S is on line CD; Q is on line BC; Q is on line SP; F is on line QE; F is on line PD. The **conclusion**: $OA \equiv OF$.

$A = (0, 0)$, $B = (u_1, 0)$, $C = (u_2, u_3)$, $O = (x_2, x_1)$, $D = (x_3, u_4)$, $E = (x_4, u_5)$, $P = (u_6, 0)$, $S = (x_6, x_5)$, $Q = (x_8, x_7)$, $F = (x_{10}, x_9)$.

The **nondegenerate** conditions: Points A, B and C are not collinear; $A \neq B$; Line CD intersects line EA; Line SP intersects line BC; Line PD intersects line QE.

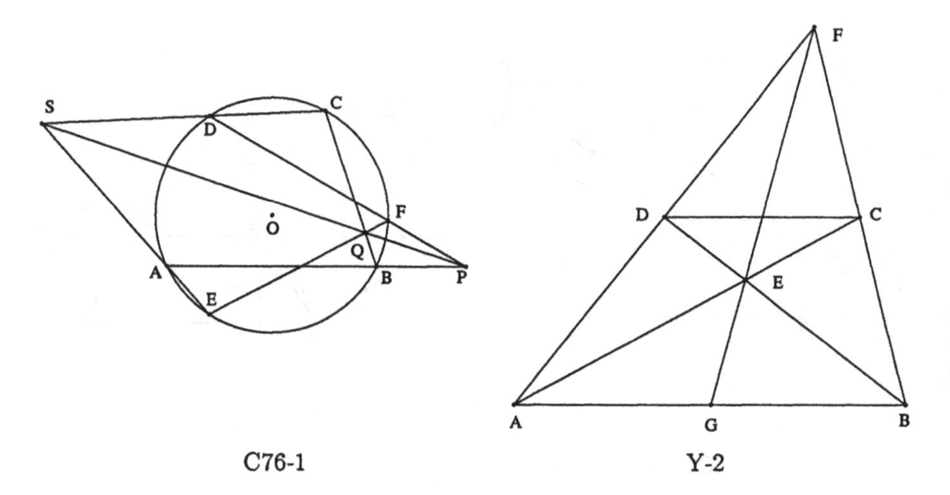

C76-1 Y-2

Example 49 (Y-2: 0.66s, 5). *Prove that the line joining the point of intersection of the extensions of the nonparallel sides of a trapezoid to the point of intersection of its diagonals bisects the base of the trapezoid.*

Points A, B, C are arbitrarily chosen. Points D, E, F, G are constructed (in order) as follows: $DC \parallel BA$; E is on line BD; E is on line AC; F is on line BC; F is on line AD; G is on line AB; G is on line EF. The **conclusion**: G is the midpoint of A and B.

$A = (0, 0)$, $B = (u_1, 0)$, $C = (u_2, u_3)$, $D = (u_4, u_3)$, $E = (x_2, x_1)$, $F = (x_4, x_3)$, $G = (x_5, 0)$.

The **nondegenerate** conditions: $B \neq A$; Line AC intersects line BD; Line AD intersects line BC; Line EF intersects line AB.

Example 50 (Y-4-a: 1.83s, 23). *Let M, N, and P be three points on the sides AB, BC and AC of a triangle ABC such that $AM/MB = BN/NC = CP/PA$. Show that the point of intersection of the medians of $\triangle MNP$ coincides with the point of intersection of the medians of $\triangle ABC$.*

Points A, B, C are arbitrarily chosen. Points M, N, P, E, F, G, L are constructed (in order) as follows: M is on line AB; N is on line BC; $\frac{\overline{AM}}{\overline{MB}} - \frac{\overline{BN}}{\overline{NC}} = 0$; P is on line CA; $\frac{\overline{AM}}{\overline{MB}} - \frac{\overline{CP}}{\overline{PA}} = 0$; E is the midpoint of A and C; F is the midpoint of A and B; G is on line BE; G is on line CF; L is on line NP; L is on line MG. The **conclusion**: L is the midpoint of N and P.

$A = (0, 0)$, $B = (u_1, 0)$, $C = (u_2, u_3)$, $M = (u_4, 0)$, $N = (x_2, x_1)$, $P = (x_4, x_3)$,

$E=(x_5,x_6)$, $F=(x_7,0)$, $G=(x_9,x_8)$, $L=(x_{11},x_{10})$.

The **nondegenerate** conditions: $A \neq B$; $u_1u_2 - u_1^2 \neq 0$; $-u_1 \neq 0$; $-u_1u_2 \neq 0$; $-u_1 \neq 0$; Line CF intersects line BE; Line MG intersects line NP.

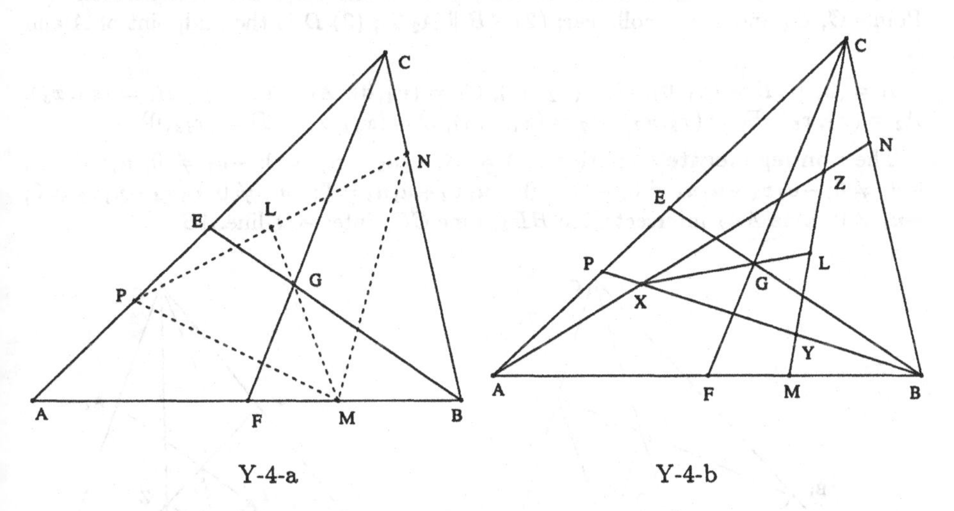

Y-4-a Y-4-b

Example 51 (Y-4-b: 10.85s, 277; 6.38s, 76). *Let M, N, and P be the same as in Y-4-A. Show that the point of intersection of the medians of the triangle formed by lines AN, BP and CM coincides with the point of intersection of the medians of $\triangle ABC$.*

Points A, B, C are arbitrarily chosen. Points M, N, P, E, F, G, X, Y, Z, L are constructed (in order) as follows: M is on line AB; N is on line BC; $\frac{\overline{AM}}{\overline{MB}} - \frac{\overline{BN}}{\overline{NC}} = 0$; P is on line CA; $\frac{\overline{AM}}{\overline{MB}} - \frac{\overline{CP}}{\overline{PA}} = 0$; E is the midpoint of A and C; F is the midpoint of A and B; G is on line BE; G is on line CF; X is on line BP; X is on line AN; Y is on line CM; Y is on line BP; Z is on line AN; Z is on line CM; L is on line YZ; L is on line XG. The **conclusion**: L is the midpoint of Y and Z.

$A=(0,0)$, $B=(u_1,0)$, $C=(u_2,u_3)$, $M=(u_4,0)$, $N=(x_2,x_1)$, $P=(x_4,x_3)$, $E=(x_5,x_6)$, $F=(x_7,0)$, $G=(x_9,x_8)$, $X=(x_{11},x_{10})$, $Y=(x_{13},x_{12})$, $Z=(x_{15},x_{14})$, $L=(x_{17},x_{16})$.

The **nondegenerate** conditions: $A \neq B$; $u_1u_2 - u_1^2 \neq 0$; $-u_1 \neq 0$; $-u_1u_2 \neq 0$; $-u_1 \neq 0$; Line CF intersects line BE; Line AN intersects line BP; Line BP intersects line CM; Line CM intersects line AN; Line XG intersects line YZ.

Example 52 (Y-6: 4.97s, 66; 4.95s, 16). *Let A_1, B_1, C_1 be points on the sides BC, CA, AB of a triangle ABC such that $BA_1/A_1C = CB_1/B_1A = AC_1/C_1B = k$. Furthermore, let A_2, B_2, C_2 be points on the sides B_1C_1, C_1A_1, A_1C_1 of a triangle $A_1B_1C_1$ such that $C_1A_2/A_2B_1 = A_1B_2/B_2C_1 = B_1C_2/C_2A_1 = k$. Show that triangles ABC and $A_2B_2C_2$ are similar.*

Points A, B, C are arbitrarily chosen. Points C_1, A_1, B_1, A_2, B_2, C_2, O, D are constructed (in order) as follows: C_1 is on line AB; A_1 is on line BC; $\frac{\overline{AC_1}}{\overline{C_1B}} - \frac{\overline{BA_1}}{\overline{A_1C}}$

$= 0$; B_1 is on line AC; $\frac{\overline{AC_1}}{\overline{C_1B}} - \frac{\overline{CB_1}}{\overline{B_1A}} = 0$; A_2 is on line B_1C_1; $\frac{\overline{AC_1}}{\overline{C_1B}} - \frac{\overline{C_1A_2}}{\overline{A_2B_1}} = 0$; B_2 is on line A_1C_1; $\frac{\overline{AC_1}}{\overline{C_1B}} - \frac{\overline{A_1B_2}}{\overline{B_2C_1}} = 0$; C_2 is on line A_1B_1; $\frac{\overline{AC_1}}{\overline{C_1B}} - \frac{\overline{B_1C_2}}{\overline{C_2A_1}} = 0$; O is on line BB_2; O is on line AA_2; D is on line AB; D is on line CC_2. The **conclusions**: (1) Points C, C_2 and O are collinear; (2) $AB \parallel A_2B_2$; (3) D is the midpoint of A and B.

$A = (0,0)$, $B = (u_1,0)$, $C = (u_2,u_3)$, $C_1 = (u_4,0)$, $A_1 = (x_2,x_1)$, $B_1 = (x_4,x_3)$, $A_2 = (x_6,x_5)$, $B_2 = (x_8,x_7)$, $C_2 = (x_{10},x_9)$, $O = (x_{12},x_{11})$, $D = (x_{13},0)$.

The **nondegenerate** conditions: $A \neq B$; $u_1u_2 - u_1^2 \neq 0$; $-u_1 \neq 0$; $u_1u_2 \neq 0$; $-u_1 \neq 0$; $-u_1x_4 + u_1u_4 \neq 0$; $-u_1 \neq 0$; $-u_1x_2 + u_1u_4 \neq 0$; $-u_1 \neq 0$; $u_1x_4 - u_1x_2 \neq 0$; $-u_1 \neq 0$; Line AA_2 intersects line BB_2; Line CC_2 intersects line AB.

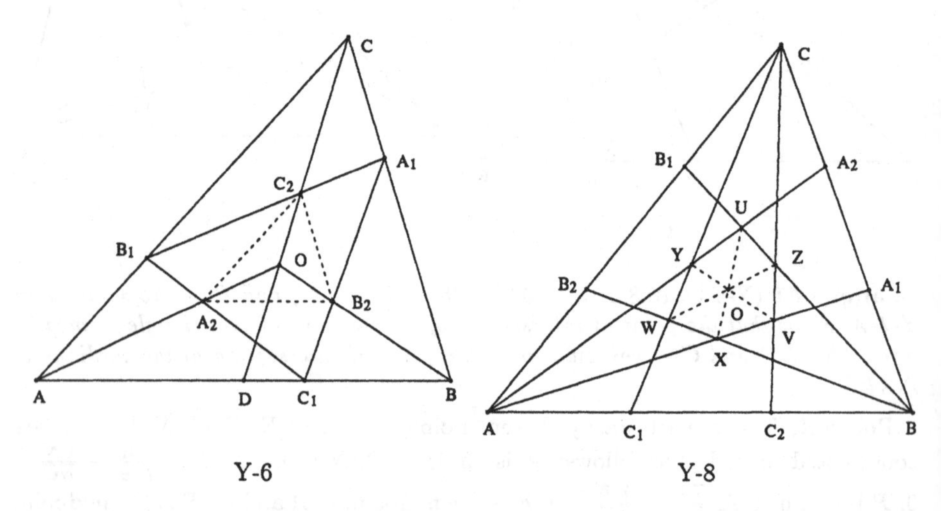

Y-6 Y-8

Example 53 (Y-8: 3.83s, 22). *Through each of the vertices of a triangle ABC we draw two lines dividing the opposite side into three equal parts. These six lines determine a hexagon. Prove that the diagonals joining opposite vertices of this hexagon meet in a point.*

Points A, B, C are arbitrarily chosen. Points A_1, A_2, B_1, B_2, C_1, C_2, X, Y, Z, U, V, W, O are constructed (in order) as follows: B, A_1, C are collinear, and $BA_1/A_1C = 1/2$; B, A_2, C are collinear, and $BA_2/A_2C = 2/1$; C, B_1, A are collinear, and $CB_1/B_1A = 1/2$; C, B_2, A are collinear, and $CB_2/B_2A = 2/1$; A, C_1, B are collinear, and $AC_1/C_1B = 1/2$; A, C_2, B are collinear, and $AC_2/C_2B = 2/1$; X is on line AA_1; X is on line BB_2; Y is on line CC_1; Y is on line AA_2; Z is on line BB_1; Z is on line CC_2; U is on line AA_2; U is on line BB_1; V is on line CC_2; V is on line AA_1; W is on line BB_2; W is on line CC_1; O is on line YV; O is on line UX. The **conclusion**: Points W, Z and O are collinear.

$A = (0,0)$, $B = (u_1,0)$, $C = (u_2,u_3)$, $A_1 = (x_1,x_2)$, $A_2 = (x_3,x_4)$, $B_1 = (x_5,x_6)$, $B_2 = (x_7,x_8)$, $C_1 = (x_9,0)$, $C_2 = (x_{10},0)$, $X = (x_{12},x_{11})$, $Y = (x_{14},x_{13})$, $Z = (x_{16},x_{15})$, $U = (x_{18},x_{17})$, $V = (x_{20},x_{19})$, $W = (x_{22},x_{21})$, $O = (x_{24},x_{23})$.

The **nondegenerate** conditions: Line BB_2 intersects line AA_1; Line AA_2 intersects line CC_1; Line CC_2 intersects line BB_1; Line BB_1 intersects line AA_2; Line AA_1 intersects line CC_2; Line CC_1 intersects line BB_2; Line UX intersects line YV.

Example 54 (Y-8-c: 58.1s, 1286; 16.25s, 219). *This is our generalization of Y-8: we only need conditions that A_1 and A_2 are isotomic; B_1 and B_2 are isotomic; C_1 and C_2 are isotomic. (See C49-1.)*

Points A, B, C are arbitrarily chosen. Points D, E, F, A_1, A_2, B_1, B_2, C_1, C_2, X, Y, Z, U, V, W, O are constructed (in order) as follows: D is the midpoint of B and C; E is the midpoint of A and C; F is the midpoint of A and B; A_1 is on line BC; D is the midpoint of A_1 and A_2; B_1 is on line CA; E is the midpoint of B_1 and B_2; C_1 is on line AB; F is the midpoint of C_1 and C_2; X is on line AA_1; X is on line BB_2; Y is on line CC_1; Y is on line AA_2; Z is on line BB_1; Z is on line CC_2; U is on line AA_2; U is on line BB_1; V is on line CC_2; V is on line AA_1; W is on line BB_2; W is on line CC_1; O is on line YV; O is on line UX. The **conclusion**: Points W, Z and O are collinear.

$A = (0,0)$, $B = (u_1,0)$, $C = (u_2,u_3)$, $D = (x_1,x_2)$, $E = (x_3,x_4)$, $F = (x_5,0)$, $A_1 = (x_6,u_4)$, $A_2 = (x_7,x_8)$, $B_1 = (x_9,u_5)$, $B_2 = (x_{10},x_{11})$, $C_1 = (u_6,0)$, $C_2 = (x_{12},0)$, $X = (x_{14},x_{13})$, $Y = (x_{16},x_{15})$, $Z = (x_{18},x_{17})$, $U = (x_{20},x_{19})$, $V = (x_{22},x_{21})$, $W = (x_{24},x_{23})$, $O = (x_{26},x_{25})$.

The **nondegenerate** conditions: $B \neq C$; $C \neq A$; $A \neq B$; Line BB_2 intersects line AA_1; Line AA_2 intersects line CC_1; Line CC_2 intersects line BB_1; Line BB_1 intersects line AA_2; Line AA_1 intersects line CC_2; Line CC_1 intersects line BB_2; Line UX intersects line YV.

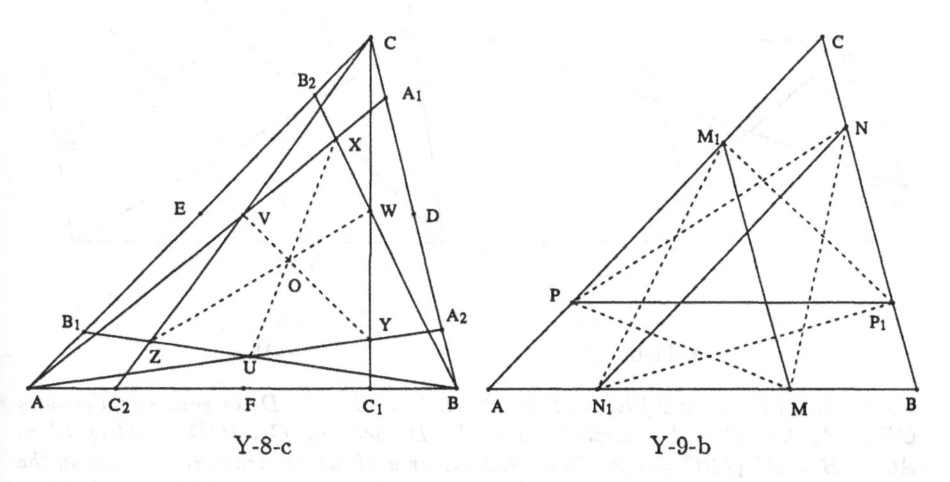

Y-8-c Y-9-b

Example 55 (Y-9-b: 0.97s, 10). *Let M, N, P be points on the sides AB, BC and AC of a triangle ABC. Show that if M_1, N_1 and P_1 are points on sides AC, BA, and BC of a triangle ABC such that $MM_1 \parallel BC$, $NN_1 \parallel CA$ and $PP_1 \parallel AB$, then triangles MNP and $M_1N_1P_1$ have equal areas.*

Points A, B, C are arbitrarily chosen. Points M, N, P, M_1, N_1, P_1 are constructed (in order) as follows: M is on line AB; N is on line BC; P is on line AC; M_1 is on line AC; $M_1M \parallel BC$; N_1 is on line AB; $N_1N \parallel AC$; P_1 is on line BC; $P_1P \parallel AB$. The **conclusion**: $\nabla(MNP) - \nabla(M_1N_1P_1) = 0$.

$A = (0,0)$, $B = (u_1,0)$, $C = (u_2,u_3)$, $M = (u_4,0)$, $N = (x_1,u_5)$, $P = (x_2,u_6)$, $M_1 = (x_4,x_3)$, $N_1 = (x_5,0)$, $P_1 = (x_6,u_6)$.

The **nondegenerate** conditions: $A \neq B$; $B \neq C$; $A \neq C$; Points B, C and A are not collinear; Points A, C and B are not collinear; Points A, B and C are not collinear.

Example 56 (Y-11-b: 0.85s, 7). *Let l be a line passing through the vertex of M of a parallelogram $MNPQ$ and intersecting the lines NP, PQ, NQ in points R, S, T. Show that $1/MR + 1/MS = 1/MT$.*

Points M, N, P are arbitrarily chosen. Points Q, R, S, T are constructed (in order) as follows: $QM \parallel PN$; $QP \parallel NM$; R is on line NP; S is on line MR; S is on line PQ; T is on line MR; T is on line NQ. The **conclusion**: $1/\overline{MR} + 1/\overline{MS} - 1/\overline{MT} = 0$.

$M = (0,0)$, $N = (u_1,0)$, $P = (u_2,u_3)$, $Q = (x_1,u_3)$, $R = (x_2,u_4)$, $S = (x_3,u_3)$, $T = (x_5,x_4)$.

The **nondegenerate** conditions: Points N, M and P are not collinear; $N \neq P$; Line PQ intersects line MR; Line NQ intersects line MR.

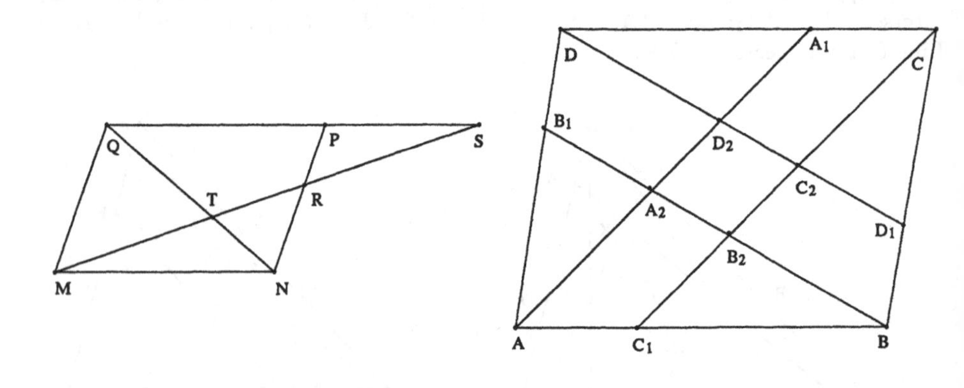

Y-11-b Y-15-a

Example 57 (Y-15-a: 2.12s, 13; 1.9s, 13). *Let A_1, B_1, C_1, D_1 be points on the sides CD, DA, AB, BC of a parallelogram $ABCD$ such that $CA_1/CD = DB_1/DA = AC_1/AB = BD_1/BC = 1/3$. Show that the area of the quadrilateral formed by the lines AA_1 BB_1, CC_1, DD_1 is one thirteenth of the area of parallelogram $ABCD$.*

Points A, B, C are arbitrarily chosen. Points D, A_1, B_1, C_1, D_1, A_2, B_2, C_2, D_2 are constructed (in order) as follows: $DA \parallel CB$; $DC \parallel BA$; C, A_1, D are collinear, and $CA_1/A_1D = 1/2$; D, B_1, A are collinear, and $DB_1/B_1A = 1/2$; A, C_1, B are collinear, and $AC_1/C_1B = 1/2$; B, D_1, C are collinear, and $BD_1/D_1C = 1/2$; A_2 is on line BB_1; A_2 is on line AA_1; B_2 is on line CC_1; B_2 is on line BB_1; C_2 is on line

DD_1; C_2 is on line CC_1; D_2 is on line AA_1; D_2 is on line DD_1. The **conclusion**: $13\ \Diamond(A_2B_2C_2D_2) - \Diamond(ABCD) = 0$.

$A = (0,0)$, $B = (u_1,0)$, $C = (u_2,u_3)$, $D = (x_1,u_3)$, $A_1 = (x_2,u_3)$, $B_1 = (x_3,x_4)$, $C_1 = (x_5,0)$, $D_1 = (x_6,x_7)$, $A_2 = (x_9,x_8)$, $B_2 = (x_{11},x_{10})$, $C_2 = (x_{13},x_{12})$, $D_2 = (x_{15},x_{14})$.

The **nondegenerate** conditions: Points B, A and C are not collinear; Line AA_1 intersects line BB_1; Line BB_1 intersects line CC_1; Line CC_1 intersects line DD_1; Line DD_1 intersects line AA_1.

Example 58 (Y-15-b: 1.18s, 9). *Let A_1, B_1, C_1 be points on the sides BC, CA, AB of a triangle ABC such that $BA_1/BC = CB_1/CA = AC_1/AB = 1/3$. Show that the area of the triangle determined by lines AA_1, BB_1 and CC_1 is one seventh of the area of triangle ABC.*

Points A, B, C are arbitrarily chosen. Points A_1, B_1, C_1, A_2, B_2, C_2 are constructed (in order) as follows: B, A_1, C are collinear, and $BA_1/A_1C = 1/2$; C, B_1, A are collinear, and $CB_1/B_1A = 1/2$; A, C_1, B are collinear, and $AC_1/C_1B = 1/2$; A_2 is on line BB_1; A_2 is on line AA_1; B_2 is on line CC_1; B_2 is on line BB_1; C_2 is on line AA_1; C_2 is on line CC_1. The **conclusion**: $7\ \nabla(A_2B_2C_2) - \nabla(ABC) = 0$.

$A = (0,0)$, $B = (u_1,0)$, $C = (u_2,u_3)$, $A_1 = (x_1,x_2)$, $B_1 = (x_3,x_4)$, $C_1 = (x_5,0)$, $A_2 = (x_7,x_6)$, $B_2 = (x_9,x_8)$, $C_2 = (x_{11},x_{10})$.

The **nondegenerate** conditions: Line AA_1 intersects line BB_1; Line BB_1 intersects line CC_1; Line CC_1 intersects line AA_1.

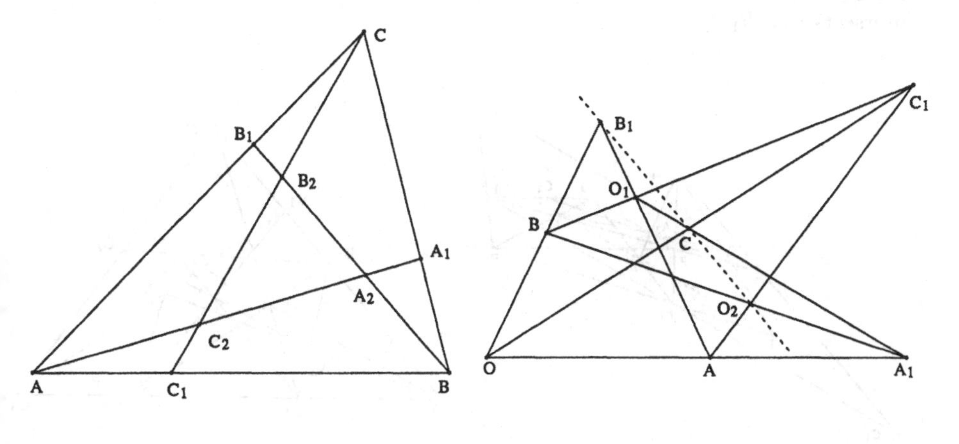

Y-15-b Y-25

Example 59 (Y-25: 1.53s, 40). *Two doubly perspective triangles are in fact triply perspective.*

Points O, A, B, C, O_1 are arbitrarily chosen. Points A_1, B_1, C_1, O_2 are constructed (in order) as follows: A_1 is on line CO_1; A_1 is on line AO; B_1 is on line AO_1; B_1 is on line BO; C_1 is on line BO_1; C_1 is on line CO; O_2 is on line BA_1; O_2 is on line AC_1. The **conclusion**: Points C, B_1 and O_2 are collinear.

$O = (0,0)$, $A = (u_1,0)$, $B = (u_2,u_3)$, $C = (u_4,u_5)$, $O_1 = (u_6,u_7)$, $A_1 = (x_1,0)$, $B_1 = (x_3,x_2)$, $C_1 = (x_5,x_4)$, $O_2 = (x_7,x_6)$.

The **nondegenerate** conditions: Line AO intersects line CO_1; Line BO intersects line AO_1; Line CO intersects line BO_1; Line AC_1 intersects line BA_1.

Example 60 (Y-27: 5.92s, 105). *Let three triangles ABC, $A_1B_1C_1$, $A_2B_2C_2$ be given such that lines AB, A_1B_1, A_2B_2 intersect in a point P, lines AC, A_1C_1, A_2C_2 intersect in a point Q, lines BC, B_1C_1, B_2C_2 intersect in a point R, and P, Q, R are collinear. In view of Desargues' theorem, the lines in each of the triads AA_1, BB_1, CC_1; AA_2, BB_2, CC_2; A_1A_2, B_1B_2, C_1C_2; intersect in a point. Prove that these three points are collinear.*

Points A, B, C, A_1, B_1, B_2 are arbitrarily chosen. Points P, Q, R, C_1, A_2, C_2, I, J, K are constructed (in order) as follows: P is on line A_1B_1; P is on line AB; Q is on line AC; R is on line PQ; R is on line BC; C_1 is on line B_1R; C_1 is on line A_1Q; A_2 is on line B_2P; C_2 is on line B_2R; C_2 is on line A_2Q; I is on line BB_1; I is on line AA_1; J is on line BB_2; J is on line AA_2; K is on line B_1B_2; K is on line A_1A_2. The **conclusion**: Points I, J and K are collinear.

$A = (0,0)$, $B = (u_1,0)$, $C = (u_2,u_3)$, $A_1 = (u_4,u_5)$, $B_1 = (u_6,u_7)$, $B_2 = (u_8,u_9)$, $P = (x_1,0)$, $Q = (x_2,u_{10})$, $R = (x_4,x_3)$, $C_1 = (x_6,x_5)$, $A_2 = (x_7,u_{11})$, $C_2 = (x_9,x_8)$, $I = (x_{11},x_{10})$, $J = (x_{13},x_{12})$, $K = (x_{15},x_{14})$.

The **nondegenerate** conditions: Line AB intersects line A_1B_1; $A \neq C$; Line BC intersects line PQ; Line A_1Q intersects line B_1R; $B_2 \neq P$; Line A_2Q intersects line B_2R; Line AA_1 intersects line BB_1; Line AA_2 intersects line BB_2; Line A_1A_2 intersects line B_1B_2.

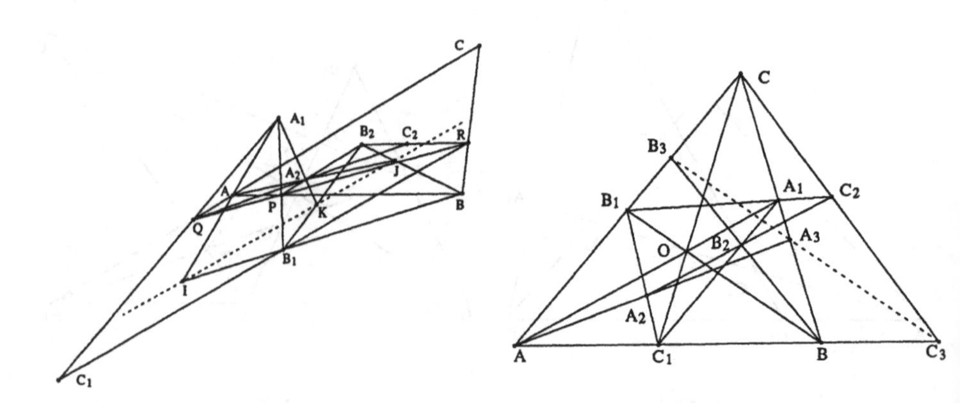

Y-27 Y-38-b

Example 61 (Y-38-b: 18.82s, 676). *Let O be a point in the plane of a triangle ABC, and let A_1, B_1, C_1 be the points of intersection of the lines AO, BO, CO with the sides of the triangle opposite A, B, C. Prove that if the points A_2, B_2, C_2 on sides B_1C_1, C_1A_1, A_1B_1 of $\triangle A_1B_1C_1$ are collinear, then the points of intersection of the liens AA_2, BB_2, CC_2 with the opposite sides of $\triangle ABC$ are collinear.*

Points A, B, C, O are arbitrarily chosen. Points A_1, B_1, C_1, A_2, B_2, C_2, A_3, B_3, C_3 are constructed (in order) as follows: A_1 is on line BC; A_1 is on line AO; B_1 is on line AC; B_1 is on line BO; C_1 is on line AB; C_1 is on line CO; A_2 is on line B_1C_1; B_2 is on line A_1C_1; C_2 is on line A_2B_2; C_2 is on line A_1B_1; A_3 is on line BC; A_3 is on line AA_2; B_3 is on line AC; B_3 is on line BB_2; C_3 is on line AB; C_3 is on line CC_2. The **conclusion**: Points A_3, B_3 and C_3 are collinear.

$A = (0,0)$, $B = (u_1,0)$, $C = (u_2,u_3)$, $O = (u_4,u_5)$, $A_1 = (x_2,x_1)$, $B_1 = (x_4,x_3)$, $C_1 = (x_5,0)$, $A_2 = (x_6,u_6)$, $B_2 = (x_7,u_7)$, $C_2 = (x_9,x_8)$, $A_3 = (x_{11},x_{10})$, $B_3 = (x_{13},x_{12})$, $C_3 = (x_{14},0)$.

The **nondegenerate** conditions: Line AO intersects line BC; Line BO intersects line AC; Line CO intersects line AB; $B_1 \neq C_1$; $A_1 \neq C_1$; Line A_1B_1 intersects line A_2B_2; Line AA_2 intersects line BC; Line BB_2 intersects line AC; Line CC_2 intersects line AB.

Example 62 (Y-page-83: 8.93s, 238; 8.78s, 69).

Points A, D, B, C are arbitrarily chosen. Points L, K, F, E, O, T, X, Y, Z are constructed (in order) as follows: L is on line AD; L is on line BC; K is on line DC; K is on line AB; F is on line BD; F is on line LK; E is on line CA; E is on line LK; O is on line BD; O is on line AC; T is on line EB; T is on line FA; X is on line EB; X is on line FC; Y is on line ED; Y is on line FC; Z is on line ED; Z is on line FA. The **conclusion**: Points T, Y and O are collinear.

$A = (0,0)$, $D = (u_1,0)$, $B = (u_2,u_3)$, $C = (u_4,u_5)$, $L = (x_1,0)$, $K = (x_3,x_2)$, $F = (x_5,x_4)$, $E = (x_7,x_6)$, $O = (x_9,x_8)$, $T = (x_{11},x_{10})$, $X = (x_{13},x_{12})$, $Y = (x_{15},x_{14})$, $Z = (x_{17},x_{16})$.

The **nondegenerate** conditions: Line BC intersects line AD; Line AB intersects line DC; Line LK intersects line BD; Line LK intersects line CA; Line AC intersects line BD; Line FA intersects line EB; Line FC intersects line EB; Line FC intersects line ED; Line FA intersects line ED.

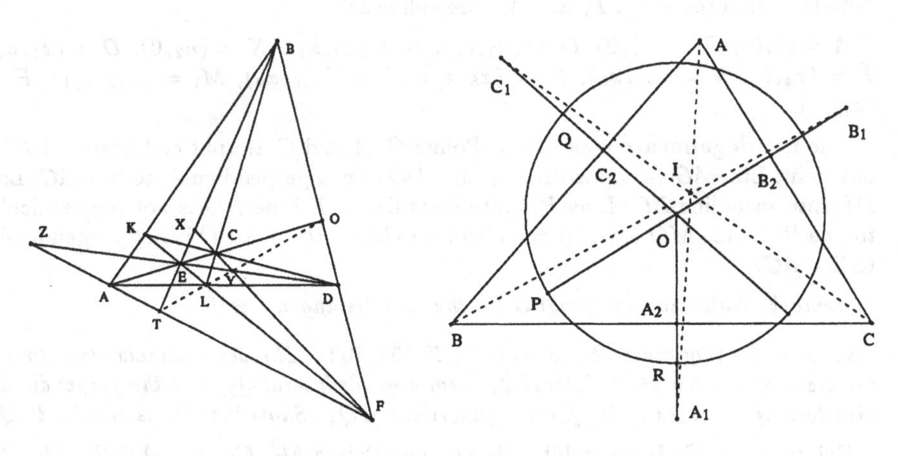

Y-page-83 Y-57

Example 63 (Y-57: 87.07s, 2408; 29.78s, 539). *Show that the lines joining the vertices of a triangle ABC with the poles A_1, B_1, C_1 of the opposite sides of the triangle relative to some circle are concurrent.*

Points B, C, A, O are arbitrarily chosen. Points A_2, C_2, B_2, R, P, Q, A_1, B_1, C_1, I are constructed (in order) as follows: A_2 is on line CB; $A_2O \perp CB$; C_2 is on line AB; $C_2O \perp AB$; B_2 is on line AC; $B_2O \perp AC$; R is on line OA_2; $PO \equiv OR$; $PO \perp AC$; $QO \equiv OR$; $QO \perp AB$; A_2 and A_1 are inversive wrpt circle (O, OR); A_1 is on line OA_2; B_2 and B_1 are inversive wrpt circle (O, OP); B_1 is on line OP; C_2 and C_1 are inversive wrpt circle (O, OQ); C_1 is on line OQ; I is on line BB_1; I is on line AA_1. The **conclusion**: Points C, C_1 and I are collinear.

$B = (0,0)$, $C = (u_1,0)$, $A = (u_2,u_3)$, $O = (u_4,u_5)$, $A_2 = (u_4,0)$, $C_2 = (x_2,x_1)$, $B_2 = (x_4,x_3)$, $R = (u_4,u_6)$, $P = (x_6,x_5)$, $Q = (x_8,x_7)$, $A_1 = (u_4,x_9)$, $B_1 = (x_{11},x_{10})$, $C_1 = (x_{13},x_{12})$, $I = (x_{15},x_{14})$.

The **nondegenerate** conditions: Line CB is non-isotropic; Line AB is non-isotropic; Line AC is non-isotropic; $O \neq A_2$; Line AC is non-isotropic; Line AB is non-isotropic; $u_5 \neq 0$; $O \neq P$; $B_2 \neq O$; $O \neq Q$; $C_2 \neq O$; Line AA_1 intersects line BB_1.

Example 64 (Y-68: 7.9s, 260). *Let l be a line tangent to the circle (O) inscribed in the triangle ABC, and let M, N, P be the points of intersection of l with the sides of that triangle. At the center O of the circle erect perpendiculars to the lines OM, ON, OP, and denote their points of intersection with the corresponding sides of the triangle by M_1, N_1, P_1. Prove that the points M_1, N_1, P_1 lie on a line.*

Points A, C, O are arbitrarily chosen. Points B, X, D, P, N, M, N_1, M_1, P_1 are constructed (in order) as follows: $\tan(ACO) = \tan(OCB)$; $\tan(CAO) = \tan(OAB)$; X is on line AC; $XO \perp AC$; $DO \equiv OX$; P is on line AC; $PD \perp DO$; N is on line BC; N is on line DP; M is on line AB; M is on line PD; N_1 is on line BC; $N_1O \perp NO$; M_1 is on line AB; $M_1O \perp MO$; P_1 is on line AC; $P_1O \perp PO$. The **conclusion**: Points N_1, P_1 and M_1 are collinear.

$A = (0,0)$, $C = (u_1,0)$, $O = (u_2,u_3)$, $B = (x_2,x_1)$, $X = (u_2,0)$, $D = (x_3,u_4)$, $P = (x_4,0)$, $N = (x_6,x_5)$, $M = (x_8,x_7)$, $N_1 = (x_{10},x_9)$, $M_1 = (x_{12},x_{11})$, $P_1 = (x_{13},0)$.

The **nondegenerate** conditions: Points O, A and C are not collinear; $\angle AOC$ is not right; Line AC is non-isotropic; Line DO is not perpendicular to line AC; Line DP intersects line BC; Line PD intersects line AB; Line NO is not perpendicular to line BC; Line MO is not perpendicular to line AB; Line PO is not perpendicular to line AC.

Remark. Note that our proof is for the four tritangent circles.

Example 65 (Ogilvy-1: 247.88s, 4553; 73.55s, 702). *The perpendicular bisectors of the sides of a given quadrilateral Q_1 form a quadrilateral Q_2, and the perpendicular bisectors of the sides of Q_2 form a quadrilateral Q_3. Show that Q_3 is similar to Q_1.*

Points A, B, C, D are arbitrarily chosen. Points M_1, C_1, D_1, A_1, B_1, M_2, A_2, D_2, B_2, C_2 are constructed (in order) as follows: M_1 is the midpoint of A and

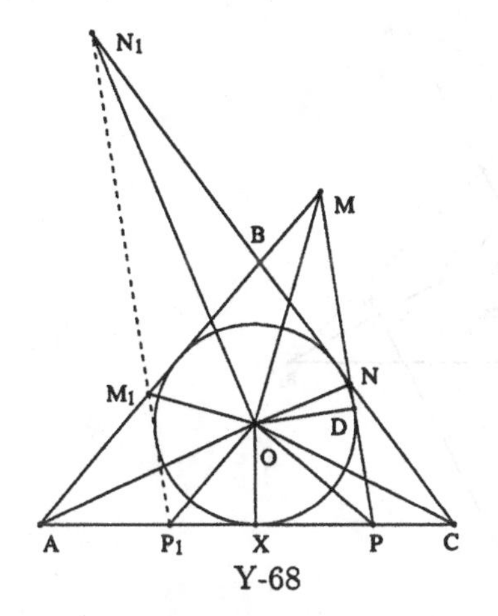

Y-68

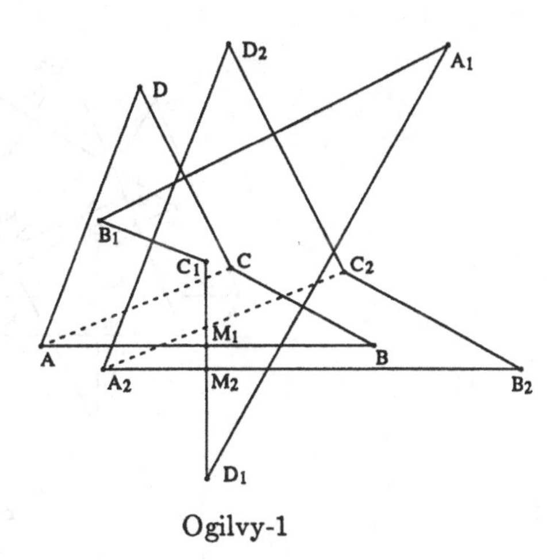

Ogilvy-1

B; $C_1A \equiv C_1D$; $C_1M_1 \perp AB$; $D_1B \equiv D_1C$; D_1 is on line C_1M_1; $A_1C \equiv A_1B$; $A_1D \equiv A_1C$; $B_1D \equiv B_1C$; $B_1A \equiv B_1D$; M_2 is the midpoint of C_1 and D_1; $A_2B_1 \equiv A_2C_1$; $A_2M_2 \parallel AB$; $D_2B_1 \equiv D_2A_1$; $D_2A_2 \parallel AD$; $B_2A_1 \equiv B_2D_1$; B_2 is on line A_2M_2; $C_2B_2 \parallel BC$; $C_2D_2 \parallel DC$. The **conclusions**: (1) $AC \parallel A_2C_2$; (2) $\frac{\overline{AB}}{\overline{A_2B_2}} - \frac{\overline{AD}}{\overline{A_2D_2}} = 0$.

$A = (0,0)$, $B = (u_1,0)$, $C = (u_2,u_3)$, $D = (u_4,u_5)$, $M_1 = (x_1,0)$, $C_1 = (x_1,x_2)$, $D_1 = (x_1,x_3)$, $A_1 = (x_5,x_4)$, $B_1 = (x_7,x_6)$, $M_2 = (x_1,x_8)$, $A_2 = (x_9,x_8)$, $D_2 = (x_{11},x_{10})$, $B_2 = (x_{12},x_8)$, $C_2 = (x_{14},x_{13})$.

The **nondegenerate** conditions: Points A, B and D are not collinear; Line C_1M_1 is not perpendicular to line BC; Points D, C and B are not collinear; Points A, D and C are not collinear; Line AB is not perpendicular to line B_1C_1; Line AD is not perpendicular to line B_1A_1; Line A_2M_2 is not perpendicular to line A_1D_1; Points D, C and B are not collinear.

Example 66 (Ogilvy-2: 8.87s, 383). *Let Q_4 be obtained from Q_3 using the iteration of the above problem. Then Q_4 and Q_2 bear the same relation as Q_3 and Q_1.*

Points A, B, C, D are arbitrarily chosen. Points M_1, C_1, D_1, A_1, B_1, M_2, M_3, M_4, A_2, B_2, D_2, C_2, M_5, M_6, M_7, C_3, D_3 are constructed (in order) as follows: M_1 is the midpoint of A and B; $C_1A \equiv C_1D$; $C_1M_1 \perp AB$; $D_1B \equiv D_1C$; D_1 is on line C_1M_1; $A_1C \equiv A_1B$; $A_1D \equiv A_1C$; $B_1D \equiv B_1C$; $B_1A \equiv B_1D$; M_2 is the midpoint of C_1 and D_1; M_3 is the midpoint of A_1 and D_1; M_4 is the midpoint of B_1 and C_1; $A_2M_2 \parallel AB$; $A_2M_4 \parallel AD$; $B_2M_2 \parallel AB$; $B_2M_3 \parallel CB$; $D_2B_2 \parallel DB$; $D_2M_4 \parallel AD$; $C_2A_2 \parallel AC$; $C_2M_3 \parallel BC$; M_5 is the midpoint of A_2 and B_2; M_6 is the midpoint of A_2 and D_2; M_7 is the midpoint of B_2 and C_2; $C_3M_5 \perp AB$; $C_3M_6 \perp AD$; $D_3M_7 \perp BC$; $D_3M_5 \perp AB$. The **conclusion**: $\frac{\overline{AB}}{\overline{A_2B_2}} - \frac{\overline{C_1D_1}}{\overline{C_3D_3}} = 0$.

$A = (0,0)$, $B = (u_1,0)$, $C = (u_2,u_3)$, $D = (u_4,u_5)$, $M_1 = (x_1,0)$, $C_1 = (x_1,x_2)$, $D_1 = (x_1,x_3)$, $A_1 = (x_5,x_4)$, $B_1 = (x_7,x_6)$, $M_2 = (x_1,x_8)$, $M_3 = (x_9,x_{10})$, $M_4 = (x_{11},x_{12})$, $A_2 = (x_{13},x_8)$, $B_2 = (x_{14},x_8)$, $D_2 = (x_{16},x_{15})$, $C_2 = (x_{18},x_{17})$, $M_5 = (x_{19},x_8)$, $M_6 = (x_{20},x_{21})$, $M_7 = (x_{22},x_{23})$, $C_3 = (x_{19},x_{24})$, $D_3 = (x_{19},x_{25})$.

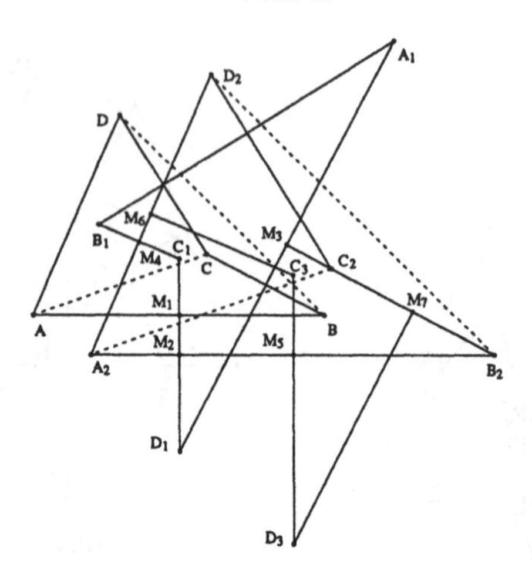

Ogilvy-2

The **nondegenerate** conditions: Points A, B and D are not collinear; Line C_1M_1 is not perpendicular to line BC; Points D, C and B are not collinear; Points A, D and C are not collinear; Points A, D and B are not collinear; Points C, B and A are not collinear; Points A, D and B are not collinear; Points B, C and A are not collinear; Points A, D and B are not collinear; Points A, B and C are not collinear.

Remark. In specifying this problem, we used the results in the previous problem. The above two problems are from C.S. Ogilvy's book "Tomrrow's math: unsolved problems for the amateur". Since we found an easy traditional proof for the first problem and the second problem seems not easy, it is possibly an open conjecture confirmed by our prover.

Example 67 (Harmonic-set: 1.28s, 12).

Points A, B, R are arbitrarily chosen. Points P, Q, S, C, F are constructed (in order) as follows: P is on line RB; Q is on line AR; S is on line QB; S is on line AP; C is on line AB; C is on line SR; F is on line AB; F is on line QP. The **conclusion**: Points A, B, C and F form a harmonic set.

$A = (0,0)$, $B = (u_1, 0)$, $R = (u_2, u_3)$, $P = (x_1, u_4)$, $Q = (x_2, u_5)$, $S = (x_4, x_3)$, $C = (x_5, 0)$, $F = (x_6, 0)$.

The **nondegenerate** conditions: $R \neq B$; $A \neq R$; Line AP intersects line QB; Line SR intersects line AB; Line QP intersects line AB.

Remark. The property proved here can serve as a definition of harmonic set in projective geometry.

Example 68 (Quadrangular-set: 3.1s, 108).

Points A, B, R, P, R_1, Q are arbitrarily chosen. Points S, D, C, E, F, P_1, S_1, Q_1 are constructed (in order) as follows: S is on line BQ; S is on line AP; D is on line AB; D is on line QR; C is on line AB; C is on line SR; E is on line AB; E is on line PR; F is on line AB; F is on line QP; P_1 is on line ER_1; S_1 is on line CR_1;

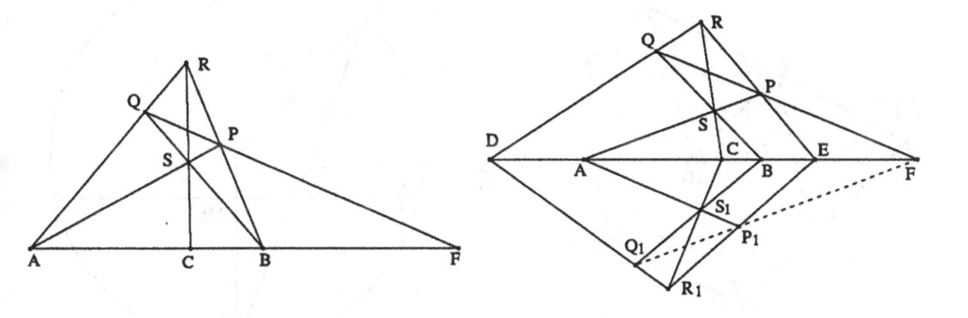

Harmonic-set Quadrangular-set

S_1 is on line AP_1; Q_1 is on line BS_1; Q_1 is on line DR_1. The **conclusion**: Points F, Q_1 and P_1 are collinear.

$A = (0,0)$, $B = (u_1,0)$, $R = (u_2,u_3)$, $P = (u_4,u_5)$, $R_1 = (u_6,u_7)$, $Q = (u_8,u_9)$, $S = (x_2,x_1)$, $D = (x_3,0)$, $C = (x_4,0)$, $E = (x_5,0)$, $F = (x_6,0)$, $P_1 = (x_7,u_{10})$, $S_1 = (x_9,x_8)$, $Q_1 = (x_{11},x_{10})$.

The **nondegenerate** conditions: Line AP intersects line BQ; Line QR intersects line AB; Line SR intersects line AB; Line PR intersects line AB; Line QP intersects line AB; $E \neq R_1$; Line AP_1 intersects line CR_1; Line DR_1 intersects line BS_1.

Remark. This theorem shows the invariance of a quadrangular set (hence a Harmonic set)

Example 69 (H-28: 2.58s, 42; 2.12s, 42). *The diagonals of a parallelogram and those of its inscribed parallelogram are concurrent.*

Points A, B, C, E are arbitrarily chosen. Points D, F, G, H, O are constructed (in order) as follows: $DA \parallel BC$; $DC \parallel BA$; $FD \parallel EB$; F is on line EA; $GC \parallel AE$; G is on line EB; H is on line GC; H is on line FD; O is on line BD; O is on line AC. The **conclusion**: Points O, H and E are collinear.

$A = (0,0)$, $B = (u_1,0)$, $C = (u_2,u_3)$, $E = (u_4,u_5)$, $D = (x_1,u_3)$, $F = (x_3,x_2)$, $G = (x_5,x_4)$, $H = (x_7,x_6)$, $O = (x_9,x_8)$.

The **nondegenerate** conditions: Points B, A and C are not collinear; Points E, A and B are not collinear; Points E, B and A are not collinear; Line FD intersects line GC; Line AC intersects line BD.

Example 70 (H-64: 43.65s, 1897; 7.72s, 250). *In a circle, the lines joining the midpoints of two arcs AB and AC meet line AB and AC at D and E. show that $AD \equiv AE$.*

Points A, B, C are arbitrarily chosen. Points O, M_1, M_2, N_1, N_2, D, E are constructed (in order) as follows: $OA \equiv OC$; $OA \equiv OB$; M_1 is the midpoint of A and B; M_2 is the midpoint of A and C; $N_1O \equiv OA$; N_1 is on line OM_1; $N_2O \equiv OA$;

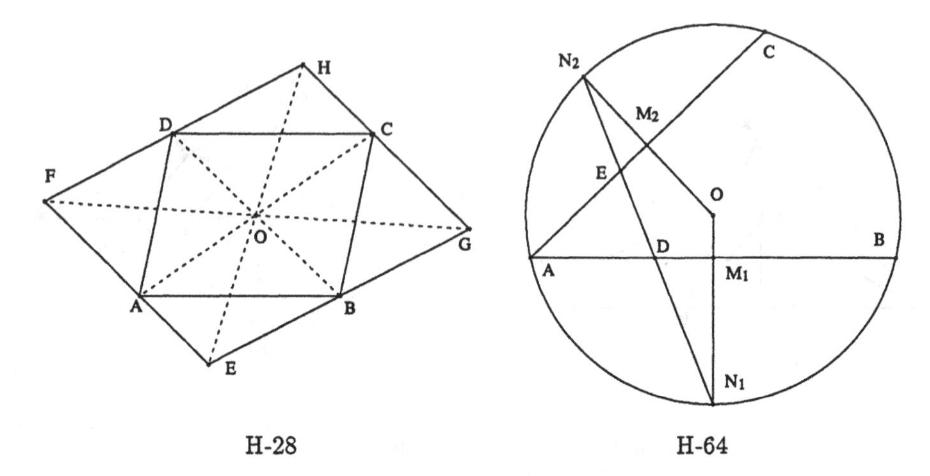

H-28 H-64

N_2 is on line OM_2; D is on line N_1N_2; D is on line AB; E is on line N_1N_2; E is on line AC. The **conclusion**: $AD \equiv AE$.

$A = (0,0)$, $B = (u_1, 0)$, $C = (u_2, u_3)$, $O = (x_2, x_1)$, $M_1 = (x_3, 0)$, $M_2 = (x_4, x_5)$, $N_1 = (x_7, x_6)$, $N_2 = (x_9, x_8)$, $D = (x_{10}, 0)$, $E = (x_{12}, x_{11})$.

The **nondegenerate** conditions: Points A, B and C are not collinear; Line OM_1 is non-isotropic; Line OM_2 is non-isotropic; Line AB intersects line N_1N_2; Line AC intersects line N_1N_2.

Remark. Note there are four pairs (D, E). Since this is an irreducible problem (as our prover checked), if the conclusion is true for one pair, it is true for all four pairs. See Section 5 of Chapter 4 of Part I.

Example 71 (H-67: 1.25s, 28). *From the midpoint C of arc AB of a circle, two secants are drawn meeting line AB at F, G, and the circle at D and E. Show that F, D, E, and G are on the same circle.*

Points C, O are arbitrarily chosen. Points A, M, B, D, E, F, G are constructed (in order) as follows: $AO \equiv OC$; M is on line OC; $MA \perp OC$; M is the midpoint of A and B; $DO \equiv OC$; $EO \equiv OC$; F is on line AM; F is on line DC; G is on line AM; G is on line CE. The **conclusion**: Points D, E, F and G are on the same circle.

$C = (0,0)$, $O = (u_1, 0)$, $A = (x_1, u_2)$, $M = (x_1, 0)$, $B = (x_1, x_2)$, $D = (x_3, u_3)$, $E = (x_4, u_4)$, $F = (x_1, x_5)$, $G = (x_1, x_6)$.

The **nondegenerate** conditions: Line OC is non-isotropic; Line DC intersects line AM; Line CE intersects line AM.

Example 72 (H-68: 251.07s, 6453). *Let Q, S and S_1 be three collinear points and (O, OP) be a circle. Circles SPQ and S_1PQ meet circle (O, OP) again at points R and R_1, respectively. Show that R_1S_1 and RS meet on the circle (O, OP).*

Points G, P, S, O are arbitrarily chosen. Points Q, A, R, B, S_1, R_1, I are constructed (in order) as follows: G is the midpoint of P and Q; $AQ \equiv AS$; $AG \perp$

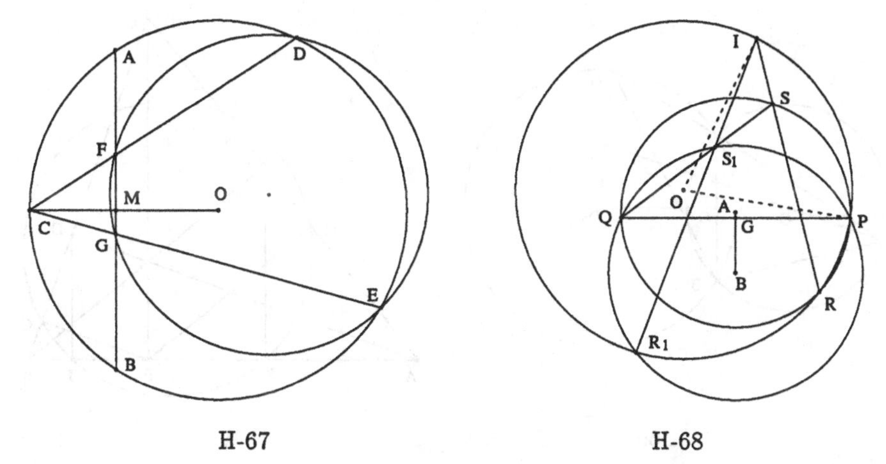

H-67 H-68

GP; $RO \equiv OP$; $RA \equiv AP$; B is on line AG; S_1 is on line QS; $S_1B \equiv BQ$; $R_1B \equiv BP$; $R_1O \equiv OP$; I is on line S_1R_1; I is on line SR. The **conclusion**: $OP \equiv OI$.

$G = (0,0)$, $P = (u_1, 0)$, $S = (u_2, u_3)$, $O = (u_4, u_5)$, $Q = (x_1, 0)$, $A = (0, x_2)$, $R = (x_4, x_3)$, $B = (0, u_6)$, $S_1 = (x_6, x_5)$, $R_1 = (x_8, x_7)$, $I = (x_{10}, x_9)$.

The **nondegenerate** conditions: Line GP intersects line QS; Line AO is non-isotropic; $A \neq G$; Line QS is non-isotropic; Line OB is non-isotropic; Line SR intersects line S_1R_1. In addition, the following nondegenerate conditions, which come from reducibility and have been detected by our prover, should be also added: $R_1 \neq P$; $S_1 \neq Q$; $R \neq P$.

Example 73 (H-69: 12.1s, 131). *Let A and B be the intersections of two circles O_1 and O_2. Through A a secant is drawn meeting the two circles at C and D, respectively. Show that angle CBD is equal to the angle formed by lines O_1C and O_2D.*

Points O, O_1 are arbitrarily chosen. Points A, O_2, B, D, C, I are constructed (in order) as follows: $AO \perp OO_1$; O_2 is on line O_1O; O is the midpoint of A and B; $DO_2 \equiv O_2A$; $CO_1 \equiv O_1A$; C is on line DA; I is on line O_2D; I is on line O_1C. The **conclusion**: $\tan(CBD) = \tan(O_1IO_2)$.

$O = (0,0)$, $O_1 = (u_1, 0)$, $A = (0, u_2)$, $O_2 = (u_3, 0)$, $B = (0, x_1)$, $D = (x_2, u_4)$, $C = (x_4, x_3)$, $I = (x_6, x_5)$.

The **nondegenerate** conditions: $O \neq O_1$; $O_1 \neq O$; Line DA is non-isotropic; Line O_1C intersects line O_2D. In addition, the following nondegenerate conditions, which come from reducibility and have been detected by our prover, should be also added: $C \neq A$.

Example 74 (H-102: 3.33s, 98). *Let ABC be a triangle. Show that the six feet obtained by drawing perpendiculars through the foot of each altitude upon the other two sides are co-circle.*

Points A, B, C are arbitrarily chosen. Points F, E, D, G, H, I, K are constructed

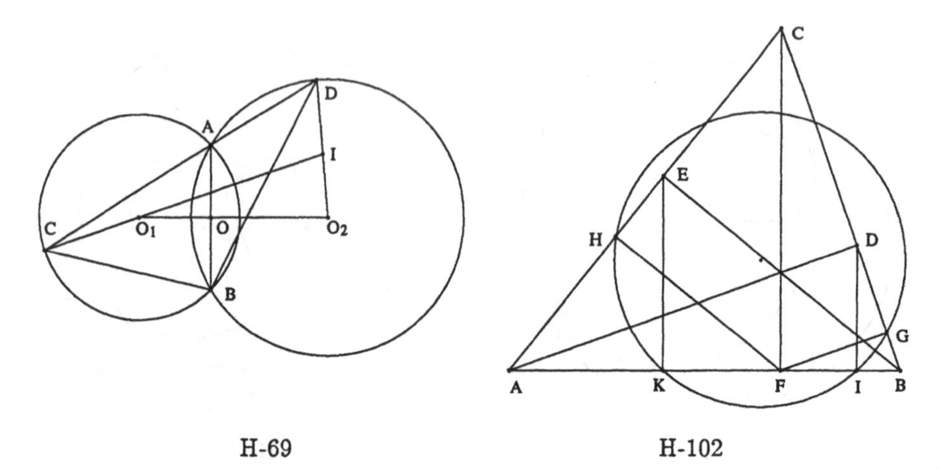

H-69 H-102

(in order) as follows: $FC \perp AB$; F is on line AB; $EB \perp AC$; E is on line AC; $DA \perp BC$; D is on line CB; $GF \perp BC$; G is on line BC; $HF \perp AC$; H is on line AC; $ID \perp AF$; I is on line AF; $KE \perp AF$; K is on line AF. The **conclusion**: Points K, I, H and G are on the same circle.

$A = (0,0)$, $B = (u_1,0)$, $C = (u_2,u_3)$, $F = (u_2,0)$, $E = (x_2,x_1)$, $D = (x_4,x_3)$, $G = (x_6,x_5)$, $H = (x_8,x_7)$, $I = (x_4,0)$, $K = (x_2,0)$.

The **nondegenerate** conditions: Line AB is non-isotropic; Line AC is non-isotropic; Line CB is non-isotropic; Line BC is non-isotropic; Line AC is non-isotropic; Line AF is non-isotropic; Line AF is non-isotropic.

Example 75 (H-104: 0.77s, 6). *Let C be a point on a chord AB of circle O. Let D and E be the intersections of perpendicular of OC through C with the two tangents of the circle at A and B, respectively. Show that $CE \equiv CD$.*

Points O, A are arbitrarily chosen. Points B, C, D, E are constructed (in order) as follows: $BO \equiv OA$; C is on line AB; $DC \perp CO$; $DA \perp AO$; $EC \perp CO$; $EB \perp BO$. The **conclusion**: C is the midpoint of D and E.

$O = (0,0)$, $A = (u_1,0)$, $B = (x_1,u_2)$, $C = (x_2,u_3)$, $D = (u_1,x_3)$, $E = (x_5,x_4)$.

The **nondegenerate** conditions: $A \neq B$; Points A, O and C are not collinear; Points B, O and C are not collinear.

Example 76 (H189-200: 6.65s, 192; 3.25s, 172). *The cross ratio of four points on a line is unchanged under projection.*

Points A, B, O are arbitrarily chosen. Points C, D, A_1, B_1, C_1, D_1 are constructed (in order) as follows: C is on line AB; D is on line AB; A_1 is on line AO; B_1 is on line BO; C_1 is on line A_1B_1; C_1 is on line CO; D_1 is on line A_1B_1; D_1 is on line DO. The **conclusion**: $crossratio(ABCD) - crossratio(A_1B_1C_1D_1) = 0$.

$A = (0,0)$, $B = (u_1,0)$, $O = (u_2,u_3)$, $C = (u_4,0)$, $D = (u_5,0)$, $A_1 = (x_1,u_6)$, $B_1 = (x_2,u_7)$, $C_1 = (x_4,x_3)$, $D_1 = (x_6,x_5)$.

The **nondegenerate** conditions: $A \neq B$; $A \neq B$; $A \neq O$; $B \neq O$; Line CO

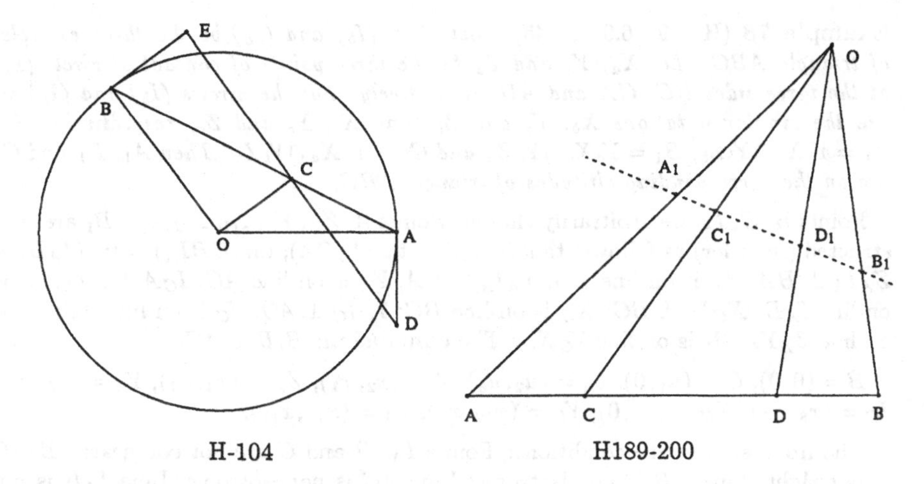

H-104 H189-200

intersects line A_1B_1; Line DO intersects line A_1B_1.

Example 77 (H199-212a: 2.23s, 36). ***The cross ratio of four points on a circles with respect to any points on the circle is constant.***

Points A, B, C are arbitrarily chosen. Points O, D, E, G, F, E_1, G_1, F_1 are constructed (in order) as follows: $OA \equiv OC$; $OA \equiv OB$; $DO \equiv OA$; $EO \equiv OA$; G is on line AB; G is on line EC; F is on line AB; F is on line ED; $E_1O \equiv OA$; G_1 is on line AB; G_1 is on line E_1C; F_1 is on line AB; F_1 is on line E_1D. The **conclusion**: $crossratio(AFGB) - crossratio(AF_1G_1B) = 0$.

$A = (0,0)$, $B = (u_1,0)$, $C = (u_2,u_3)$, $O = (x_2,x_1)$, $D = (x_3,u_4)$, $E = (x_4,u_5)$, $G = (x_5,0)$, $F = (x_6,0)$, $E_1 = (x_7,u_6)$, $G_1 = (x_8,0)$, $F_1 = (x_9,0)$.

The **nondegenerate** conditions: Points A, B and C are not collinear; Line EC intersects line AB; Line ED intersects line AB; Line E_1C intersects line AB; Line E_1D intersects line AB.

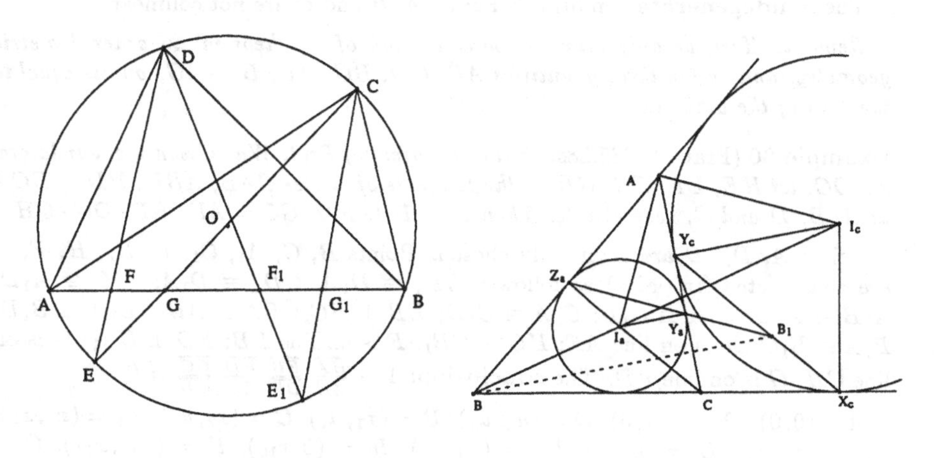

H199-212a H-379

Example 78 (H-379: 6.93s, 245). *Let (I_a), (I_b) and (I_c) be the three excircles of triangle ABC. Let X_a, Y_a and Z_a be the three points of contact of circle (I_a) at the three sides BC, CA and AB, respectively. For the circles (I_b) and (I_c) we use the similar notations X_b, Y_b and Z_b, and X_c, Y_c and Z_c, respectively. Let $A_1 = Y_cX_c \cap Y_bZ_b$, $B_1 = Y_cX_c \cap Y_aZ_a$ and $C_1 = Y_aX_a \cap Y_bZ_b$. Then A_1, B_1 and C_1 are on the corresponding altitudes of triangle ABC.*

Points B, C, I_A are arbitrarily chosen. Points A, Z_A, Y_A, I_C, X_C, Y_C, B_1 are constructed (in order) as follows: $\tan(BCI_A) = \tan(I_ACA)$; $\tan(CBI_A) = \tan(I_ABA)$; $Z_AI_A \perp BA$; Z_A is on line AB; $Y_AI_A \perp CA$; Y_A is on line AC; $I_CA \perp AI_A$; I_C is on line I_AB; $X_CI_C \perp BC$; X_C is on line BC; $Y_CI_C \perp AC$; Y_C is on line AC; B_1 is on line Z_AY_A; B_1 is on line Y_CX_C. The **conclusion:** $B_1B \perp AC$.

$B = (0,0)$, $C = (u_1,0)$, $I_A = (u_2,u_3)$, $A = (x_2,x_1)$, $Z_A = (x_4,x_3)$, $Y_A = (x_6,x_5)$, $I_C = (x_8,x_7)$, $X_C = (x_8,0)$, $Y_C = (x_{10},x_9)$, $B_1 = (x_{12},x_{11})$.

The **nondegenerate** conditions: Points I_A, B and C are not collinear; $\angle BI_AC$ is not right; Line AB is non-isotropic; Line AC is non-isotropic; Line I_AB is not perpendicular to line AI_A; Line BC is non-isotropic; Line AC is non-isotropic; Line Y_CX_C intersects line Z_AY_A.

Remark. Actually we prove a more general version of the theorem. Point $B_2 = Y_bX_b \cap YZ$ has the same property as B_1, where X, Y, Z are similar points for the incircle (I). Our figure is for this case.

Example 79 (Ptolemy: 1.3s, 41). *The product of two diagonals of a cyclic quadrilateral is equal to sum of the products of opposite sides.*

Points A, B, C are arbitrarily chosen. Points O, D are constructed (in order) as follows: $OA \equiv OC$; $OA \equiv OB$; $DO \equiv OA$; $\overline{AB}^2\ \overline{CD}^2 - x_4\ x_4 = 0$; $\overline{BC}^2\ \overline{AD}^2 - x_5\ x_5 = 0$; $\overline{BD}^2\ \overline{AC}^2 - x_6\ x_6 = 0$. The **conclusion:** $(x_4 + x_5 - x_6)\ (x_4 + x_5 + x_6)\ (x_4 - (x_5 - x_6))\ (x_4 - (x_5 + x_6)) = 0$.

$A = (0,0)$, $B = (u_1,0)$, $C = (u_2,u_3)$, $O = (x_2,x_1)$, $D = (x_3,u_4)$.

The **nondegenerate** conditions: Points A, B and C are not collinear.

Remark. Here we only proved a weak version of the theorem: in ordered metric geometry, among the three quantities $AB \cdot CD$, $BC \cdot AD$, $BD \cdot AC$, one is equal to the sum of the other two.

Example 80 (Pratt-4: 285.28s, 4657). *Theorem of Pratt-Wu. Given a quadrilateral $ABDC$, let HE, EF, FG, GH be the tangents of circles CAB, ABD, BDC, DCA at A, B, D and C, respectively. Then $HA \cdot EB \cdot FD \cdot GC = AE \cdot BF \cdot DG \cdot CH$.*

Points A, D_1, D are arbitrarily chosen. Points B, C, A_1, C_1, E, B_1, H, F, G are constructed (in order) as follows: $BD_1 \equiv D_1A$; $CD_1 \equiv D_1A$; $A_1C \equiv A_1D$; $A_1B \equiv A_1D$; $C_1A \equiv C_1B$; $C_1A \equiv C_1D$; $EB \perp BC_1$; $EA \perp AD_1$; $B_1A \equiv B_1D$; $B_1A \equiv B_1C$; H is on line AE; $HC \perp CB_1$; F is on line EB; $FD \perp DA_1$; G is on line CH; G is on line FD. The **conclusion:** $1 - \frac{\overline{HA}}{\overline{AE}} \frac{\overline{EB}}{\overline{BF}} \frac{\overline{FD}}{\overline{DG}} \frac{\overline{GC}}{\overline{CH}} = 0$.

$A = (0,0)$, $D_1 = (u_1,0)$, $D = (u_2,u_3)$, $B = (x_1,u_4)$, $C = (x_2,u_5)$, $A_1 = (x_4,x_3)$, $C_1 = (x_6,x_5)$, $E = (0,x_7)$, $B_1 = (x_9,x_8)$, $H = (0,x_{10})$, $F = (x_{12},x_{11})$, $G =$

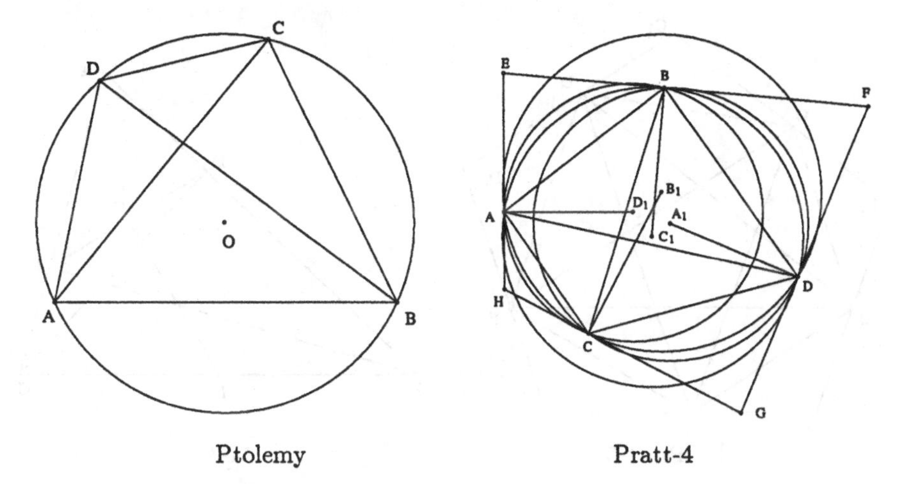

Ptolemy　　　　Pratt-4

(x_{14}, x_{13}).

The **nondegenerate** conditions: Points B, D and C are not collinear; Points A, D and B are not collinear; Line AD_1 intersects line BC_1; Points A, C and D are not collinear; Line CB_1 is not perpendicular to line AE; Line DA_1 is not perpendicular to line EB; Line FD intersects line CH.

Remark. Actually, this property is valid for any polygon, so is the next problem.

Example 81 (Pratt-5: 10.82s, 391; 5.5s, 284). *Theorem of Pratt-Kasapi. Let $ABCDE$ be a pentagon. $A_1B_1 \parallel AC$, $B_1C_1 \parallel BD$, $C_1D_1 \parallel CE$, $D_1E_1 \parallel AD$, $E_1A_1 \parallel EB$. Show that $A_1B \cdot B_1C \cdot C_1D \cdot D_1E \cdot E_1A = BB_1 \cdot CC_1 \cdot DD_1 \cdot EE_1 \cdot AA_1$.*

Points B, E, A, C, D are arbitrarily chosen. Points A_1, B_1, C_1, D_1, E_1 are constructed (in order) as follows: $A_1B \parallel CA$; $A_1A \parallel EB$; $B_1C \parallel BD$; $B_1B \parallel AC$; $C_1D \parallel CE$; $C_1C \parallel BD$; $D_1E \parallel AD$; $D_1D \parallel CE$; $E_1A \parallel BE$; $E_1E \parallel AD$. The **conclusion**: $\frac{\overline{AA_1}}{\overline{A_1B}} \frac{\overline{BB_1}}{\overline{B_1C}} \frac{\overline{CC_1}}{\overline{C_1D}} \frac{\overline{DD_1}}{\overline{D_1E}} \frac{\overline{EE_1}}{\overline{E_1A}} - 1 = 0$.

$B = (0,0)$, $E = (u_1, 0)$, $A = (u_2, u_3)$, $C = (u_4, u_5)$, $D = (u_6, u_7)$, $A_1 = (x_1, u_3)$, $B_1 = (x_3, x_2)$, $C_1 = (x_5, x_4)$, $D_1 = (x_7, x_6)$, $E_1 = (x_8, u_3)$.

The **nondegenerate** conditions: Line EB intersects line CA; Line AC intersects line BD; Line BD intersects line CE; Line CE intersects line AD; Line AD intersects line BE.

Example 82 (M-3: 2.43s, 81). *In triangle ABC, let F the midpoint of the side BC, D and E the feet of the altitudes on AB and BC, respectively. FG is perpendicular to DE at G. Show that G is the midpoint of DE.*

Points B, C, A are arbitrarily chosen. Points D, E, F, G are constructed (in order) as follows: $DC \perp AB$; D is on line BA; $EB \perp AC$; E is on line AC; F is the midpoint of B and C; $GF \perp ED$; G is on line ED. The **conclusion**: G is the midpoint of D and E.

$B = (0,0)$, $C = (u_1, 0)$, $A = (u_2, u_3)$, $D = (x_2, x_1)$, $E = (x_4, x_3)$, $F = (x_5, 0)$, $G = (x_7, x_6)$.

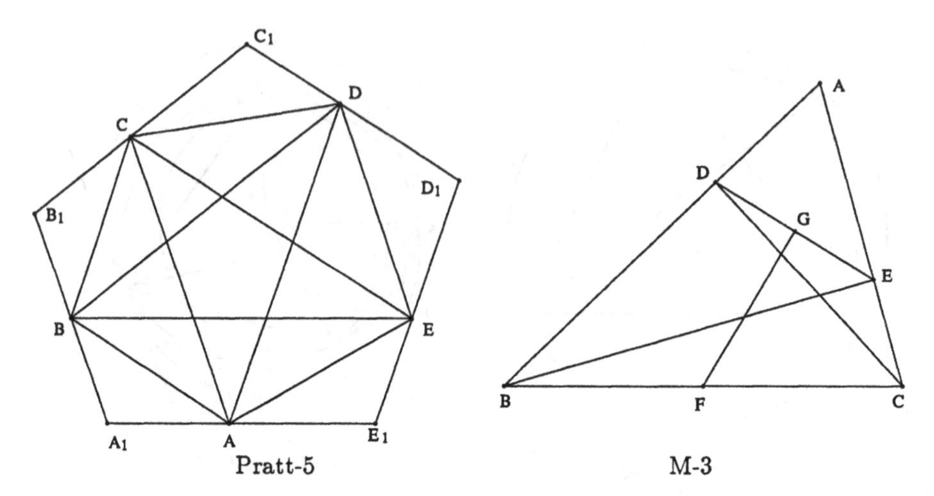

Pratt-5

M-3

The **nondegenerate** conditions: Line BA is non-isotropic; Line AC is non-isotropic; Line ED is non-isotropic.

Example 83 (M-8: 1.5s, 16). *Let $ABCD$ be a quadrilateral with $AB \equiv AD$, F and E be the midpoints of AD and BC respectively. $M = BA \cap EF$, $N = CD \cap EF$. Show that $\angle BME = \angle ENC$.*

Points B, C, A are arbitrarily chosen. Points D, E, F, M, N are constructed (in order) as follows: $DC \equiv AB$; E is the midpoint of B and C; F is the midpoint of A and D; M is on line AB; M is on line EF; N is on line EF; N is on line DC. The **conclusion**: $\tan(BME) = \tan(ENC)$.

$B = (0,0)$, $C = (u_1, 0)$, $A = (u_2, u_3)$, $D = (x_1, u_4)$, $E = (x_2, 0)$, $F = (x_3, x_4)$, $M = (x_6, x_5)$, $N = (x_8, x_7)$.

The **nondegenerate** conditions: Line EF intersects line AB; Line DC intersects line EF.

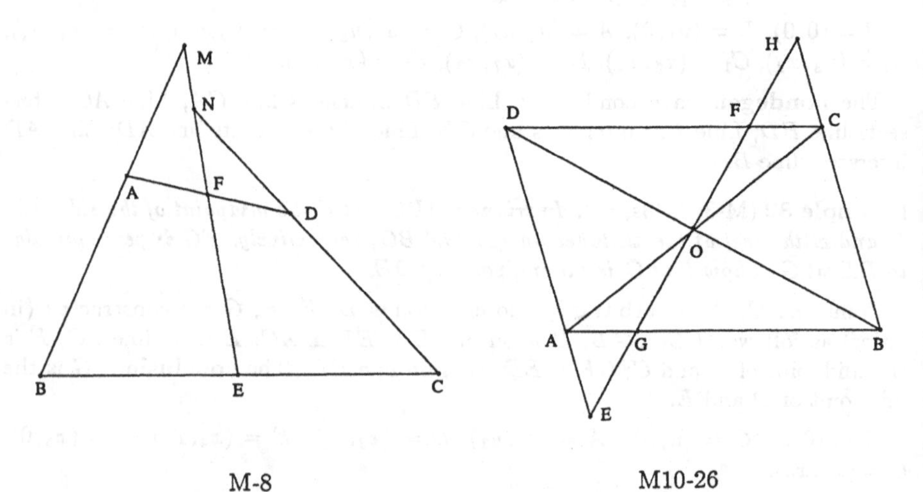

M-8

M10-26

Example 84 (M10-26: 1.17s, 6). *A line passing through the intersection O of the diagonals of parallelogram ABCD meets the four sides at E, F, G, H. Show that* $EF \equiv GH$.

Points A, B, C are arbitrarily chosen. Points D, O, G, F, E, H are constructed (in order) as follows: $DA \parallel CB$; $DC \parallel BA$; O is on line BD; O is on line AC; G is on line AB; F is on line GO; F is on line CD; E is on line GO; E is on line AD; H is on line GO; H is on line BC. The **conclusion**: $\frac{\overline{EF}}{\overline{GH}} - 1 = 0$.

$A = (0,0)$, $B = (u_1,0)$, $C = (u_2,u_3)$, $D = (x_1,u_3)$, $O = (x_3,x_2)$, $G = (u_4,0)$, $F = (x_4,u_3)$, $E = (x_6,x_5)$, $H = (x_8,x_7)$.

The **nondegenerate** conditions: Points B, A and C are not collinear; Line AC intersects line BD; $A \neq B$; Line CD intersects line GO; Line AD intersects line GO; Line BC intersects line GO.

Example 85 (M10-32: 2.48s, 24; 2.15s, 16). *Let P and Q be two points on side BC and AD of a parallelogram such that* $PQ \parallel AB$; $M = AP \cap BQ$, $N = DP \cap QC$. *Show that* $MN \parallel AD$ *and* $MN = 1/2 \ldots AD$.

Points A, B, C are arbitrarily chosen. Points D, P, Q, M, N are constructed (in order) as follows: $DA \parallel CB$; $DC \parallel BA$; P is on line BC; Q is on line AD; $QP \parallel AB$; M is on line BQ; M is on line AP; N is on line CQ; N is on line DP. The **conclusions**: (1) $NM \parallel AD$; (2) $\frac{\overline{AD}}{\overline{MN}} - 2 = 0$.

$A = (0,0)$, $B = (u_1,0)$, $C = (u_2,u_3)$, $D = (x_1,u_3)$, $P = (x_2,u_4)$, $Q = (x_3,u_4)$, $M = (x_5,x_4)$, $N = (x_7,x_6)$.

The **nondegenerate** conditions: Points B, A and C are not collinear; $B \neq C$; Points A, B and D are not collinear; Line AP intersects line BQ; Line DP intersects line CQ.

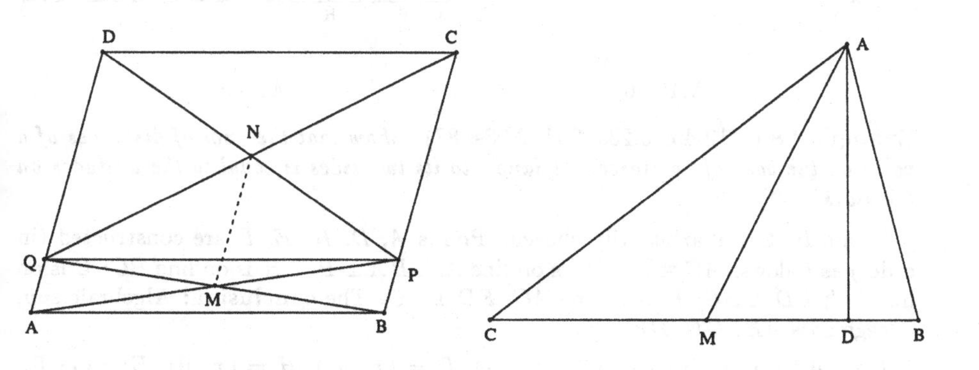

M10-32 M10-34

Example 86 (M10-34: 0.4s, 6). *Let ABC be a triangle with* $\angle B = 2\angle C$, *D the foot of the altitude on CB and M the midpoint of B and C. Show that* $AB = 2DM$.

Points C, B are arbitrarily chosen. Points A, D, M are constructed (in order) as follows: $\tan(2\angle BCA) - \tan(\angle ABC) = 0$; D is on line BC; $DA \perp BC$; M is the midpoint of B and C. The **conclusion**: $2DM = AB$.

$C = (0,0)$, $B = (u_1, 0)$, $A = (x_1, u_2)$, $D = (x_1, 0)$, $M = (x_2, 0)$.

The **nondegenerate** conditions: $3u_2 \neq 0$; Line BC is non-isotropic.

Example 87 (M10-40: 1.1s, 14). *Three equilateral triangles A_1BC, AB_1C, ABC_1 are erected on the three sides of triangle ABC. Show that $CA_1C_1B_1$ is a parallelogram.*

Points A, B, C are arbitrarily chosen. Points C_1, A_1, B_1 are constructed (in order) as follows: $C_1A \equiv AB$; $C_1A \equiv C_1B$; $\tan(BAC_1) = \tan(A_1BC)$; $A_1B \equiv A_1C$; $\tan(BAC_1) = \tan(CAB_1)$; $B_1A \equiv B_1C$. The **conclusion**: $B_1C_1 \parallel CA_1$.

$A = (0,0)$, $B = (u_1, 0)$, $C = (u_2, u_3)$, $C_1 = (x_2, x_1)$, $A_1 = (x_4, x_3)$, $B_1 = (x_6, x_5)$.

The **nondegenerate** conditions: Line AB is non-isotropic; $(-2u_1u_3^2 - 2u_1u_2^2 + 4u_1^2u_2 - 2u_1^3)x_2 \neq 0$; $(2u_1u_3^2 + 2u_1u_2^2)x_2 \neq 0$.

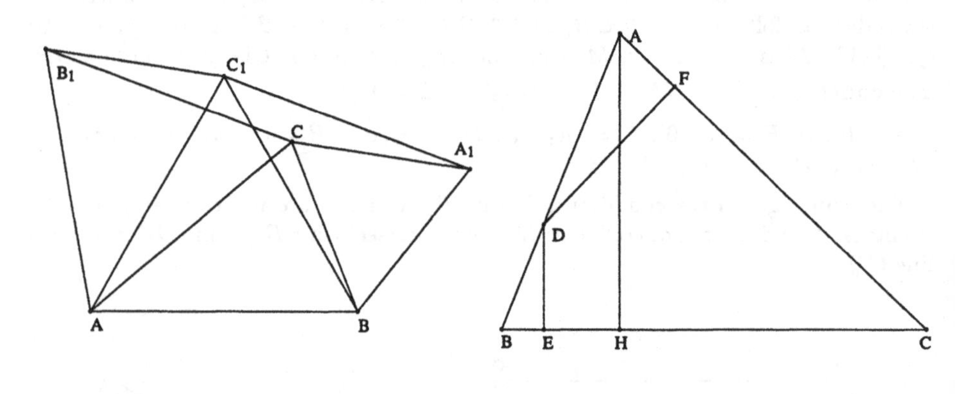

M10-40 M12-46

Example 88 (M12-46: 5.23s, 204; 2.08s, 87). *Show that the sum of distances of a point on the base of an isosceles triangle to its two sides is equal to the altitudes on the sides.*

Points B, C are arbitrarily chosen. Points A, D, H, E, F are constructed (in order) as follows: $AC \equiv CB$; D is on line AB; $HA \perp BC$; H is on line BC; E is on line BC; $ED \perp BC$; F is on line AC; $FD \perp AC$. The **conclusion**: Algebraic sum of segments AH, DE, $DF = 0$.

$B = (0,0)$, $C = (u_1, 0)$, $A = (x_1, u_2)$, $D = (x_2, u_3)$, $H = (x_1, 0)$, $E = (x_2, 0)$, $F = (x_4, x_3)$.

The **nondegenerate** conditions: $A \neq B$; Line BC is non-isotropic; Line BC is non-isotropic; Line AC is non-isotropic.

Example 89 (M12-47: 3.95s, 78; 3.25s, 73). *Show that the sum of the distances* *of a point to the three sides of an equilateral triangle is equal to the altitude of that triangle.*

Points B, C, P are arbitrarily chosen. Points A, D, E, F, G, H are constructed (in order) as follows: $AB \equiv AC$; $AB \equiv BC$; D is on line BC; $DA \perp BC$; E is on line BC; $EP \perp BC$; F is on line AC; $FP \perp AC$; G is on line AB; $GP \perp AB$; H is on line AD; $HP \perp AD$. The **conclusion**: Algebraic sum of segments AH, PF, $PG = 0$.

$B = (0,0)$, $C = (u_1,0)$, $P = (u_2,u_3)$, $A = (x_2,x_1)$, $D = (x_2,0)$, $E = (u_2,0)$, $F = (x_4,x_3)$, $G = (x_6,x_5)$, $H = (x_2,u_3)$.

The **nondegenerate** conditions: Line BC is non-isotropic; Line BC is non-isotropic; Line BC is non-isotropic; Line AC is non-isotropic; Line AB is non-isotropic; Line AD is non-isotropic.

Remark. For problem M12-46 and this problem, we only confirmed their week versions. A more precise and general version proved is problem A97-5.

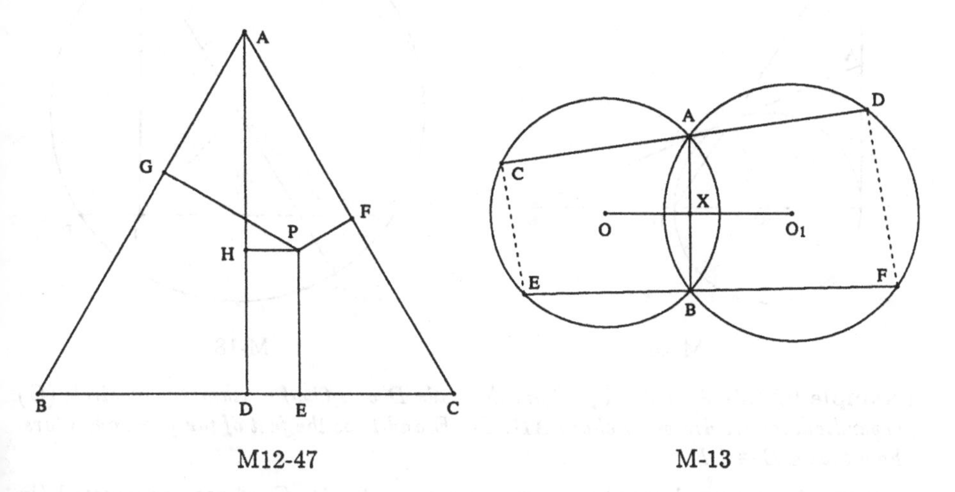

M12-47 M-13

Example 90 (M-13: 2.33s, 64). *Through the two common points A, B of two circles (O) and (O_1) two lines are drawn meeting the circles at C and D, E and F, respectively. Show that $CE \parallel DF$.*

Points X, O are arbitrarily chosen. Points A, B, O_1, C, D, E, F are constructed (in order) as follows: $AX \perp OX$; X is the midpoint of A and B; O_1 is on line OX; $CO \equiv OA$; $DO_1 \equiv O_1A$; D is on line CA; $EO \equiv OA$; $FO_1 \equiv O_1B$; F is on line BE. The **conclusion**: $EC \parallel DF$.

$X = (0,0)$, $O = (u_1,0)$, $A = (0,u_2)$, $B = (0,x_1)$, $O_1 = (u_3,0)$, $C = (x_2,u_4)$, $D = (x_4,x_3)$, $E = (x_5,u_5)$, $F = (x_7,x_6)$.

The **nondegenerate** conditions: $O \neq X$; $O \neq X$; Line CA is non-isotropic; Line BE is non-isotropic. In addition, the following nondegenerate conditions, which come from reducibility and have been detected by our prover, should be also added: $F \neq B$; $D \neq A$.

Example 91 (M-16: 0.8s, 7). *Let D be a point on the side CB of a right triangle ABC such that the circle (O) with diameter CD touches the hypotenuse AB at E. Let $F = AC \cap DE$. Show that $AF \equiv AE$.*

Points C, D are arbitrarily chosen. Points O, E, A, B, F are constructed (in order) as follows: O is the midpoint of D and C; $EO \equiv OC$; $AE \perp OE$; $AC \perp DC$; B is on line DC; B is on line AE; F is on line DE; F is on line AC. The **conclusion**: $AF \equiv AE$.

$C = (0,0)$, $D = (u_1,0)$, $O = (x_1,0)$, $E = (x_2,u_2)$, $A = (0,x_3)$, $B = (x_4,0)$, $F = (0,x_5)$.

The **nondegenerate** conditions: Line DC intersects line OE; Line AE intersects line DC; Line AC intersects line DE.

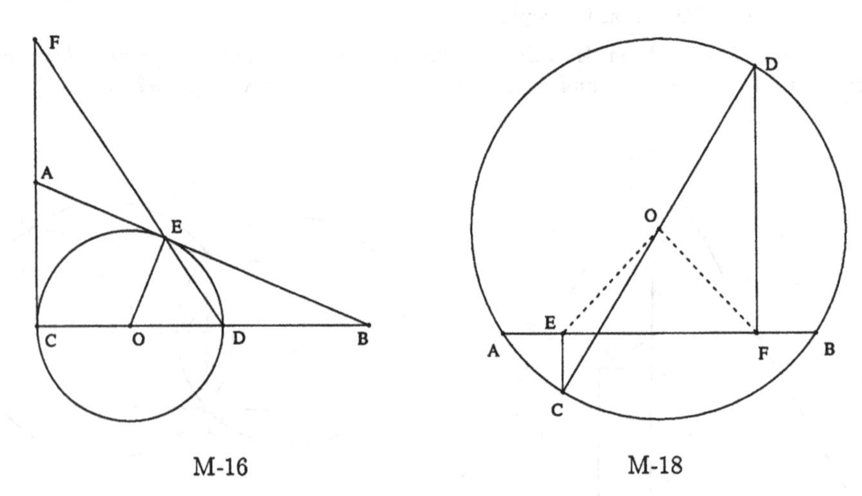

M-16 M-18

Example 92 (M-18: 0.53s, 5). *From the ends D and C of a diameter of circle (O) perpendiculars are drawn to chord AB. Let E and F be the feet of the perpendiculars. Show that $OE \equiv OF$.*

Points A, B are arbitrarily chosen. Points O, C, D, E, F are constructed (in order) as follows: $OA \equiv OB$; $CO \equiv OA$; O is the midpoint of D and C; E is on line AB; $EC \perp AB$; F is on line AB; $FD \perp AB$. The **conclusion**: $OE \equiv OF$.

$A = (0,0)$, $B = (u_1,0)$, $O = (x_1,u_2)$, $C = (x_2,u_3)$, $D = (x_3,x_4)$, $E = (x_2,0)$, $F = (x_3,0)$.

The **nondegenerate** conditions: $A \neq B$; Line AB is non-isotropic; Line AB is non-isotropic.

Example 93 (M-19: 1.42s, 34). *Let AD be the altitude on the hypotenuse BC of right triangle ABC. A circle passing through C and D meets AC at E. BE meets the circle at another point F. Show that $AF \perp BE$.*

Points A, B are arbitrarily chosen. Points C, D, O, E, F are constructed (in order) as follows: $CA \perp BA$; D is on line BC; $DA \perp BC$; $OD \equiv OC$; E is on line AC; $EO \equiv OC$; F is on line BE; $FO \equiv OC$. The **conclusion**: $AF \perp BE$.

$A = (0,0)$, $B = (u_1,0)$, $C = (0,u_2)$, $D = (x_2,x_1)$, $O = (x_3,u_3)$, $E = (0,x_4)$, $F = (x_6,x_5)$.

The **nondegenerate** conditions: $B \neq A$; Line BC is non-isotropic; $D \neq C$; Line AC is non-isotropic; Line BE is non-isotropic. In addition, the following nondegenerate conditions, which come from reducibility and have been detected by our prover, should be also added: $F \neq E$; $E \neq C$.

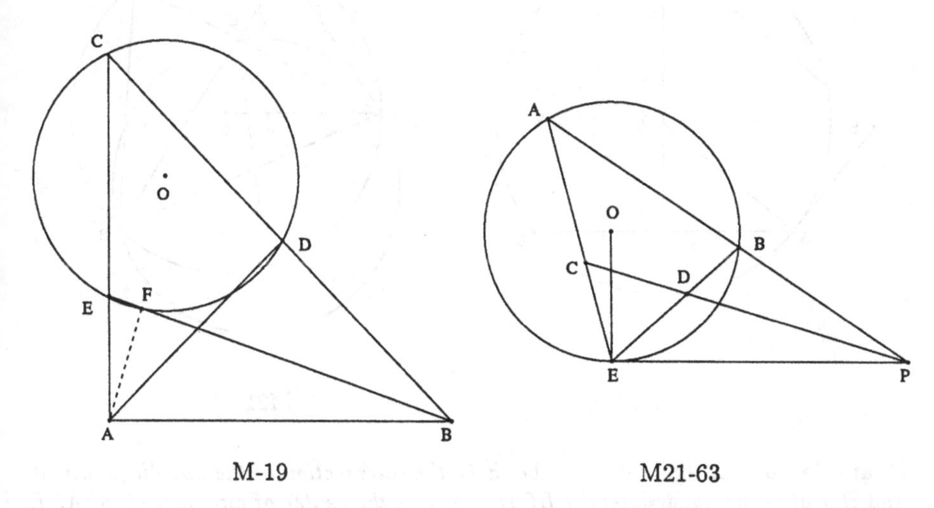

M-19 M21-63

Example 94 (M21-63: 1.65s, 21). *Through P a tangent PE and a secant PAB of circle (O) are drawn. The bisector of angle APE meets AE and BE at C and D. Show that $EC \equiv ED$.*

Points E, P are arbitrarily chosen. Points O, A, B, C, D are constructed (in order) as follows: $OE \perp PE$; $AO \equiv OE$; B is on line PA; $BO \equiv OE$; C is on line AE; $\tan(EPC) = \tan(CPA)$; D is on line BE; D is on line PC. The **conclusion**: $EC \equiv ED$.

$E = (0,0)$, $P = (u_1,0)$, $O = (0,u_2)$, $A = (x_1,u_3)$, $B = (x_3,x_2)$, $C = (x_5,x_4)$, $D = (x_7,x_6)$.

The **nondegenerate** conditions: $P \neq E$; Line PA is non-isotropic; $u_1u_3x_1^2 - 2u_1^2u_3x_1+u_1u_3^3 \neq 0$; $-u_3 \neq 0$; Line PC intersects line BE. In addition, the following nondegenerate conditions, which come from reducibility and have been detected by our prover, should be also added: $B \neq A$.

Remark. Actually, there are two cases for this theorem: the internal and the external bisector of APE. Our proof is for both cases.

Example 95 (M21-67: 0.63s, 6). *Let M be the midpoint of the arc AB of circle (O), D be the midpoint of AB. The perpendicular through M is drawn to the tangent of the circle at A meeting that tangent at E. Show $ME \equiv MD$.*

Points D, A are arbitrarily chosen. Points B, O, M, E are constructed (in order) as follows: D is the midpoint of A and B; $OD \perp DA$; $MO \equiv OA$; M is on line OD; $EM \parallel OA$; $EA \perp OA$. The **conclusion**: $MD \equiv ME$.

$D = (0,0)$, $A = (u_1, 0)$, $B = (x_1, 0)$, $O = (0, u_2)$, $M = (0, x_2)$, $E = (x_4, x_3)$.

The **nondegenerate** conditions: $D \neq A$; Line OD is non-isotropic; Line OA is non-isotropic.

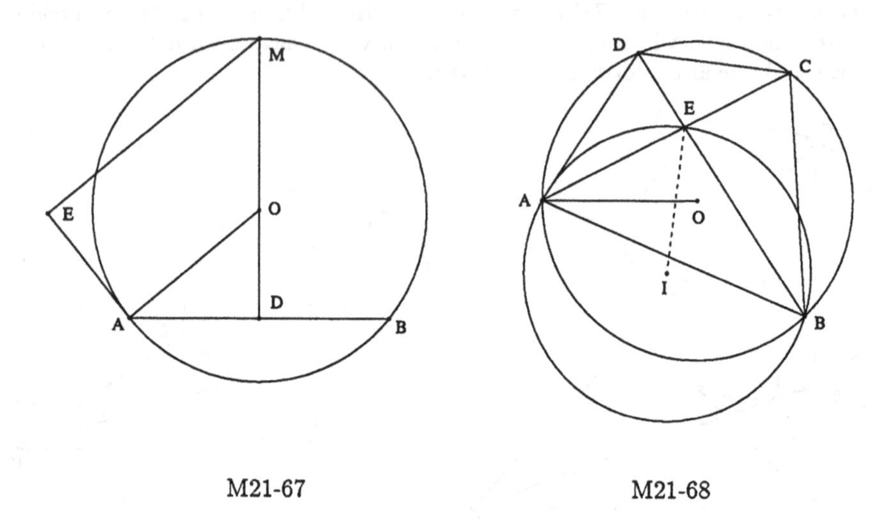

M21-67 M21-68

Example 96 (M21-68: 1.4s, 24). *Let E be the intersection of the two diagonals AC and BD of cyclic quadrilateral $ABCD$. Let I be the center of circumcircle of ABE. Show the $IE \perp DC$.*

Points A, O are arbitrarily chosen. Points B, C, D, E, I are constructed (in order) as follows: $BO \equiv OA$; $CO \equiv OA$; $DO \equiv OA$; E is on line BD; E is on line AC; $IA \equiv IB$; $IA \equiv IE$. The **conclusion**: $IE \perp CD$.

$A = (0,0)$, $O = (u_1, 0)$, $B = (x_1, u_2)$, $C = (x_2, u_3)$, $D = (x_3, u_4)$, $E = (x_5, x_4)$, $I = (x_7, x_6)$.

The **nondegenerate** conditions: Line AC intersects line BD; Points A, E and B are not collinear.

Example 97 (M22-77: 2.33s, 14; 2.3s, 25). *Let G be a point on the circle (O) with diameter BC, A be the midpoint of the arc BG. $AD \perp BC$. $E = AD \cap BG$ and $F = AC \cap BG$. Show that $AE \equiv BE \equiv EF$.*

Points O, B are arbitrarily chosen. Points C, G, M, A, D, E, F are constructed (in order) as follows: O is the midpoint of B and C; $GO \equiv OB$; M is the midpoint of B and G; A is on line OM; $AO \equiv OB$; D is on line BC; $DA \perp BC$; E is on line AD; E is on line BG; F is on line AC; F is on line BG. The **conclusions**: (1) $AE \equiv BE$; (2) $BE \equiv EF$.

$O = (0,0)$, $B = (u_1, 0)$, $C = (x_1, 0)$, $G = (x_2, u_2)$, $M = (x_3, x_4)$, $A = (x_6, x_5)$, $D = (x_6, 0)$, $E = (x_6, x_7)$, $F = (x_9, x_8)$.

The **nondegenerate** conditions: Line OM is non-isotropic; Line BC is non-isotropic; Line BG intersects line AD; Line BG intersects line AC.

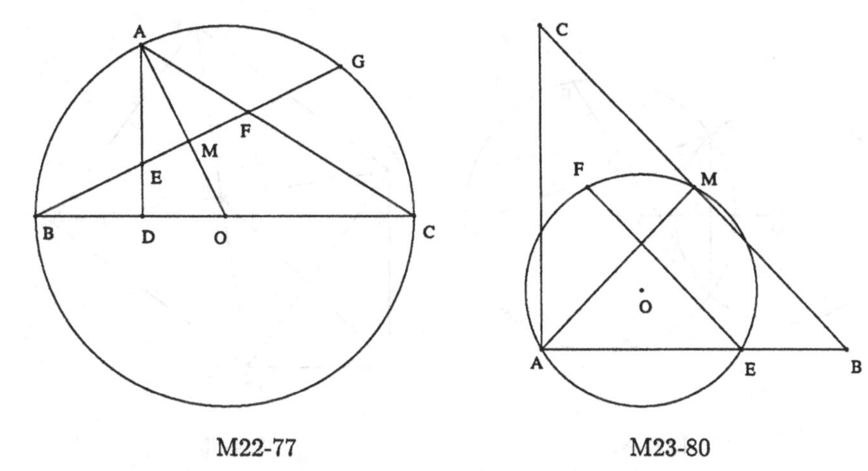

M22-77 M23-80

Example 98 (M23-80: 0.95s, 11). *Let M be the midpoint of the hypotenuse of the right triangle ABC. A circle passing through A and M meet AB at E. F is the point on the circle such that $EF \parallel BC$. Show that $BC = 2EF$.*

Points A, B are arbitrarily chosen. Points C, M, O, E, F are constructed (in order) as follows: $CA \perp BA$; M is the midpoint of B and C; $OA \equiv OM$; E is on line AB; $EO \equiv OA$; $FO \equiv OA$; $FE \parallel BC$. The **conclusion**: $2EF = BC$.

$A = (0,0)$, $B = (u_1,0)$, $C = (0,u_2)$, $M = (x_1,x_2)$, $O = (x_3,u_3)$, $E = (x_4,0)$, $F = (x_6,x_5)$.

The **nondegenerate** conditions: $B \neq A$; $A \neq M$; Line AB is non-isotropic; Line BC is non-isotropic. In addition, the following nondegenerate conditions, which come from reducibility and have been detected by our prover, should be also added: $F \neq E$; $E \neq A$.

Example 99 (M23-81: 59.7s, 1472). *Let A and B be the two common points of two circles (O) and (O_1). Through A a line is drawn meeting the circles at C and D respectively. G is the midpoint of CD. Line BG intersects circles (O) and (O_1) at E and F, respectively. Show that $EG = GF$.*

Points X, O are arbitrarily chosen. Points O_1, A, B, C, D, G, E, F are constructed (in order) as follows: O_1 is on line XO; $AX \perp XO$; X is the midpoint of A and B; $CO \equiv OA$; D is on line CA; $DO_1 \equiv O_1A$; G is the midpoint of C and D; $EO \equiv OB$; E is on line BG; $FO_1 \equiv O_1B$; F is on line BG. The **conclusion**: G is the midpoint of F and E.

$X = (0,0)$, $O = (u_1,0)$, $O_1 = (u_2,0)$, $A = (0,u_3)$, $B = (0,x_1)$, $C = (x_2,u_4)$, $D = (x_4,x_3)$, $G = (x_5,x_6)$, $E = (x_8,x_7)$, $F = (x_{10},x_9)$.

The **nondegenerate** conditions: $X \neq O$; $X \neq O$; Line CA is non-isotropic; Line BG is non-isotropic; Line BG is non-isotropic. In addition, the following nondegenerate conditions, which come from reducibility and have been detected by our prover, should be also added: $F \neq B$; $E \neq B$; $D \neq A$.

Example 100 (M24-95: 2.37s, 59; 1.7s, 15). *On the hypotenuse AB of right triangle*

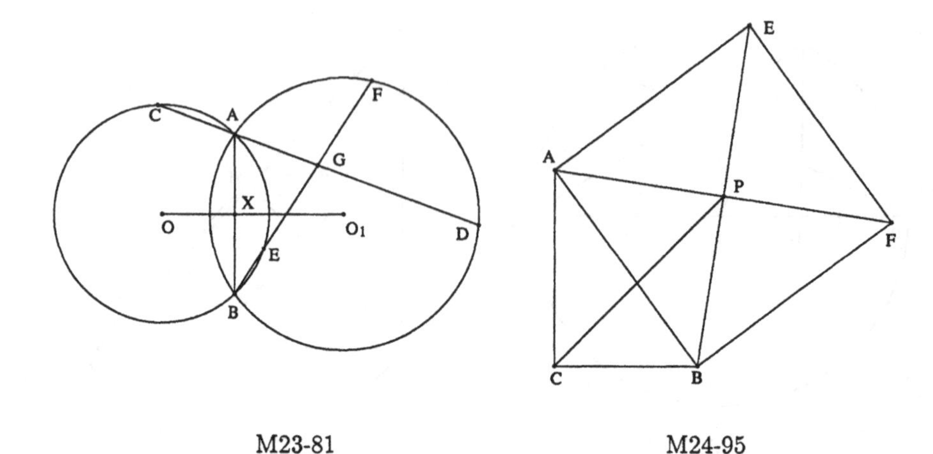

M23-81 M24-95

ABC a square ABFE is erected. Let P be the intersection of the diagonals AF and BE of ABFE. Show that $\angle ACP \equiv \angle PCB$.

Points C, B are arbitrarily chosen. Points A, E, F, P are constructed (in order) as follows: $AC \perp BC$; $EA \equiv AB$; $EA \perp AB$; $FB \parallel AE$; $FE \parallel AB$; P is on line BE; P is on line AF. The **conclusion**: $\tan(ACP) = \tan(PCB)$.

$C = (0,0)$, $B = (u_1, 0)$, $A = (0, u_2)$, $E = (x_2, x_1)$, $F = (x_4, x_3)$, $P = (x_6, x_5)$.

The **nondegenerate** conditions: $B \neq C$; Line AB is non-isotropic; Points A, B and E are not collinear; Line AF intersects line BE.

Example 101 (M25-96: 1.1s, 18). *The circle with the altitude AD of triangle ABC as a diameter meets AB and AC at E and F, respectively. Show that B, C, E and F are on the same circle.*

Points B, C, A are arbitrarily chosen. Points D, O, E, F are constructed (in order) as follows: D is on line BC; $DA \perp BC$; O is the midpoint of A and D; E is on line AB; $EO \equiv OD$; F is on line AC; $FO \equiv OD$. The **conclusion**: Points E, F, C and B are on the same circle.

$B = (0,0)$, $C = (u_1, 0)$, $A = (u_2, u_3)$, $D = (u_2, 0)$, $O = (u_2, x_1)$, $E = (x_3, x_2)$, $F = (x_5, x_4)$.

The **nondegenerate** conditions: Line BC is non-isotropic; Line AB is non-isotropic; Line AC is non-isotropic. In addition, the following nondegenerate conditions, which come from reducibility and have been detected by our prover, should be also added: $F \neq A$; $E \neq A$.

Example 102 (M25-98: 1.07s, 17). *The two tangents to the circumcircle of ABC at A and C meet at E. The mediator of BC meet AB at D. Show that* $DE \parallel BC$.

Points A, B, C are arbitrarily chosen. Points O, H, D, E are constructed (in order) as follows: $OA \equiv OC$; $OA \equiv OB$; H is the midpoint of B and C; D is on line AB; D is on line OH; $EC \perp OC$; $EA \perp OA$. The **conclusion**: $DE \parallel BC$.

$A = (0,0)$, $B = (u_1, 0)$, $C = (u_2, u_3)$, $O = (x_2, x_1)$, $H = (x_3, x_4)$, $D = (x_5, 0)$,

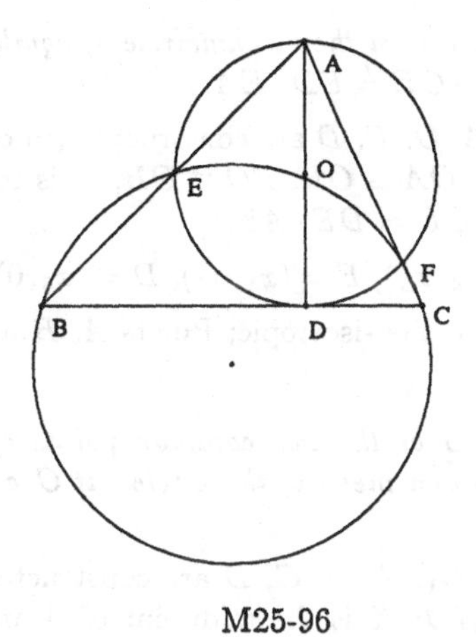

M25-96

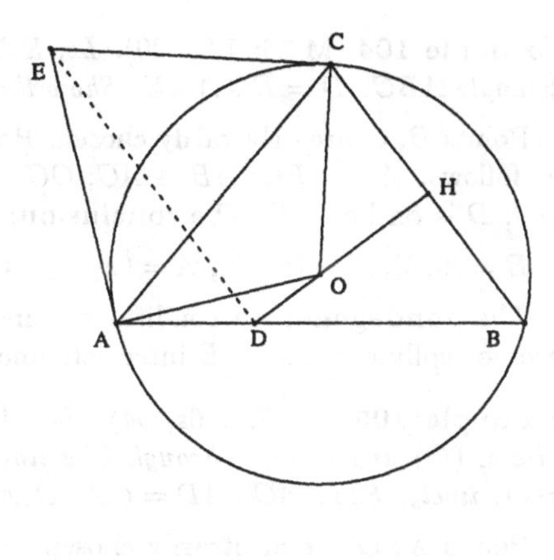

M25-98

$E = (x_7, x_6)$.

The **nondegenerate** conditions: Points A, B and C are not collinear; Line OH intersects line AB; Points O, A and C are not collinear.

Example 103 (M-24: 0.77s, 13). *A line parallel to the base of trapezoid $ABCD$ meet its two sides and two diagonals at H, G, F, E. Show that $EF \equiv GH$.*

Points A, B, C are arbitrarily chosen. Points D, E, H, F, G are constructed (in order) as follows: $DC \parallel BA$; E is on line BC; H is on line AD; $HE \parallel BA$; F is on line EH; F is on line BD; G is on line EF; G is on line AC. The **conclusion**: $EF \equiv GH$.

$A = (0,0)$, $B = (u_1, 0)$, $C = (u_2, u_3)$, $D = (u_4, u_3)$, $E = (x_1, u_5)$, $H = (x_2, u_5)$, $F = (x_3, u_5)$, $G = (x_4, u_5)$.

The **nondegenerate** conditions: $B \neq A$; $B \neq C$; Points B, A and D are not collinear; Line BD intersects line EH; Line AC intersects line EF.

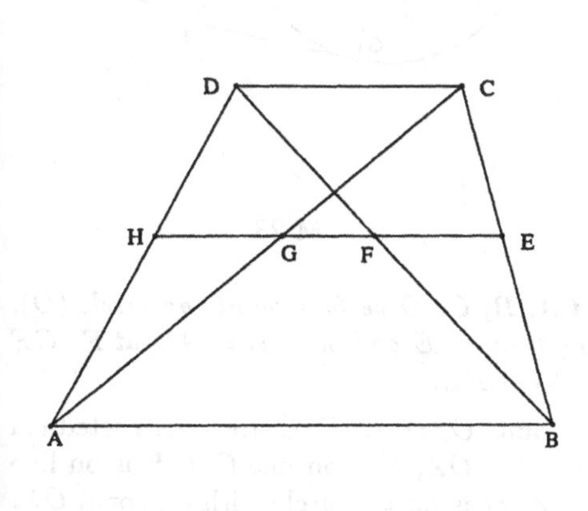

M-24

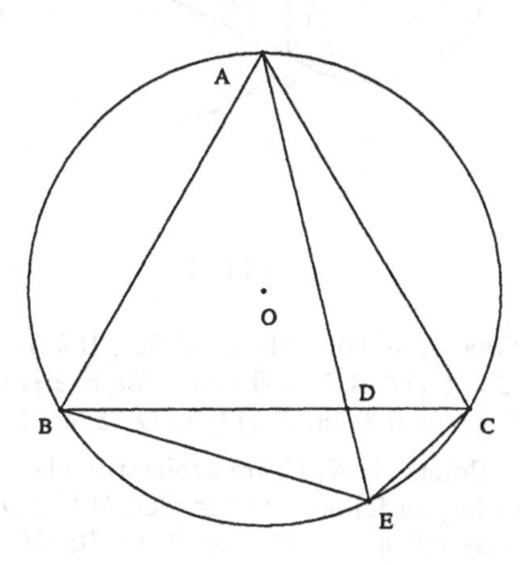

M-26

Example 104 (M-26: 1.2s, 30). *Let E be a point on the circumcircle of equilateral triangle ABC. $D = BC \cap AE$. Show that $BE \cdot CE = ED \cdot EA$.*

Points B, C are arbitrarily chosen. Points A, O, E, D are constructed (in order) as follows: $AB \equiv BC$; $AB \equiv AC$; $OC \equiv OB$; $OA \equiv OB$; $EO \equiv OB$; D is on line BC; D is on line AE. The **conclusion**: $BE \cdot CE = DE \cdot AE$.

$B = (0,0)$, $C = (u_1,0)$, $A = (x_2,x_1)$, $O = (x_4,x_3)$, $E = (x_5,u_2)$, $D = (x_6,0)$.

The **nondegenerate** conditions: Line BC is non-isotropic; Points A, B and C are not collinear; Line AE intersects line BC.

Example 105 (M-27: 1.6s, 94). *Let A and B be the two common points of two circles (O) and (O_1). Through B a line is drawn meeting the circles at C and D respectively. Show $AC : AD = OA : O_1A$.*

Points X, O are arbitrarily chosen. Points O_1, A, B, C, D are constructed (in order) as follows: O_1 is on line OX; $AX \perp XO$; X is the midpoint of A and B; $CO \equiv OB$; $DO_1 \equiv O_1B$; D is on line CB. The **conclusion**: $AC \cdot O_1A = AD \cdot OA$.

$X = (0,0)$, $O = (u_1,0)$, $O_1 = (u_2,0)$, $A = (0,u_3)$, $B = (0,x_1)$, $C = (x_2,u_4)$, $D = (x_4,x_3)$.

The **nondegenerate** conditions: $O \neq X$; $X \neq O$; Line CB is non-isotropic. In addition, the following nondegenerate conditions, which come from reducibility and have been detected by our prover, should be also added: $D \neq B$.

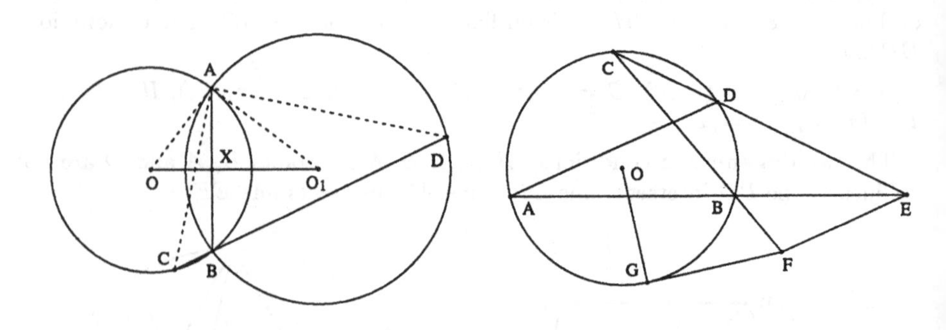

M-27 M-28

Example 106 (M-28: 51.82s, 1594). *Let A, B, C, D be four points on circle (O). $E = CD \cap AB$. CB meets the line passing through E and parallel to AD at F. GF is tangent to circle (O) at G. Show that $FG \equiv FE$.*

Points A, B, C are arbitrarily chosen. Points O, D, E, F, G are constructed (in order) as follows: $OA \equiv OC$; $OA \equiv OB$; $DO \equiv OA$; E is on line CD; E is on line AB; $FE \parallel AD$; F is on line CB; $GO \equiv OA$; G is on the circle with diagonal OF. The **conclusion**: $FG \equiv FE$.

$A = (0,0)$, $B = (u_1,0)$, $C = (u_2,u_3)$, $O = (x_2,x_1)$, $D = (x_3,u_4)$, $E = (x_4,0)$, $F = (x_6,x_5)$, $G = (x_8,x_7)$.

The **nondegenerate** conditions: Points A, B and C are not collinear; Line AB intersects line CD; Line CB intersects line AD; The line joining the midpoint of O and F and the point O is non-isotropic.

Example 107 (M34-121: 1.0s, 21). *Let ABC be a triangle with $AC \equiv AB$. D is a point on BC. Line AD meets the circumcircle of ABC at E. Show that $AB^2 = AD \cdot AE$.*

Points A, B are arbitrarily chosen. Points C, O, E, D are constructed (in order) as follows: $CA \equiv AB$; $OA \equiv OC$; $OA \equiv OB$; $EO \equiv OA$; D is on line AE; D is on line BC. The **conclusion**: $AB \cdot AB = AD \cdot AE$.

$A = (0,0)$, $B = (u_1,0)$, $C = (x_1,u_2)$, $O = (x_3,x_2)$, $E = (x_4,u_3)$, $D = (x_6,x_5)$.

The **nondegenerate** conditions: Points A, B and C are not collinear; Line BC intersects line AE.

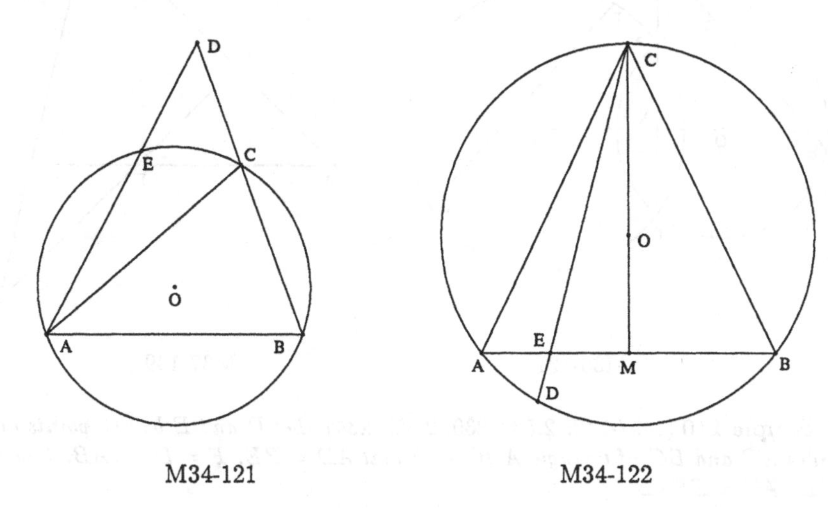

M34-121 M34-122

Example 108 (M34-122: 1.75s, 20). *Let C be the midpoint of the arc AB of circle (O). D is a point on the circle. $E = AB \cap CD$. Show that $CA^2 = CE \cdot CD$.*

Points A, B are arbitrarily chosen. Points O, M, C, D, E are constructed (in order) as follows: $OA \equiv OB$; M is the midpoint of A and B; $CO \equiv OA$; C is on line OM; $DO \equiv OA$; E is on line AB; E is on line CD. The **conclusion**: $CA \cdot CA = CE \cdot CD$.

$A = (0,0)$, $B = (u_1,0)$, $O = (x_1,u_2)$, $M = (x_2,0)$, $C = (x_4,x_3)$, $D = (x_5,u_3)$, $E = (x_6,0)$.

The **nondegenerate** conditions: $A \neq B$; Line OM is non-isotropic; Line CD intersects line AB.

Example 109 (M35-127: 31.55s, 1015). *From a point P on the line joining the two common points A and B of two circles (O) and (O_1) two secants PCE and PFD*

are drawn to the circles respectively. Show that $PC \cdot PE = PF \cdot PD$.

Points X, O are arbitrarily chosen. Points O_1, A, B, P, E, C, F, D are constructed (in order) as follows: O_1 is on line OX; $AX \perp XO$; X is the midpoint of A and B; P is on line AB; $EO \equiv OA$; $CO \equiv OA$; C is on line PE; $FO_1 \equiv O_1A$; $DO_1 \equiv O_1A$; D is on line PF. The **conclusion**: $PC \cdot PE = PD \cdot PF$.

$X = (0,0)$, $O = (u_1,0)$, $O_1 = (u_2,0)$, $A = (0,u_3)$, $B = (0,x_1)$, $P = (0,u_4)$, $E = (x_2,u_5)$, $C = (x_4,x_3)$, $F = (x_5,u_6)$, $D = (x_7,x_6)$.

The **nondegenerate** conditions: $O \neq X$; $X \neq O$; $A \neq B$; Line PE is non-isotropic; Line PF is non-isotropic. In addition, the following nondegenerate conditions, which come from reducibility and have been detected by our prover, should be also added: $D \neq F$; $C \neq E$.

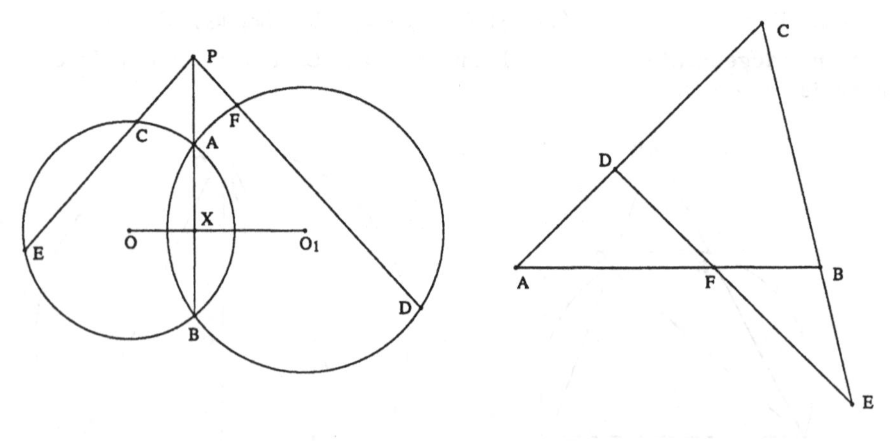

M35-127 M37-149

Example 110 (M37-149: 2.53s, 230; 2.35s, 234). *Let D and E be two points on two sides AC and BC of triangle ABC such that $AD = BE$. $F = DE \cap AB$. Show that $FD \cdot AC = EF \cdot BC$.*

Points A, B, C are arbitrarily chosen. Points D, E, F are constructed (in order) as follows: D is on line AC; $EB \equiv AD$; E is on line CB; F is on line ED; F is on line AB. The **conclusion**: $FD \cdot AC = EF \cdot BC$.

$A = (0,0)$, $B = (u_1,0)$, $C = (u_2,u_3)$, $D = (x_1,u_4)$, $E = (x_3,x_2)$, $F = (x_4,0)$.

The **nondegenerate** conditions: $A \neq C$; Line CB is non-isotropic; Line AB intersects line ED.

Example 111 (M38-156: 2.28s, 91). *Let D be the intersection of one of the bisectors of $\angle A$ of triangle ABC with side BC, E be the intersection of AD with the circumcircle of ABC. Show that $AB \cdot AC = AD \cdot AE$.*

Points B, C, A are arbitrarily chosen. Points D, O, E are constructed (in order) as follows: D is on line BC; $\tan(BAD) = \tan(DAC)$; $OC \equiv OB$; $OA \equiv OB$; E is on line AD; $EO \equiv OB$. The **conclusion**: $AB \cdot AC = AE \cdot AD$.

$B = (0,0)$, $C = (u_1,0)$, $A = (u_2,u_3)$, $D = (x_1,0)$, $O = (x_3,x_2)$, $E = (x_5,x_4)$.

The **nondegenerate** conditions: $(2u_2 - u_1)u_3 \neq 0$; Points A, B and C are not collinear; Line AD is non-isotropic. In addition, the following nondegenerate conditions, which come from reducibility and have been detected by our prover, should be also added: $E \neq A$.

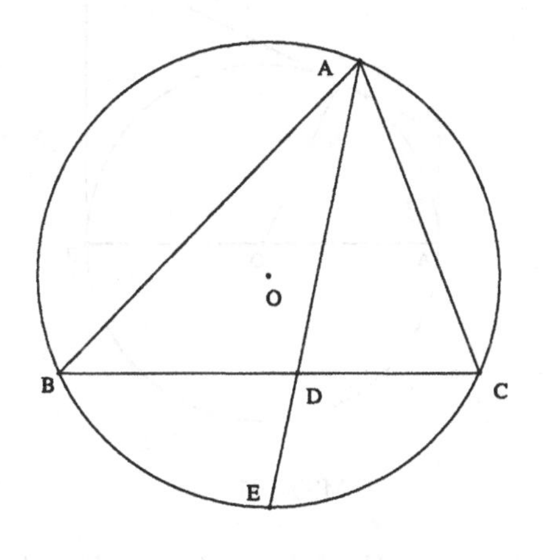

M38-156

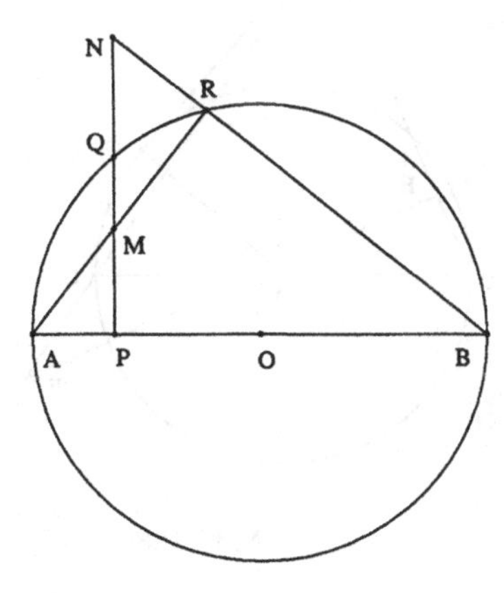

M38-157

Example 112 (M38-157: 0.78s, 16). *Let R be a point on the circle with diameter AB. At a point P of AB a perpendicular is drawn meeting BR at N, AR at M, and meeting the circle at Q. Show that $PQ^2 = PM \cdot PN$.*

Points O, A are arbitrarily chosen. Points B, P, Q, R, M, N are constructed (in order) as follows: O is the midpoint of A and B; P is on line AB; $QO \equiv OA$; $QP \perp AB$; $RO \equiv OA$; M is on line PQ; M is on line AR; N is on line PQ; N is on line BR. The **conclusion**: $PM \cdot PN = PQ \cdot PQ$.

$O = (0,0)$, $A = (u_1,0)$, $B = (x_1,0)$, $P = (u_2,0)$, $Q = (u_2,x_2)$, $R = (x_3,u_3)$, $M = (u_2,x_4)$, $N = (u_2,x_5)$.

The **nondegenerate** conditions: $A \neq B$; Line AB is non-isotropic; Line AR intersects line PQ; Line BR intersects line PQ.

Example 113 (M39-158: 8.62s, 336). *Let ABC be a triangle. Through A a line is drawn tangent to the circle with diameter BC at D. Let E be a point on AB such that $AD \equiv AE$. The perpendicular to AB at E meets AC at F. Show that $AE/AB = AC/AF$.*

Points B, C, A are arbitrarily chosen. Points O, D, E, F are constructed (in order) as follows: O is the midpoint of B and C; $DO \equiv OB$; D is on the circle with diagonal AO; E is on line AB; $EA \equiv AD$; F is on line AC; $FE \perp AB$. The **conclusion**: $AB \cdot AC = AE \cdot AF$.

$B = (0,0)$, $C = (u_1,0)$, $A = (u_2,u_3)$, $O = (x_1,0)$, $D = (x_3,x_2)$, $E = (x_5,x_4)$, $F = (x_7,x_6)$.

The **nondegenerate** conditions: The line joining the midpoint of A and O and the point O is non-isotropic; Line AB is non-isotropic; Line AB is not perpendicular to line AC.

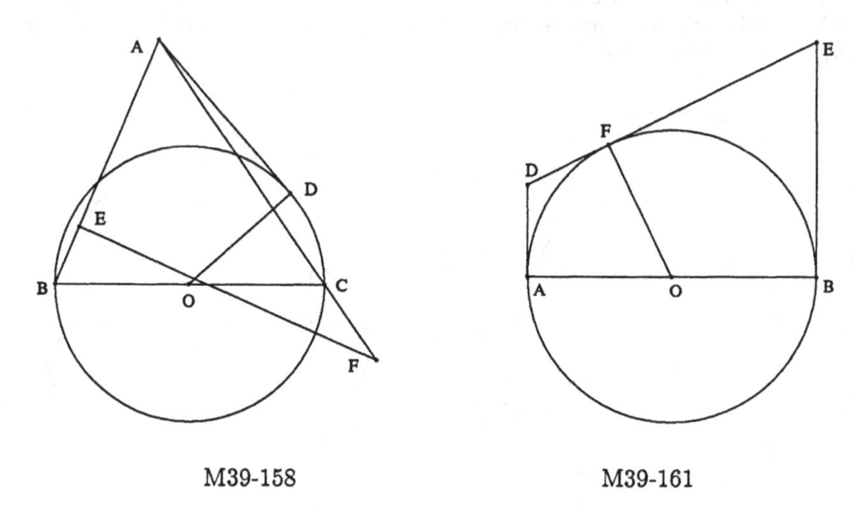

M39-158 M39-161

Example 114 (M39-161: 0.75s, 15). *Through point F on the circle with diameter AB a tangent to the circle is drawn meeting the two lines, perpendicular to AB at A and B, at D and E. Show the* $OA^2 = DF \cdot EF$.

Points O, A are arbitrarily chosen. Points B, F, D, E are constructed (in order) as follows: O is the midpoint of A and B; $FO \equiv OA$; $DF \perp OF$; $DA \perp AB$; E is on line FD; $EB \perp AB$. The **conclusion**: $DF \cdot EF = OA \cdot OA$.

$O = (0,0)$, $A = (u_1, 0)$, $B = (x_1, 0)$, $F = (x_2, u_2)$, $D = (u_1, x_3)$, $E = (x_1, x_4)$.

The **nondegenerate** conditions: Line AB intersects line OF; Line AB is not perpendicular to line FD.

Example 115 (M40-167: 1.78s, 23). *Let PA tangent to circle (O) at point A. M is the midpoint of PA. C is a point on the circle. PC and MC meet the circle at points E and B, respectively. PB meets the circle at D. Show that ED is parallel to AP.*

Points A, P are arbitrarily chosen. Points O, M, B, C, D, E are constructed (in order) as follows: $OA \perp AP$; M is the midpoint of A and P; $BO \equiv OA$; C is on line MB; $CO \equiv OA$; D is on line PB; $DO \equiv OA$; E is on line PC; $EO \equiv OA$. The **conclusion**: $ED \parallel PA$.

$A = (0,0)$, $P = (u_1, 0)$, $O = (0, u_2)$, $M = (x_1, 0)$, $B = (x_2, u_3)$, $C = (x_4, x_3)$, $D = (x_6, x_5)$, $E = (x_8, x_7)$.

The **nondegenerate** conditions: $A \neq P$; Line MB is non-isotropic; Line PB is non-isotropic; Line PC is non-isotropic. In addition, the following nondegenerate conditions, which come from reducibility and have been detected by our prover, should be also added: $E \neq C$; $D \neq B$; $C \neq B$.

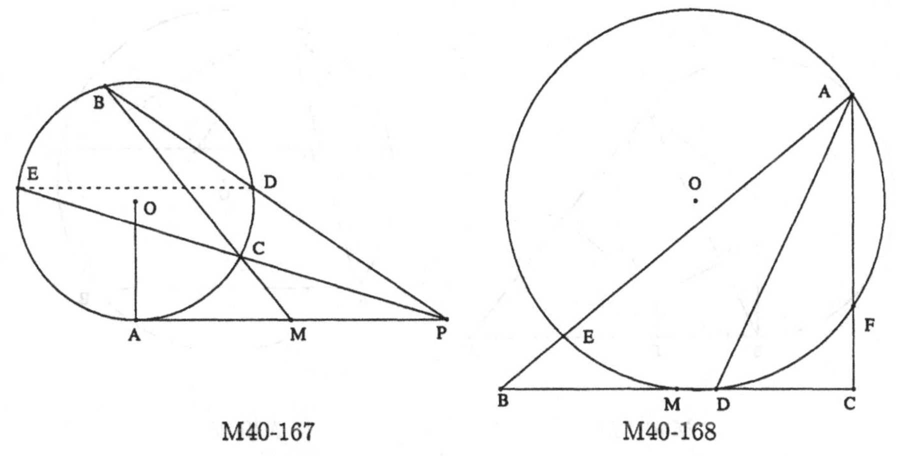

M40-167 M40-168

Example 116 (M40-168: 2.4s, 134). *In triangle ABC, M is the midpoint of BC; the bisector of $\angle A$ meets BC at D. The circle passing through A, D, M meets AB and AC at E and F, respectively. Show that $BE \equiv CF$.*

Points M, B, A are arbitrarily chosen. Points C, D, O, E, F are constructed (in order) as follows: M is the midpoint of B and C; D is on line BC; $\tan(BAD) = \tan(DAC)$; $OM \equiv OA$; $OM \equiv OD$; E is on line AB; $EO \equiv OM$; F is on line AC; $FO \equiv OM$. The **conclusion**: $BE \equiv CF$.

$M = (0,0)$, $B = (u_1,0)$, $A = (u_2,u_3)$, $C = (x_1,0)$, $D = (x_2,0)$, $O = (x_4,x_3)$, $E = (x_6,x_5)$, $F = (x_8,x_7)$.

The **nondegenerate** conditions: $-u_3x_1 + (2u_2 - u_1)u_3 \neq 0$; Points M, D and A are not collinear; Line AB is non-isotropic; Line AC is non-isotropic. In addition, the following nondegenerate conditions, which come from reducibility and have been detected by our prover, should be also added: $F \neq A$; $E \neq A$.

Example 117 (M40-169: 10.02s, 277). *From a point A two lines are drawn tangent to circle (O) at B and C. From a point P on the circle perpendiculars are drawn to BC, AB, and AC. Let D, F, E be the feet. Show that $PD^2 = PE \cdot PF$.*

Points B, A are arbitrarily chosen. Points O, C, P, D, E, F are constructed (in order) as follows: $OB \perp BA$; $CO \equiv OB$; C is on the circle with diagonal AO; $PO \equiv OB$; D is on line BC; $DP \perp BC$; E is on line AC; $EP \perp AC$; F is on line AB; $FP \perp AB$. The **conclusion**: $PD \cdot PD = PE \cdot PF$.

$B = (0,0)$, $A = (u_1,0)$, $O = (0,u_2)$, $C = (x_2,x_1)$, $P = (x_3,u_3)$, $D = (x_5,x_4)$, $E = (x_7,x_6)$, $F = (x_3,0)$.

The **nondegenerate** conditions: $B \neq A$; The line joining the midpoint of A and O and the point O is non-isotropic; Line BC is non-isotropic; Line AC is non-isotropic; Line AB is non-isotropic.

Example 118 (A30-28: 1.53s, 77). *The circle through the vertices A, B, C of a parallelogram $ABCD$ meets DA, DC in the points A_1, C_1. Prove that $A_1D : A_1C_1 = A_1C : A_1B$.*

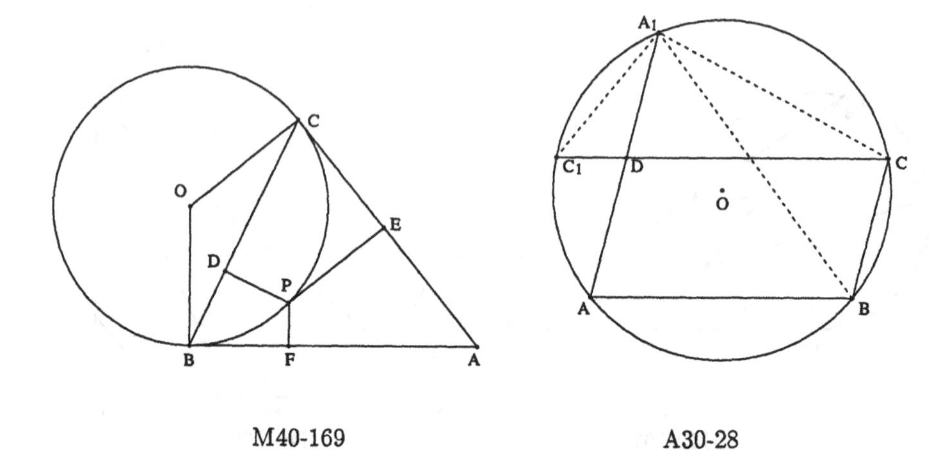

M40-169 A30-28

Points A, B, C are arbitrarily chosen. Points D, O, A_1, C_1 are constructed (in order) as follows: $DA \parallel CB$; $DC \parallel BA$; $OA \equiv OC$; $OA \equiv OB$; A_1 is on line DA; $A_1O \equiv OA$; C_1 is on line DC; $C_1O \equiv OA$. The **conclusion**: $A_1D \cdot A_1B = A_1C_1 \cdot A_1C$.

$A = (0,0)$, $B = (u_1, 0)$, $C = (u_2, u_3)$, $D = (x_1, u_3)$, $O = (x_3, x_2)$, $A_1 = (x_5, x_4)$, $C_1 = (x_6, u_3)$.

The **nondegenerate** conditions: Points B, A and C are not collinear; Points A, B and C are not collinear; Line DA is non-isotropic; Line DC is non-isotropic. In addition, the following nondegenerate conditions, which come from reducibility and have been detected by our prover, should be also added: $C_1 \neq C$; $A_1 \neq A$.

Example 119 (A30-29: 0.88s, 33). *Of the three lines joining the vertices of an equilateral triangle to a point on its circumcircle, one is equal to the sum of the other two.*

Points B, C are arbitrarily chosen. Points A, O, D are constructed (in order) as follows: $AB \equiv AC$; $AB \equiv BC$; $OA \equiv OB$; $OB \equiv OC$; $DO \equiv OB$. The **conclusion**: Algebraic sum of segments DA, DB, $DC = 0$.

$B = (0,0)$, $C = (u_1, 0)$, $A = (x_2, x_1)$, $O = (x_4, x_3)$, $D = (x_5, u_2)$.

The **nondegenerate** conditions: Line BC is non-isotropic; Points B, C and A are not collinear.

Example 120 (A30-30: 0.62s, 7). *Three parallel lines drawn through the vertices of a triangle ABC meet the respectively opposite sides in the points X, Y, Z. Show that area XYZ : area ABC = 2 : 1.*

Points A, B, C are arbitrarily chosen. Points X, Y, Z are constructed (in order) as follows: X is on line BC; Y is on line AC; $YB \parallel AX$; Z is on line AB; $ZC \parallel AX$. The **conclusion**: $2\ \nabla(BAC) - \nabla(XYZ) = 0$.

$A = (0,0)$, $B = (u_1, 0)$, $C = (u_2, u_3)$, $X = (x_1, u_4)$, $Y = (x_3, x_2)$, $Z = (x_4, 0)$.

The **nondegenerate** conditions: $B \neq C$; Points A, X and C are not collinear;

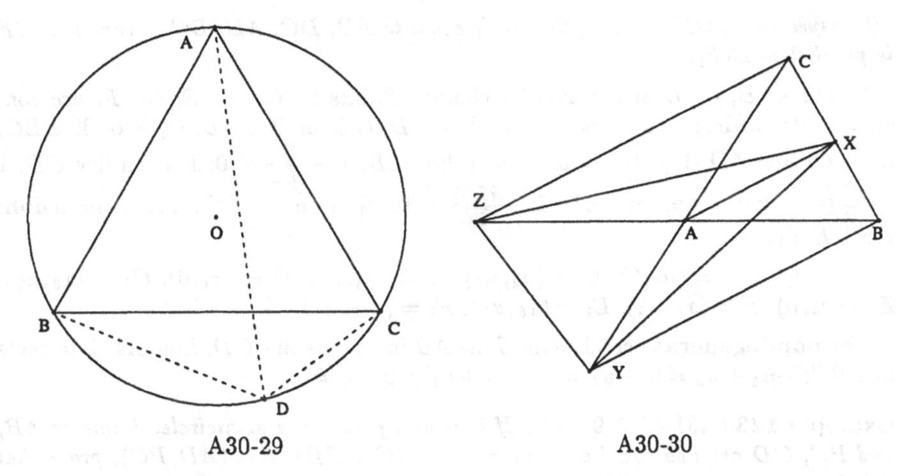

A30-29 A30-30

Points A, X and B are not collinear.

Example 121 (A30-32: 0.55s, 6). *Two parallel lines AE, BD through the vertices A, B of the triangle ABC meet a line through the vertex C in the points E, D. If the parallel through E to BC meets AB in F, show that DF is parallel to AC.*

Points A, B, C, D are arbitrarily chosen. Points E, F are constructed (in order) as follows: E is on line CD; $EA \parallel BD$; F is on line AB; $FE \parallel BC$. The **conclusion**: $DF \parallel AC$.

$A = (0,0)$, $B = (u_1, 0)$, $C = (u_2, u_3)$, $D = (u_4, u_5)$, $E = (x_2, x_1)$, $F = (x_3, 0)$.

The **nondegenerate** conditions: Points B, D and C are not collinear; Points B, C and A are not collinear.

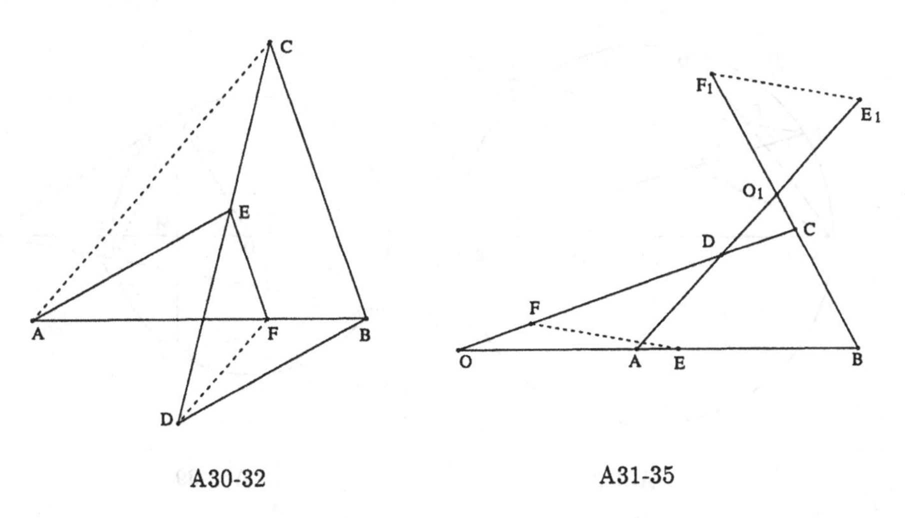

A30-32 A31-35

Example 122 (A31-35: 1.7s, 44). *The sides BA, CD of the quadrilateral $ABCD$ meet in O, and the sides DA, CB meet O_1. Along OA, OC, O_1A, O_1C are measured* *off, respectively, OE, OF, O_1E_1, O_1F_1 equal to AB, DC, AD, BC. Prove that EF is parallel to E_1F_1.*

Points A, B, C, D are arbitrarily chosen. Points O, O_1, E, F, E_1, F_1 are constructed (in order) as follows: O is on line CD; O is on line AB; O_1 is on line BC; O_1 is on line AD; $1 - \frac{\overline{OE}}{\overline{AB}} = 0$; E is on line AB; $1 - \frac{\overline{OF}}{\overline{DC}} = 0$; F is on line CD; $1 - \frac{\overline{O_1E_1}}{\overline{AD}} = 0$; E_1 is on line AD; $1 - \frac{\overline{O_1F_1}}{\overline{BC}} = 0$; F_1 is on line BC. The **conclusion:** $EF \parallel E_1F_1$.

$A = (0,0)$, $B = (u_1,0)$, $C = (u_2,u_3)$, $D = (u_4,u_5)$, $O = (x_1,0)$, $O_1 = (x_3,x_2)$, $E = (x_4,0)$, $F = (x_6,x_5)$, $E_1 = (x_8,x_7)$, $F_1 = (x_{10},x_9)$.

The **nondegenerate** conditions: Line AB intersects line CD; Line AD intersects line BC; $-u_4 + u_2 \neq 0$; $-u_4 \neq 0$; $-u_2 + u_1 \neq 0$; $x_5 \neq 0$.

Example 123 (A31-37: 1.9s, 43). *If P is any point on a semicircle, diameter AB, and BC, CD are two equal arcs, then if $E = (CA, PB)$, $F = (AD, PC)$, prove that AD is perpendicular to EF.*

Points O, A are arbitrarily chosen. Points B, C, D, P, E, F are constructed (in order) as follows: O is the midpoint of A and B; $CO \equiv OA$; $DC \equiv BC$; $DO \equiv OA$; $PO \equiv OA$; E is on line PB; E is on line CA; F is on line PC; F is on line AD. The **conclusion:** $AD \perp EF$.

$O = (0,0)$, $A = (u_1,0)$, $B = (x_1,0)$, $C = (x_2,u_2)$, $D = (x_4,x_3)$, $P = (x_5,u_3)$, $E = (x_7,x_6)$, $F = (x_9,x_8)$.

The **nondegenerate** conditions: Line OC is non-isotropic; Line CA intersects line PB; Line AD intersects line PC. In addition, the following nondegenerate conditions, which come from reducibility and have been detected by our prover, should be also added: $D \neq B$.

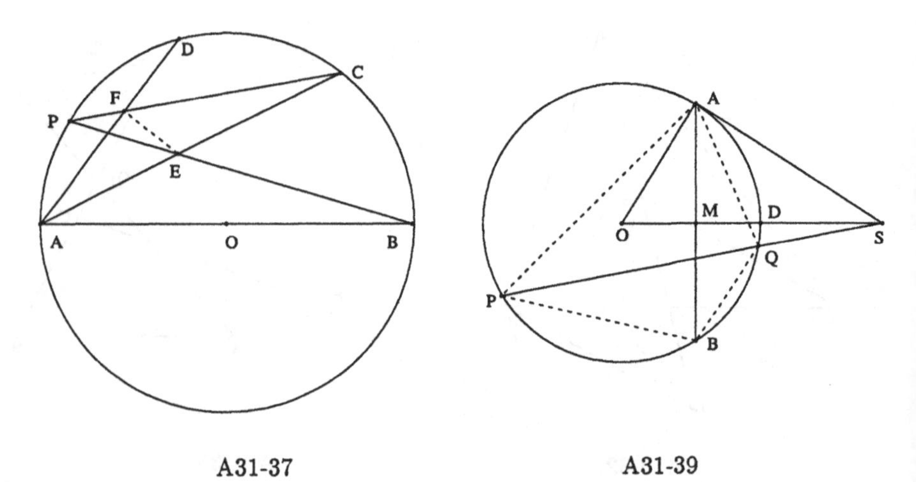

A31-37 A31-39

Example 124 (A31-39: 1.1s, 28). *From the point S the two tangents SA, SB and the secant SPQ are drawn to the same circle. Prove that $AP : AQ = BP : BQ$.*

Points O, D are arbitrarily chosen. Points A, S, M, B, P, Q are constructed (in order) as follows: $AO \equiv OD$; S is on line OD; $SA \perp AO$; M is on line OD; $MA \perp OD$; M is the midpoint of A and B; $PO \equiv OD$; Q is on line SP; $QO \equiv OD$. The **conclusion**: $AQ \cdot BP = AP \cdot BQ$.

$O = (0,0)$, $D = (u_1,0)$, $A = (x_1,u_2)$, $S = (x_2,0)$, $M = (x_1,0)$, $B = (x_1,x_3)$, $P = (x_4,u_3)$, $Q = (x_6,x_5)$.

The **nondegenerate** conditions: Line AO is not perpendicular to line OD; Line OD is non-isotropic; Line SP is non-isotropic. In addition, the following nondegenerate conditions, which come from reducibility and have been detected by our prover, should be also added: $Q \neq P$.

Example 125 (A31-40: 0.73s, 5). *On the radius OA, produced, take any point P and draw a tangent PT; produce OP to Q, making $PQ = PT$, and draw a tangent QV; if VR be drawn perpendicular to OA, meeting OA at R, prove that $PR = PQ = PT$.*

Points O, A are arbitrarily chosen. Points T, P, Q, V, R are constructed (in order) as follows: $TO \equiv OA$; P is on line OA; $PT \perp TO$; Q is on line AO; $QP \equiv PT$; $VO \equiv OA$; V is on the circle with diagonal QO; R is on line AO; $RV \perp AO$. The **conclusion**: $PQ \equiv PR$.

$O = (0,0)$, $A = (u_1,0)$, $T = (x_1,u_2)$, $P = (x_2,0)$, $Q = (x_3,0)$, $V = (x_5,x_4)$, $R = (x_5,0)$.

The **nondegenerate** conditions: Line TO is not perpendicular to line OA; Line AO is non-isotropic; The line joining the midpoint of Q and O and the point O is non-isotropic; Line AO is non-isotropic.

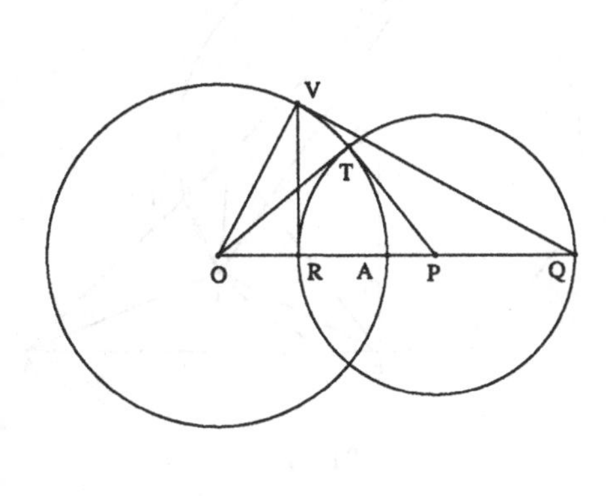

A31-40

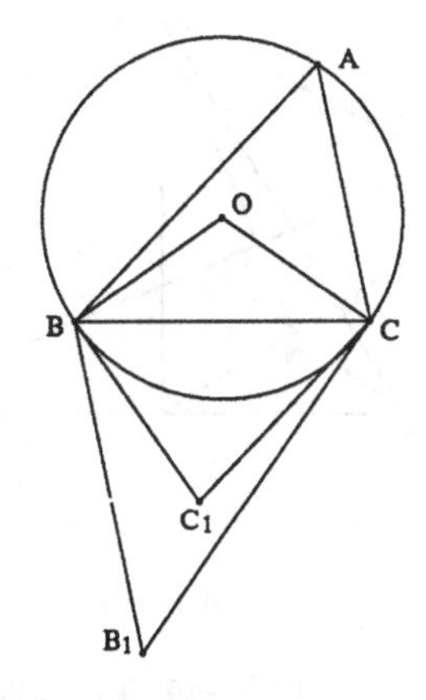

A31-41

Example 126 (A31-41: 1.07s, 20). *The parallel to the side AC through the vertex B of the triangle ABC meets the tangent to the circumcircle (O) of ABC at C in B_1, and the parallel through C to AB meets the tangent to (O) at B in C_1. Prove that $BC^2 = BC_1 \cdot B_1C$.*

Points B, C, A are arbitrarily chosen. Points O, B_1, C_1 are constructed (in order) as follows: $OB \equiv OC$; $OA \equiv OB$; $B_1C \perp CO$; $B_1B \parallel CA$; $C_1B \perp BO$; $C_1C \parallel BA$. The **conclusion**: $BC \cdot BC = BC_1 \cdot B_1C$.

$B = (0,0)$, $C = (u_1, 0)$, $A = (u_2, u_3)$, $O = (x_2, x_1)$, $B_1 = (x_4, x_3)$, $C_1 = (x_6, x_5)$.

The **nondegenerate** conditions: Points A, B and C are not collinear; Line CA is not perpendicular to line CO; Line BA is not perpendicular to line BO.

Example 127 (A31-43: 1.03s, 9). *If Q, R are the projections of a point M of the internal bisector AM of the angle A of the triangle ABC upon the sides AC, AB, show that the perpendicular MP from M upon BC meets QR in the point N on the median AA_1 of ABC.*

Points A, B, M are arbitrarily chosen. Points C, Q, R, A_1, N are constructed (in order) as follows: $\tan(BAM) = \tan(MAC)$; Q is on line AC; $QM \perp AC$; R is on line AB; $RM \perp AB$; A_1 is the midpoint of B and C; N is on line AA_1; $NM \perp BC$. The **conclusion**: Points Q, R and N are collinear.

$A = (0,0)$, $B = (u_1, 0)$, $M = (u_2, u_3)$, $C = (x_1, u_4)$, $Q = (x_3, x_2)$, $R = (u_2, 0)$, $A_1 = (x_4, x_5)$, $N = (x_7, x_6)$.

The **nondegenerate** conditions: $A \neq B$, $A \neq M$; Line AC is non-isotropic; Line AB is non-isotropic; Line BC is not perpendicular to line AA_1.

Remark. In fact, our proof is for both the internal and external bisectors of angle A.

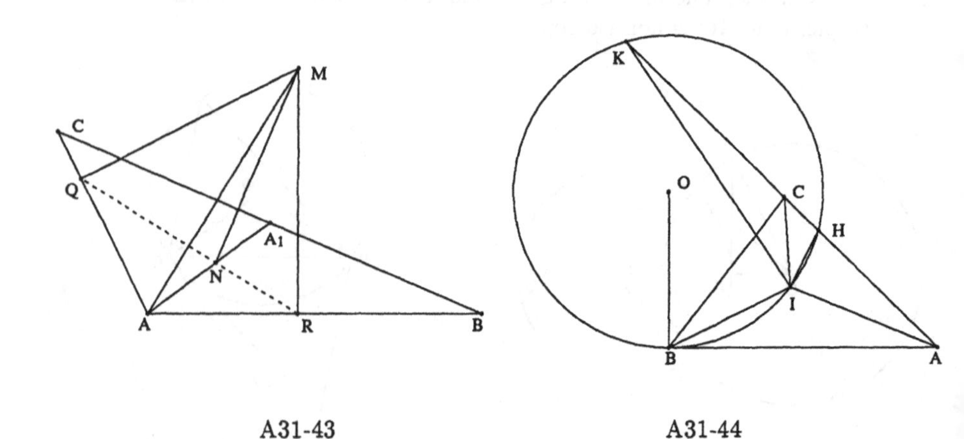

A31-43 A31-44

Example 128 (A31-44: 4.67s, 238). *A circle touching AB at B and passing through the incenter I (i.e., the center I of the inscribed circle) of the triangle ABC meets AC in H, K. Prove that IC bisects the angle HIK.*

Points B, A, I are arbitrarily chosen. Points C, O, H, K are constructed (in order) as follows: $\tan(BAI) = \tan(IAC)$; $\tan(ABI) = \tan(IBC)$; $OB \equiv OI$; $OB \perp AB$; H is on line AC; $HO \equiv OB$; $KO \equiv OB$; K is on line AC. The **conclusion**: $\tan(HIC) = \tan(CIK)$.

$B = (0,0)$, $A = (u_1, 0)$, $I = (u_2, u_3)$, $C = (x_2, x_1)$, $O = (0, x_3)$, $H = (x_5, x_4)$, $K = (x_7, x_6)$.

The **nondegenerate** conditions: Points I, B and A are not collinear; $\angle BIA$ is not right; Points A, B and I are not collinear; Line AC is non-isotropic; Line AC is non-isotropic. In addition, the following nondegenerate conditions, which come from reducibility and have been detected by our prover, should be also added: $K \neq H$.

Remark. Our proof is also for the three excenters.

Example 129 (A31-45: 2.05s, 30; 1.63s, 42). *AB, CD are two chords of the same circle, and the lines joining A, B to the midpoint of CD make equal angles with CD. Show that the lines joining C, D to the midpoint of AB make equal angles with AB.*

Points A, B are arbitrarily chosen. Points M, O, C, D, N are constructed (in order) as follows: M is the midpoint of A and B; $OM \perp AB$; $CO \equiv OA$; $\tan(BMC) = \tan(DMA)$; $DO \equiv OA$; N is the midpoint of C and D. The **conclusion**: $\tan(AND) = \tan(CNB)$.

$A = (0,0)$, $B = (u_1, 0)$, $M = (x_1, 0)$, $O = (x_1, u_2)$, $C = (x_2, u_3)$, $D = (x_4, x_3)$, $N = (x_5, x_6)$.

The **nondegenerate** conditions: $A \neq B$; $(-x_1^4 + 2u_1x_1^3 - u_1^2x_1^2)x_2^2 + (2x_1^5 - 4u_1x_1^4 + 2u_1^2x_1^3)x_2 - x_1^6 + 2u_1x_1^5 + (-u_3^2 - u_1^2)x_1^4 + 2u_1u_3^2x_1^3 - u_1^2u_3^2x_1^2 \neq 0$; $-u_3x_1^2 + u_1u_3x_1 \neq 0$.

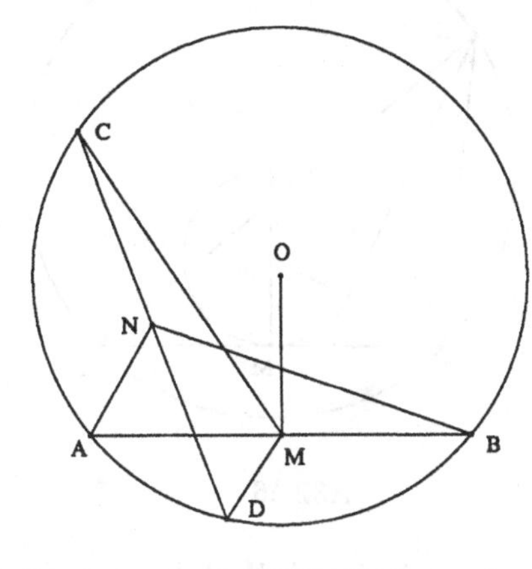

A31-45

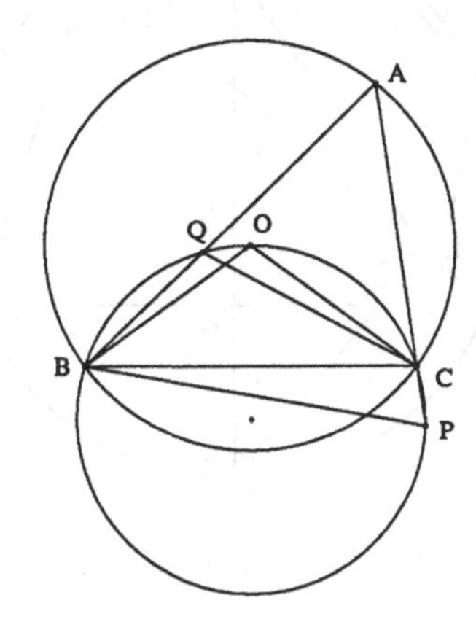

A31-47

Example 130 (A31-47: 2.23s, 28). *The mediators of the sides AC, AB of the triangle ABC meet the sides AB, AC in P and Q. Prove that the points B, C, P, Q lie on a circle which passes through the circumcenter (i.e., the center of the circumscribed circle) of ABC.*

Points B, C, A are arbitrarily chosen. Points P, Q, O are constructed (in order) as follows: $PA \equiv PB$; P is on line AC; $QA \equiv QC$; Q is on line AB; $OB \equiv OC$; $OB \equiv OA$. The **conclusions**: (1) Points B, C, P and Q are on the same circle; (2) Points B, C, P and O are on the same circle.

$B = (0,0)$, $C = (u_1, 0)$, $A = (u_2, u_3)$, $P = (x_2, x_1)$, $Q = (x_4, x_3)$, $O = (x_6, x_5)$.

The **nondegenerate** conditions: Line AC is not perpendicular to line AB; Line AB is not perpendicular to line AC; Points B, A and C are not collinear.

Example 131 (A32-49: 5.45s, 130; 2.9s, 104). *ABC is triangle inscribed in a circle; DE is the diameter bisecting BC at G; from E a perpendicular EK is drawn to one of the sides, and the perpendicular from the vertex A on DE meets DE in H. Show that EK touches the circle GHK.*

Points B, C, A are arbitrarily chosen. Points G, O, E, K, H, N are constructed (in order) as follows: G is the midpoint of B and C; $OG \perp BC$; $OB \equiv OA$; $EO \equiv OB$; E is on line GO; K is on line AB; $KE \perp AB$; H is on line OE; $HA \perp OE$; $NG \equiv NK$; $NG \equiv NH$. The **conclusion**: Line EK is tangent to circle (N, GN).

$B = (0,0)$, $C = (u_1,0)$, $A = (u_2,u_3)$, $G = (x_1,0)$, $O = (x_1,x_2)$, $E = (x_1,x_3)$, $K = (x_5,x_4)$, $H = (x_1,u_3)$, $N = (x_7,x_6)$.

The **nondegenerate** conditions: Points B, A and C are not collinear; Line GO is non-isotropic; Line AB is non-isotropic; Line OE is non-isotropic; Points G, H and K are not collinear.

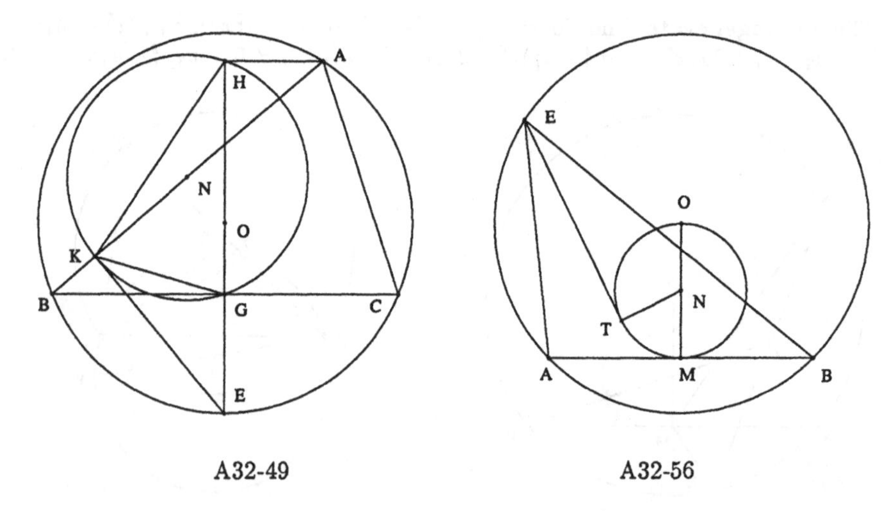

A32-49 A32-56

Example 132 (A32-56: 0.87s, 17). *Let M be the midpoint of chord AB of a circle, center O; on OM as diameter draw another circle, and at any point T of this circle draw a tangent to it meeting the outer circle in E. Prove that $AE^2 + BE^2 = 4ET^2$.*

Points M, A are arbitrarily chosen. Points B, O, N, T, E are constructed (in order) as follows: M is the midpoint of A and B; $OM \perp MA$; N is the midpoint of M and O; $TN \equiv NM$; $EO \equiv OA$; $ET \perp TN$. The **conclusion**: $4\,\overline{ET}^2 - (\overline{AE}^2 + \overline{BE}^2) = 0$.

$M = (0,0)$, $A = (u_1,0)$, $B = (x_1,0)$, $O = (0,u_2)$, $N = (0,x_2)$, $T = (x_3,u_3)$, $E = (x_5,x_4)$.

The **nondegenerate** conditions: $M \neq A$; Line TN is non-isotropic.

Example 133 (A32-58: 0.82s, 6). *If M, N are points on the sides AC, AB of a triangle ABC and the lines BM, CN intersect on the altitude AD, show that AD is the bisector of the angle MDN.*

Points B, C, A are arbitrarily chosen. Points D, J, M, N are constructed (in order) as follows: D is on line BC; $DA \perp BC$; J is on line AD; M is on line AB; M is on line CJ; N is on line AC; N is on line BJ. The **conclusion**: $\tan(MDA) = \tan(ADN)$.

$B = (0,0)$, $C = (u_1,0)$, $A = (u_2,u_3)$, $D = (u_2,0)$, $J = (u_2,u_4)$, $M = (x_2,x_1)$, $N = (x_4,x_3)$.

The **nondegenerate** conditions: Line BC is non-isotropic; $A \neq D$; Line CJ intersects line AB; Line BJ intersects line AC.

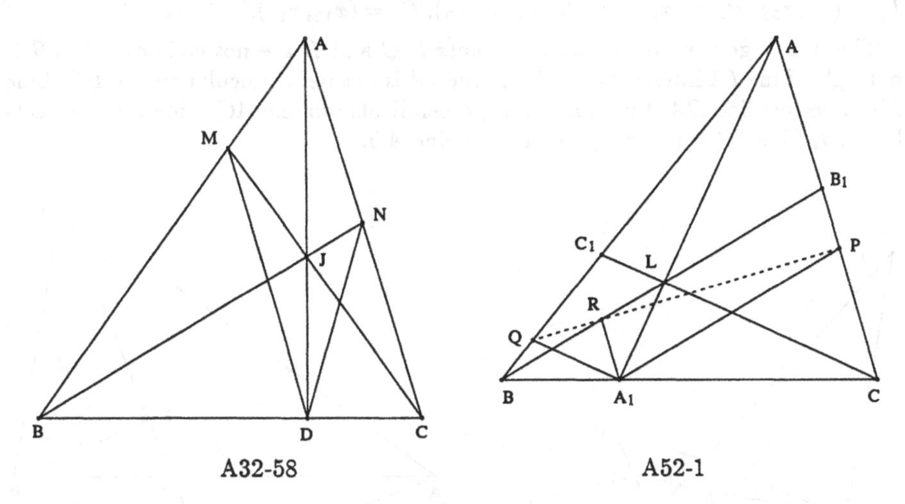

A32-58 A52-1

Example 134 (A52-1: 1.85s, 18). *The lines AL, BL, CL joining the vertices of a triangle ABC to a point L meet the respectively opposite sides in A_1, B_1, C_1. The parallels through A_1 to BB_1 CC_1 meet AC, AB in P, Q, and the parallels through A_1 to AC, AB meet BB_1, CC_1 in R, S. Show that the four points P, Q, R, S are collinear.*

Points B, C, A, L are arbitrarily chosen. Points A_1, B_1, C_1, P, Q, R are constructed (in order) as follows: A_1 is on line BC; A_1 is on line AL; B_1 is on line AC; B_1 is on line BL; C_1 is on line AB; C_1 is on line CL; P is on line AC; $PA_1 \parallel BB_1$; Q is on line AB; $QA_1 \parallel CC_1$; R is on line BB_1; $RA_1 \parallel AC$. The **conclusion**: Points P, Q and R are collinear.

$B = (0,0)$, $C = (u_1,0)$, $A = (u_2,u_3)$, $L = (u_4,u_5)$, $A_1 = (x_1,0)$, $B_1 = (x_3,x_2)$, $C_1 = (x_5,x_4)$, $P = (x_7,x_6)$, $Q = (x_9,x_8)$, $R = (x_{11},x_{10})$.

The **nondegenerate** conditions: Line AL intersects line BC; Line BL intersects line AC; Line CL intersects line AB; Line BB_1 intersects line AC; Line CC_1 intersects line AB; Line AC intersects line BB_1.

Example 135 (A57-2: > 14400s, > 50000; 112.27s, 976). *Show that the pairs of* *bisectors of the angles of a triangle determine on the respectively opposite sides three segments such that the reciprocal of one is equal to the sum of the reciprocals of the other two.*

Points B, C, I are arbitrarily chosen. Points A, A_1, A_2, B_1, B_2, C_1, C_2 are constructed (in order) as follows: $\tan(CBI) = \tan(IBA)$; $\tan(BCI) = \tan(ICA)$; A_1 is on line BC; A_1 is on line IA; A_2 is on line BC; $A_2A \perp IA$; B_1 is on line CA; B_1 is on line IB; B_2 is on line AC; $B_2B \perp IB$; C_1 is on line AB; C_1 is on line CI; C_2 is on line AB; $C_2C \perp IC$; $\overline{A_1A_2}^2\ x_{13}\ x_{13} - 1 = 0$; $\overline{B_1B_2}^2\ x_{14}\ x_{14} - 1 = 0$; $\overline{C_1C_2}^2\ x_{15}\ x_{15} - 1 = 0$. The **conclusion**: $(x_{13} + x_{14} - x_{15})\ (x_{13} + x_{14} + x_{15})\ (x_{13} - (x_{14} - x_{15}))\ (x_{13} - (x_{14} + x_{15})) = 0$.

$B = (0,0)$, $C = (u_1, 0)$, $I = (u_2, u_3)$, $A = (x_2, x_1)$, $A_1 = (x_3, 0)$, $A_2 = (x_4, 0)$, $B_1 = (x_6, x_5)$, $B_2 = (x_8, x_7)$, $C_1 = (x_{10}, x_9)$, $C_2 = (x_{12}, x_{11})$.

The **nondegenerate** conditions: Points I, C and B are not collinear; $\angle CIB$ is not right; Line IA intersects line BC; Line IA is not perpendicular to line BC; Line IB intersects line CA; Line IB is not perpendicular to line AC; Line CI intersects line AB; Line IC is not perpendicular to line AB.

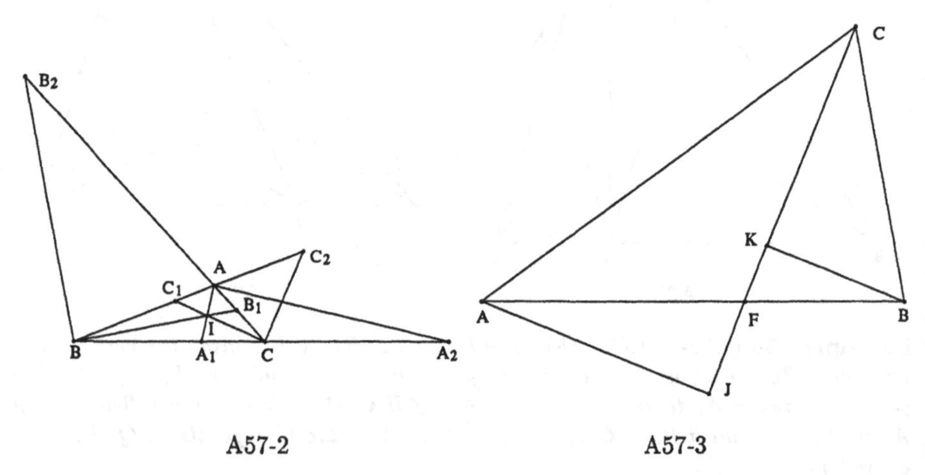

A57-2 A57-3

Example 136 (A57-3: 2.23s, 25). *Show that the internal (or external) bisector of an angle of a triangle is divided harmonically by the feet of the perpendiculars dropped upon it from the two other vertices of the triangle.*

Points A, B, C are arbitrarily chosen. Points F, J, K are constructed (in order) as follows: F is on line AB; $\tan(ACF) = \tan(FCB)$; J is on line CF; $JA \perp CF$; K is on line CF; $KB \perp CF$. The **conclusion**: Points C, F, J and K form a harmonic set.

$A = (0,0)$, $B = (u_1, 0)$, $C = (u_2, u_3)$, $F = (x_1, 0)$, $J = (x_3, x_2)$, $K = (x_5, x_4)$.

The **nondegenerate** conditions: $(2u_2 - u_1)u_3 \neq 0$; Line CF is non-isotropic; Line CF is non-isotropic.

Example 137 (A-73: 0.6s, 14). *The angle between the circumdiameter and the* *altitude issued from the same vertex of a triangle is bisected by the bisector of angle of the triangle at the vertex considered.*

Points A, B, C are arbitrarily chosen. Points O, D, F are constructed (in order) as follows: $OA \equiv OC$; $OA \equiv OB$; $DC \perp BA$; D is on line AB; $\tan(ACF) = \tan(FCB)$; F is on line AB. The **conclusion**: $\tan(DCF) = \tan(FCO)$.

$A = (0,0)$, $B = (u_1,0)$, $C = (u_2,u_3)$, $O = (x_2,x_1)$, $D = (u_2,0)$, $F = (x_3,0)$.

The **nondegenerate** conditions: Points A, B and C are not collinear; Line AB is non-isotropic; $(2u_2 - u_1)u_3 \neq 0$.

Remark. The bisector can be internal or external.

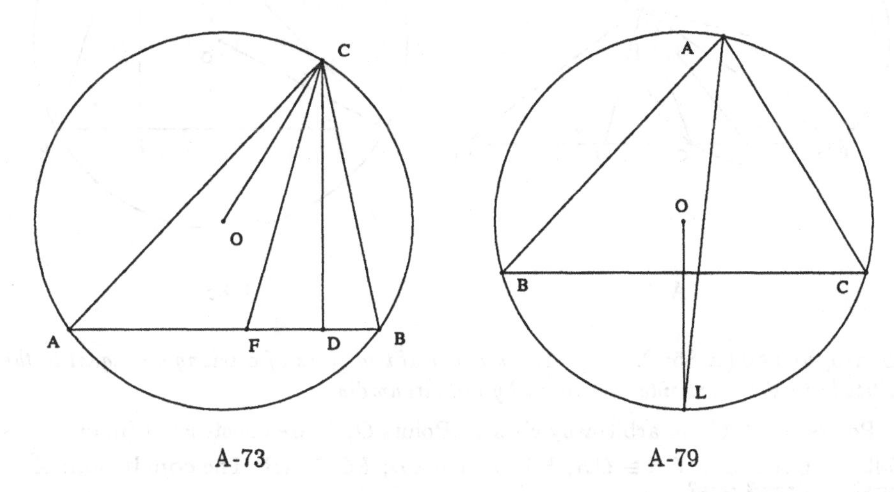

A-73 A-79

Example 138 (A-79: 0.67s, 17). *The internal and external bisectors of an angle of a triangle pass through the ends of the circumdiameters which is perpendicular to the side opposite the vertex considered.*

Points B, C, A are arbitrarily chosen. Points O, L are constructed (in order) as follows: $OB \equiv OC$; $OA \equiv OB$; $LO \equiv OB$; $LO \perp BC$. The **conclusion**: $\tan(BAL) = \tan(LAC)$.

$B = (0,0)$, $C = (u_1,0)$, $A = (u_2,u_3)$, $O = (x_2,x_1)$, $L = (x_2,x_3)$.

The **nondegenerate** conditions: Points A, B and C are not collinear; Line BC is non-isotropic.

Example 139 (A-81: 1.43s, 11). *Let U and U_1 be the intersections of the bisectors of angle A of a triangle ABC with side BC. If the tangent at A to the circumcircle meets BC in T, we have $TA \equiv TU \equiv TU_1$.*

Points B, C, A are arbitrarily chosen. Points U, U_1, O, T are constructed (in order) as follows: U is on line BC; $\tan(BAU) = \tan(UAC)$; $U_1A \perp UA$; U_1 is on line BC; $OB \equiv OA$; $OB \equiv OC$; T is on line BC; $TA \perp AO$. The **conclusions**: (1) $TA \equiv TU$; (2) $TA \equiv TU_1$.

$B = (0,0)$, $C = (u_1,0)$, $A = (u_2,u_3)$, $U = (x_1,0)$, $U_1 = (x_2,0)$, $O = (x_4,x_3)$,

$T = (x_5, 0)$.

The **nondegenerate** conditions: $(2u_2 - u_1)u_3 \neq 0$; Line BC is not perpendicular to line UA; Points B, C and A are not collinear; Line AO is not perpendicular to line BC.

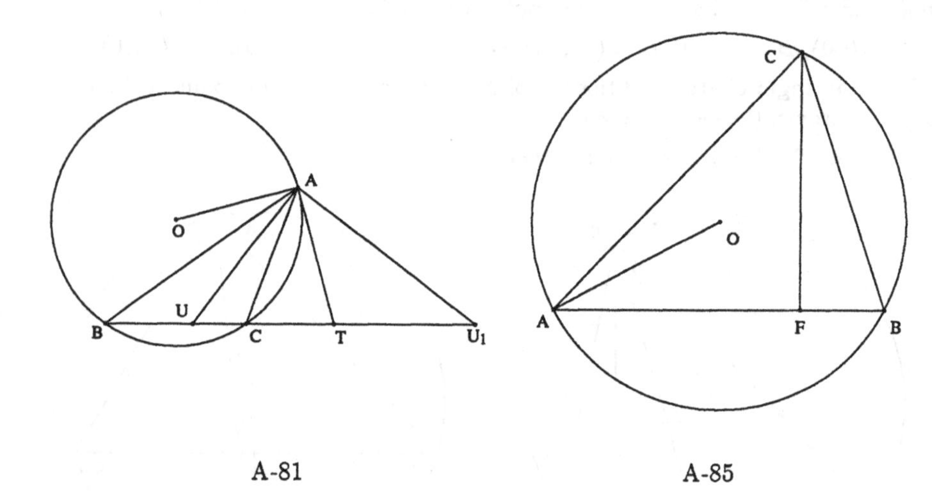

A-81 A-85

Example 140 (A-85: 0.5s, 9). *The product of two sides of a triangle is equal to the altitude to the third side multiplied by the circumdiameter.*

Points A, B, C are arbitrarily chosen. Points O, F are constructed (in order) as follows: $OA \equiv OC$; $OA \equiv OB$; F is on line AB; $FC \perp AB$. The **conclusion**: $\overline{AC}^2\ \overline{BC}^2 - 4\ \overline{OA}^2\ \overline{CF}^2 = 0$.

$A = (0,0)$, $B = (u_1, 0)$, $C = (u_2, u_3)$, $O = (x_2, x_1)$, $F = (u_2, 0)$.

The **nondegenerate** conditions: Points A, B and C are not collinear; Line AB is non-isotropic.

Example 141 (A-86: 0.43s, 9). *The area of a triangle is equal to the product of its three sides divided by the double circumdiameter of the triangle.*

Points A, B, C are arbitrarily chosen. Points O are constructed (in order) as follows: $OA \equiv OC$; $OA \equiv OB$. The **conclusion**: $\overline{AB}^2\ \overline{AC}^2\ \overline{CB}^2 - 4\ \nabla(ABC)\ \nabla(ABC)\ \overline{OA}^2 = 0$.

$A = (0,0)$, $B = (u_1, 0)$, $C = (u_2, u_3)$, $O = (x_2, x_1)$.

The **nondegenerate** conditions: Points A, B and C are not collinear.

Example 142 (A-88: 0.7s, 5). *Theorem of Centroid.*

Points A, B, C are arbitrarily chosen. Points D, E, F, M are constructed (in order) as follows: D is the midpoint of B and C; E is the midpoint of A and C; F is the midpoint of A and B; M is on line BE; M is on line AD. The **conclusion**: Points C, F and M are collinear.

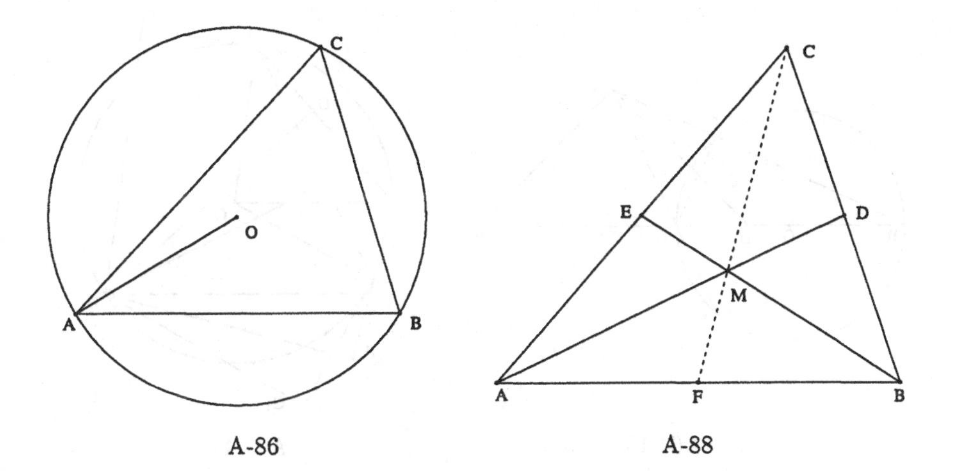

A-86 A-88

$A = (0,0)$, $B = (u_1, 0)$, $C = (u_2, u_3)$, $D = (x_1, x_2)$, $E = (x_3, x_4)$, $F = (x_5, 0)$, $M = (x_7, x_6)$.

The **nondegenerate** conditions: Line AD intersects line BE.

Example 143 (A64-10-a: 2.53s, 78). *If the two bisectors of the angle A of the triangle ABC are equal, and the circle having BC for diameter cuts the sides AB, AC in the points P, Q, show that $CP \equiv CQ$.*

Points U, V are arbitrarily chosen. Points B, A, C, O, P, Q are constructed (in order) as follows: B is on line UV; $AU \equiv AV$; A is on the circle with diagonal UV; C is on line UV; $\tan(CAU) = \tan(UAB)$; O is the midpoint of B and C; $PO \equiv OB$; P is on line AB; $QO \equiv OB$; Q is on line AC. The **conclusion**: $CQ \equiv CP$.

$U = (0,0)$, $V = (u_1, 0)$, $B = (u_2, 0)$, $A = (x_2, x_1)$, $C = (x_3, 0)$, $O = (x_4, 0)$, $P = (x_6, x_5)$, $Q = (x_8, x_7)$.

The **nondegenerate** conditions: $U \neq V$; Line UV is non-isotropic; $x_1x_2^2 - 2u_2x_1x_2 + x_1^3 \neq 0$; Line AB is non-isotropic; Line AC is non-isotropic. In addition, the following nondegenerate conditions, which come from reducibility and have been detected by our prover, should be also added: $Q \neq C$; $P \neq B$.

Example 144 (A65-4: 883.08s, 16537; 620.28s, 13017). *Show that the product of the distances of a point of the circumcircle of a triangle from the sides of the triangle is equal to the product of the distances of the same point from the sides of the tangential triangle of the given triangle.*

Points A, O are arbitrarily chosen. Points B, C, D, E, F, G, E_1, F_1, G_1 are constructed (in order) as follows: $BO \equiv OA$; $CO \equiv OA$; $DO \equiv OA$; E is on line BC; $ED \perp BC$; F is on line AC; $FD \perp AC$; G is on line AB; $GD \perp AB$; $E_1D \parallel AO$; $E_1A \perp AO$; $F_1D \parallel BO$; $F_1B \perp BO$; $G_1D \parallel CO$; $G_1C \perp CO$. The **conclusion**: $\overline{DE}^2\ \overline{DF}^2\ \overline{DG}^2 - \overline{DE_1}^2\ \overline{DF_1}^2\ \overline{DG_1}^2 = 0$.

$A = (0,0)$, $O = (u_1, 0)$, $B = (x_1, u_2)$, $C = (x_2, u_3)$, $D = (x_3, u_4)$, $E = (x_5, x_4)$, $F = (x_7, x_6)$, $G = (x_9, x_8)$, $E_1 = (0, u_4)$, $F_1 = (x_{11}, x_{10})$, $G_1 = (x_{13}, x_{12})$.

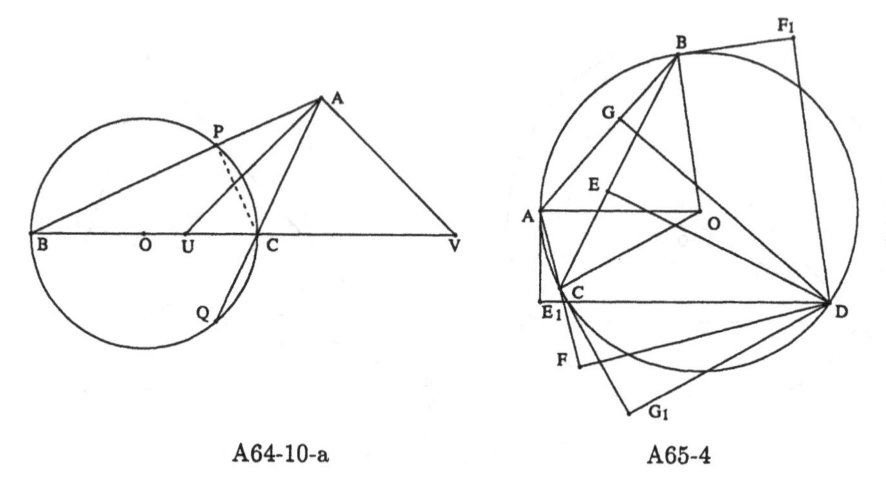

A64-10-a A65-4

The **nondegenerate** conditions: Line BC is non-isotropic; Line AC is non-isotropic; Line AB is non-isotropic; Line AO is non-isotropic; Line BO is non-isotropic; Line CO is non-isotropic.

Example 145 (A68-8: 2.17s, 148). *Show that the distances of a point on a median of triangle from the sides including the median are inversely proportional to these sides.*

Points A, B, C are arbitrarily chosen. Points F, N, K, J are constructed (in order) as follows: F is the midpoint of A and B; N is on line CF; $KN \perp AC$; K is on line AC; $JN \perp BC$; J is on line BC. The **conclusion**: $NK \cdot AC = NJ \cdot BC$.

$A = (0,0)$, $B = (u_1, 0)$, $C = (u_2, u_3)$, $F = (x_1, 0)$, $N = (x_2, u_4)$, $K = (x_4, x_3)$, $J = (x_6, x_5)$.

The **nondegenerate** conditions: $C \neq F$; Line AC is non-isotropic; Line BC is non-isotropic.

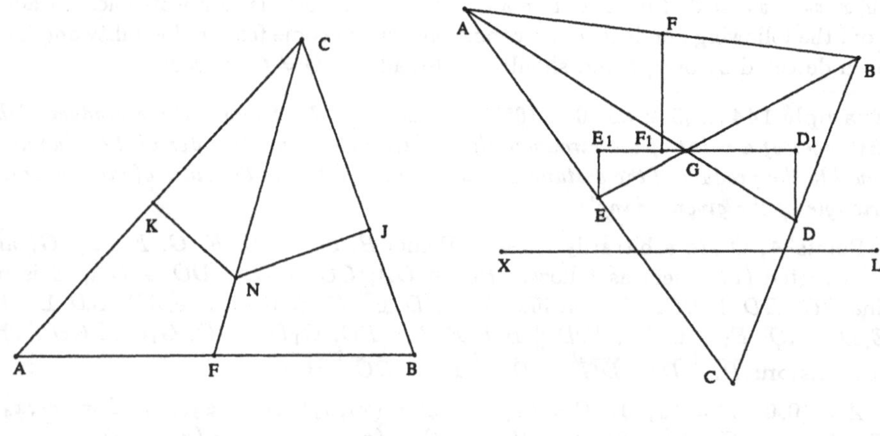

A68-8 A68-9

Example 146 (A68-9: 1.45s, 26). *A line is dawn through the centroid of a triangle. Show that the sum of the distances of the line from the two vertices of the triangle situated on the same side of the line is equal to the distance of the line from the third vertex.*

Points X, L, A, B, C are arbitrarily chosen. Points D, E, F, G, D_1, E_1, F_1 are constructed (in order) as follows: D is the midpoint of B and C; E is the midpoint of A and C; F is the midpoint of A and B; G is on line BE; G is on line AD; $D_1D \perp XL$; $D_1G \parallel XL$; $E_1E \perp XL$; $E_1G \parallel XL$; $F_1F \perp XL$; $F_1G \parallel XL$. The **conclusion**: $\overline{DD_1} + \overline{EE_1} + \overline{FF_1} = 0$.

$X = (0,0)$, $L = (u_1,0)$, $A = (u_2,u_3)$, $B = (u_4,u_5)$, $C = (u_6,u_7)$, $D = (x_1,x_2)$, $E = (x_3,x_4)$, $F = (x_5,x_6)$, $G = (x_8,x_7)$, $D_1 = (x_1,x_7)$, $E_1 = (x_3,x_7)$, $F_1 = (x_5,x_7)$.

The **nondegenerate** conditions: Line AD intersects line BE; Line XL is non-isotropic; Line XL is non-isotropic; Line XL is non-isotropic.

Example 147 (A70-9: 2.88s, 19). *The centroids of the four triangles determined by the vertices of a quadrilateral taken three at a time form a quadrilateral homothetic to the given quadrilateral. Find the ratio of this homothecy.*

Points B, C, D, A are arbitrarily chosen. Points Q, P, R, A_1, D_1, J are constructed (in order) as follows: Q is the midpoint of B and C; P is the midpoint of A and B; R is the midpoint of C and D; A_1 is on line BR; A_1 is on line DQ; D_1 is on line CP; D_1 is on line AQ; J is on line D_1D; J is on line A_1A. The **conclusions**: (1) $\frac{\overline{JA}}{\overline{JA_1}} - \frac{\overline{JD}}{\overline{JD_1}} = 0$; (2) $\frac{\overline{JA}}{\overline{JA_1}} - {-3} = 0$.

$B = (0,0)$, $C = (u_1,0)$, $D = (u_2,u_3)$, $A = (u_4,u_5)$, $Q = (x_1,0)$, $P = (x_2,x_3)$, $R = (x_4,x_5)$, $A_1 = (x_7,x_6)$, $D_1 = (x_9,x_8)$, $J = (x_{11},x_{10})$.

The **nondegenerate** conditions: Line DQ intersects line BR; Line AQ intersects line CP; Line A_1A intersects line D_1D.

Remark. Our system can calculate the ratio mechanically: -3.

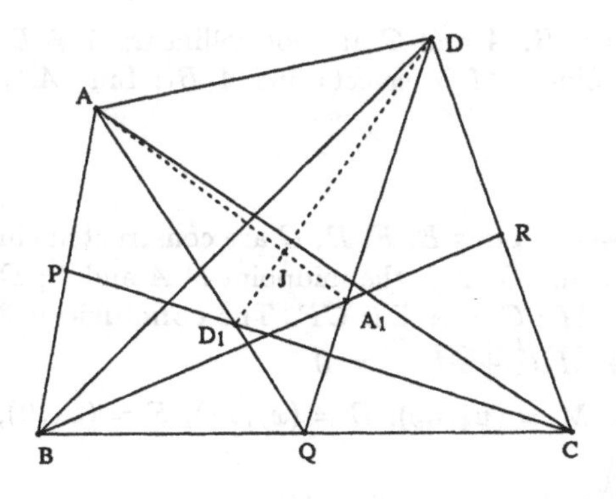

A70-9

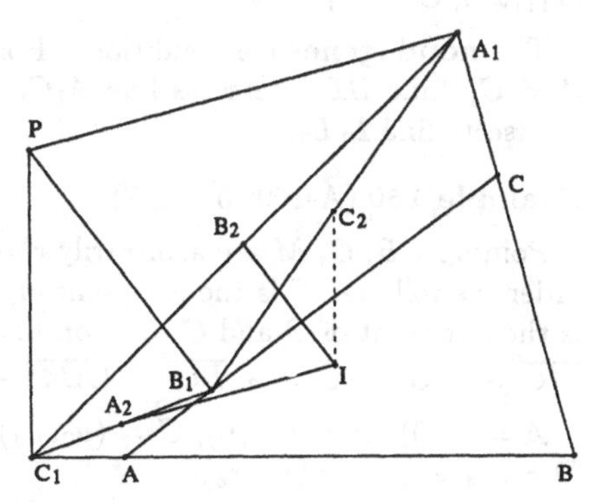

A70-10

Example 148 (A70-10: 1.5s, 10). *From the midpoints A_2, B_2, C_2 of the lines B_1C_1, C_1A_1, A_1B_1 joining the projections A_1, B_1, C_1 of a point P upon the sides of the triangle ABC, perpendiculars are dropped upon the sides BC, CA, AB of ABC. Show that these perpendiculars are concurrent.*

Points A, B, C, P are arbitrarily chosen. Points A_1, B_1, C_1, A_2, B_2, C_2, I are constructed (in order) as follows: A_1 is on line BC; $A_1P \perp BC$; $B_1P \perp AC$; B_1 is on line AC; $C_1P \perp AB$; C_1 is on line AB; A_2 is the midpoint of B_1 and C_1; B_2 is the midpoint of A_1 and C_1; C_2 is the midpoint of A_1 and B_1; $IA_2 \perp BC$; $IB_2 \perp AC$. The **conclusion**: $C_2I \perp AB$.

$A = (0,0)$, $B = (u_1,0)$, $C = (u_2,u_3)$, $P = (u_4,u_5)$, $A_1 = (x_2,x_1)$, $B_1 = (x_4,x_3)$, $C_1 = (u_4,0)$, $A_2 = (x_5,x_6)$, $B_2 = (x_7,x_8)$, $C_2 = (x_9,x_{10})$, $I = (x_{12},x_{11})$.

The **nondegenerate** conditions: Line BC is non-isotropic; Line AC is non-isotropic; Line AB is non-isotropic; Points A, C and B are not collinear.

Example 149 (A70-12: 3.4s, 32; 3.28s, 10). *Let L, L_1 and M, M_1 be two pairs of isotomic points on the two sides AC, AB of the triangle ABC, and L_2, M_2 the traces of the lines BL, CM on the sides A_1C_1, A_1B_1 of the complementary triangle $A_1B_1C_1$ of ABC. Show that the triangles AL_1M_1, $A_1L_2M_2$ are inversely homothetic.*

Points A, B, C are arbitrarily chosen. Points B_1, A_1, C_1, M, L, L_1, M_1, L_2, M_2, O are constructed (in order) as follows: B_1 is the midpoint of A and C; A_1 is on line BC; $A_1B_1 \parallel BA$; C_1 is the midpoint of A and B; M is on line AB; L is on line AC; B_1 is the midpoint of L and L_1; C_1 is the midpoint of M and M_1; L_2 is on line A_1C_1; L_2 is on line BL; M_2 is on line A_1B_1; M_2 is on line CM; O is on line L_1L_2; O is on line AA_1. The **conclusions**: (1) Points O, M_1 and M_2 are collinear; (2) $\frac{\overline{OA}}{\overline{OA_1}} - \frac{\overline{OM_1}}{\overline{OM_2}} = 0$.

$A = (0,0)$, $B = (u_1,0)$, $C = (u_2,u_3)$, $B_1 = (x_1,x_2)$, $A_1 = (x_3,x_2)$, $C_1 = (x_4,0)$, $M = (u_4,0)$, $L = (x_5,u_5)$, $L_1 = (x_6,x_7)$, $M_1 = (x_8,0)$, $L_2 = (x_{10},x_9)$, $M_2 = (x_{11},x_2)$, $O = (x_{13},x_{12})$.

The **nondegenerate** conditions: Points B, A and C are not collinear; $A \neq B$; $A \neq C$; Line BL intersects line A_1C_1; Line CM intersects line A_1B_1; Line AA_1 intersects line L_1L_2.

Example 150 (A-109: 0.83s, 7).

Points A, B, C, M are arbitrarily chosen. Points E, F, D, G are constructed (in order) as follows: E is the midpoint of A and C; F is the midpoint of A and B; D is the midpoint of B and C; G is on line AD; G is on line CF. The **conclusion**: $3\,\overline{MG}^2 + \overline{AG}^2 + \overline{GC}^2 + \overline{BG}^2 - (\overline{AM}^2 + \overline{MB}^2 + \overline{MC}^2) = 0$.

$A = (0,0)$, $B = (u_1,0)$, $C = (u_2,u_3)$, $M = (u_4,u_5)$, $E = (x_1,x_2)$, $F = (x_3,0)$, $D = (x_4,x_5)$, $G = (x_7,x_6)$.

The **nondegenerate** conditions: Line CF intersects line AD.

Example 151 (A-110: 1.03s, 8).

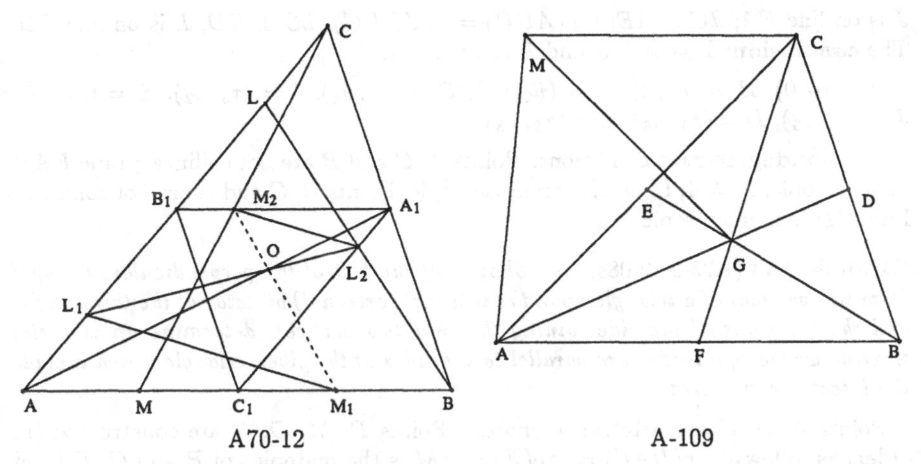

A70-12 A-109

Points A, B, C are arbitrarily chosen. Points O, E, F, D, G are constructed (in order) as follows: $OA \equiv OC$; $OA \equiv OB$; E is the midpoint of A and C; F is the midpoint of A and B; D is the midpoint of B and C; G is on line AD; G is on line CF. The **conclusion**: $3\,\overline{OG}^2 + \overline{AG}^2 + \overline{GC}^2 + \overline{BG}^2 - 3\,\overline{OA}^2 = 0$.

$A = (0,0)$, $B = (u_1,0)$, $C = (u_2,u_3)$, $O = (x_2,x_1)$, $E = (x_3,x_4)$, $F = (x_5,0)$, $D = (x_6,x_7)$, $G = (x_9,x_8)$.

The **nondegenerate** conditions: Points A, B and C are not collinear; Line CF intersects line AD.

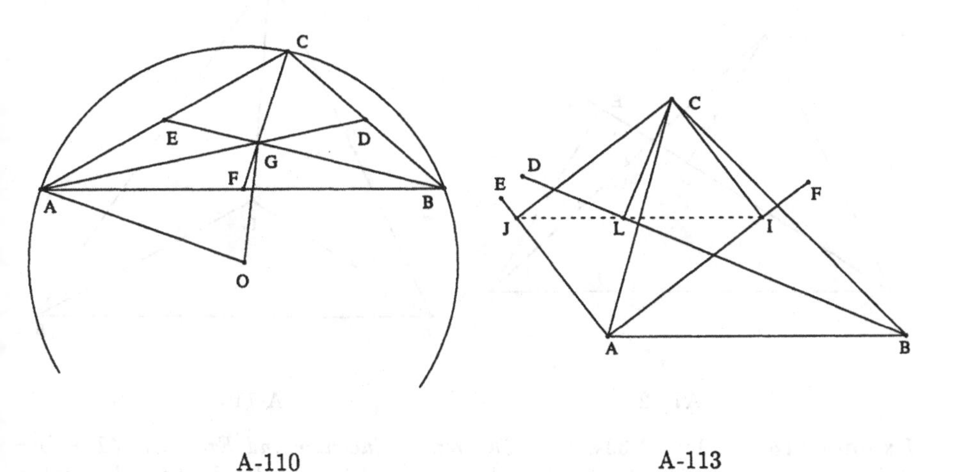

A-110 A-113

Example 152 (A-113: 2.18s, 36; 1.42s, 10). *The feet of the four perpendiculars dropped from a vertex of a triangle upon the four bisectors of the other two angles are collinear.*

Points A, B, C are arbitrarily chosen. Points F, I, E, J, D, L are constructed (in order) as follows: $\tan(BAF) = \tan(FAC)$; $IC \perp AF$; I is on line FA; $EA \perp FA$;

J is on line EA; $JC \perp AE$; $\tan(ABD) = \tan(DBC)$; $LC \perp BD$; L is on line BD. The **conclusion**: Points I, J and L are collinear.

$A = (0,0)$, $B = (u_1, 0)$, $C = (u_2, u_3)$, $F = (x_1, u_4)$, $I = (x_3, x_2)$, $E = (x_4, u_5)$, $J = (x_6, x_5)$, $D = (x_7, u_6)$, $L = (x_9, x_8)$.

The **nondegenerate** conditions: Points A, C and B are not collinear; Line FA is non-isotropic; $F \neq A$; Line AE is non-isotropic; Points B, C and A are not collinear; Line BD is non-isotropic.

Example 153 (A73-2: 2.08s, 19). *Show that the feet of the perpendiculars dropped from two vertices of a triangle upon the internal (external) bisector of the third angle, and the midpoint of the side joining the first two vertices, determine an isosceles triangle whose equal sides are parallel to the sides of the given triangle which include the bisector considered.*

Points A, B, C are arbitrarily chosen. Points F, M, D, E are constructed (in order) as follows: $\tan(BAF) = \tan(FAC)$; M is the midpoint of B and C; D is on line AF; $DC \perp AF$; E is on line FA; $EB \perp AF$. The **conclusions**: (1) $DM \equiv EM$; (2) $DM \parallel AB$.

$A = (0,0)$, $B = (u_1, 0)$, $C = (u_2, u_3)$, $F = (x_1, u_4)$, $M = (x_2, x_3)$, $D = (x_5, x_4)$, $E = (x_7, x_6)$.

The **nondegenerate** conditions: Points A, C and B are not collinear; Line AF is non-isotropic; Line AF is non-isotropic.

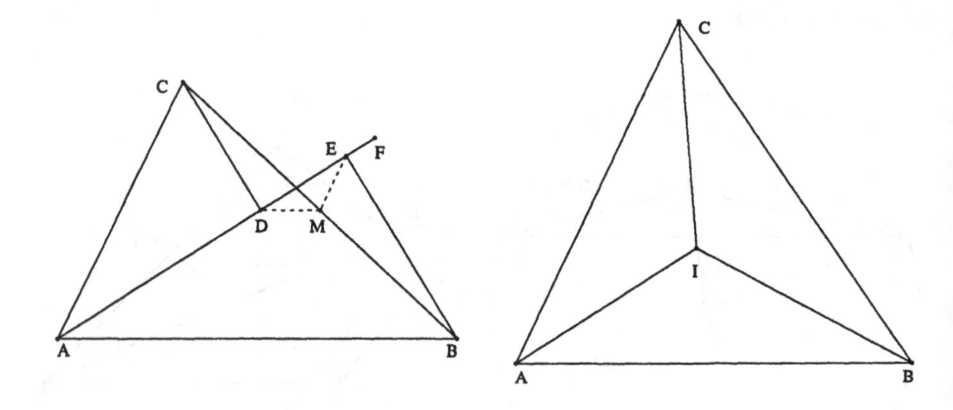

A73-2 A-116

Example 154 (A-116: 1.33s, 16). *Theorem of Incenter and Excenters (The four Tritangent Center). The three internal bisectors of the angels of a triangle meet in a point, the incenter I of the triangle.*

Points A, B, I are arbitrarily chosen. Points C are constructed (in order) as follows: $\tan(ABI) = \tan(IBC)$; $\tan(BAI) = \tan(IAC)$. The **conclusion**: $\tan(ACI) = \tan(ICB)$.

$A = (0,0)$, $B = (u_1, 0)$, $I = (u_2, u_3)$, $C = (x_2, x_1)$.

The **nondegenerate** conditions: Points I, A and B are not collinear; $\angle AIB$ is not right.

Remark. In unordered geometry there are no differences among the incenter and the three excenters. This theorem is for these **four** *tritangent centers. We will recall this fact in some examples occasionally.*

Example 155 (A-120: 0.93s, 13). *Two tritangent centers divide the bisector on which they are located, harmonically.*

Points B, C, I are arbitrarily chosen. Points A, D, I_A are constructed (in order) as follows: $\tan(BCI) = \tan(ICA)$; $\tan(CBI) = \tan(IBA)$; D is on line BC; D is on line AI; I_A is on line AI; $I_AB \perp BI$. The **conclusion**: Points A, D, I and I_A form a harmonic set.

$B = (0,0)$, $C = (u_1,0)$, $I = (u_2,u_3)$, $A = (x_2,x_1)$, $D = (x_3,0)$, $I_A = (x_5,x_4)$.

The **nondegenerate** conditions: Points I, B and C are not collinear; $\angle BIC$ is not right; Line AI intersects line BC; Line BI is not perpendicular to line AI.

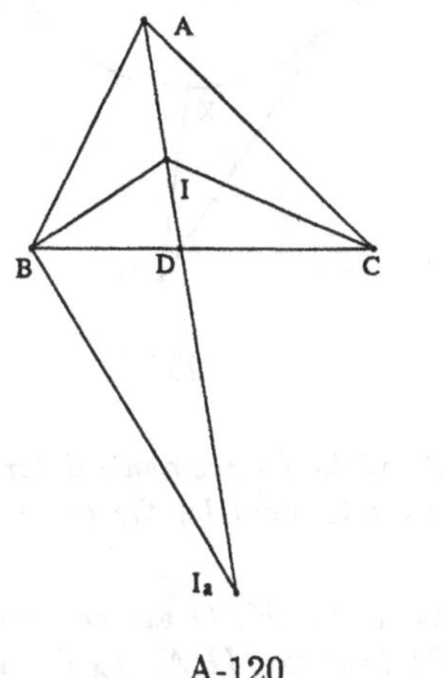

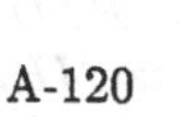

A-120

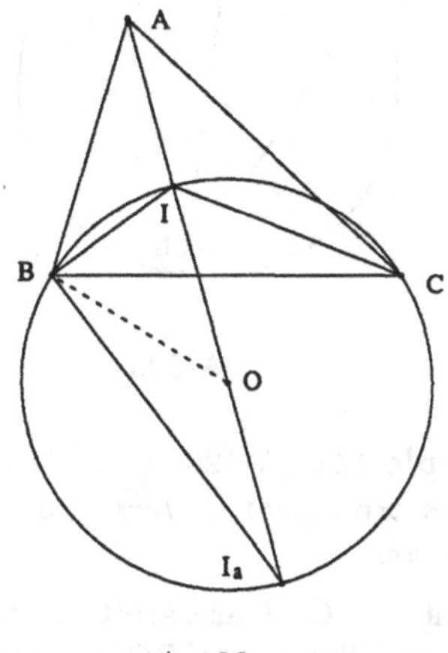

A-122

Example 156 (A-122: 0.82s, 6). *Two tritangent centers of a triangle are the ends of a diameter of a circle passing through the two vertices of the triangle which are not collinear with the centers considered.*

Points B, C, I are arbitrarily chosen. Points A, I_A, O are constructed (in order) as follows: $\tan(BCI) = \tan(ICA)$; $\tan(CBI) = \tan(IBA)$; I_A is on line AI; $I_AB \perp BI$; O is the midpoint of I and I_A. The **conclusion**: $OB \equiv OI$.

$B = (0,0)$, $C = (u_1,0)$, $I = (u_2,u_3)$, $A = (x_2,x_1)$, $I_A = (x_4,x_3)$, $O = (x_5,x_6)$.

The **nondegenerate** conditions: Points I, B and C are not collinear; $\angle BIC$ is not right; Line BI is not perpendicular to line AI.

Example 157 (A-123: 2.73s, 17; 1.63s, 14). *The four tritangent centers of a triangle lie on six circles which pass through the pairs of vertices of the triangle and have for their centers the midpoints of the arcs subtended by the respective sides of the* *triangle on its circumcircle.*

Points B, C, I are arbitrarily chosen. Points A, I_A, K are constructed (in order) as follows: $\tan(BCI) = \tan(ICA)$; $\tan(CBI) = \tan(IBA)$; I_A is on line AI; $I_AB \perp BI$; K is on line AI; $KB \equiv KC$. The **conclusions**: (1) $KB \equiv KI$; (2) $KB \equiv KI_A$.

$B = (0,0)$, $C = (u_1,0)$, $I = (u_2,u_3)$, $A = (x_2,x_1)$, $I_A = (x_4,x_3)$, $K = (x_6,x_5)$.

The **nondegenerate** conditions: Points I, B and C are not collinear; $\angle BIC$ is not right; Line BI is not perpendicular to line AI; Line BC is not perpendicular to line AI.

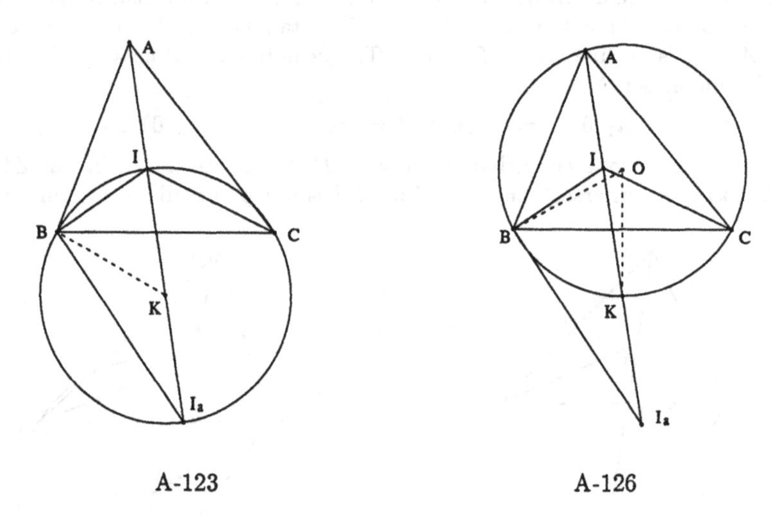

A-123 A-126

Example 158 (A-126: 1.42s, 50). *The midpoints of the six segments determined by the four tritangent centers of a triangle lie on a circle, namely, the circumcircle of the triangle.*

Points B, C, I are arbitrarily chosen. Points A, I_A, K, O are constructed (in order) as follows: $\tan(BCI) = \tan(ICA)$; $\tan(CBI) = \tan(IBA)$; I_A is on line AI; $I_AB \perp BI$; K is the midpoint of I_A and I; $OB \equiv OA$; $OB \equiv OC$. The **conclusion**: $OB \equiv OK$.

$B = (0,0)$, $C = (u_1,0)$, $I = (u_2,u_3)$, $A = (x_2,x_1)$, $I_A = (x_4,x_3)$, $K = (x_5,x_6)$, $O = (x_8,x_7)$.

The **nondegenerate** conditions: Points I, B and C are not collinear; $\angle BIC$ is not right; Line BI is not perpendicular to line AI; Points B, C and A are not collinear.

Example 159 (A-127: 1.8s, 76). *The product of the distances of two tritangent centers of a triangle from the vertex of the triangle collinear with them is equal to the product of the two sides of the triangle passing through the vertex considered.*

Points B, C, I are arbitrarily chosen. Points A, I_A are constructed (in order) as follows: $\tan(BCI) = \tan(ICA)$; $\tan(CBI) = \tan(IBA)$; I_A is on line AI; $I_AB \perp BI$. The **conclusion**: $AI \cdot AI_A = AB \cdot AC$.

$B = (0,0)$, $C = (u_1, 0)$, $I = (u_2, u_3)$, $A = (x_2, x_1)$, $I_A = (x_4, x_3)$.

The **nondegenerate** conditions: Points I, B and C are not collinear; $\angle BIC$ is not right; Line BI is not perpendicular to line AI.

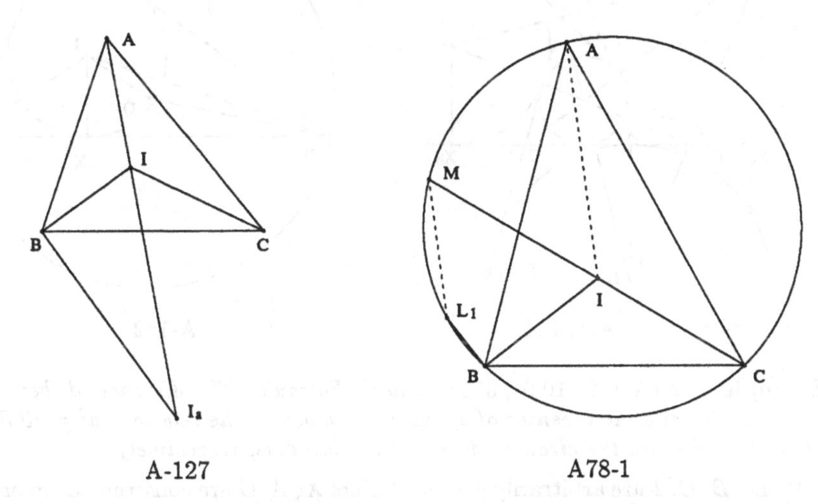

A-127 A78-1

Example 160 (A78-1: 1.63s, 42). *Show that an external bisector of an angle of a triangle is parallel to the line joining the points where the circumcircle is met by the external (internal) bisectors of the other two angles of the triangle.*

Points B, C, I are arbitrarily chosen. Points A, M, L_1 are constructed (in order) as follows: $\tan(BCI) = \tan(ICA)$; $\tan(CBI) = \tan(IBA)$; M is on line IC; $MA \equiv MB$; $L_1A \equiv L_1C$; $L_1B \perp BI$. The **conclusion**: $ML_1 \parallel AI$.

$B = (0,0)$, $C = (u_1, 0)$, $I = (u_2, u_3)$, $A = (x_2, x_1)$, $M = (x_4, x_3)$, $L_1 = (x_6, x_5)$.

The **nondegenerate** conditions: Points I, B and C are not collinear; $\angle BIC$ is not right; Line AB is not perpendicular to line IC; Line BI intersects line AC.

Example 161 (A-134: 5.5s, 331; 3.1s, 56). *The product of the four tritangent radii of a triangle is equal to the square of its area.*

Points B, C, I are arbitrarily chosen. Points A, I_A, I_C, I_B, X, X_A, X_B, X_C are constructed (in order) as follows: $\tan(CBI) = \tan(IBA)$; $\tan(BCI) = \tan(ICA)$; $I_AB \perp IB$; I_A is on line AI; I_C is on line BI_A; I_C is on line CI; I_B is on line AI_C; I_B is on line BI; X is on line BC; $XI \perp BC$; X_A is on line BC; $X_AI_A \perp BC$; X_B is on line BC; $X_BI_B \perp BC$; X_C is on line BC; $X_CI_C \perp BC$. The **conclusion**: $16\ \overline{IX}^2\ \overline{I_AX_A}^2\ \overline{I_BX_B}^2\ \overline{I_CX_C}^2 - \nabla(ABC)\ \nabla(ABC)\ \nabla(ABC)\ \nabla(ABC) = 0$.

$B = (0,0)$, $C = (u_1, 0)$, $I = (u_2, u_3)$, $A = (x_2, x_1)$, $I_A = (x_4, x_3)$, $I_C = (x_6, x_5)$, $I_B = (x_8, x_7)$, $X = (u_2, 0)$, $X_A = (x_4, 0)$, $X_B = (x_8, 0)$, $X_C = (x_6, 0)$.

The **nondegenerate** conditions: Points I, C and B are not collinear; $\angle CIB$ is not right; Line AI is not perpendicular to line IB; Line CI intersects line BI_A; Line BI intersects line AI_C; Line BC is non-isotropic; Line BC is non-isotropic; Line BC is non-isotropic; Line BC is non-isotropic.

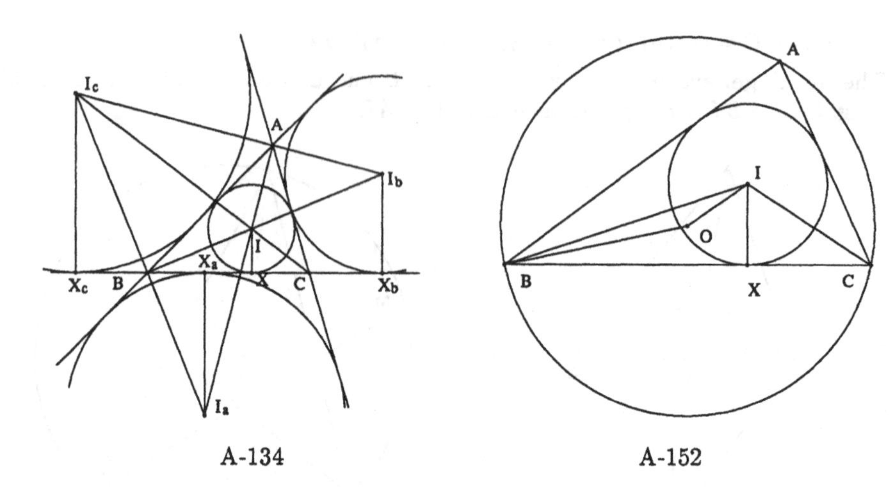

A-134 A-152

Example 162 (A-152: 10.9s, 351). *Euler's Formula. The distance, d, between the circumcenter and the incenter of a triangle is given by the relation:* $d^2 = R(R-2r)$, *where R and r are the circumradius and the inradius, respectively.*

Points B, C, I are arbitrarily chosen. Points X, A, O are constructed (in order) as follows: X is on line BC; $XI \perp BC$; $\tan(CBI) = \tan(IBA)$; $\tan(BCI) = \tan(ICA)$; $OA \equiv OB$; $OB \equiv OC$; $\overline{OI}^2\ \overline{OI}^2 - x_5\ x_5 = 0$; $\overline{OB}^2\ \overline{OB}^2 - x_6\ x_6 = 0$; $4\ \overline{XI}^2\ \overline{OB}^2 - x_7\ x_7 = 0$. The **conclusion**: $(x_5 + x_6 - x_7)\ (x_5 + x_6 + x_7)\ (x_5 - (x_6 - x_7))\ (x_5 - (x_6 + x_7)) = 0$.

$B = (0,0)$, $C = (u_1,0)$, $I = (u_2,u_3)$, $X = (u_2,0)$, $A = (x_2,x_1)$, $O = (x_4,x_3)$.

The **nondegenerate** conditions: Line BC is non-isotropic; Points I, C and B are not collinear; $\angle CIB$ is not right; Points B, C and A are not collinear.

Remark. Our proof proved a weak form of Euler's formula. But it for the four tritangents.

Example 163 (A-154: 145.07s, 3564; 12.57s, 353). $II_a{}^2 = 4R(r_a - r)$, $I_bI_c{}^2 = 4R(r_a + r_b)$.

Points B, C, I are arbitrarily chosen. Points X, A, O, I_A, X_A are constructed (in order) as follows: X is on line BC; $XI \perp BC$; $\tan(CBI) = \tan(IBA)$; $\tan(BCI) = \tan(ICA)$; $OA \equiv OB$; $OB \equiv OC$; I_A is on line AI; $I_AB \perp BI$; X_A is on line BC; $X_AI_A \perp BC$; $\overline{I_AI}^2\ \overline{I_AI}^2 - x_7\ x_7 = 0$; $16\ \overline{X_AI_A}^2\ \overline{OB}^2 - x_8\ x_8 = 0$; $16\ \overline{XI}^2\ \overline{OB}^2 -$ [illegible]0. The **conclusion**: $(x_7 + x_8 - x_9)\ (x_7 + x_8 + x_9)\ (x_7 - (x_8 - x_9))\ (x_7 + x_9)) = 0$.

$B = (0,0)$, $C = (u_1,0)$, $I = (u_2,u_3)$, $X = (u_2,0)$, $A = (x_2,x_1)$, $O = (x_4,x_3)$, $I_A = (x_6,x_5)$, $X_A = (x_6,0)$.

The **nondegenerate** conditions: Line BC is non-isotropic; Points I, C and B are not collinear; $\angle CIB$ is not right; Points B, C and A are not collinear; Line BI is not perpendicular to line AI; Line BC is non-isotropic.

Remark. Again, we only proved a weak form; but a single proof is for both formu-

las.

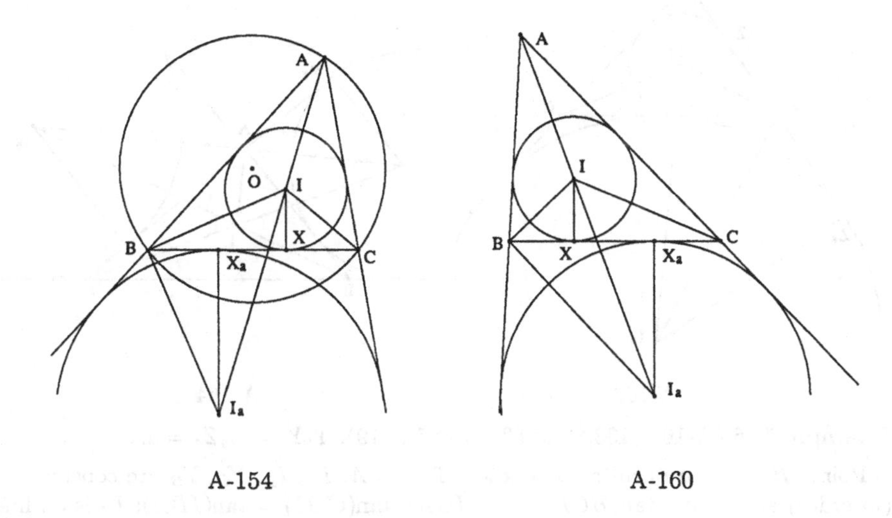

A-154 A-160

Example 164 (A-160: 1.03s, 30). *The points of contact of a side of a triangle with the incircle and the excircle relative to this side are two isotomic points.*

Points B, C, I are arbitrarily chosen. Points A, I_A, X, X_A are constructed (in order) as follows: $\tan(BCI) = \tan(ICA)$; $\tan(CBI) = \tan(IBA)$; I_A is on line AI; $I_AB \perp BI$; $XI \perp BC$; X is on line BC; X_A is on line BC; $X_AI_A \perp BC$. The **conclusion:** $BX \equiv CX_A$.

$B = (0,0)$, $C = (u_1, 0)$, $I = (u_2, u_3)$, $A = (x_2, x_1)$, $I_A = (x_4, x_3)$, $X = (u_2, 0)$, $X_A = (x_4, 0)$.

The **nondegenerate** conditions: Points I, B and C are not collinear; $\angle BIC$ is not right; Line BI is not perpendicular to line AI; Line BC is non-isotropic; Line BC is non-isotropic.

Remark. This proof is also for the excenters, i.e., article 163 – The two points of contact of a side of a triangle with the two excircles relative to the other two sides are two isotomic points – is a repetition of A-160.

Example 165 (A-162: 5.08s, 129; 3.83s, 43). $ZZ_a = YY_a = a$.

Points B, C, I are arbitrarily chosen. Points A, I_A, Z, Z_A are constructed (in order) as follows: $\tan(BCI) = \tan(ICA)$; $\tan(CBI) = \tan(IBA)$; I_A is on line AI; $I_AB \perp BI$; Z is on line AB; $ZI \perp AB$; Z_A is on line AB; $Z_AI_A \perp AB$. The **conclusion:** $ZZ_A \equiv BC$.

$B = (0,0)$, $C = (u_1, 0)$, $I = (u_2, u_3)$, $A = (x_2, x_1)$, $I_A = (x_4, x_3)$, $Z = (x_6, x_5)$, $Z_A = (x_8, x_7)$.

The **nondegenerate** conditions: Points I, B and C are not collinear; $\angle BIC$ is not right; Line BI is not perpendicular to line AI; Line AB is non-isotropic; Line AB is non-isotropic.

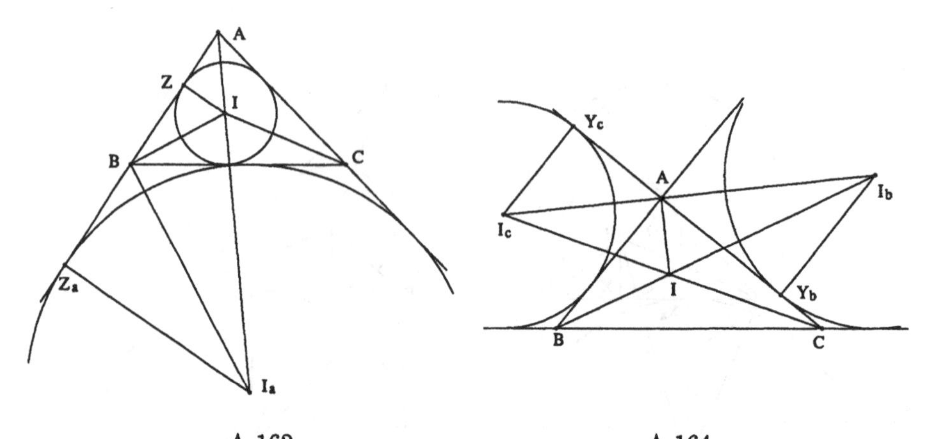

A-162 A-164

Example 166 (A-164: 133.83s, 2839; 15.87s, 319). *$Y_bY_c = Z_bZ_c = a$.*

Points B, C, I are arbitrarily chosen. Points A, I_C, I_B, Y_C, Y_B are constructed (in order) as follows: $\tan(BCI) = \tan(ICA)$; $\tan(CBI) = \tan(IBA)$; I_C is on line IC; $I_CA \perp AI$; I_B is on line IB; I_B is on line AI_C; Y_C is on line AC; $Y_CI_C \perp AC$; Y_B is on line AC; $Y_BI_B \perp AC$. The **conclusion**: $Y_CY_B \equiv BC$.

$B = (0,0)$, $C = (u_1,0)$, $I = (u_2,u_3)$, $A = (x_2,x_1)$, $I_C = (x_4,x_3)$, $I_B = (x_6,x_5)$, $Y_C = (x_8,x_7)$, $Y_B = (x_{10},x_9)$.

The **nondegenerate** conditions: Points I, B and C are not collinear; $\angle BIC$ is not right; Line AI is not perpendicular to line IC; Line AI_C intersects line IB; Line AC is non-isotropic; Line AC is non-isotropic.

Remark. This is a repetition of A-162.

Example 167 (A87-4-a: 35.9s, 1233; 6.27s, 280). *Prove the formulas: $OI^2 + OI_a^2 + OI_b^2 + OI_c^2 = 12R^2$..*

Points B, C, I are arbitrarily chosen. Points A, O, I_A, I_C, I_B are constructed (in order) as follows: $\tan(CBI) = \tan(IBA)$; $\tan(BCI) = \tan(ICA)$; $OB \equiv OC$; $OA \equiv OB$; $I_AB \perp IB$; I_A is on line AI; I_C is on line BI_A; I_C is on line CI; I_B is on line AI_C; I_B is on line BI. The **conclusion**: $12\,\overline{OB}^2 - (\overline{OI}^2 + \overline{OI_A}^2 + \overline{OI_B}^2 + \overline{OI_C}^2) = 0$.

$B = (0,0)$, $C = (u_1,0)$, $I = (u_2,u_3)$, $A = (x_2,x_1)$, $O = (x_4,x_3)$, $I_A = (x_6,x_5)$, $I_C = (x_8,x_7)$, $I_B = (x_{10},x_9)$.

The **nondegenerate** conditions: Points I, C and B are not collinear; $\angle CIB$ is not right; Points A, B and C are not collinear; Line AI is not perpendicular to line IB; Line CI intersects line BI_A; Line BI intersects line AI_C.

Example 168 (A87-4b: 9.78s, 568; 6.42s, 224). *$II_a{}^2 + II_b{}^2 + II_c{}^2 + I_aI_b{}^2 + I_bI_c{}^2 + I_cI_a{}^2 = 48R^2$.*

Points B, C, I are arbitrarily chosen. Points A, O, I_A, I_B, I_C are constructed (in order) as follows: $\tan(CBI) = \tan(IBA)$; $\tan(BCI) = \tan(ICA)$; $OA \equiv OB$;

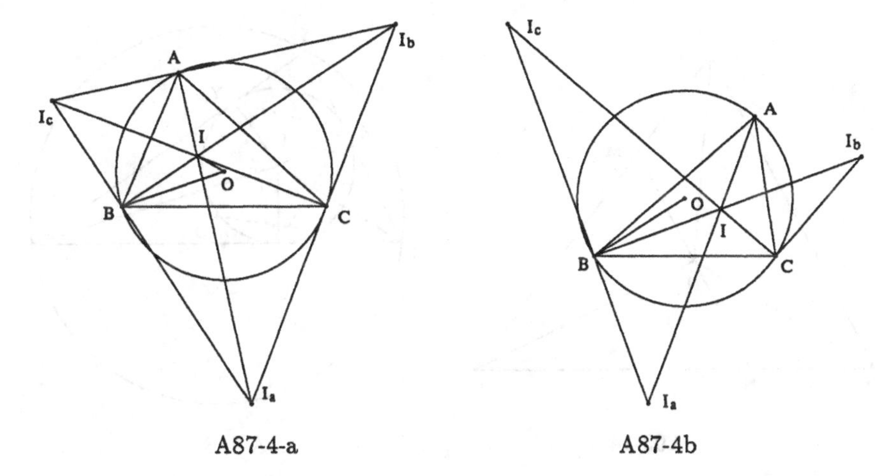

A87-4-a A87-4b

$OB \equiv OC$; I_A is on line AI; $I_AB \perp BI$; $I_BC \perp CI$; I_B is on line BI; $I_CB \perp BI$; I_C is on line CI. The **conclusion**: $48\,\overline{OB}^2 - (\overline{II_A}^2 + \overline{II_B}^2 + \overline{II_C}^2 + \overline{I_AI_B}^2 + \overline{I_AI_C}^2 + \overline{I_BI_C}^2) = 0$.

$B = (0,0)$, $C = (u_1,0)$, $I = (u_2,u_3)$, $A = (x_2,x_1)$, $O = (x_4,x_3)$, $I_A = (x_6,x_5)$, $I_B = (x_8,x_7)$, $I_C = (x_{10},x_9)$.

The **nondegenerate** conditions: Points I, C and B are not collinear; $\angle CIB$ is not right; Points B, C and A are not collinear; Line BI is not perpendicular to line AI; Line BI is not perpendicular to line CI; Line CI is not perpendicular to line BI.

Example 169 (A93-14: 9.33s, 190). *Show that a parallel through a tritangent center to a side of a triangle is equal to the sum, or difference, of the two segments on the other two sides of the triangle between the two parallel lines considered.*

Points B, C, I are arbitrarily chosen. Points A, M, N are constructed (in order) as follows: $\tan(BCI) = \tan(ICA)$; $\tan(CBI) = \tan(IBA)$; M is on line AC; $MI \parallel AB$; N is on line BC; N is on line IM. The **conclusion**: Algebraic sum of segments $MN, AM, BN = 0$.

$B = (0,0)$, $C = (u_1,0)$, $I = (u_2,u_3)$, $A = (x_2,x_1)$, $M = (x_4,x_3)$, $N = (x_5,0)$.

The **nondegenerate** conditions: Points I, B and C are not collinear; $\angle BIC$ is not right; Points A, B and C are not collinear; Line IM intersects line BC.

Example 170 (A-168: 7.72s, 237). *The ratio of the area of a triangle to the area of the triangle determined by the points of contact of the sides with the incircle is equal to the ratio of the circumdiameter of the given triangle to its inradius.*

Points B, C, I are arbitrarily chosen. Points A, X, Y, Z, O are constructed (in order) as follows: $\tan(BCI) = \tan(ICA)$; $\tan(CBI) = \tan(IBA)$; X is on line BC; $XI \perp BC$; Y is on line AC; $YI \perp AC$; Z is on line AB; $ZI \perp AB$; $OB \equiv OA$; $OB \equiv OC$. The **conclusion**: $\nabla(XYZ)\,\nabla(XYZ)\,4\,\overline{OB}^2 - \nabla(ABC)\,\nabla(ABC)\,\overline{IX}^2 = 0$.

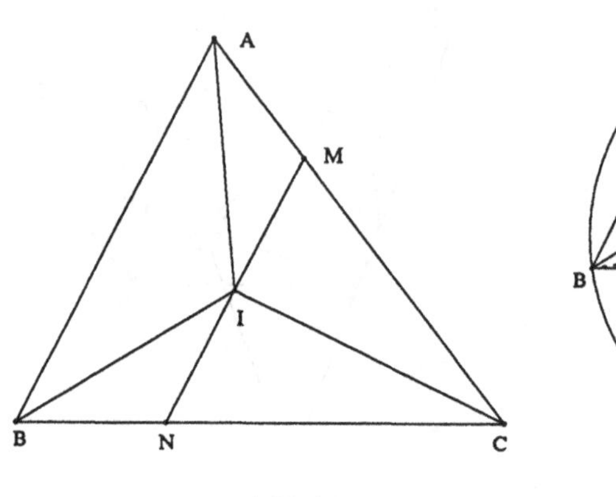

A93-14

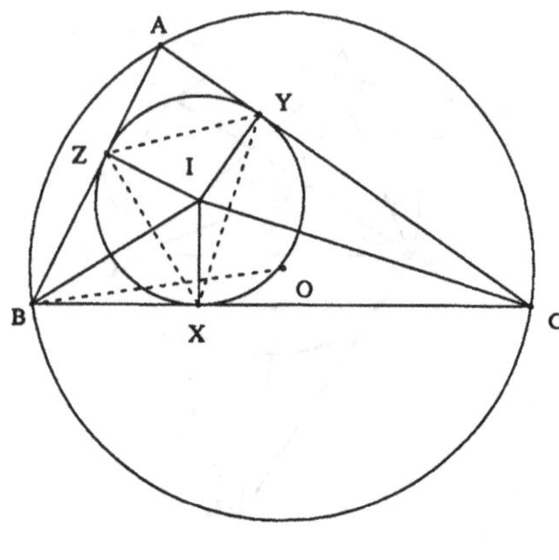

A-168

$B = (0,0)$, $C = (u_1,0)$, $I = (u_2,u_3)$, $A = (x_2,x_1)$, $X = (u_2,0)$, $Y = (x_4,x_3)$, $Z = (x_6,x_5)$, $O = (x_8,x_7)$.

The **nondegenerate** conditions: Points I, B and C are not collinear; $\angle BIC$ is not right; Line BC is non-isotropic; Line AC is non-isotropic; Line AB is non-isotropic; Points B, C and A are not collinear.

Remark. I can be any of the four tritangent centers.

Example 171 (A93-17: 8.13s, 290). *Prove the formula:* $AZ \cdot BX \cdot CY = rS$.

Points B, C, I are arbitrarily chosen. Points A, X, Y, Z are constructed (in order) as follows: $\tan(CBI) = \tan(IBA)$; $\tan(BCI) = \tan(ICA)$; X is on line BC; $XI \perp BC$; Y is on line AC; $YI \perp AC$; Z is on line AB; $ZI \perp AB$. The **conclusion**: $4\,\overline{AZ}^2\,\overline{BX}^2\,\overline{CY}^2 - \overline{XI}^2\,\nabla(ABC)\,\nabla(ABC) = 0$.

$B = (0,0)$, $C = (u_1,0)$, $I = (u_2,u_3)$, $A = (x_2,x_1)$, $X = (u_2,0)$, $Y = (x_4,x_3)$, $Z = (x_6,x_5)$.

The **nondegenerate** conditions: Points I, C and B are not collinear; $\angle CIB$ is not right; Line BC is non-isotropic; Line AC is non-isotropic; Line AB is non-isotropic.

Remark. Our proof is for the four tritangents.

Example 172 (A93-20: 4.22s, 190). *Show that the area of a right triangle is equal to the product of the two segments into which the hypotenuse is divided by its point of contact with the incircle.*

Points B, C are arbitrarily chosen. Points A, D, F, I, Y are constructed (in order) as follows: $AB \perp CB$; $\tan(BAD) = \tan(DAC)$; D is on line BC; $\tan(BCF) = \tan(FCA)$; F is on line BA; I is on line AD; I is on line CF; Y is on line AC; $YI \perp AC$. The **conclusion**: $\nabla(ABC)\,\nabla(ABC) - 4\,\overline{AY}^2\,\overline{CY}^2 = 0$.

$B = (0,0)$, $C = (u_1,0)$, $A = (0,u_2)$, $D = (x_1,0)$, $F = (0,x_2)$, $I = (x_4,x_3)$, $Y = (x_6,x_5)$.

The **nondegenerate** conditions: $C \neq B$; $-u_1u_2 \neq 0$; $u_1u_2 \neq 0$; Line CF inter-

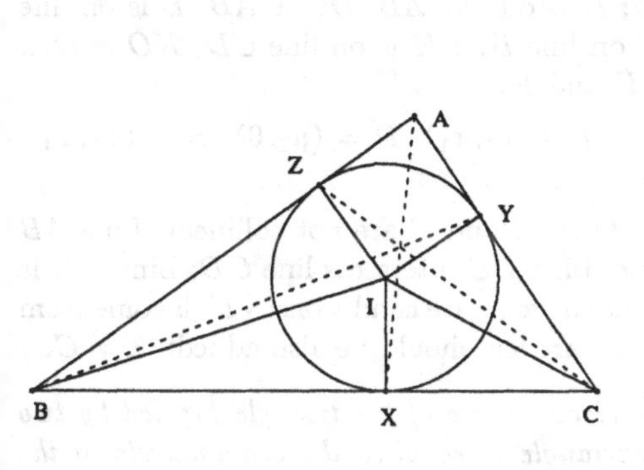

A93-17

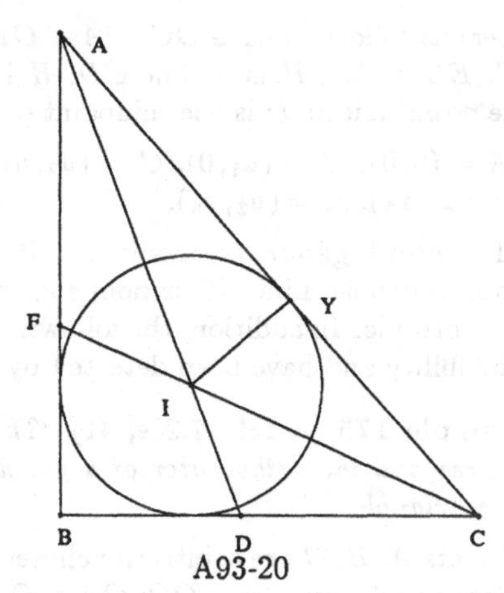

A93-20

sects line AD; Line AC is non-isotropic.

Remark. Our proof is also for I_b.

Example 173 (A-175: 2.08s, 0; 0.37s, 3). *The theorem of orthocenter. The three altitudes of a triangle are concurrent.*

Points A, B, C are arbitrarily chosen. Points H are constructed (in order) as follows: $HA \perp BC$; $HC \perp AB$. The **conclusion**: $BH \perp AC$.

$A = (0,0)$, $B = (u_1, 0)$, $C = (u_2, u_3)$, $H = (u_2, x_1)$.

The **nondegenerate** conditions: Points A, B and C are not collinear.

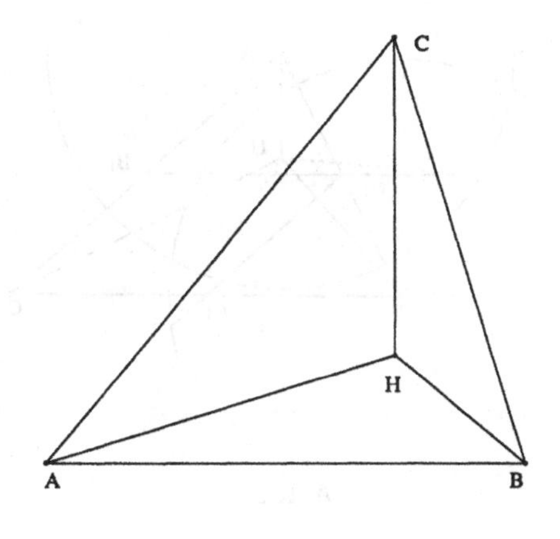

A-175

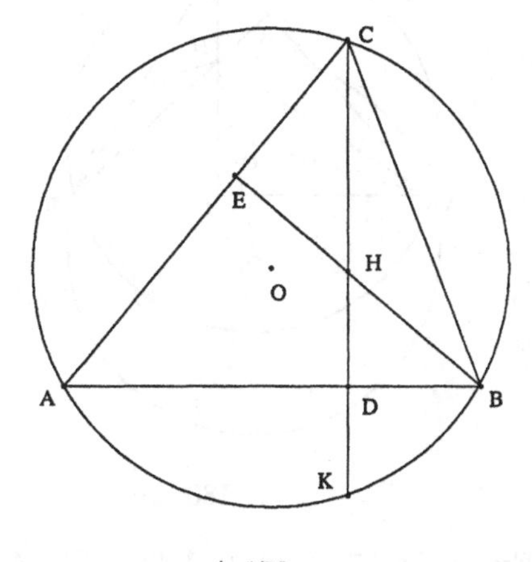

A-178

Example 174 (A-178: 1.05s, 6). *The segment of the altitude extended between the orthocenter and the second point of intersection with the circumcircle is bisected by the corresponding side of the triangle.*

Points A, B, C are arbitrarily chosen. Points O, D, E, H, K are constructed (in order) as follows: $OA \equiv OC$; $OA \equiv OB$; D is on line AB; $DC \perp AB$; E is on line AC; $EB \perp AC$; H is on line CD; H is on line BE; K is on line CD; $KO \equiv OA$. The **conclusion**: D is the midpoint of H and K.

$A = (0,0)$, $B = (u_1,0)$, $C = (u_2,u_3)$, $O = (x_2,x_1)$, $D = (u_2,0)$, $E = (x_4,x_3)$, $H = (u_2,x_5)$, $K = (u_2,x_6)$.

The **nondegenerate** conditions: Points A, B and C are not collinear; Line AB is non-isotropic; Line AC is non-isotropic; Line BE intersects line CD; Line CD is non-isotropic. In addition, the following nondegenerate conditions, which come from reducibility and have been detected by our prover, should be also added: $K \neq C$.

Example 175 (A-180: 1.28s, 41). *The circumcircle of the triangle formed by two vertices and the orthocenter of a given triangle is equal to the circumcircle of the given triangle.*

Points A, B, C are arbitrarily chosen. Points O, D, E, H, O_1 are constructed (in order) as follows: $OA \equiv OC$; $OA \equiv OB$; D is on line AB; $DC \perp AB$; E is on line AC; $EB \perp AC$; H is on line CD; H is on line BE; $O_1A \equiv O_1H$; $O_1A \equiv O_1B$. The **conclusion**: $OA \equiv O_1H$.

$A = (0,0)$, $B = (u_1,0)$, $C = (u_2,u_3)$, $O = (x_2,x_1)$, $D = (u_2,0)$, $E = (x_4,x_3)$, $H = (u_2,x_5)$, $O_1 = (x_7,x_6)$.

The **nondegenerate** conditions: Points A, B and C are not collinear; Line AB is non-isotropic; Line AC is non-isotropic; Line BE intersects line CD; Points A, B and H are not collinear.

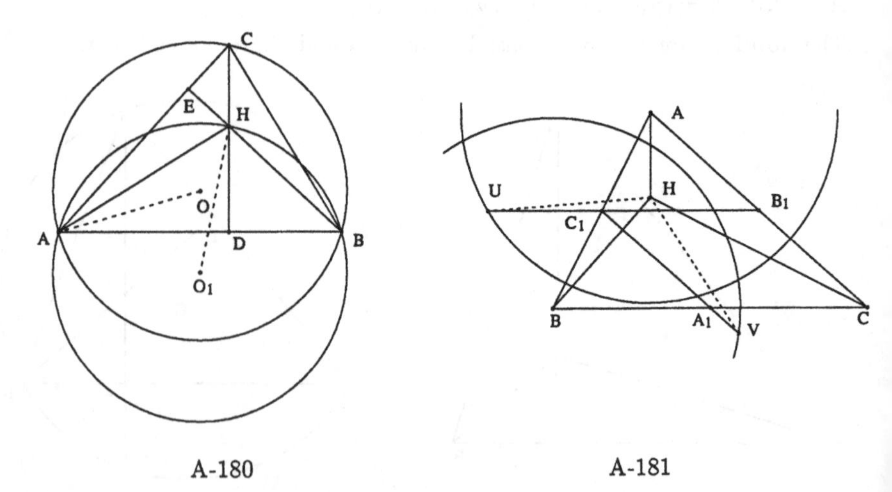

A-180 A-181

Example 176 (A-181: 1.42s, 36). *Any three equal circles having for centers the vertices of a given triangle cut the respective sides of the medial triangle in six points equidistant from the orthocenter of the given triangle.*

Points B, C, A are arbitrarily chosen. Points H, A_1, B_1, C_1, V, U are constructed (in order) as follows: $HA \perp BC$; $HB \perp AC$; A_1 is the midpoint of B and C; B_1 is the midpoint of A and C; C_1 is the midpoint of A and B; V is on line A_1C_1;

$UA \equiv BV$; U is on line B_1C_1. The **conclusion**: $HV \equiv HU$.

$B = (0,0)$, $C = (u_1, 0)$, $A = (u_2, u_3)$, $H = (u_2, x_1)$, $A_1 = (x_2, 0)$, $B_1 = (x_3, x_4)$, $C_1 = (x_5, x_6)$, $V = (u_4, x_7)$, $U = (x_8, x_9)$.

The **nondegenerate** conditions: Points A, C and B are not collinear; $A_1 \neq C_1$; Line B_1C_1 is non-isotropic.

Example 177 (A97-5: 2.12s, 53; 1.85s, 50). *If p, q, r are the distances of a point inside a triangle ABC from the sides of the triangle, show that $(p : h_a)+(q : h_b)+(r : h_c) = 1$.*

Points A, B, C, O are arbitrarily chosen. Points D, E, F, D_1, E_1, F_1 are constructed (in order) as follows: $DA \perp CB$; D is on line BC; $EB \perp CA$; E is on line AC; $FC \perp BA$; F is on line AB; $D_1O \perp BC$; D_1 is on line BC; $E_1O \perp AC$; E_1 is on line CA; $F_1O \perp AB$; F_1 is on line AB. The **conclusion**: $\frac{\overline{OD_1}}{\overline{AD}} + \frac{\overline{OE_1}}{\overline{BE}} + \frac{\overline{OF_1}}{\overline{CF}} - 1 = 0$.

$A = (0,0)$, $B = (u_1, 0)$, $C = (u_2, u_3)$, $O = (u_4, u_5)$, $D = (x_2, x_1)$, $E = (x_4, x_3)$, $F = (u_2, 0)$, $D_1 = (x_6, x_5)$, $E_1 = (x_8, x_7)$, $F_1 = (u_4, 0)$.

The **nondegenerate** conditions: Line BC is non-isotropic; Line AC is non-isotropic; Line AB is non-isotropic; Line BC is non-isotropic; Line CA is non-isotropic; Line AB is non-isotropic.

Remark. Here we consider the directed segments; we proved a more general version. Condition "inside" is not necessary.

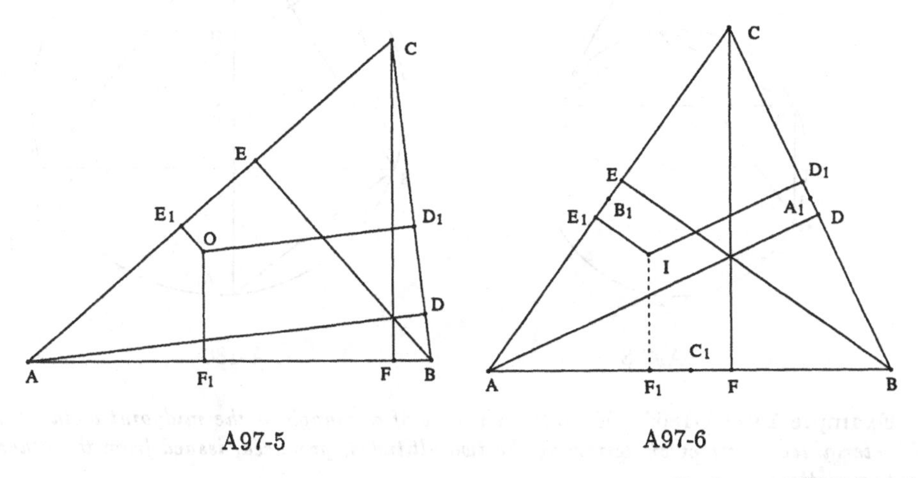

A97-5 A97-6

Example 178 (A97-6: 2.8s, 12). *Show that the three perpendiculars to the sides of a triangle at the points isotomic to the feet of the respective altitudes are concurrent.*

Points A, B, C are arbitrarily chosen. Points A_1, B_1, C_1, D, E, F, D_1, E_1, F_1, I are constructed (in order) as follows: A_1 is the midpoint of B and C; B_1 is the midpoint of A and C; C_1 is the midpoint of A and B; D is on line BC; $DA \perp BC$; E is on line AC; $EB \perp AC$; F is on line AB; $FC \perp AB$; A_1 is the midpoint of D_1

and D; B_1 is the midpoint of E_1 and E; C_1 is the midpoint of F_1 and F; $IE_1 \perp AC$; $ID_1 \perp BC$. The **conclusion**: $IF_1 \perp AB$.

$A = (0,0)$, $B = (u_1,0)$, $C = (u_2,u_3)$, $A_1 = (x_1,x_2)$, $B_1 = (x_3,x_4)$, $C_1 = (x_5,0)$, $D = (x_7,x_6)$, $E = (x_9,x_8)$, $F = (u_2,0)$, $D_1 = (x_{10},x_{11})$, $E_1 = (x_{12},x_{13})$, $F_1 = (x_{14},0)$, $I = (x_{16},x_{15})$.

The **nondegenerate** conditions: Line BC is non-isotropic; Line AC is non-isotropic; Line AB is non-isotropic; Points B, C and A are not collinear.

Example 179 (A97-8: 0.9s, 27). *The perpendicular at the orthocenter H to the altitude HC of the triangle ABC meets the circumcircle of HBC in P. Show that $ABPH$ is a parallelogram.*

Points B, C, A are arbitrarily chosen. Points H, O, P are constructed (in order) as follows: $HA \perp BC$; $HC \perp AB$; $OB \equiv OH$; $OB \equiv OC$; $PO \equiv OB$; $PH \parallel AB$. The **conclusion**: $HP \equiv AB$.

$B = (0,0)$, $C = (u_1,0)$, $A = (u_2,u_3)$, $H = (u_2,x_1)$, $O = (x_3,x_2)$, $P = (x_5,x_4)$.

The **nondegenerate** conditions: Points A, B and C are not collinear; Points B, C and H are not collinear; Line AB is non-isotropic. In addition, the following nondegenerate conditions, which come from reducibility and have been detected by our prover, should be also added: $P \neq H$.

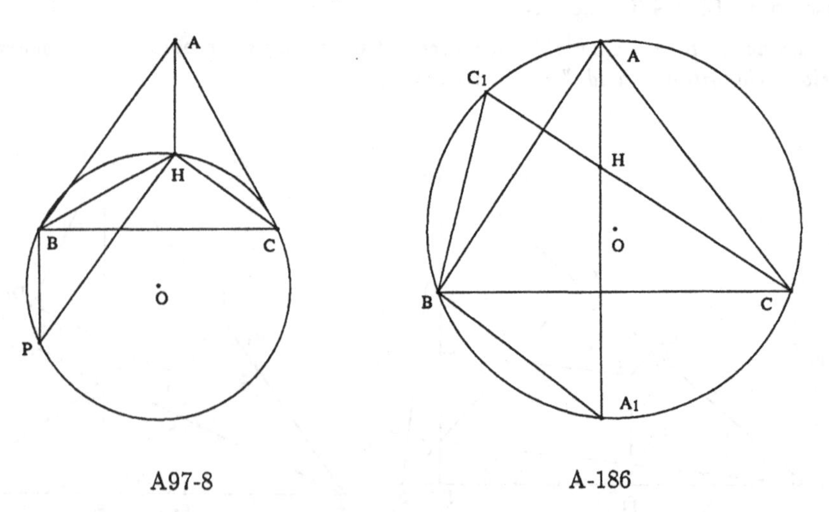

A97-8 A-186

Example 180 (A-186: 1.08s, 40). *A vertex of a triangle is the midpoint of the arc determined on its circumcircle by the two altitudes, produced, issued from the other two vertices.*

Points B, C, A are arbitrarily chosen. Points H, O, C_1, A_1 are constructed (in order) as follows: $HA \perp BC$; $HC \perp AB$; $OB \equiv OC$; $OB \equiv OA$; $C_1O \equiv OB$; C_1 is on line CH; $A_1O \equiv OB$; A_1 is on line AH. The **conclusion**: $BC_1 \equiv BA_1$.

$B = (0,0)$, $C = (u_1,0)$, $A = (u_2,u_3)$, $H = (u_2,x_1)$, $O = (x_3,x_2)$, $C_1 = (x_5,x_4)$, $A_1 = (u_2,x_6)$.

The **nondegenerate** conditions: Points A, B and C are not collinear; Points B, A and C are not collinear; Line CH is non-isotropic; Line AH is non-isotropic. In addition, the following nondegenerate conditions, which come from reducibility and have been detected by our prover, should be also added: $A_1 \neq A$; $C_1 \neq C$.

Example 181 (A-188: 0.58s, 6). *The radii of the circumcircle passing through the vertices of a triangle are perpendicular to the corresponding sides of the orthic triangle.*

Points A, B, C are arbitrarily chosen. Points F, E, O are constructed (in order) as follows: F is on line AB; $FC \perp AB$; E is on line AC; $EB \perp AC$; $OA \equiv OC$; $OA \equiv OB$. The **conclusion**: $OA \perp EF$.

$A = (0,0)$, $B = (u_1, 0)$, $C = (u_2, u_3)$, $F = (u_2, 0)$, $E = (x_2, x_1)$, $O = (x_4, x_3)$.

The **nondegenerate** conditions: Line AB is non-isotropic; Line AC is non-isotropic; Points A, B and C are not collinear.

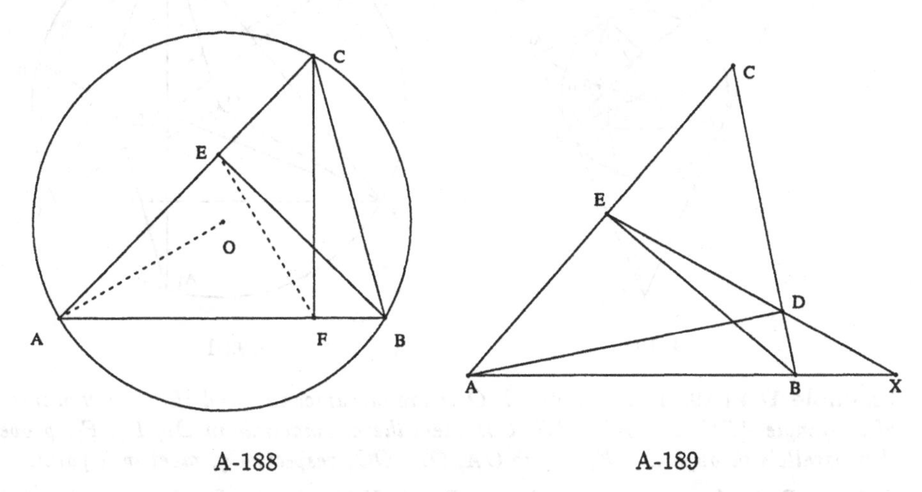

A-188 A-189

Example 182 (A-189: 1.08s, 11). *The angle which a side of a triangle makes with the corresponding side of the orthic triangle is equal to the difference of the angles of the given triangle adjacent to the side considered.*

Points A, B, C are arbitrarily chosen. Points D, E, X are constructed (in order) as follows: $DA \perp CB$; D is on line BC; $EB \perp CA$; E is on line AC; X is on line DE; X is on line AB. The **conclusion**: $\tan(\angle CAB + \angle CBA) - \tan(\angle DXA) = 0$.

$A = (0,0)$, $B = (u_1, 0)$, $C = (u_2, u_3)$, $D = (x_2, x_1)$, $E = (x_4, x_3)$, $X = (x_5, 0)$.

The **nondegenerate** conditions: Line BC is non-isotropic; Line AC is non-isotropic; Line AB intersects line DE.

Example 183 (A-191: 205.35s, 5505; 9.55s, 61). *The tangential and the orthic triangles of a given triangle are homothetic.*

Points A, B, C are arbitrarily chosen. Points O, D, E, F, C_1, A_1, B_1, I are constructed (in order) as follows: $OA \equiv OC$; $OA \equiv OB$; $DA \perp CB$; D is on line BC; $EB \perp CA$; E is on line AC; $FC \perp BA$; F is on line AB; $C_1B \perp BO$; $C_1A \perp AO$; A_1 is on line C_1B; $A_1C \perp CO$; B_1 is on line CA_1; B_1 is on line AC_1; I is on line B_1E; I is on line A_1D. The **conclusions**: (1) Points C_1, F and I are collinear; (2) $\frac{\overline{IE}}{\overline{IB_1}} - \frac{\overline{IF}}{\overline{IC_1}} = 0$.

$A = (0,0)$, $B = (u_1,0)$, $C = (u_2,u_3)$, $O = (x_2,x_1)$, $D = (x_4,x_3)$, $E = (x_6,x_5)$, $F = (u_2,0)$, $C_1 = (x_8,x_7)$, $A_1 = (x_{10},x_9)$, $B_1 = (x_{12},x_{11})$, $I = (x_{14},x_{13})$.

The **nondegenerate** conditions: Points A, B and C are not collinear; Line BC is non-isotropic; Line AC is non-isotropic; Line AB is non-isotropic; Points A, O and B are not collinear; Line CO is not perpendicular to line C_1B; Line AC_1 intersects line CA_1; Line A_1D intersects line B_1E.

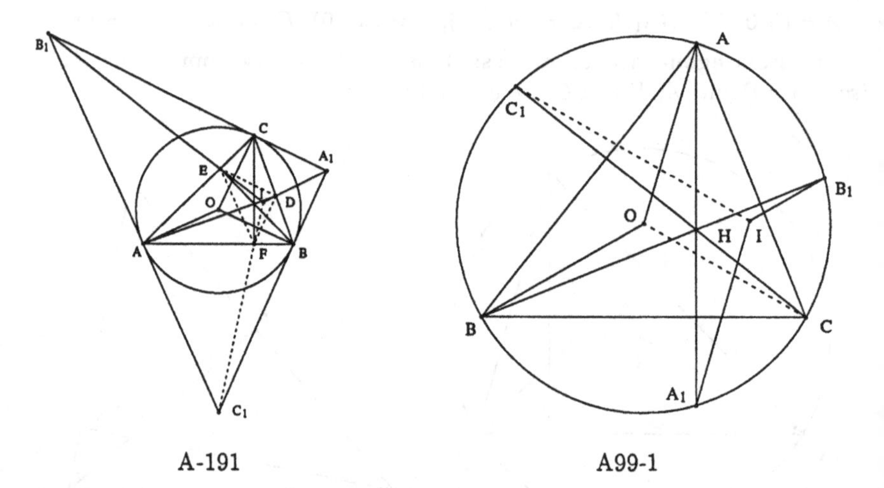

A-191 A99-1

Example 184 (A99-1: 1.75s, 49). *If O is the circumcenter and H the orthocenter of a triangle ABC, and AH, BH, CH meet the circumcircle in D_1, E_1, F_1, prove that parallels through D_1, E_1, F_1 to OA, OB, OC, respectively, meet in a point.*

Points B, C, A are arbitrarily chosen. Points H, O, C_1, A_1, B_1, I are constructed (in order) as follows: $HA \perp BC$; $HC \perp AB$; $OB \equiv OC$; $OB \equiv OA$; $C_1O \equiv OB$; C_1 is on line CH; $A_1O \equiv OB$; A_1 is on line AH; $B_1O \equiv OB$; B_1 is on line BH; $IA_1 \parallel OA$; $IB_1 \parallel OB$. The **conclusion**: $IC_1 \parallel OC$.

$B = (0,0)$, $C = (u_1,0)$, $A = (u_2,u_3)$, $H = (u_2,x_1)$, $O = (x_3,x_2)$, $C_1 = (x_5,x_4)$, $A_1 = (u_2,x_6)$, $B_1 = (x_8,x_7)$, $I = (x_{10},x_9)$.

The **nondegenerate** conditions: Points A, B and C are not collinear; Points B, A and C are not collinear; Line CH is non-isotropic; Line AH is non-isotropic; Line BH is non-isotropic; Points O, B and A are not collinear. In addition, the following nondegenerate conditions, which come from reducibility and have been detected by our prover, should be also added: $B_1 \neq B$; $A_1 \neq A$; $C_1 \neq C$.

Example 185 (A99-2: 1.47s, 56). *Show that the product of the segments into which a side of a triangle is divided by the corresponding vertex of the orthic triangle is equal to the product of the sides of the orthic triangle passing through the vertex* *considered.*

Points B, C, A are arbitrarily chosen. Points F, E, D are constructed (in order) as follows: F is on line AB; $FC \perp AB$; E is on line AC; $EB \perp AC$; D is on line BC; $DA \perp BC$. The **conclusion**: $BD \cdot DC = ED \cdot FD$.

$B = (0,0)$, $C = (u_1, 0)$, $A = (u_2, u_3)$, $F = (x_2, x_1)$, $E = (x_4, x_3)$, $D = (u_2, 0)$.

The **nondegenerate** conditions: Line AB is non-isotropic; Line AC is non-isotropic; Line BC is non-isotropic.

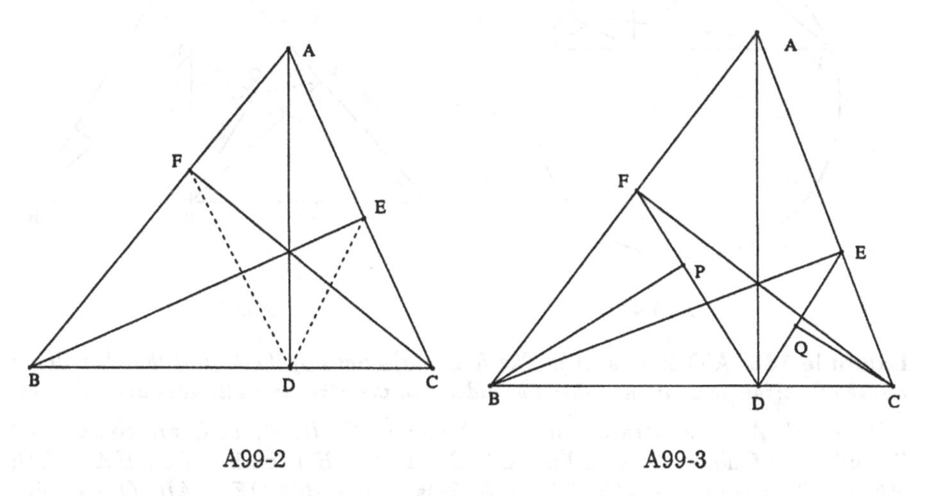

A99-2 A99-3

Example 186 (A99-3: 51.75s, 2097; 41.92s, 1079). *If P, Q are the feet of the perpendiculars from the vertices B, C of the triangle ABC upon th sides DF, DE, respectively, of the orthic triangle DEF, show that $EQ = FP$.*

Points B, C, A are arbitrarily chosen. Points F, E, D, Q, P are constructed (in order) as follows: F is on line AB; $FC \perp AB$; E is on line AC; $EB \perp AC$; D is on line BC; $DA \perp BC$; Q is on line DE; $QC \perp DE$; P is on line DF; $PB \perp DF$. The **conclusion**: $EQ \equiv FP$.

$B = (0,0)$, $C = (u_1, 0)$, $A = (u_2, u_3)$, $F = (x_2, x_1)$, $E = (x_4, x_3)$, $D = (u_2, 0)$, $Q = (x_6, x_5)$, $P = (x_8, x_7)$.

The **nondegenerate** conditions: Line AB is non-isotropic; Line AC is non-isotropic; Line BC is non-isotropic; Line DE is non-isotropic; Line DF is non-isotropic.

Example 187 (A99-4: 7.38s, 42). *DP, DQ are the perpendiculars from the foot D of the altitude AD of the triangle ABC upon the sides AC, AB. Prove that the points B, C, P, Q are concyclic, and that angle $DPB = CQD$.*

Points B, C, A are arbitrarily chosen. Points D, Q, P are constructed (in order) as follows: D is on line BC; $DA \perp BC$; Q is on line AB; $QD \perp AB$; P is on line AC; $PD \perp AC$. The **conclusions**: (1) Points B, C, P and Q are on the same circle; (2) $\tan(DPB) = \tan(CQD)$.

$B = (0,0)$, $C = (u_1,0)$, $A = (u_2,u_3)$, $D = (u_2,0)$, $Q = (x_2,x_1)$, $P = (x_4,x_3)$.

The **nondegenerate** conditions: Line BC is non-isotropic; Line AB is non-isotropic; Line AC is non-isotropic.

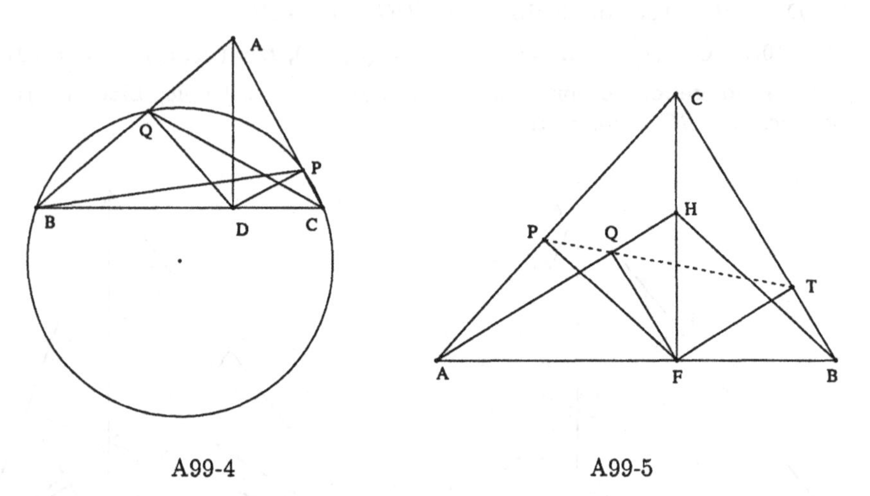

A99-4 A99-5

Example 188 (A99-5: 1.1s, 10). *The four projections of the foot of the altitude on a side of a triangle upon the other two sides and the other two altitudes are collinear.*

Points F, A are arbitrarily chosen. Points B, C, H, P, T, Q are constructed (in order) as follows: B is on line AF; $CF \perp AF$; H is on line FC; $HA \perp CB$; $PF \perp AC$; P is on line AC; $TF \perp CB$; T is on line BC; $QF \perp AH$; Q is on line AH. The **conclusion**: Points P, Q and T are collinear.

$F = (0,0)$, $A = (u_1,0)$, $B = (u_2,0)$, $C = (0,u_3)$, $H = (0,x_1)$, $P = (x_3,x_2)$, $T = (x_5,x_4)$, $Q = (x_7,x_6)$.

The **nondegenerate** conditions: $A \neq F$; $A \neq F$; Line CB is not perpendicular to line FC; Line AC is non-isotropic; Line BC is non-isotropic; Line AH is non-isotropic.

Example 189 (A-195: 0.98s, 8). *The distance of a side of a triangle from the circumcenter is equal to half the distance of the opposite vertex from the orthocenter.*

Points B, C, A are arbitrarily chosen. Points D, H, O, A_1 are constructed (in order) as follows: D is on line BC; $DA \perp BC$; H is on line AD; $HB \perp AC$; $OB \equiv OC$; $OA \equiv OB$; $A_1O \perp BC$; A_1 is on line BC. The **conclusion**: $HA = 2OA_1$.

$B = (0,0)$, $C = (u_1,0)$, $A = (u_2,u_3)$, $D = (u_2,0)$, $H = (u_2,x_1)$, $O = (x_3,x_2)$, $A_1 = (x_3,0)$.

The **nondegenerate** conditions: Line BC is non-isotropic; Line AC is not perpendicular to line AD; Points A, B and C are not collinear; Line BC is non-isotropic.

Example 190 (A-197: 1.7s, 105). *The ratio of a side of a triangle to the corresponding side of the orthic triangle is equal to the ratio of the circumradius to the*

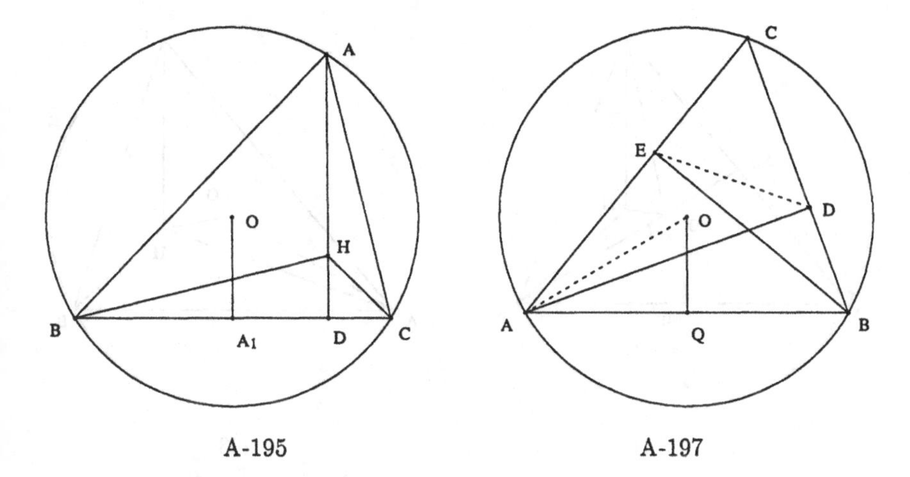

A-195 A-197

distance of the side considered from the circumcenter.

Points A, B, C are arbitrarily chosen. Points E, D, O, Q are constructed (in order) as follows: E is on line AC; $EB \perp AC$; D is on line BC; $DA \perp BC$; $OA \equiv OC$; $OA \equiv OB$; Q is on line AB; $QO \perp AB$. The **conclusion**: $AB \cdot OQ = OA \cdot ED$.

$A = (0,0)$, $B = (u_1, 0)$, $C = (u_2, u_3)$, $E = (x_2, x_1)$, $D = (x_4, x_3)$, $O = (x_6, x_5)$, $Q = (x_6, 0)$.

The **nondegenerate** conditions: Line AC is non-isotropic; Line BC is non-isotropic; Points A, B and C are not collinear; Line AB is non-isotropic.

Example 191 (A-201: 0.97s, 6). *Euler Line. The circumcenter, the orthocenter, and the centroid of a triangle are collinear, and the distance from the centroid to the orthocenter is equal to twice the distance from the centroid to the circumcenter.*

Points A, B, C are arbitrarily chosen. Points H, O, M_1, M_2, M are constructed (in order) as follows: $HB \perp AC$; $HC \perp AB$; $OA \equiv OC$; $OA \equiv OB$; M_1 is the midpoint of A and B; M_2 is the midpoint of A and C; M is on line CM_1; M is on line BM_2. The **conclusion**: Points O, M, H are collinear, and $\overline{OM}/\overline{MH} = 1/2$.

$A = (0,0)$, $B = (u_1, 0)$, $C = (u_2, u_3)$, $H = (u_2, x_1)$, $O = (x_3, x_2)$, $M_1 = (x_4, 0)$, $M_2 = (x_5, x_6)$, $M = (x_8, x_7)$.

The **nondegenerate** conditions: Points A, B and C are not collinear; Points A, B and C are not collinear; Line BM_2 intersects line CM_1.

Example 192 (A-203: 0.8s, 9).

Points A, B, C are arbitrarily chosen. Points H, O are constructed (in order) as follows: $HB \perp AC$; $HC \perp AB$; $OA \equiv OC$; $OA \equiv OB$. The **conclusion**: $9\,\overline{AO}^2 - (\overline{AB}^2 + \overline{AC}^2 + \overline{BC}^2 + \overline{OH}^2) = 0$.

$A = (0,0)$, $B = (u_1, 0)$, $C = (u_2, u_3)$, $H = (u_2, x_1)$, $O = (x_3, x_2)$.

The **nondegenerate** conditions: Points A, B and C are not collinear; Points A,

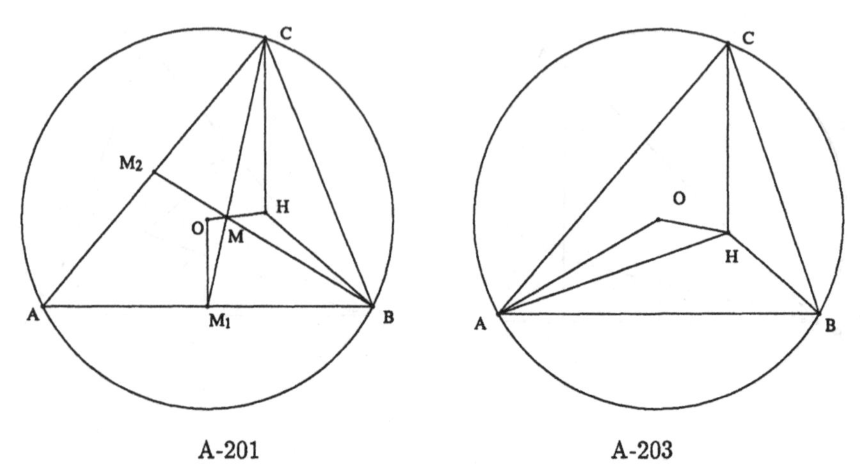

B and C are not collinear.

Example 193 (A-204: 0.52s, 8).

Points A, B, C are arbitrarily chosen. Points H, O are constructed (in order) as follows: $HB \perp AC$; $HC \perp AB$; $OA \equiv OC$; $OA \equiv OB$. The **conclusion**: $12\ \overline{AO}^2 - (\overline{AB}^2 + \overline{AC}^2 + \overline{BC}^2 + \overline{AH}^2 + \overline{BH}^2 + \overline{CH}^2) = 0$.

$A = (0,0)$, $B = (u_1,0)$, $C = (u_2,u_3)$, $H = (u_2,x_1)$, $O = (x_3,x_2)$.

The **nondegenerate** conditions: Points A, B and C are not collinear; Points A, B and C are not collinear.

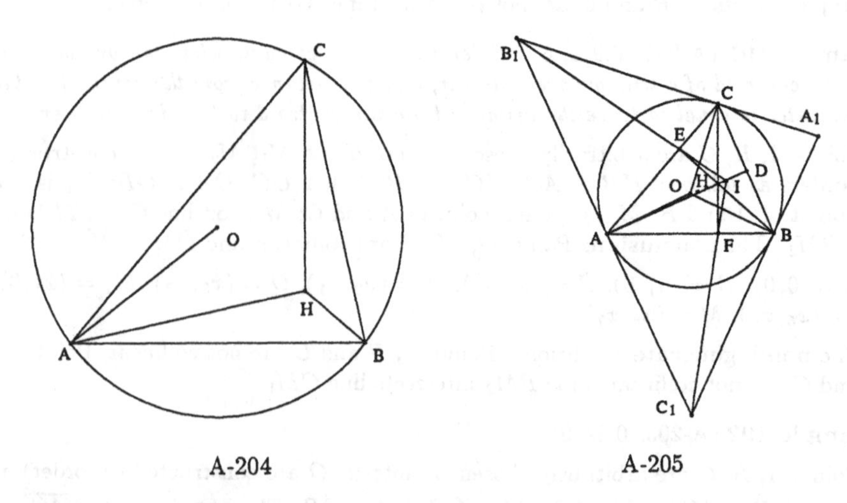

Example 194 (A-205: 3.27s, 61). *The homothetic center of the orthic and the tangential triangles of a given triangle lies on the Euler line of the given triangle.*

Points A, B, C are arbitrarily chosen. Points H, O, D, E, F, B_1, C_1, A_1, I are constructed (in order) as follows: $HB \perp AC$; $HC \perp AB$; $OA \equiv OC$; $OA \equiv OB$; D

is on line BC; D is on line AH; E is on line AC; E is on line BH; F is on line AB; F is on line CH; $B_1C \perp CO$; $B_1A \perp AO$; $C_1B \perp OB$; C_1 is on line AB_1; A_1 is on line C_1B; A_1 is on line B_1C; I is on line FC_1; I is on line B_1E. The **conclusion**: Points O, H and I are collinear.

$A = (0,0)$, $B = (u_1,0)$, $C = (u_2,u_3)$, $H = (u_2,x_1)$, $O = (x_3,x_2)$, $D = (x_5,x_4)$, $E = (x_7,x_6)$, $F = (u_2,0)$, $B_1 = (x_9,x_8)$, $C_1 = (x_{11},x_{10})$, $A_1 = (x_{13},x_{12})$, $I = (x_{15},x_{14})$.

The **nondegenerate** conditions: Points A, B and C are not collinear; Points A, B and C are not collinear; Line AH intersects line BC; Line BH intersects line AC; Line CH intersects line AB; Points A, O and C are not collinear; Line AB_1 is not perpendicular to line OB; Line B_1C intersects line C_1B; Line B_1E intersects line FC_1.

Example 195 (A103-5: 4.85s, 30; 1.98s, 35). *The line joining the centroid of a triangle to a point P on the circumcircle bisects the line joining the diametric opposite of P to the orthocenter.*

Points A, B, C are arbitrarily chosen. Points M_3, M_2, G, H, O, P, Q, I are constructed (in order) as follows: M_3 is the midpoint of A and B; M_2 is the midpoint of A and C; G is on line BM_2; G is on line CM_3; $HB \perp AC$; $HC \perp AB$; $OA \equiv OC$; $OA \equiv OB$; $PO \equiv OA$; O is the midpoint of P and Q; I is on line HQ; I is on line PG. The **conclusion**: I is the midpoint of Q and H.

$A = (0,0)$, $B = (u_1,0)$, $C = (u_2,u_3)$, $M_3 = (x_1,0)$, $M_2 = (x_2,x_3)$, $G = (x_5,x_4)$, $H = (u_2,x_6)$, $O = (x_8,x_7)$, $P = (x_9,u_4)$, $Q = (x_{10},x_{11})$, $I = (x_{13},x_{12})$.

The **nondegenerate** conditions: Line CM_3 intersects line BM_2; Points A, B and C are not collinear; Points A, B and C are not collinear; Line PG intersects line HQ.

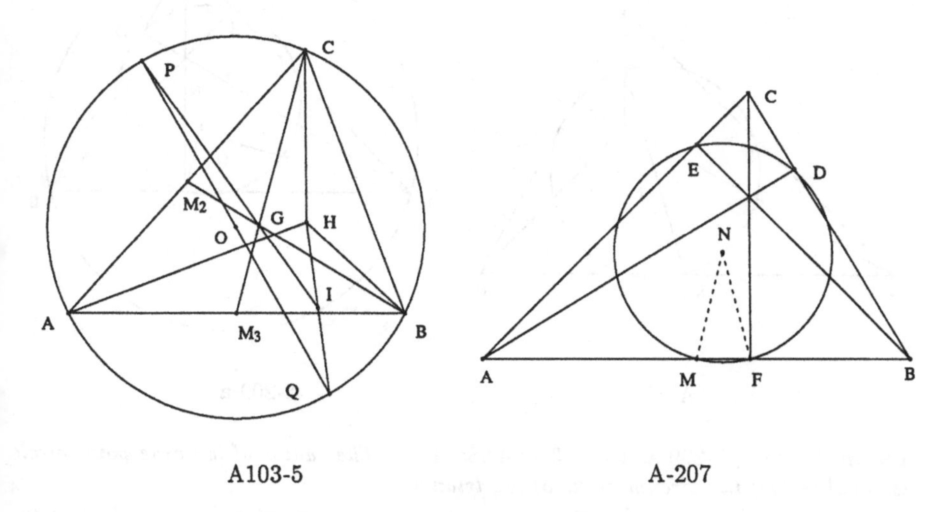

A103-5 A-207

Example 196 (A-207: 1.37s, 31). *The Nine-Point Circle Theorem. In a triangle the midpoints of the sides, the feet of altitudes, and the Euler points lie on the same* *circle. (The midpoints of the segments joining the orthocenter of a triangle to its vertices are called the Euler points of the triangle. The three Euler points determine the Euler triangle.)*

Points A, B, C are arbitrarily chosen. Points D, E, F, M, N are constructed (in order) as follows: D is on line BC; $DA \perp CB$; E is on line AC; $EB \perp CA$; F is on line AB; $FC \perp BA$; M is the midpoint of A and B; $NF \equiv NE$; $NF \equiv ND$. The **conclusion**: $NF \equiv NM$.

$A = (0,0)$, $B = (u_1,0)$, $C = (u_2,u_3)$, $D = (x_2,x_1)$, $E = (x_4,x_3)$, $F = (u_2,0)$, $M = (x_5,0)$, $N = (x_7,x_6)$.

The **nondegenerate** conditions: Line CB is non-isotropic; Line CA is non-isotropic; Line BA is non-isotropic; Points F, D and E are not collinear.

Example 197 (A-208: 1.07s, 14). *The center of the nine-point circle is the midpoint of a Euler point and the midpoint of the opposite side.*

Points A, B, C are arbitrarily chosen. Points M_1, M_2, M_3, H, H_1, N are constructed (in order) as follows: M_1 is the midpoint of B and C; M_2 is the midpoint of A and C; M_3 is the midpoint of A and B; $HB \perp AC$; $HC \perp AB$; H_1 is the midpoint of C and H; $NM_3 \equiv NM_2$; $NM_3 \equiv NM_1$. The **conclusion**: N is the midpoint of H_1 and M_3.

$A = (0,0)$, $B = (u_1,0)$, $C = (u_2,u_3)$, $M_1 = (x_1,x_2)$, $M_2 = (x_3,x_4)$, $M_3 = (x_5,0)$, $H = (u_2,x_6)$, $H_1 = (u_2,x_7)$, $N = (x_9,x_8)$.

The **nondegenerate** conditions: Points A, B and C are not collinear; Points M_3, M_1 and M_2 are not collinear.

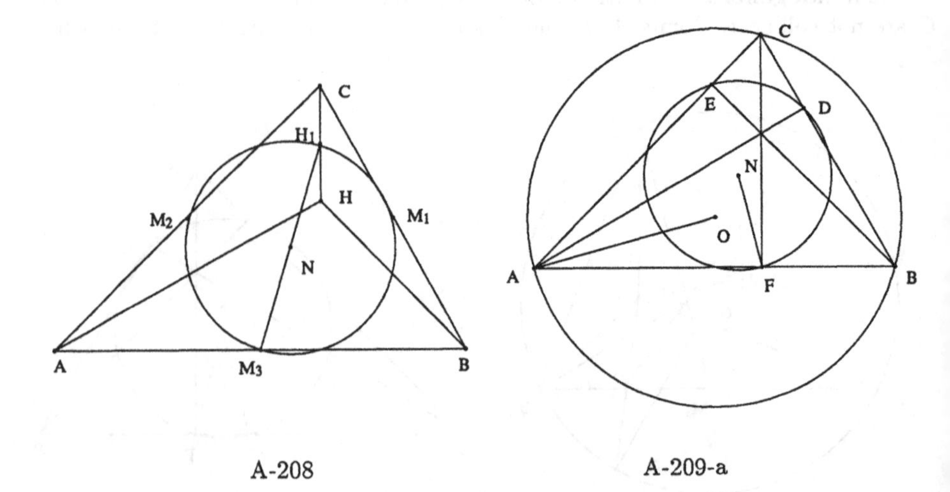

A-208

A-209-a

Example 198 (A-209-a: 11.5s, 376; 4.98s, 117). *The radius of the nine-point circle is equal to half the circumradius of the triangle.*

Points A, B, C are arbitrarily chosen. Points O, D, E, F, N are constructed (in order) as follows: $OA \equiv OC$; $OA \equiv OB$; D is on line BC; $DA \perp CB$; E is on line AC; $EB \perp CA$; F is on line AB; $FC \perp BA$; $NF \equiv NE$; $NF \equiv ND$. The **conclusion**: $4\,\overline{NF}^2 - \overline{OA}^2 = 0$.

$A = (0,0)$, $B = (u_1, 0)$, $C = (u_2, u_3)$, $O = (x_2, x_1)$, $D = (x_4, x_3)$, $E = (x_6, x_5)$, $F = (u_2, 0)$, $N = (x_8, x_7)$.

The **nondegenerate** conditions: Points A, B and C are not collinear; Line CB is non-isotropic; Line CA is non-isotropic; Line BA is non-isotropic; Points F, D and E are not collinear.

Example 199 (A-209-b: 1.45s, 24). *The nine-point center lies on the Euler line, midway between the circumcenter and the orthocenter.*

Points A, B, H are arbitrarily chosen. Points C, O, D, E, F, N are constructed (in order) as follows: $CH \perp AB$; $CB \perp AH$; $OA \equiv OC$; $OA \equiv OB$; D is on line BC; $DA \perp CB$; E is on line AC; $EB \perp CA$; F is on line AB; $FC \perp BA$; $NF \equiv NE$; $NF \equiv ND$. The **conclusion**: N is the midpoint of H and O.

$A = (0,0)$, $B = (u_1, 0)$, $H = (u_2, u_3)$, $C = (u_2, x_1)$, $O = (x_3, x_2)$, $D = (x_5, x_4)$, $E = (x_7, x_6)$, $F = (u_2, 0)$, $N = (x_9, x_8)$.

The **nondegenerate** conditions: Points A, H and B are not collinear; Points A, B and C are not collinear; Line CB is non-isotropic; Line CA is non-isotropic; Line BA is non-isotropic; Points F, D and E are not collinear.

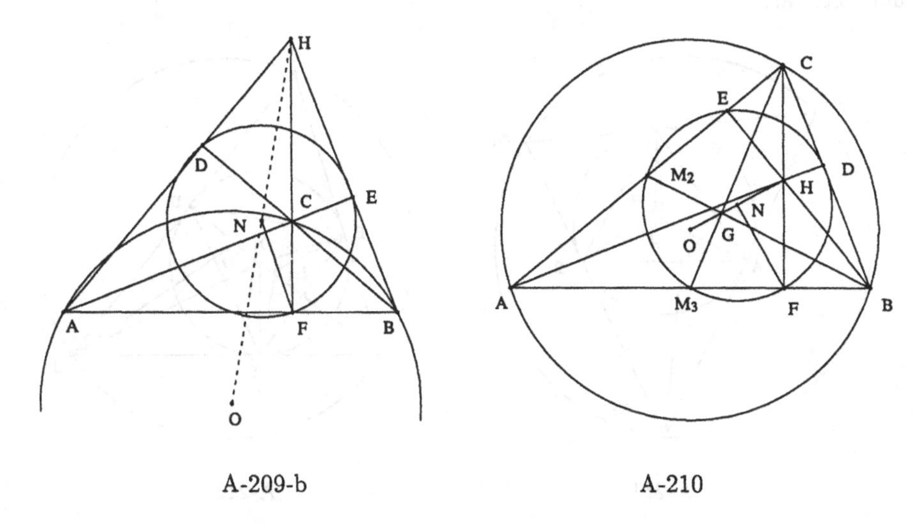

A-209-b A-210

Example 200 (A-210: 4.97s, 31). *The two pairs of points O and N, G and H separate each other harmonically.*

Points A, B, H are arbitrarily chosen. Points C, M_2, M_3, G, O, D, E, F, N are constructed (in order) as follows: $CH \perp AB$; $CB \perp AH$; M_2 is the midpoint of A and C; M_3 is the midpoint of A and B; G is on line BM_2; G is on line CM_3; $OA \equiv OC$; $OA \equiv OB$; D is on line BC; $DA \perp CB$; E is on line AC; $EB \perp CA$; F is on line AB; $FC \perp BA$; $NF \equiv NE$; $NF \equiv ND$. The **conclusion**: Points O, N, G and H form a harmonic set.

$A = (0, 0)$, $B = (u_1, 0)$, $H = (u_2, u_3)$, $C = (u_2, x_1)$, $M_2 = (x_2, x_3)$, $M_3 = (x_4, 0)$, $G = (x_6, x_5)$, $O = (x_8, x_7)$, $D = (x_{10}, x_9)$, $E = (x_{12}, x_{11})$, $F = (u_2, 0)$, $N = (x_{14}, x_{13})$.

The **nondegenerate** conditions: Points A, H and B are not collinear; Line CM_3 intersects line BM_2; Points A, B and C are not collinear; Line CB is non-isotropic; Line CA is non-isotropic; Line BA is non-isotropic; Points F, D and E are not collinear.

Example 201 (A-211: 21.38s, 690; 8.42s, 93). *The circumcenter of the tangential triangle of a given triangle lies on the Euler line of the given triangle.*

Points A, B, C are arbitrarily chosen. Points H, O, B_1, C_1, A_1, O_1 are constructed (in order) as follows: $HB \perp AC$; $HC \perp AB$; $OA \equiv OC$; $OA \equiv OB$; $B_1C \perp CO$; $B_1A \perp AO$; $C_1B \perp OB$; C_1 is on line AB_1; A_1 is on line C_1B; A_1 is on line B_1C; $O_1A_1 \equiv O_1C_1$; $O_1A_1 \equiv O_1B_1$. The **conclusion**: Points O, H and O_1 are collinear.

$A = (0, 0)$, $B = (u_1, 0)$, $C = (u_2, u_3)$, $H = (u_2, x_1)$, $O = (x_3, x_2)$, $B_1 = (x_5, x_4)$, $C_1 = (x_7, x_6)$, $A_1 = (x_9, x_8)$, $O_1 = (x_{11}, x_{10})$.

The **nondegenerate** conditions: Points A, B and C are not collinear; Points A, B and C are not collinear; Points A, O and C are not collinear; Line AB_1 is not perpendicular to line OB; Line B_1C intersects line C_1B; Points A_1, B_1 and C_1 are not collinear.

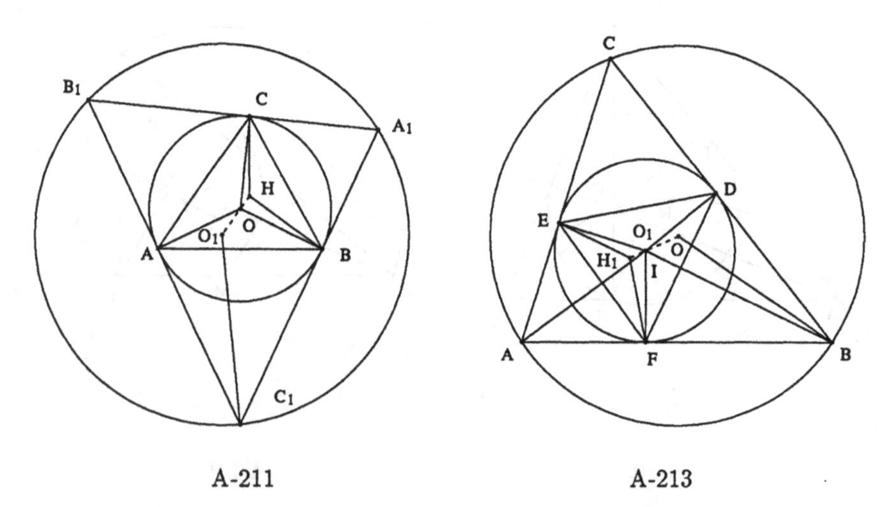

A-211 A-213

Example 202 (A-213: 61.88s, 2110; 9.8s, 104). *The Circumcenter of a triangle lies on the Euler line of the triangle determined by the points of contact of the sides of the given triangle with its inscribed circle.*

Points A, B, I are arbitrarily chosen. Points C, O, D, E, F, O_1, H_1 are constructed (in order) as follows: $\tan(ABI) = \tan(IBC)$; $\tan(BAI) = \tan(IAC)$; $OA \equiv OC$; $OA \equiv OB$; D is on line BC; $DI \perp BC$; E is on line AC; $EI \perp AC$; F is on line AB; $FI \perp AB$; $O_1F \equiv O_1D$; $O_1F \equiv O_1E$; $H_1E \perp FD$; $H_1F \perp ED$. The **conclusion**: Points O, O_1 and H_1 are collinear.

$A = (0,0)$, $B = (u_1,0)$, $I = (u_2,u_3)$, $C = (x_2,x_1)$, $O = (x_4,x_3)$, $D = (x_6,x_5)$, $E = (x_8,x_7)$, $F = (u_2,0)$, $O_1 = (x_{10},x_9)$, $H_1 = (x_{12},x_{11})$.

The **nondegenerate** conditions: Points I, A and B are not collinear; $\angle AIB$ is not right; Points A, B and C are not collinear; Line BC is non-isotropic; Line AC is non-isotropic; Line AB is non-isotropic; Points F, E and D are not collinear; Points E, D and F are not collinear.

Remark. I can be any of the four tritangent centers.

Example 203 (A-214: 1.24s, 14). *The projections of the orthocenter of a triangle upon the two bisectors of an angle of the triangle lie on the line joining the midpoint of the side opposite the vertex considered to the nine-point center of the triangle.*

Points A, B, C are arbitrarily chosen. Points H, I, I_1, D, E, F, N are constructed (in order) as follows: $HC \perp AB$; $HA \perp BC$; $\tan(BAI) = \tan(IAC)$; I_1 is on line IA; $I_1H \perp AI$; D is the midpoint of B and C; E is the midpoint of A and C; F is the midpoint of A and B; $NF \equiv ND$; $NF \equiv NE$. The **conclusion**: Points D, N and I_1 are collinear.

$A = (0,0)$, $B = (u_1,0)$, $C = (u_2,u_3)$, $H = (u_2,x_1)$, $I = (x_2,u_4)$, $I_1 = (x_4,x_3)$, $D = (x_5,x_6)$, $E = (x_7,x_8)$, $F = (x_9,0)$, $N = (x_{11},x_{10})$.

The **nondegenerate** conditions: Points B, C and A are not collinear; Points A, C and B are not collinear; Line AI is non-isotropic; Points F, E and D are not collinear.

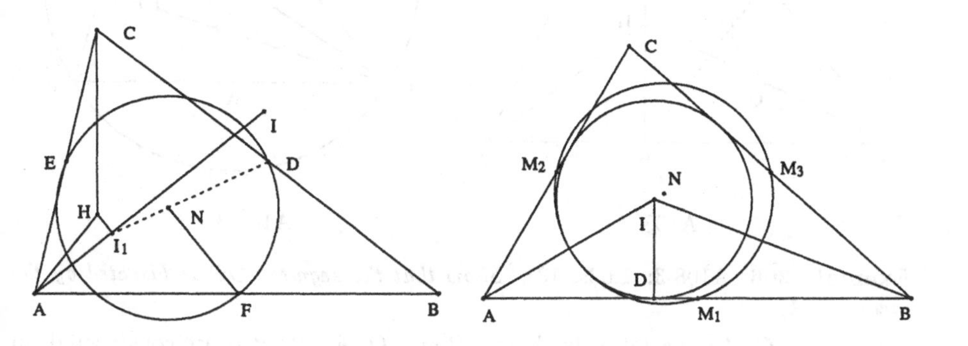

A-214 A-215

Example 204 (A-215: 4.4s, 80). *Feuerbach's Theorem. The nine-point circle of a triangle touches each of the four tritangent circles of the triangle.*

Points D, A are arbitrarily chosen. Points I, B, C, M_1, M_2, M_3, N are constructed (in order) as follows: $ID \perp AD$; B is on line AD; $\tan(ABI) = \tan(IBC)$; $\tan(BAI) = \tan(IAC)$; M_1 is the midpoint of A and B; M_2 is the midpoint of A and C; M_3 is the midpoint of B and C; $NM_1 \equiv NM_3$; $NM_1 \equiv NM_2$. The **conclusion**: Circle (I, DI) is tangent to circle (N, M_1N).

$D = (0,0)$, $A = (u_1,0)$, $I = (0,u_2)$, $B = (u_3,0)$, $C = (x_2,x_1)$, $M_1 = (x_3,0)$, $M_2 = (x_4,x_5)$, $M_3 = (x_6,x_7)$, $N = (x_9,x_8)$.

The **nondegenerate** conditions: $A \neq D$; $A \neq D$; Points I, A and B are not collinear; $\angle AIB$ is not right; Points M_1, M_2 and M_3 are not collinear.

Example 205 (A108-2: 0.9s, 8). *Show that the triangle DB_1C_1 is congruent to the Euler triangle.*

Points B, C, A are arbitrarily chosen. Points D, H, P, Q, C_1 are constructed (in order) as follows: D is on line BC; $DA \perp BC$; $HB \perp AC$; H is on line AD; P is the midpoint of A and H; Q is the midpoint of B and H; C_1 is the midpoint of B and A. The **conclusion**: $C_1D \equiv PQ$.

$B = (0,0)$, $C = (u_1,0)$, $A = (u_2,u_3)$, $D = (u_2,0)$, $H = (u_2,x_1)$, $P = (u_2,x_2)$, $Q = (x_3,x_4)$, $C_1 = (x_5,x_6)$.

The **nondegenerate** conditions: Line BC is non-isotropic; Line AD is not perpendicular to line AC.

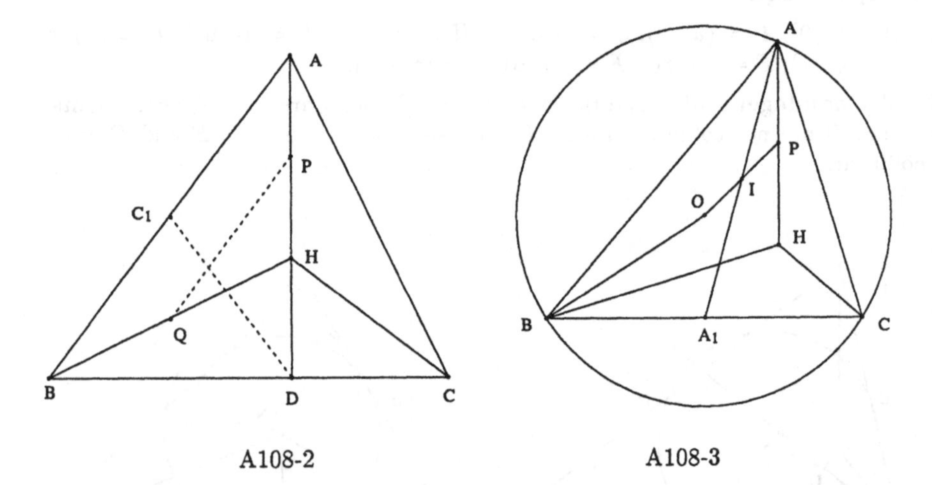

A108-2 A108-3

Example 206 (A108-3: 1.03s, 13). *Show that the segment OP is bisected by the median AA_1.*

Points B, C, A are arbitrarily chosen. Points O, A_1, H, P, I are constructed (in order) as follows: $OB \equiv OC$; $OA \equiv OB$; A_1 is the midpoint of B and C; $HB \perp AC$; $HA \perp BC$; P is the midpoint of A and H; I is on line OP; I is on line AA_1. The **conclusion**: I is the midpoint of O and P.

$B = (0,0)$, $C = (u_1,0)$, $A = (u_2,u_3)$, $O = (x_2,x_1)$, $A_1 = (x_3,0)$, $H = (u_2,x_4)$, $P = (u_2,x_5)$, $I = (x_7,x_6)$.

The **nondegenerate** conditions: Points A, B and C are not collinear; Points B, C and A are not collinear; Line AA_1 intersects line OP.

Example 207 (A108-4: 1.87s, 14). *If P is the symmetric of the vertex A with respect to the opposite side BC, show that HP is equal to four times the distance of* *the nine-point center from BC.*

Points B, C, A are arbitrarily chosen. Points A_1, B_1, C_1, H, D, N, P, K are constructed (in order) as follows: A_1 is the midpoint of B and C; B_1 is the midpoint of A and C; C_1 is the midpoint of B and A; $HB \perp AC$; $HA \perp BC$; D is on line BC; $DA \perp BC$; $NA_1 \equiv NC_1$; $NA_1 \equiv NB_1$; D is the midpoint of A and P; K is on line BC; $KN \perp BC$. The **conclusion**: $\overline{PH}^2 - 16\,\overline{NK}^2 = 0$.

$B = (0,0)$, $C = (u_1, 0)$, $A = (u_2, u_3)$, $A_1 = (x_1, 0)$, $B_1 = (x_2, x_3)$, $C_1 = (x_4, x_5)$, $H = (u_2, x_6)$, $D = (u_2, 0)$, $N = (x_8, x_7)$, $P = (u_2, x_9)$, $K = (x_8, 0)$.

The **nondegenerate** conditions: Points B, C and A are not collinear; Line BC is non-isotropic; Points A_1, B_1 and C_1 are not collinear; Line BC is non-isotropic.

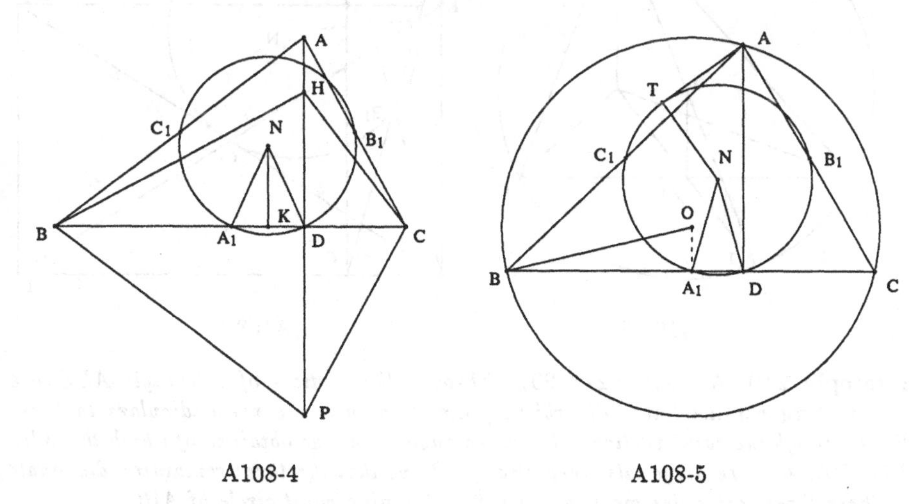

A108-4 A108-5

Example 208 (A108-5: 83.78s, 1605; 8.15s, 356). *Show that the square of the tangent from a vertex of a triangle to the nine-point circle is equal to the altitude issued from that vertex multiplied by the distance of the opposite side from the circumcenter.*

Points B, C, A are arbitrarily chosen. Points O, A_1, B_1, C_1, D, N, T are constructed (in order) as follows: $OA \equiv OB$; $OB \equiv OC$; A_1 is the midpoint of B and C; B_1 is the midpoint of A and C; C_1 is the midpoint of B and A; D is on line BC; $DA \perp BC$; $NA_1 \equiv NC_1$; $NA_1 \equiv NB_1$; $TN \equiv NA_1$; T is on the circle with diagonal NA. The **conclusion**: $AT \cdot AT = AD \cdot OA_1$.

$B = (0,0)$, $C = (u_1, 0)$, $A = (u_2, u_3)$, $O = (x_2, x_1)$, $A_1 = (x_3, 0)$, $B_1 = (x_4, x_5)$, $C_1 = (x_6, x_7)$, $D = (u_2, 0)$, $N = (x_9, x_8)$, $T = (x_{11}, x_{10})$.

The **nondegenerate** conditions: Points B, C and A are not collinear; Line BC is non-isotropic; Points A_1, B_1 and C_1 are not collinear; The line joining the midpoint of N and A and the point N is non-isotropic.

Example 209 (A108-6: 6.7s, 16; 1.1s, 16). *Prove that HA_1 passes through the diametric opposite of A on the circumcircle.*

Points B, C, A are arbitrarily chosen. Points O, A_1, H, I are constructed (in order) as follows: $OB \equiv OC$; $OA \equiv OB$; A_1 is the midpoint of B and C; $HB \perp AC$; $HA \perp BC$; I is on line AO; I is on line HA_1. The **conclusion**: $OB \equiv OI$.

$B = (0,0)$, $C = (u_1, 0)$, $A = (u_2, u_3)$, $O = (x_2, x_1)$, $A_1 = (x_3, 0)$, $H = (u_2, x_4)$, $I = (x_6, x_5)$.

The **nondegenerate** conditions: Points A, B and C are not collinear; Points B, C and A are not collinear; Line HA_1 intersects line AO.

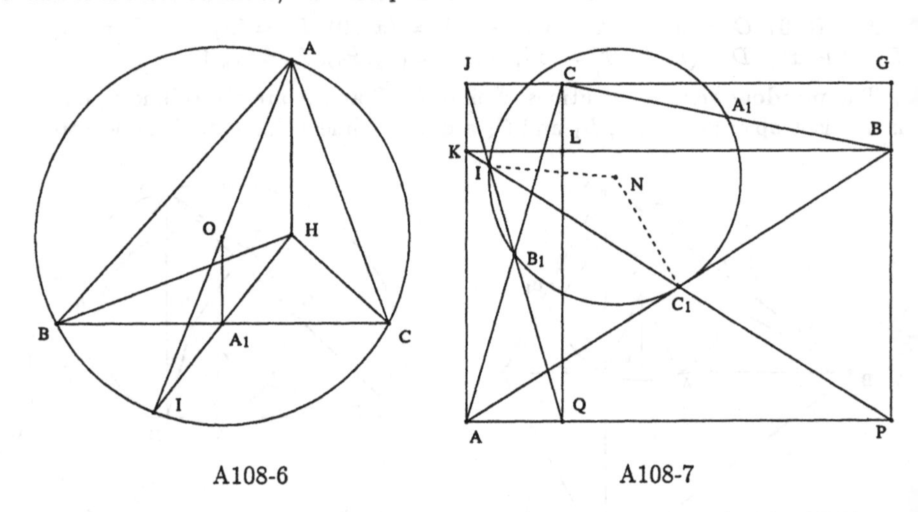

A108-6 A108-7

Example 210 (A108-7: 2.22s, 20). *Through the vertices of a triangle ABC are drawn three parallel lines, of arbitrary direction, and the perpendiculars to these lines through the same vertices. Three rectangles are thus obtained of which the sides BC, CA, AB are diagonals, respectively. Prove that the three remaining diagonals of these three rectangles meet in a point on the nine point circle of ABC.*

Points A, P are arbitrarily chosen. Points Q, B, C, A_1, B_1, C_1, N, K, L, J, G, I are constructed (in order) as follows: Q is on line AP; $BP \perp AP$; $CQ \perp AP$; A_1 is the midpoint of B and C; B_1 is the midpoint of A and C; C_1 is the midpoint of B and A; $NA_1 \equiv NC_1$; $NA_1 \equiv NB_1$; $KA \perp AP$; $KB \parallel AP$; L is on line KB; L is on line QC; J is on line KA; $JC \parallel AP$; G is on line BP; G is on line JC; I is on line JQ; I is on line KP. The **conclusion**: $NI \equiv NC_1$.

$A = (0,0)$, $P = (u_1, 0)$, $Q = (u_2, 0)$, $B = (u_1, u_3)$, $C = (u_2, u_4)$, $A_1 = (x_1, x_2)$, $B_1 = (x_3, x_4)$, $C_1 = (x_5, x_6)$, $N = (x_8, x_7)$, $K = (0, u_3)$, $L = (u_2, u_3)$, $J = (0, u_4)$, $G = (u_1, u_4)$, $I = (x_{10}, x_9)$.

The **nondegenerate** conditions: $A \neq P$; $A \neq P$; $A \neq P$; Points A_1, B_1 and C_1 are not collinear; Line AP is non-isotropic; Line QC intersects line KB; Points A, P and K are not collinear; Line JC intersects line BP; Line KP intersects line JQ.

Example 211 (A109-18: 3.72s, 25; 2.5s, 57). *The tangent to the nine-point circle at the midpoint of a side of the given triangle is antiparallel to this side with respect to the two other sides of the triangle.*

Points B, C, A are arbitrarily chosen. Points A_1, B_1, C_1, N, K, J are constructed (in order) as follows: A_1 is the midpoint of B and C; B_1 is the midpoint of A and C; C_1 is the midpoint of B and A; $NA_1 \equiv NC_1$; $NA_1 \equiv NB_1$; K is on line AC; $KA_1 \perp A_1N$; J is on line AB; J is on line KA_1. The **conclusion**: Points K, J, B and C are on the same circle.

$B = (0,0)$, $C = (u_1,0)$, $A = (u_2,u_3)$, $A_1 = (x_1,0)$, $B_1 = (x_2,x_3)$, $C_1 = (x_4,x_5)$, $N = (x_7,x_6)$, $K = (x_9,x_8)$, $J = (x_{11},x_{10})$.

The **nondegenerate** conditions: Points A_1, B_1 and C_1 are not collinear; Line A_1N is not perpendicular to line AC; Line KA_1 intersects line AB.

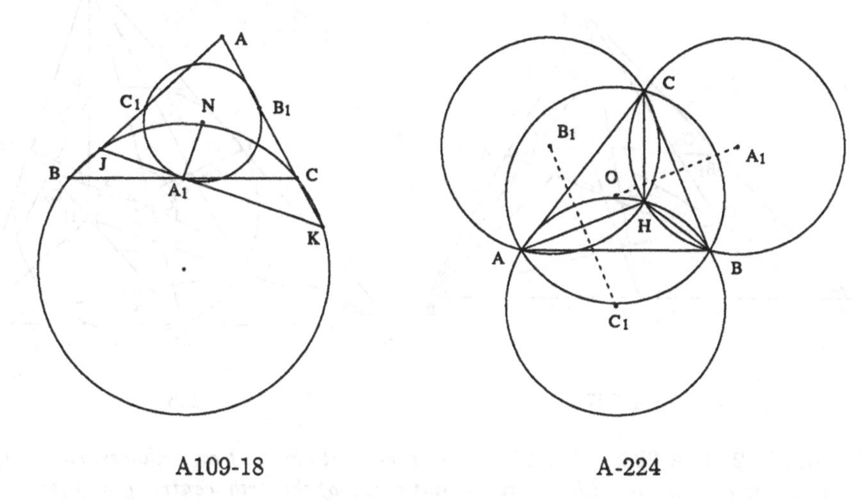

A109-18 A-224

Example 212 (A-224: 1.28s, 13). *The circumcenters of an orthocentric group of triangles form an orthocentric quadrilateral.*

Points A, B, C are arbitrarily chosen. Points H, O, A_1, B_1, C_1 are constructed (in order) as follows: $HC \perp AB$; $HA \perp BC$; $OA \equiv OC$; $OA \equiv OB$; $A_1C \equiv A_1B$; $A_1B \equiv A_1H$; $B_1A \equiv B_1C$; $B_1A \equiv B_1H$; $C_1A \equiv C_1B$; $C_1A \equiv C_1H$. The **conclusion**: $OA_1 \perp B_1C_1$.

$A = (0,0)$, $B = (u_1,0)$, $C = (u_2,u_3)$, $H = (u_2,x_1)$, $O = (x_3,x_2)$, $A_1 = (x_5,x_4)$, $B_1 = (x_7,x_6)$, $C_1 = (x_9,x_8)$.

The **nondegenerate** conditions: Points B, C and A are not collinear; Points A, B and C are not collinear; Points B, H and C are not collinear; Points A, H and C are not collinear; Points A, H and B are not collinear.

Example 213 (A-227: 2.18s, 15). *The four centroids of an orthocentric group of triangles form an orthocentric group.*

Points A, B, C are arbitrarily chosen. Points H, C_1, B_1, A_1, G, P, Q, G_A, G_B, G_C are constructed (in order) as follows: $HC \perp AB$; $HA \perp BC$; C_1 is the midpoint of A and B; B_1 is the midpoint of A and C; A_1 is the midpoint of B and C; G is on line CC_1; G is on line AA_1; P is the midpoint of C and H; Q is the midpoint of A and H; G_A is on line HA_1; G_A is on line PB; G_B is on line HB_1; G_B is on line CQ; G_C is on line HC_1; G_C is on line BQ. The **conclusion**: $GG_A \perp G_BG_C$.

$A = (0,0)$, $B = (u_1,0)$, $C = (u_2,u_3)$, $H = (u_2,x_1)$, $C_1 = (x_2,0)$, $B_1 = (x_3,x_4)$, $A_1 = (x_5,x_6)$, $G = (x_8,x_7)$, $P = (u_2,x_9)$, $Q = (x_{10},x_{11})$, $G_A = (x_{13},x_{12})$, $G_B = (x_{15},x_{14})$, $G_C = (x_{17},x_{16})$.

The **nondegenerate** conditions: Points B, C and A are not collinear; Line AA_1 intersects line CC_1; Line PB intersects line HA_1; Line CQ intersects line HB_1; Line BQ intersects line HC_1.

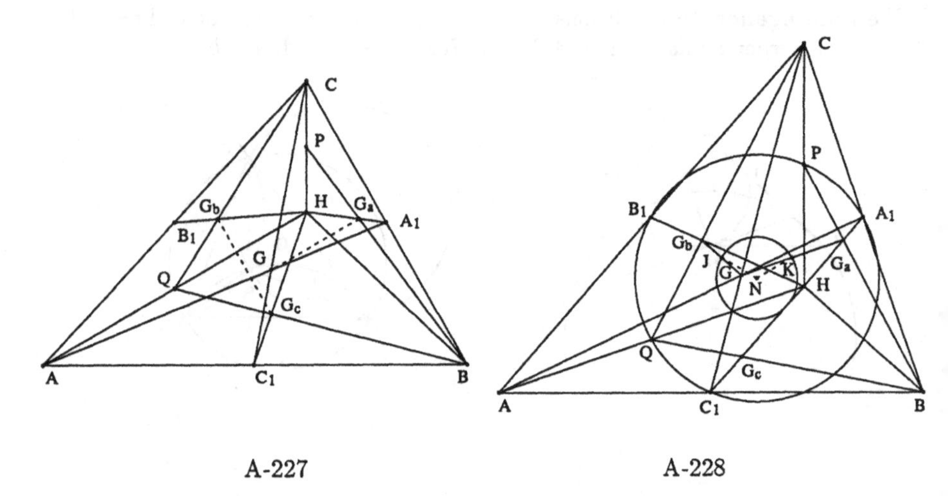

A-227 A-228

Example 214 (A-228: 3.02s, 22). *The nine-point circle of an orthocentric group of triangles is concentric with the nine-point circle of the orthocentric group formed by the centroids of the given group of triangles.*

Points A, B, C are arbitrarily chosen. Points H, C_1, B_1, A_1, G, P, Q, G_A, G_B, G_C, N, K, J are constructed (in order) as follows: $HC \perp AB$; $HA \perp BC$; C_1 is the midpoint of A and B; B_1 is the midpoint of A and C; A_1 is the midpoint of B and C; G is on line CC_1; G is on line AA_1; P is the midpoint of C and H; Q is the midpoint of A and H; G_A is on line HA_1; G_A is on line PB; G_B is on line HB_1; G_B is on line CQ; G_C is on line HC_1; G_C is on line BQ; $NC_1 \equiv NB_1$; $NC_1 \equiv NA_1$; K is the midpoint of G and G_A; J is the midpoint of G and G_B. The **conclusion**: $NK \equiv NJ$.

$A = (0,0)$, $B = (u_1,0)$, $C = (u_2,u_3)$, $H = (u_2,x_1)$, $C_1 = (x_2,0)$, $B_1 = (x_3,x_4)$, $A_1 = (x_5,x_6)$, $G = (x_8,x_7)$, $P = (u_2,x_9)$, $Q = (x_{10},x_{11})$, $G_A = (x_{13},x_{12})$, $G_B = (x_{15},x_{14})$, $G_C = (x_{17},x_{16})$, $N = (x_{19},x_{18})$, $K = (x_{20},x_{21})$, $J = (x_{22},x_{23})$.

The **nondegenerate** conditions: Points B, C and A are not collinear; Line AA_1 intersects line CC_1; Line PB intersects line HA_1; Line CQ intersects line HB_1; Line BQ intersects line HC_1; Points C_1, A_1 and B_1 are not collinear.

Example 215 (A111-1: 1.42s, 17). *The Euler lines of the four triangles of an orthocentric group are concurrent.*

Points A, B, C are arbitrarily chosen. Points H, O, A_1, B_1, I are constructed (in order) as follows: $HC \perp AB$; $HA \perp BC$; $OA \equiv OC$; $OA \equiv OB$; $A_1C \equiv A_1B$;

$A_1B \equiv A_1H$; $B_1A \equiv B_1C$; $B_1A \equiv B_1H$; I is on line BB_1; I is on line AA_1. The **conclusion**: Points O, H and I are collinear.

$A = (0,0)$, $B = (u_1, 0)$, $C = (u_2, u_3)$, $H = (u_2, x_1)$, $O = (x_3, x_2)$, $A_1 = (x_5, x_4)$, $B_1 = (x_7, x_6)$, $I = (x_9, x_8)$.

The **nondegenerate** conditions: Points B, C and A are not collinear; Points A, B and C are not collinear; Points B, H and C are not collinear; Points A, H and C are not collinear; Line AA_1 intersects line BB_1.

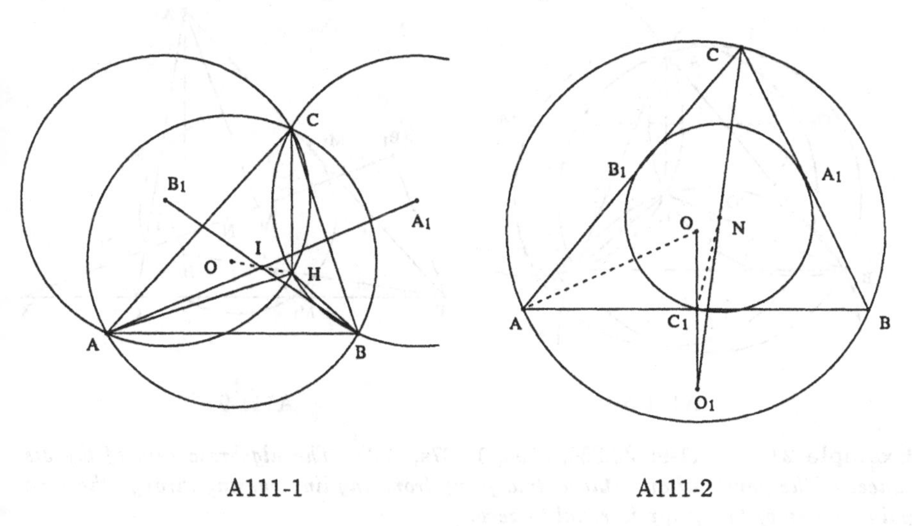

A111-1 A111-2

Example 216 (A111-2: 1.4s, 22). *Show that the symmetric of the circumcenter of a triangle with respect to a side coincides with the symmetric of the vertex opposite the side considered with respect to the nine-point center of the triangle.*

Points A, B, C are arbitrarily chosen. Points O, C_1, B_1, A_1, N, O_1 are constructed (in order) as follows: $OA \equiv OC$; $OA \equiv OB$; C_1 is the midpoint of A and B; B_1 is the midpoint of A and C; A_1 is the midpoint of B and C; $NC_1 \equiv NB_1$; $NC_1 \equiv NA_1$; C_1 is the midpoint of O and O_1. The **conclusion**: N is the midpoint of O_1 and C.

$A = (0,0)$, $B = (u_1, 0)$, $C = (u_2, u_3)$, $O = (x_2, x_1)$, $C_1 = (x_3, 0)$, $B_1 = (x_4, x_5)$, $A_1 = (x_6, x_7)$, $N = (x_9, x_8)$, $O_1 = (x_{10}, x_{11})$.

The **nondegenerate** conditions: Points A, B and C are not collinear; Points C_1, A_1 and B_1 are not collinear.

Example 217 (A111-5: 3.05s, 13). *Show that the circumcenters of the triangles HBC, HCA, HAB of an orthocentric group $HABC$ form a triangle congruent to ABC; the sides of the two triangles are parallel, and the point H is the circumcenter of the new triangle; the circumcenter of ABC is the orthocenter of the new triangle.*

Points B, C, A are arbitrarily chosen. Points H, O_A, O_B, O_C, O are constructed (in order) as follows: $HB \perp AC$; $HA \perp BC$; $O_AB \equiv O_AH$; $O_AB \equiv O_AC$; $O_BC \equiv O_BH$; $O_BA \equiv O_BC$; $O_CB \equiv O_CH$; $O_CB \equiv O_CA$; $OB \equiv OA$; $OB \equiv OC$. The **conclusions**: (1) $O_BO_C \parallel BC$; (2) $O_BO_C \equiv BC$; (3) $OO_A \perp O_BO_C$.

$B = (0,0)$, $C = (u_1, 0)$, $A = (u_2, u_3)$, $H = (u_2, x_1)$, $O_A = (x_3, x_2)$, $O_B = (x_5, x_4)$, $O_C = (x_7, x_6)$, $O = (x_9, x_8)$.

The **nondegenerate** conditions: Points B, C and A are not collinear; Points B, C and H are not collinear; Points A, C and H are not collinear; Points B, A and H are not collinear; Points B, C and A are not collinear.

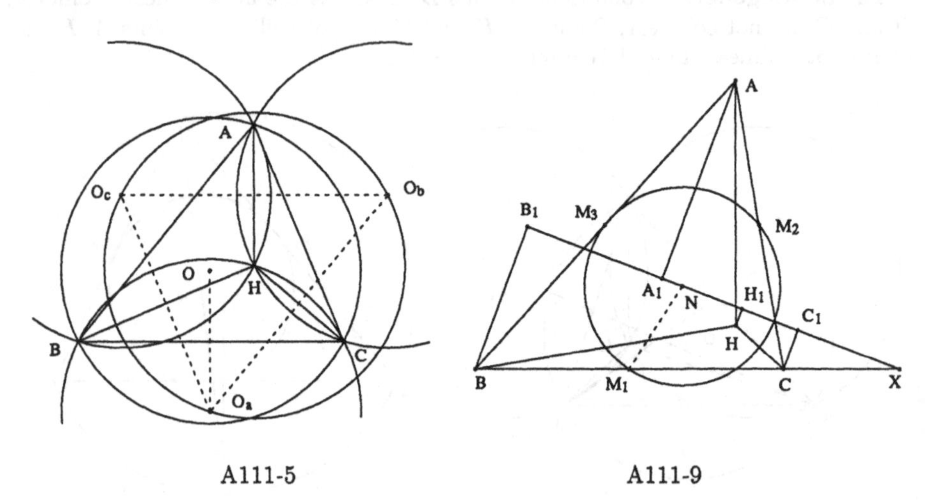

A111-5 A111-9

Example 218 (A111-9: 33.25s, 2195; 12.37s, 223). *The algebraic sum of the distances of the points of an orthocentric group from any line passing through the nine-point center of the group is equal to zero.*

Points B, C, A are arbitrarily chosen. Points H, M_1, M_2, M_3, N, X, A_1, B_1, C_1, H_1 are constructed (in order) as follows: $HB \perp AC$; $HA \perp BC$; M_1 is the midpoint of B and C; M_2 is the midpoint of C and A; M_3 is the midpoint of A and B; $NM_3 \equiv NM_2$; $NM_3 \equiv NM_1$; X is on line BC; A_1 is on line XN; $A_1A \perp XN$; B_1 is on line XN; $B_1B \perp XN$; C_1 is on line XN; $C_1C \perp XN$; H_1 is on line XN; $H_1H \perp XN$. The **conclusion**: $\overline{AA_1} + \overline{BB_1} + \overline{CC_1} + \overline{HH_1} = 0$.

$B = (0,0)$, $C = (u_1, 0)$, $A = (u_2, u_3)$, $H = (u_2, x_1)$, $M_1 = (x_2, 0)$, $M_2 = (x_3, x_4)$, $M_3 = (x_5, x_6)$, $N = (x_8, x_7)$, $X = (u_4, 0)$, $A_1 = (x_{10}, x_9)$, $B_1 = (x_{12}, x_{11})$, $C_1 = (x_{14}, x_{13})$, $H_1 = (x_{16}, x_{15})$.

The **nondegenerate** conditions: Points B, C and A are not collinear; Points M_3, M_1 and M_2 are not collinear; $B \neq C$; Line XN is non-isotropic; Line XN is non-isotropic; Line XN is non-isotropic; Line XN is non-isotropic.

Example 219 (A115-4: 6.47s, 16; 2.75s, 41). *Show that the mediator of the bisector AU of the triangle ABC, the perpendicular to BC at U, and the circumdiameter of ABC passing through A are concurrent.*

Points B, C, A are arbitrarily chosen. Points U, O, I are constructed (in order) as follows: U is on line BC; $\tan(BAU) = \tan(UAC)$; $OB \equiv OC$; $OA \equiv OB$; I is on line OA; $IA \equiv IU$. The **conclusion**: $IU \perp BC$.

$B = (0,0)$, $C = (u_1, 0)$, $A = (u_2, u_3)$, $U = (x_1, 0)$, $O = (x_3, x_2)$, $I = (x_5, x_4)$.

The **nondegenerate** conditions: $(2u_2 - u_1)u_3 \neq 0$; Points A, B and C are not collinear; Line AU is not perpendicular to line OA.

Remark. This is for the both bisectors of angle A.

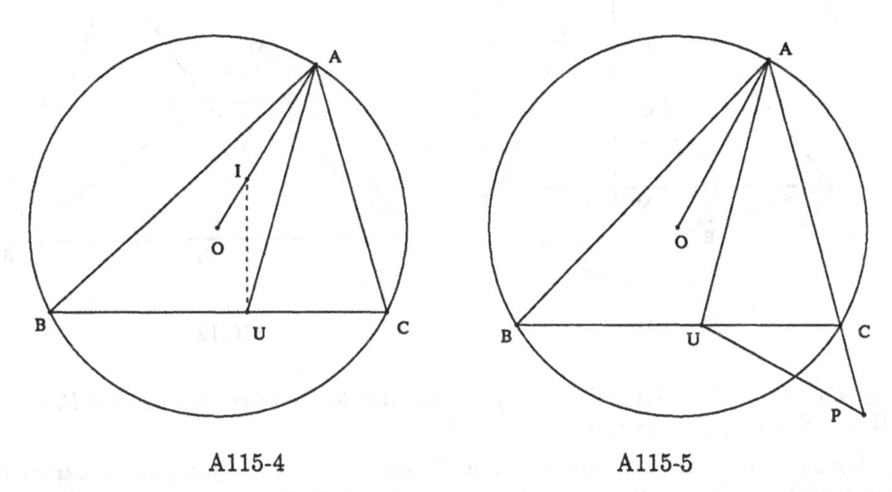

A115-4 A115-5

Example 220 (A115-5: 1.2s, 34). *From the foot U of the bisector AU of the triangle ABC a perpendicular UQ is drawn to the circumradius AO of ABCmeeting AC in P. Prove that AP is equal to AB.*

Points B, C, A are arbitrarily chosen. Points U, O, P are constructed (in order) as follows: U is on line BC; $\tan(BAU) = \tan(UAC)$; $OB \equiv OC$; $OA \equiv OB$; P is on line AC; $PU \perp AO$. The **conclusion**: $AP \equiv AB$.

$B = (0,0)$, $C = (u_1, 0)$, $A = (u_2, u_3)$, $U = (x_1, 0)$, $O = (x_3, x_2)$, $P = (x_5, x_4)$.

The **nondegenerate** conditions: $(2u_2 - u_1)u_3 \neq 0$; Points A, B and C are not collinear; Line AO is not perpendicular to line AC.

Remark. This is for the both bisectors of angle A.

Example 221 (A116-11: 1.42s, 60). *Show that the foot of the altitude to the base of a triangle and the projections of the ends of the base upon the circumdiameter passing through the opposite vertex of the triangle determine a circle having for center the midpoint the base.*

Points A, B, C are arbitrarily chosen. Points O, M, D, E are constructed (in order) as follows: $OA \equiv OC$; $OA \equiv OB$; M is the midpoint of A and B; $DC \perp BA$; D is on line AB; E is on line OC; $EA \perp OC$. The **conclusion**: $ME \equiv MD$.

$A = (0,0)$, $B = (u_1, 0)$, $C = (u_2, u_3)$, $O = (x_2, x_1)$, $M = (x_3, 0)$, $D = (u_2, 0)$, $E = (x_5, x_4)$.

The **nondegenerate** conditions: Points A, B and C are not collinear; Line AB is non-isotropic; Line OC is non-isotropic.

Example 222 (A116-12: 2.4s, 26; 2.3s, 73). *The internal bisector of the angle B of the triangle ABC meets the sides B_1C_1, B_1A_1 of the medial triangle in the points*

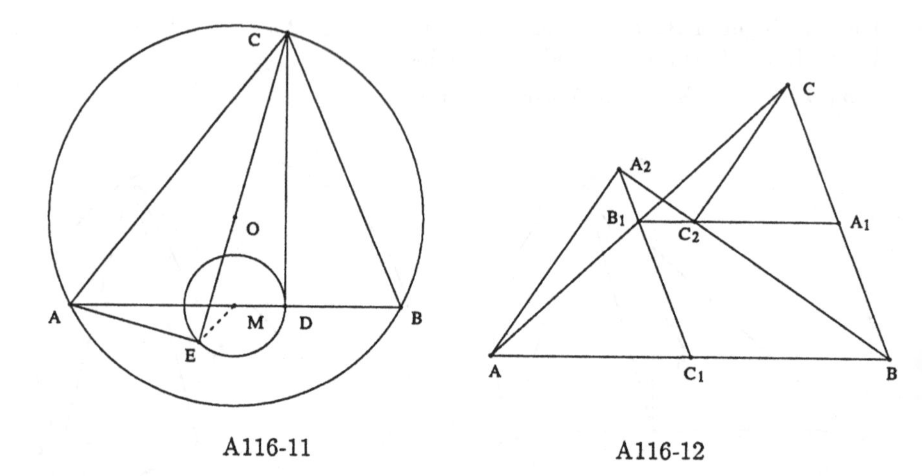

A116-11 A116-12

A_2, C_2. Prove that AA_2, CC_2 are perpendicular to the bisector, and that $B_1A_2 = B_1C_2$. Similarly for the external bisector.

Points A, B, C are arbitrarily chosen. Points A_1, B_1, C_1, A_2, C_2 are constructed (in order) as follows: A_1 is the midpoint of B and C; B_1 is the midpoint of A and C; C_1 is the midpoint of A and B; A_2 is on line B_1C_1; $\tan(ABA_2) = \tan(A_2BC)$; C_2 is on line A_1B_1; C_2 is on line BA_2. The **conclusions**: (1) $B_1A_2 \equiv B_1C_2$; (2) $AA_2 \perp A_2B$.

$A = (0,0)$, $B = (u_1,0)$, $C = (u_2,u_3)$, $A_1 = (x_1,x_2)$, $B_1 = (x_3,x_4)$, $C_1 = (x_5,0)$, $A_2 = (x_7,x_6)$, $C_2 = (x_9,x_8)$.

The **nondegenerate** conditions: $-u_1u_3x_5^2 + ((-2u_1u_2 + 2u_1^2)x_4 + 2u_1u_3x_3)x_5 + u_1u_3x_4^2 + ((2u_1u_2 - 2u_1^2)x_3)x_4 - u_1u_3x_3^2 \neq 0$; $-x_4 \neq 0$; Line BA_2 intersects line A_1B_1.

Example 223 (A116-13: 11.0s, 376). *A_1, B_1, C_1 are the traces of the medians of the triangle ABC on the circumcircle. If $B_1C_1 = a_1$, $C_1A_1 = b_1$, $A_1B_1 = c_1$, show that, in the usual notation for triangle ABC: $am_a : a_1 = bm_b : b_1 = cm_c : c_1$.*

Points A, B, C are arbitrarily chosen. Points A_M, B_M, C_M, O, A_1, B_1, C_1 are constructed (in order) as follows: A_M is the midpoint of C and B; B_M is the midpoint of A and C; C_M is the midpoint of B and A; $OA \equiv OC$; $OA \equiv OB$; A_1 is on line AA_M; $A_1O \equiv OA$; B_1 is on line BB_M; $B_1O \equiv OA$; C_1 is on line CC_M; $C_1O \equiv OA$. The **conclusion**: $\overline{AA_M}^2\ \overline{BC}^2\ \overline{A_1B_1}^2 - \overline{CC_M}^2\ \overline{AB}^2\ \overline{B_1C_1}^2 = 0$.

$A = (0,0)$, $B = (u_1,0)$, $C = (u_2,u_3)$, $A_M = (x_1,x_2)$, $B_M = (x_3,x_4)$, $C_M = (x_5,0)$, $O = (x_7,x_6)$, $A_1 = (x_9,x_8)$, $B_1 = (x_{11},x_{10})$, $C_1 = (x_{13},x_{12})$.

The **nondegenerate** conditions: Points A, B and C are not collinear; Line AA_M is non-isotropic; Line BB_M is non-isotropic; Line CC_M is non-isotropic. In addition, the following nondegenerate conditions, which come from reducibility and have been detected by our prover, should be also added: $C_1 \neq C$; $B_1 \neq B$; $A_1 \neq A$.

Example 224 (A116-14: 1.4s, 10). *Show that the parallels through the vertices A,*

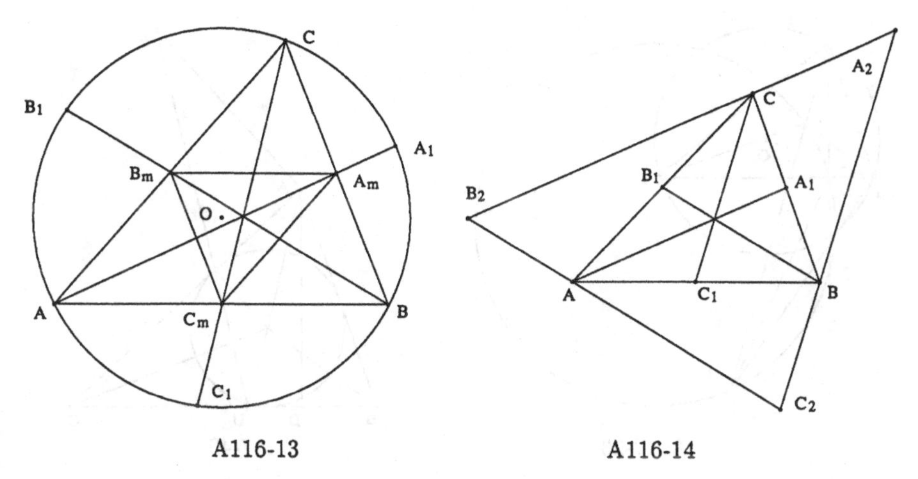

A116-13 A116-14

B, C of the triangle ABC to the medians of this triangle issued from the vertices B, C, A, respectively, form a triangle whose area is three times the area of the given triangle.

Points A, B, C are arbitrarily chosen. Points A_1, B_1, C_1, C_2, B_2, A_2 are constructed (in order) as follows: A_1 is the midpoint of C and B; B_1 is the midpoint of A and C; C_1 is the midpoint of B and A; $C_2B \parallel CC_1$; $C_2A \parallel BB_1$; $B_2C \parallel AA_1$; B_2 is on line C_2A; A_2 is on line B_2C; A_2 is on line C_2B. The **conclusion:** $\nabla(A_2B_2C_2) - 3\,\nabla(ABC) = 0$.

$A = (0,0)$, $B = (u_1,0)$, $C = (u_2,u_3)$, $A_1 = (x_1,x_2)$, $B_1 = (x_3,x_4)$, $C_1 = (x_5,0)$, $C_2 = (x_7,x_6)$, $B_2 = (x_9,x_8)$, $A_2 = (x_{11},x_{10})$.

The **nondegenerate** conditions: Line BB_1 intersects line CC_1; Points C_2, A and A_1 are not collinear; Line C_2B intersects line B_2C.

Example 225 (A117-23: 10.32s, 490; 6.03s, 65). *The circle determined by the foot D of the altitude AD and the points I, I_a of the triangle ABC meets AD again in L. Show that AL is equal to the circumdiameter of ABC. State and prove an analogous proposition involving two excenters.*

Points B, C, I are arbitrarily chosen. Points A, I_A, D, N, L, O are constructed (in order) as follows: $\tan(CBI) = \tan(IBA)$; $\tan(BCI) = \tan(ICA)$; I_A is on line AI; $I_AB \perp BI$; $DA \perp BC$; D is on line BC; $NI \equiv NI_A$; $ND \equiv NI$; L is on line AD; $LN \equiv NI$; $OB \equiv OA$; $OB \equiv OC$. The **conclusion:** $4\,\overline{OB}^2 - \overline{AL}^2 = 0$.

$B = (0,0)$, $C = (u_1,0)$, $I = (u_2,u_3)$, $A = (x_2,x_1)$, $I_A = (x_4,x_3)$, $D = (x_2,0)$, $N = (x_6,x_5)$, $L = (x_2,x_7)$, $O = (x_9,x_8)$.

The **nondegenerate** conditions: Points I, C and B are not collinear; $\angle CIB$ is not right; Line BI is not perpendicular to line AI; Line BC is non-isotropic; Points D, I and I_A are not collinear; Line AD is non-isotropic; Points B, C and A are not collinear. In addition, the following nondegenerate conditions, which come from reducibility and have been detected by our prover, should be also added: $L \neq D$.

Remark. Actually, our proof is also for the statement involving two excenters.

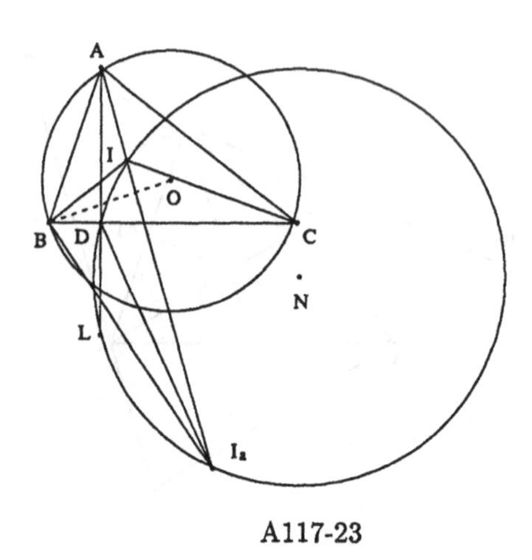

A117-23

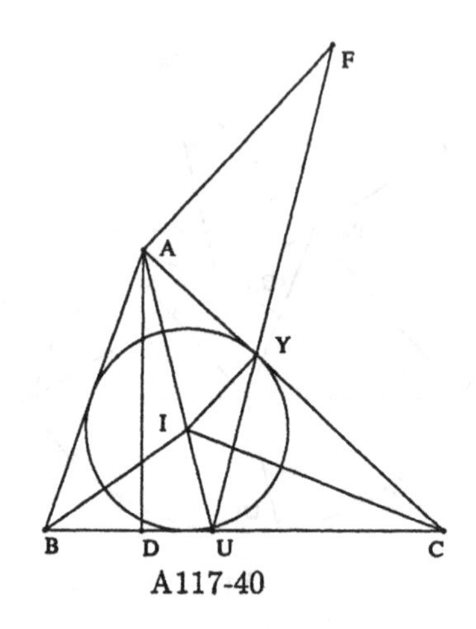

A117-40

Example 226 (A117-40: 5.95s, 220). *The line joining the foot U of the internal bisector AU of the angle A of the triangle ABC to the point of contact Y of the incircle with the side CA meets the perpendicular to CA at A in the point F. Prove that $AF = h_a$.*

Points B, C, I are arbitrarily chosen. Points A, U, D, Y, F are constructed (in order) as follows: $\tan(CBI) = \tan(IBA)$; $\tan(BCI) = \tan(ICA)$; U is on line BC; U is on line AI; $DA \perp BC$; D is on line BC; Y is on line AC; $YI \perp AC$; $FA \perp AC$; F is on line YU. The **conclusion:** $AD \equiv AF$.

$B = (0,0)$, $C = (u_1,0)$, $I = (u_2,u_3)$, $A = (x_2,x_1)$, $U = (x_3,0)$, $D = (x_2,0)$, $Y = (x_5,x_4)$, $F = (x_7,x_6)$.

The **nondegenerate** conditions: Points I, C and B are not collinear; $\angle CIB$ is not right; Line AI intersects line BC; Line BC is non-isotropic; Line AC is non-isotropic; Line YU is not perpendicular to line AC.

Example 227 (A118-42: 25.25s, 818; 4.1s, 108). *With the usual notions, show that: $AX^2 + AX_a{}^2 + AX_b{}^2 + AX_c{}^2 = 3(b^2+c^2) - a^2$.*

Points B, C, I are arbitrarily chosen. Points A, I_A, I_C, I_B, X, X_A, X_B, X_C are constructed (in order) as follows: $\tan(CBI) = \tan(IBA)$; $\tan(BCI) = \tan(ICA)$; $I_AB \perp IB$; I_A is on line AI; I_C is on line BI_A; I_C is on line CI; I_B is on line AI_C; I_B is on line BI; X is on line BC; $XI \perp BC$; X_A is on line BC; $X_AI_A \perp BC$; X_B is on line BC; $X_BI_B \perp BC$; X_C is on line BC; $X_CI_C \perp BC$. The **conclusion:** $\overline{AX}^2 + \overline{AX_A}^2 + \overline{AX_B}^2 + \overline{AX_C}^2 - (3\,(\overline{AC}^2 + \overline{AB}^2) - \overline{BC}^2) = 0$.

$B = (0,0)$, $C = (u_1,0)$, $I = (u_2,u_3)$, $A = (x_2,x_1)$, $I_A = (x_4,x_3)$, $I_C = (x_6,x_5)$, $I_B = (x_8,x_7)$, $X = (u_2,0)$, $X_A = (x_4,0)$, $X_B = (x_8,0)$, $X_C = (x_6,0)$.

The **nondegenerate** conditions: Points I, C and B are not collinear; $\angle CIB$ is not right; Line AI is not perpendicular to line IB; Line CI intersects line BI_A; Line BI intersects line AI_C; Line BC is non-isotropic; Line BC is non-isotropic; Line BC is non-isotropic; Line BC is non-isotropic.

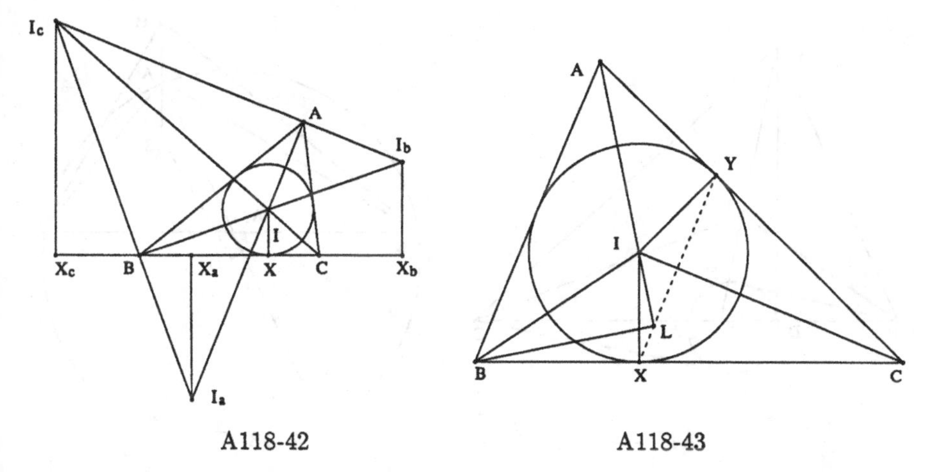

A118-42 A118-43

Example 228 (A118-43: 2.03s, 37). *The projection of the vertex B of the triangle ABC upon the internal bisector of the angle A lies on the line joining the points of contact of the incircle with the sides BC and AC. State and prove an analogous proposition for the external bisectors.*

Points B, C, I are arbitrarily chosen. Points A, X, Y, L are constructed (in order) as follows: $\tan(CBI) = \tan(IBA)$; $\tan(BCI) = \tan(ICA)$; $XI \perp BC$; X is on line BC; $YI \perp AC$; Y is on line AC; L is on line AI; $LB \perp AI$. The **conclusion**: Points X, L and Y are collinear.

$B = (0,0)$, $C = (u_1,0)$, $I = (u_2,u_3)$, $A = (x_2,x_1)$, $X = (u_2,0)$, $Y = (x_4,x_3)$, $L = (x_6,x_5)$.

The **nondegenerate** conditions: Points I, C and B are not collinear; $\angle CIB$ is not right; Line BC is non-isotropic; Line AC is non-isotropic; Line AI is non-isotropic.

Remark. Actually, our proof is for both cases.

Example 229 (A118-44: 24.82s, 665; 7.1s, 214). *The midpoint of a side of a triangle, the foot of the altitude on this side, and the projections of the ends of this side upon the internal bisector of the opposite angle are four concyclic points. Does the proposition hold for the external bisector?*

Points B, C, I are arbitrarily chosen. Points A, X, Y, A_1, D are constructed (in order) as follows: $\tan(CBI) = \tan(IBA)$; $\tan(BCI) = \tan(ICA)$; X is on line AI; $XB \perp AI$; Y is on line AI; $YC \perp AI$; A_1 is the midpoint of B and C; $DA \perp CB$; D is on line BC. The **conclusion**: Points X, Y, A_1 and D are on the same circle.

$B = (0,0)$, $C = (u_1,0)$, $I = (u_2,u_3)$, $A = (x_2,x_1)$, $X = (x_4,x_3)$, $Y = (x_6,x_5)$, $A_1 = (x_7,0)$, $D = (x_2,0)$.

The **nondegenerate** conditions: Points I, C and B are not collinear; $\angle CIB$ is not right; Line AI is non-isotropic; Line AI is non-isotropic; Line BC is non-isotropic.

Remark. Yes, it does hold for the external bisector. Our proof is for both cases.

Example 230 (A118-47: 1.75s, 8). *Show that the symmetric of the orthocenter of*

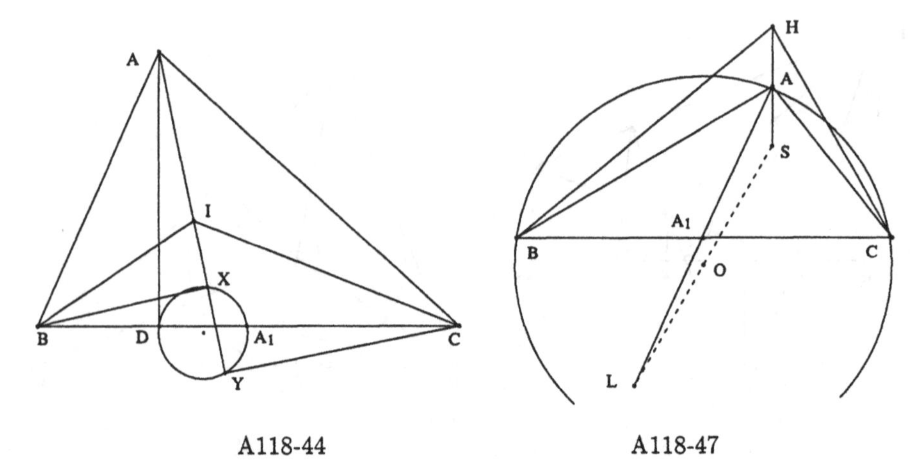

A118-44 A118-47

a triangle with respect to a vertex, and the symmetric of that vertex with respect to the midpoint of the opposite side, are collinear with the circumcenter of the triangle.

Points B, C, A are arbitrarily chosen. Points H, O, A_1, S, L are constructed (in order) as follows: $HB \perp AC$; $HA \perp BC$; $OB \equiv OA$; $OB \equiv OC$; A_1 is the midpoint of B and C; A is the midpoint of H and S; A_1 is the midpoint of A and L. The **conclusion**: Points O, L and S are collinear.

$B = (0,0)$, $C = (u_1,0)$, $A = (u_2,u_3)$, $H = (u_2,x_1)$, $O = (x_3,x_2)$, $A_1 = (x_4,0)$, $S = (u_2,x_5)$, $L = (x_6,x_7)$.

The **nondegenerate** conditions: Points B, C and A are not collinear; Points B, C and A are not collinear.

Example 231 (A118-49: 2.48s, 22). *If through the midpoints of the sides of a triangle having its vertices on the altitudes of a given triangle, perpendiculars are dropped to the respective sides of the given triangle, show that the three perpendiculars are concurrent.*

Points B, C, A are arbitrarily chosen. Points D, E, F, R, P, Q, R_1, Q_1, P_1, I are constructed (in order) as follows: D is on line BC; $DA \perp BC$; E is on line AC; $EB \perp AC$; F is on line AB; $FC \perp AB$; R is on line AD; P is on line BE; Q is on line CF; R_1 is the midpoint of P and Q; Q_1 is the midpoint of R and P; P_1 is the midpoint of Q and R; $IR_1 \perp PQ$; $IQ_1 \perp RP$. The **conclusion**: $IP_1 \perp QR$.

$B = (0,0)$, $C = (u_1,0)$, $A = (u_2,u_3)$, $D = (u_2,0)$, $E = (x_2,x_1)$, $F = (x_4,x_3)$, $R = (u_2,u_4)$, $P = (x_5,u_5)$, $Q = (x_6,u_6)$, $R_1 = (x_7,x_8)$, $Q_1 = (x_9,x_{10})$, $P_1 = (x_{11},x_{12})$, $I = (x_{14},x_{13})$.

The **nondegenerate** conditions: Line BC is non-isotropic; Line AC is non-isotropic; Line AB is non-isotropic; $A \neq D$; $B \neq E$; $C \neq F$; Points R, P and Q are not collinear.

Example 232 (A118-50: 1.03s, 11). *If D_1 is the second point of intersection of the altitude ADD_1 of the triangle ABC with the circumcircle, center O, and P is the trace on BC of the perpendicular from D_1 to AC, show that the lines AP, AO make*

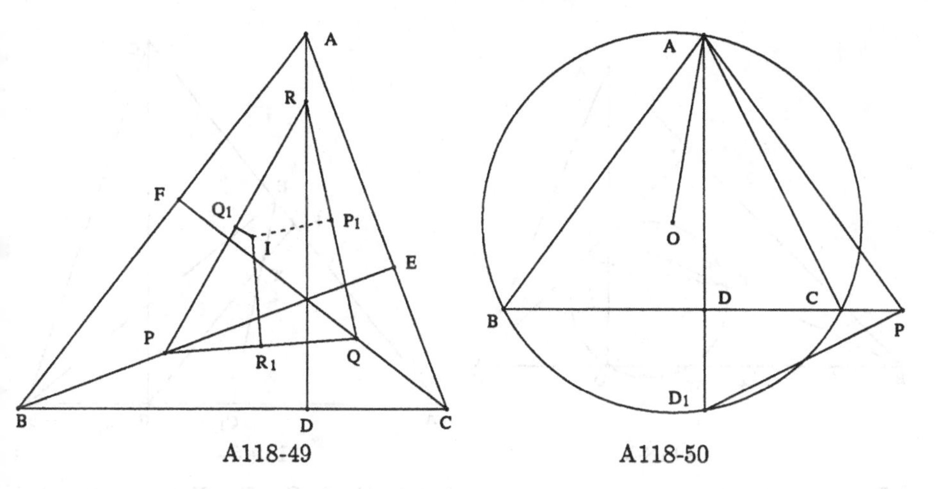

A118-49 A118-50

equal angles with the bisector of the angle DAC.

Points B, C, A are arbitrarily chosen. Points D, O, D_1, P are constructed (in order) as follows: D is on line BC; $DA \perp BC$; $OB \equiv OA$; $OB \equiv OC$; D_1 is on line AD; $D_1O \equiv OB$; $PD_1 \perp AC$; P is on line BC. The **conclusion**: $\tan(OAD) = \tan(CAP)$.

$B = (0,0)$, $C = (u_1,0)$, $A = (u_2,u_3)$, $D = (u_2,0)$, $O = (x_2,x_1)$, $D_1 = (u_2,x_3)$, $P = (x_4,0)$.

The **nondegenerate** conditions: Line BC is non-isotropic; Points B, C and A are not collinear; Line AD is non-isotropic; Line BC is not perpendicular to line AC. In addition, the following nondegenerate conditions, which come from reducibility and have been detected by our prover, should be also added: $D_1 \neq A$.

Example 233 (A118-53: 6.85s, 59). *Show that the triangle formed by the foot of the altitude to the base of a triangle and the midpoints of the altitudes to the lateral sides is similar to the given triangle; its circumcircle passes through the orthocenter of the given triangle and through the midpoint of its base.*

Points B, C, A are arbitrarily chosen. Points A_1, D, E, F, H, P, Q are constructed (in order) as follows: A_1 is the midpoint of B and C; D is on line BC; $DA \perp BC$; E is on line AC; $EB \perp AC$; F is on line AB; $FC \perp AB$; H is on line BE; H is on line AD; P is the midpoint of B and E; Q is the midpoint of C and F. The **conclusions**: (1) Points H, P, Q and D are on the same circle; (2) Points A_1, P, Q and D are on the same circle; (3) $PQ \cdot AB = PD \cdot BC$.

$B = (0,0)$, $C = (u_1,0)$, $A = (u_2,u_3)$, $A_1 = (x_1,0)$, $D = (u_2,0)$, $E = (x_3,x_2)$, $F = (x_5,x_4)$, $H = (u_2,x_6)$, $P = (x_7,x_8)$, $Q = (x_9,x_{10})$.

The **nondegenerate** conditions: Line BC is non-isotropic; Line AC is non-isotropic; Line AB is non-isotropic; Line AD intersects line BE.

Example 234 (A119-54: 1.68s, 21). *With the usual notation, show that the angle formed by the lines C_1E, B_1F is equal to $3A$, or its supplement.*

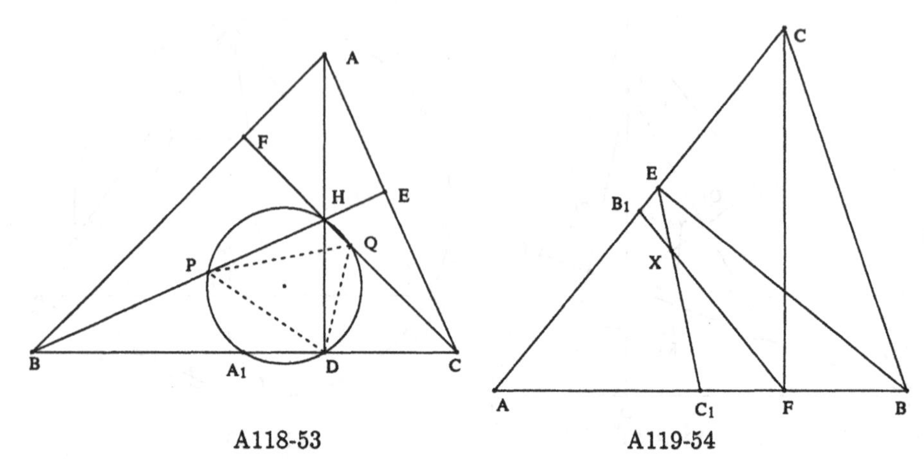

A118-53 A119-54

Points A, B, C are arbitrarily chosen. Points E, F, C_1, B_1, X are constructed (in order) as follows: $EB \perp CA$; E is on line AC; $FC \perp BA$; F is on line AB; C_1 is the midpoint of A and B; B_1 is the midpoint of A and C; X is on line B_1F; X is on line C_1E. The **conclusion**: $\tan(3\angle BAC) - \tan(\angle B_1XC_1) = 0$.

$A = (0,0)$, $B = (u_1,0)$, $C = (u_2,u_3)$, $E = (x_2,x_1)$, $F = (u_2,0)$, $C_1 = (x_3,0)$, $B_1 = (x_4,x_5)$, $X = (x_7,x_6)$.

The **nondegenerate** conditions: Line AC is non-isotropic; Line AB is non-isotropic; Line C_1E intersects line B_1F.

Example 235 (A119-55: 2.58s, 15). *Show that the symmetrics, P, Q of a given point L with respect to the sides Ox, Oy of a given angle, and the points $P_1 = (LQ, Ox)$, $Q_1 = (LP, Oy)$ lie on a circle passing through O.*

Points O, X, Y, L are arbitrarily chosen. Points A, B, P, Q, P_1, Q_1, I are constructed (in order) as follows: A is on line OX; $AL \perp OX$; B is on line OY; $BL \perp OY$; A is the midpoint of L and P; B is the midpoint of L and Q; P_1 is on line OX; P_1 is on line LQ; Q_1 is on line OY; Q_1 is on line LP; $IO \equiv IQ$; $IO \equiv IP$. The **conclusions**: (1) $IO \equiv IP_1$; (2) $IO \equiv IQ_1$.

$O = (0,0)$, $X = (u_1,0)$, $Y = (u_2,u_3)$, $L = (u_4,u_5)$, $A = (u_4,0)$, $B = (x_2,x_1)$, $P = (u_4,x_3)$, $Q = (x_4,x_5)$, $P_1 = (x_6,0)$, $Q_1 = (u_4,x_7)$, $I = (x_9,x_8)$.

The **nondegenerate** conditions: Line OX is non-isotropic; Line OY is non-isotropic; Line LQ intersects line OX; Line LP intersects line OY; Points O, P and Q are not collinear.

Example 236 (A119-56: 2.55s, 37). *The sides of the anticomplementary triangle of the triangle ABC meet the circumcircle of ABC in the points P, Q, R. Show that the area of the triangle PQR is equal to four times the area of the orthic triangle of ABC.*

Points B, C, A are arbitrarily chosen. Points D, E, F, O, P, Q, R are constructed (in order) as follows: D is on line BC; $DA \perp BC$; E is on line AC; $EB \perp AC$; F is on line AB; $FC \perp AB$; $OB \equiv OC$; $OB \equiv OA$; $PA \parallel BC$; $PO \equiv OB$; $QB \parallel AC$;

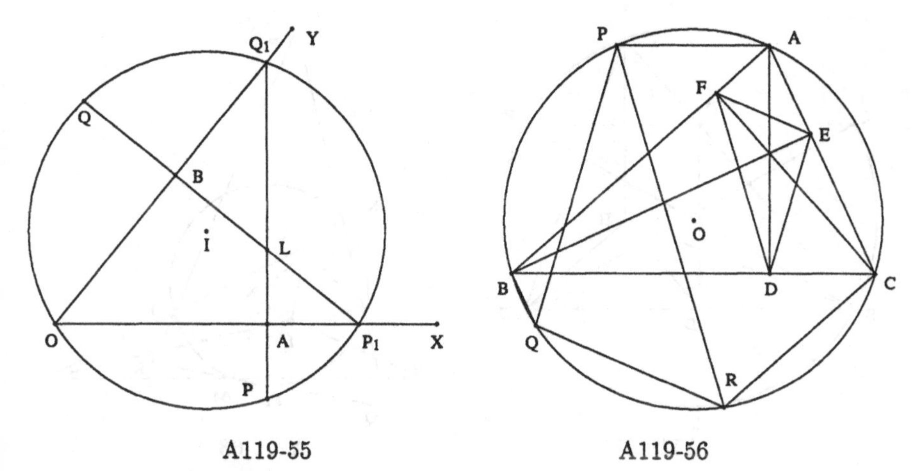

A119-55 A119-56

$QO \equiv OB$; $RC \parallel AB$; $RO \equiv OB$. The **conclusion**: $\nabla(DEF)\ 4 - \nabla(PQR) = 0$.

$B = (0,0)$, $C = (u_1,0)$, $A = (u_2,u_3)$, $D = (u_2,0)$, $E = (x_2,x_1)$, $F = (x_4,x_3)$, $O = (x_6,x_5)$, $P = (x_7,u_3)$, $Q = (x_9,x_8)$, $R = (x_{11},x_{10})$.

The **nondegenerate** conditions: Line BC is non-isotropic; Line AC is non-isotropic; Line AB is non-isotropic; Points B, A and C are not collinear; Line BC is non-isotropic; Line AC is non-isotropic; Line AB is non-isotropic. In addition, the following nondegenerate conditions, which come from reducibility and have been detected by our prover, should be also added: $R \neq C$; $Q \neq B$; $P \neq A$.

Example 237 (A119-61: 1.55s, 10). *Through the orthocenter of the triangle ABC parallels are drawn to the sides AB, AC, meeting BC in D, E. The perpendiculars to BC at D, E meet AB, AC in two points D_1, E_1 which are collinear with the diametric opposites of B, C on the circumcircle of ABC.*

Points B, C, A are arbitrarily chosen. Points H, O, D, D_1, P, Q are constructed (in order) as follows: $HB \perp AC$; $HA \perp BC$; $OB \equiv OC$; $OB \equiv OA$; D is on line BC; $DH \parallel AB$; D_1 is on line AB; $D_1D \perp BC$; P is on line BO; $PO \equiv OB$; Q is on line CO; $QO \equiv OB$. The **conclusion**: Points P, Q and D_1 are collinear.

$B = (0,0)$, $C = (u_1,0)$, $A = (u_2,u_3)$, $H = (u_2,x_1)$, $O = (x_3,x_2)$, $D = (x_4,0)$, $D_1 = (x_4,x_5)$, $P = (x_7,x_6)$, $Q = (x_9,x_8)$.

The **nondegenerate** conditions: Points B, C and A are not collinear; Points B, A and C are not collinear; Points A, B and C are not collinear; Line BC is not perpendicular to line AB; Line BO is non-isotropic; Line CO is non-isotropic. In addition, the following nondegenerate conditions, which come from reducibility and have been detected by our prover, should be also added: $Q \neq C$; $P \neq B$.

Example 238 (A119-63: 53.72s, 2340; 27.1s, 331). *Through the midpoints of the sides of a triangle parallels are drawn to the external bisectors of the respectively opposite angels. Show that the triangle thus formed has the same nine-point circle as the given triangle.*

Points B, C, I are arbitrarily chosen. Points A, A_1, B_1, C_1, N, Q, R, P, P_1, R_1 are

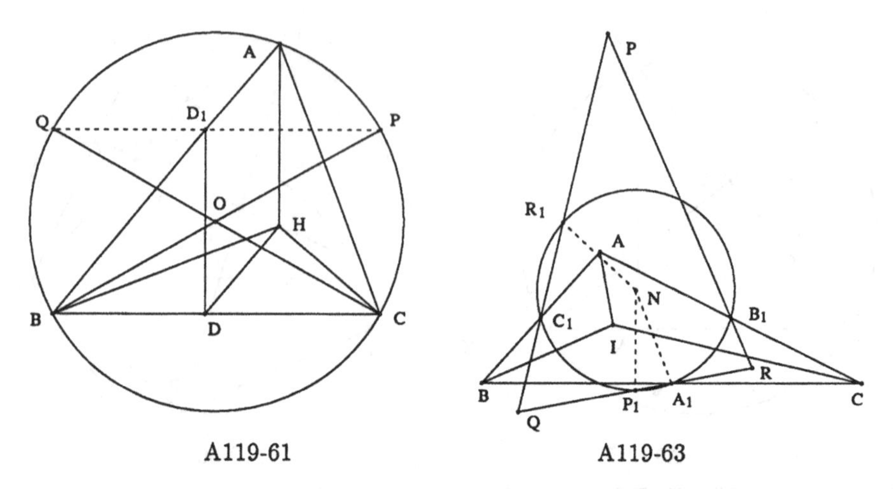

A119-61 A119-63

constructed (in order) as follows: $\tan(CBI) = \tan(IBA)$; $\tan(BCI) = \tan(ICA)$; A_1 is the midpoint of C and B; B_1 is the midpoint of A and C; C_1 is the midpoint of B and A; $NA_1 \equiv NC_1$; $NA_1 \equiv NB_1$; $QC_1 \perp CI$; $QA_1 \perp AI$; $RB_1 \perp BI$; R is on line QA_1; P is on line QC_1; P is on line RB_1; P_1 is the midpoint of Q and R; R_1 is the midpoint of Q and P. The **conclusion**: $NA_1 \equiv NP_1$.

$B = (0,0)$, $C = (u_1,0)$, $I = (u_2,u_3)$, $A = (x_2,x_1)$, $A_1 = (x_3,0)$, $B_1 = (x_4,x_5)$, $C_1 = (x_6,x_7)$, $N = (x_9,x_8)$, $Q = (x_{11},x_{10})$, $R = (x_{13},x_{12})$, $P = (x_{15},x_{14})$, $P_1 = (x_{16},x_{17})$, $R_1 = (x_{18},x_{19})$.

The **nondegenerate** conditions: Points I, C and B are not collinear; $\angle CIB$ is not right; Points A_1, B_1 and C_1 are not collinear; Points A, I and C are not collinear; Line QA_1 is not perpendicular to line BI; Line RB_1 intersects line QC_1.

Example 239 (A119-67: 7.73s, 14; 2.18s, 14). *Show that the foot of the altitude of a triangle on a side, the midpoint of the segment of the circumdiameter between this side and the opposite vertex, and the nine-point center are collinear.*

Points B, C, A are arbitrarily chosen. Points D, O, E, M, A_1, B_1, C_1, N are constructed (in order) as follows: D is on line BC; $DA \perp BC$; $OB \equiv OA$; $OB \equiv OC$; E is on line BC; E is on line AO; M is the midpoint of A and E; A_1 is the midpoint of C and B; B_1 is the midpoint of A and C; C_1 is the midpoint of B and A; $NA_1 \equiv NC_1$; $NA_1 \equiv NB_1$. The **conclusion**: Points M, D and N are collinear.

$B = (0,0)$, $C = (u_1,0)$, $A = (u_2,u_3)$, $D = (u_2,0)$, $O = (x_2,x_1)$, $E = (x_3,0)$, $M = (x_4,x_5)$, $A_1 = (x_6,0)$, $B_1 = (x_7,x_8)$, $C_1 = (x_9,x_{10})$, $N = (x_{12},x_{11})$.

The **nondegenerate** conditions: Line BC is non-isotropic; Points B, C and A are not collinear; Line AO intersects line BC; Points A_1, B_1 and C_1 are not collinear.

Example 240 (A120-68-a: 33.92s, 1686; 14.65s, 522). *If A, B, C are the centers of three equal circles (A), (B), (C) having a point L in common, and D, E, F are the other points which the circles (B) and (C), (C) and (A), (A) and (B) have in common, show that the circle DEF is equal to the given circles, and that the center*

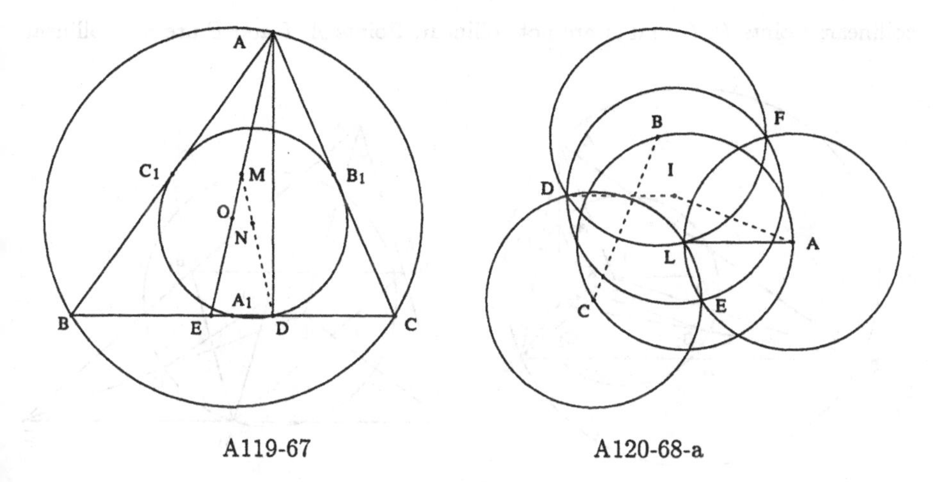

A119-67 A120-68-a

of this circle coincides with the orthocenter of the triangle ABC.

Points L, A are arbitrarily chosen. Points B, C, D, E, F, I are constructed (in order) as follows: $BL \equiv LA$; $CL \equiv LA$; $DC \equiv LA$; $DB \equiv LA$; $EA \equiv LA$; $EC \equiv LA$; $FB \equiv LA$; $FA \equiv LA$; $ID \equiv IE$; $ID \equiv IF$. The **conclusions**: (1) $ID \equiv LA$; (2) $AI \perp BC$.

$L = (0,0)$, $A = (u_1,0)$, $B = (x_1,u_2)$, $C = (x_2,u_3)$, $D = (x_4,x_3)$, $E = (x_6,x_5)$, $F = (x_8,x_7)$, $I = (x_{10},x_9)$.

The **nondegenerate** conditions: Line BC is non-isotropic; Line CA is non-isotropic; Line AB is non-isotropic; Points D, F and E are not collinear. In addition, the following nondegenerate conditions, which come from reducibility and have been detected by our prover, should be also added: $F \neq L$; $E \neq L$; $D \neq L$.

Example 241 (A120-69: 8.37s, 17). *Show that the parallels to the internal bisectors of a triangle drawn through the respective Euler points are concurrent; the line joining their common point to the nine-point center of the given triangle is parallel to th line joining the circumcenter of the given triangle to its incenter. State and prove analogous propositions.*

Points B, C, I are arbitrarily chosen. Points A, H, A_1, B_1, C_1, N, O, A_2, B_2, C_2, K are constructed (in order) as follows: $\tan(CBI) = \tan(IBA)$; $\tan(BCI) = \tan(ICA)$; $HB \perp AC$; $HA \perp BC$; A_1 is the midpoint of C and B; B_1 is the midpoint of A and C; C_1 is the midpoint of B and A; $NB_1 \equiv NA_1$; $NC_1 \equiv NA_1$; $OB \equiv OA$; $OB \equiv OC$; A_2 is the midpoint of A and H; B_2 is the midpoint of B and H; C_2 is the midpoint of C and H; $KB_2 \parallel BI$; $KA_2 \parallel AI$. The **conclusions**: (1) $C_2K \parallel CI$; (2) Points I, H and K are collinear; (3) $KN \parallel OI$.

$B = (0,0)$, $C = (u_1,0)$, $I = (u_2,u_3)$, $A = (x_2,x_1)$, $H = (x_2,x_3)$, $A_1 = (x_4,0)$, $B_1 = (x_5,x_6)$, $C_1 = (x_7,x_8)$, $N = (x_{10},x_9)$, $O = (x_{12},x_{11})$, $A_2 = (x_2,x_{13})$, $B_2 = (x_{14},x_{15})$, $C_2 = (x_{16},x_{17})$, $K = (x_{19},x_{18})$.

The **nondegenerate** conditions: Points I, C and B are not collinear; $\angle CIB$ is not right; Points B, C and A are not collinear; Points C_1, A_1 and B_1 are not collinear; Points B, C and A are not collinear; Points A, I and B are not collinear.

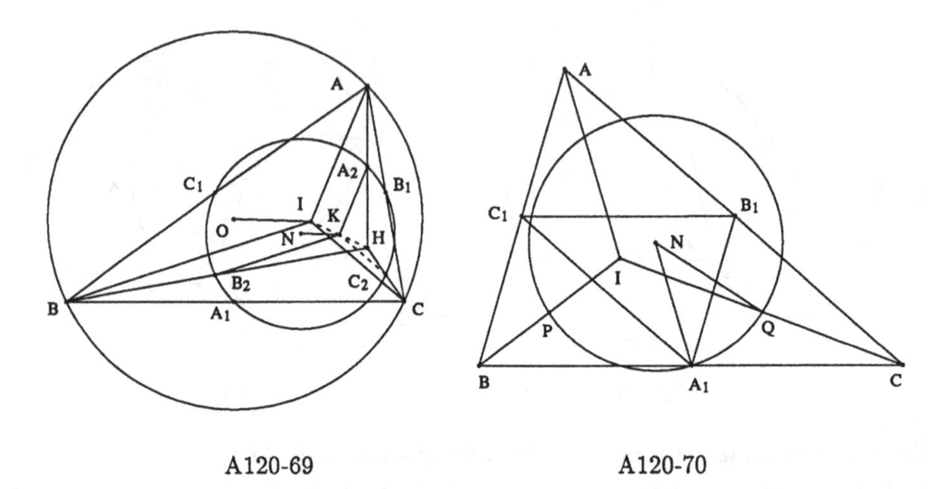

A120-69 A120-70

Example 242 (A120-70: 2.27s, 58). *Show that the nine-point center of the triangle IBC lies on the internal bisector of the angle A_1 of the complementary triangle $A_1B_1C_1$ of the given triangle ABC.*

Points B, C, I are arbitrarily chosen. Points A, A_1, B_1, C_1, P, Q, N are constructed (in order) as follows: $\tan(CBI) = \tan(IBA)$; $\tan(BCI) = \tan(ICA)$; A_1 is the midpoint of C and B; B_1 is the midpoint of C and A; C_1 is the midpoint of A and B; P is the midpoint of B and I; Q is the midpoint of C and I; $NA_1 \equiv NQ$; $NA_1 \equiv NP$. The **conclusion**: $\tan(B_1A_1N) = \tan(NA_1C_1)$.

$B = (0,0)$, $C = (u_1, 0)$, $I = (u_2, u_3)$, $A = (x_2, x_1)$, $A_1 = (x_3, 0)$, $B_1 = (x_4, x_5)$, $C_1 = (x_6, x_7)$, $P = (x_8, x_9)$, $Q = (x_{10}, x_{11})$, $N = (x_{13}, x_{12})$.

The **nondegenerate** conditions: Points I, C and B are not collinear; $\angle CIB$ is not right; Points A_1, P and Q are not collinear.

Example 243 (A120-71: 368.5s, 3833; 34.85s, 818). *If O, H are the circumcenter and the orthocenter of the triangle ABC, show that the nine-point circles of the three triangles OHA, OHB, OHC have two points in common.*

Points B, C, A are arbitrarily chosen. Points H, O, M, A_H, B_H, C_H, A_O, B_O, C_O, N_A, N_B, N_C, I are constructed (in order) as follows: $HB \perp AC$; $HA \perp BC$; $OB \equiv OC$; $OB \equiv OA$; M is the midpoint of O and H; A_H is the midpoint of A and H; B_H is the midpoint of B and H; C_H is the midpoint of C and H; A_O is the midpoint of A and O; B_O is the midpoint of B and O; C_O is the midpoint of C and O; $N_AM \equiv N_AA_O$; $N_AM \equiv N_AA_H$; $N_BM \equiv N_BB_O$; $N_BM \equiv N_BB_H$; $N_CM \equiv N_CC_O$; $N_CM \equiv N_CC_H$; $IN_C \equiv N_CC$; $IN_B \equiv N_BB$. The **conclusion**: $N_AA \equiv N_AI$.

$B = (0,0)$, $C = (u_1, 0)$, $A = (u_2, u_3)$, $H = (u_2, x_1)$, $O = (x_3, x_2)$, $M = (x_4, x_5)$, $A_H = (u_2, x_6)$, $B_H = (x_7, x_8)$, $C_H = (x_9, x_{10})$, $A_O = (x_{11}, x_{12})$, $B_O = (x_{13}, x_{14})$, $C_O = (x_{15}, x_{16})$, $N_A = (x_{18}, x_{17})$, $N_B = (x_{20}, x_{19})$, $N_C = (x_{22}, x_{21})$, $I = (x_{24}, x_{23})$.

The **nondegenerate** conditions: Points B, C and A are not collinear; Points B, A and C are not collinear; Points M, A_H and A_O are not collinear; Points M, B_H and B_O are not collinear; Points M, C_H and C_O are not collinear; Line $N_B N_C$ is non-isotropic.

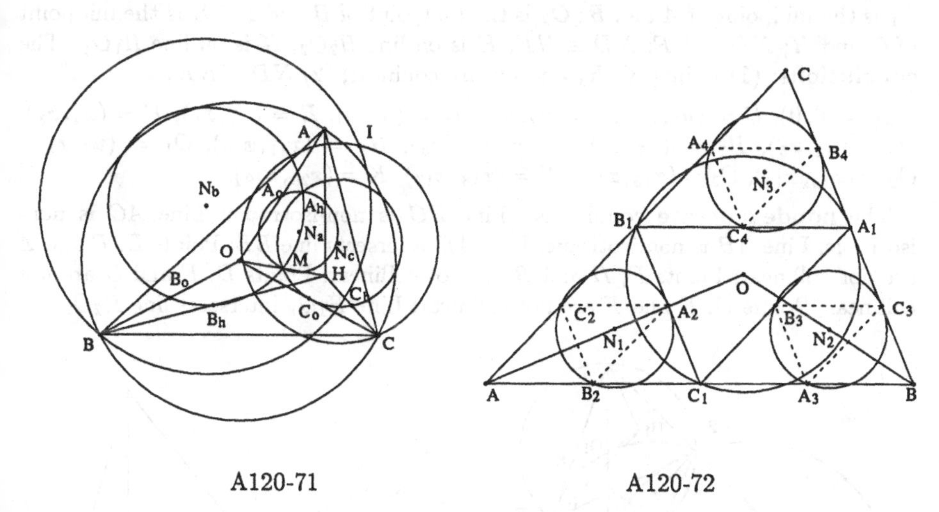

A120-71 A120-72

Example 244 (A120-72: 4.4s, 28). *If A_1, B_1, C_1 are the midpoints of the sides BC, CA, AB of a triangle ABC, respectively, show that the nine-point centers of the triangle AB_1C_1, A_1BC_1, A_1B_1C form a triangle homothetic with ABC in the ratio* $1:2$.

Points A, B, C are arbitrarily chosen. Points A_1, B_1, C_1, A_2, B_2, C_2, A_3, B_3, C_3, A_4, B_4, C_4, N_1, N_2, N_3, O are constructed (in order) as follows: A_1 is the midpoint of B and C; B_1 is the midpoint of A and C; C_1 is the midpoint of A and B; A_2 is the midpoint of B_1 and C_1; B_2 is the midpoint of A and C_1; C_2 is the midpoint of A and B_1; A_3 is the midpoint of B and C_1; B_3 is the midpoint of A_1 and C_1; C_3 is the midpoint of A_1 and B; A_4 is the midpoint of B_1 and C; B_4 is the midpoint of A_1 and C; C_4 is the midpoint of A_1 and B_1; $N_1C_2 \equiv N_1B_2$; $N_1C_2 \equiv N_1A_2$; $N_2C_3 \equiv N_2A_3$; $N_2C_3 \equiv N_2B_3$; $N_3C_4 \equiv N_3B_4$; $N_3C_4 \equiv N_3A_4$; O is on line BN_2; O is on line AN_1. The **conclusion:** $\frac{\overline{OA}}{\overline{ON_1}} - 2 = 0$.

$A = (0,0)$, $B = (u_1, 0)$, $C = (u_2, u_3)$, $A_1 = (x_1, x_2)$, $B_1 = (x_3, x_4)$, $C_1 = (x_5, 0)$, $A_2 = (x_6, x_7)$, $B_2 = (x_8, 0)$, $C_2 = (x_9, x_{10})$, $A_3 = (x_{11}, 0)$, $B_3 = (x_{12}, x_{13})$, $C_3 = (x_{14}, x_{15})$, $A_4 = (x_{16}, x_{17})$, $B_4 = (x_{18}, x_{19})$, $C_4 = (x_{20}, x_{21})$, $N_1 = (x_{23}, x_{22})$, $N_2 = (x_{25}, x_{24})$, $N_3 = (x_{27}, x_{26})$, $O = (x_{29}, x_{28})$.

The **nondegenerate** conditions: Points C_2, A_2 and B_2 are not collinear; Points C_3, B_3 and A_3 are not collinear; Points C_4, A_4 and B_4 are not collinear; Line AN_1 intersects line BN_2.

Example 245 (A120-73: 1250.22s, 10007; 37.43s, 210). *Show that th Euler lines of the three triangles cut off from a given triangle by the sides of its orthic triangle have a point in common, on the nine-point circle of the given triangle.*

Points B, C, A are arbitrarily chosen. Points D, E, F, H, H_1, H_2, H_3, O_1, O_2, O_3, N, K are constructed (in order) as follows: D is on line BC; $DA \perp BC$; E is on line AC; $EB \perp AC$; F is on line AB; $FC \perp AB$; H is on line BE; H is on line AD; $H_1E \perp AF$; $H_1A \perp EF$; $H_2D \perp BF$; $H_2B \perp FD$; $H_3D \perp CE$; $H_3C \perp DE$; O_1 is the midpoint of A and H; O_2 is the midpoint of B and H; O_3 is the midpoint of C and H; $ND \equiv NF$; $ND \equiv NE$; K is on line H_2O_2; K is on line H_1O_1. The **conclusions**: (1) Points K, H_3 and O_3 are collinear; (2) $ND \equiv NK$.

$B = (0,0)$, $C = (u_1,0)$, $A = (u_2,u_3)$, $D = (u_2,0)$, $E = (x_2,x_1)$, $F = (x_4,x_3)$, $H = (u_2,x_5)$, $H_1 = (x_7,x_6)$, $H_2 = (x_9,x_8)$, $H_3 = (x_{11},x_{10})$, $O_1 = (u_2,x_{12})$, $O_2 = (x_{13},x_{14})$, $O_3 = (x_{15},x_{16})$, $N = (x_{18},x_{17})$, $K = (x_{20},x_{19})$.

The **nondegenerate** conditions: Line BC is non-isotropic; Line AC is non-isotropic; Line AB is non-isotropic; Line AD intersects line BE; Points E, F and A are not collinear; Points F, D and B are not collinear; Points D, E and C are not collinear; Points D, E and F are not collinear; Line H_1O_1 intersects line H_2O_2.

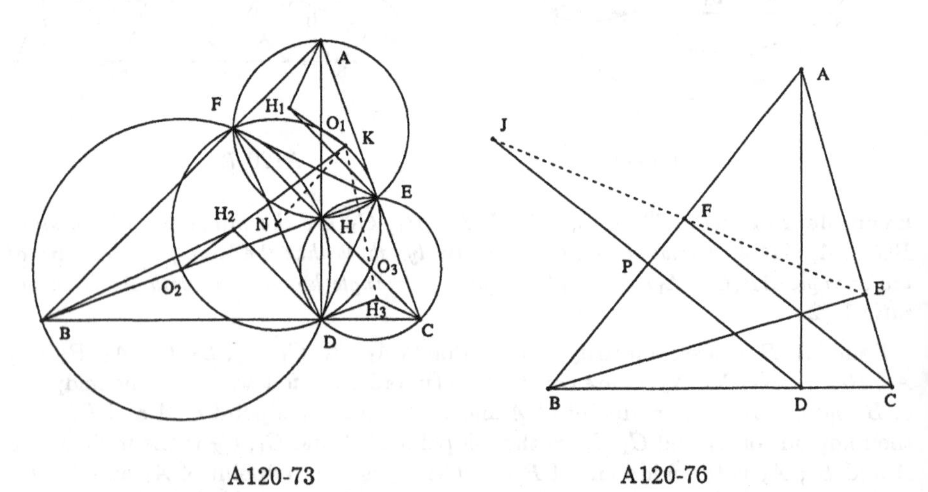

A120-73 A120-76

Example 246 (A120-76: 7.28s, 12; 2.0s, 25). *Show that the symmetrics of the foot of the altitude to the base of a triangle with respect to the other two sides lie on the side of the orthic triangle relative to the base.*

Points B, C, A are arbitrarily chosen. Points D, E, F, P, J are constructed (in order) as follows: D is on line BC; $DA \perp BC$; E is on line AC; $EB \perp AC$; F is on line AB; $FC \perp AB$; P is on line BA; $PD \perp BA$; P is the midpoint of D and J. The **conclusion**: Points E, F and J are collinear.

$B = (0,0)$, $C = (u_1,0)$, $A = (u_2,u_3)$, $D = (u_2,0)$, $E = (x_2,x_1)$, $F = (x_4,x_3)$, $P = (x_6,x_5)$, $J = (x_7,x_8)$.

The **nondegenerate** conditions: Line BC is non-isotropic; Line AC is non-isotropic; Line AB is non-isotropic; Line BA is non-isotropic.

Example 247 (A120-78: 1.6s, 6). *Show that the midpoint of an altitude of a triangle, the point of contact of the corresponding side with the excircle relative* *to that side, and the incenter of the triangle are collinear.*

Points B, C, I are arbitrarily chosen. Points A, D, I_A, X_A, M are constructed (in order) as follows: $\tan(CBI) = \tan(IBA)$; $\tan(BCI) = \tan(ICA)$; D is on line BC; $DA \perp BC$; I_A is on line IA; $I_AB \perp BI$; X_A is on line BC; $X_AI_A \perp BC$; M is the midpoint of A and D. The **conclusion**: Points X_A, M and I are collinear.

$B = (0,0)$, $C = (u_1,0)$, $I = (u_2,u_3)$, $A = (x_2,x_1)$, $D = (x_2,0)$, $I_A = (x_4,x_3)$, $X_A = (x_4,0)$, $M = (x_2,x_5)$.

The **nondegenerate** conditions: Points I, C and B are not collinear; $\angle CIB$ is not right; Line BC is non-isotropic; Line BI is not perpendicular to line IA; Line BC is non-isotropic.

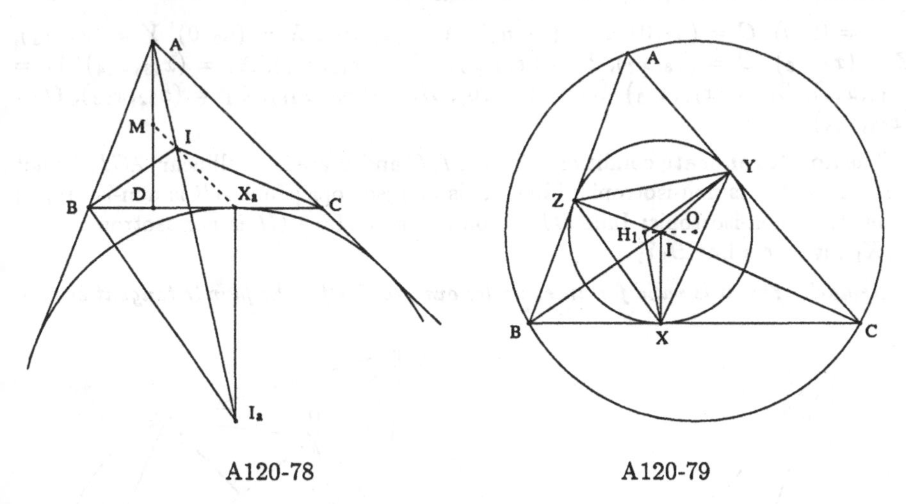

A120-78 A120-79

Example 248 (A120-79: 120.77s, 3858; 12.93s, 288). *Show that the line joining the circumcenter O of a triangle to its incenter I passes through the orthocenter H_1 of the triangle formed by the points of contact of the sides of the triangle with the incircle, and that, moreover, $H_1I : OI = r : R$. Is the proposition valid for the excenters?*

Points B, C, I are arbitrarily chosen. Points A, O, X, Y, Z, H_1 are constructed (in order) as follows: $\tan(CBI) = \tan(IBA)$; $\tan(BCI) = \tan(ICA)$; $OB \equiv OC$; $OB \equiv OA$; X is on line BC; $XI \perp BC$; Y is on line AC; $YI \perp AC$; Z is on line AB; $ZI \perp AB$; $H_1Y \perp XZ$; $H_1X \perp YZ$. The **conclusions**: (1) Points H_1, I and O are collinear; (2) $H_1I \cdot OB = OI \cdot IX$.

$B = (0,0)$, $C = (u_1,0)$, $I = (u_2,u_3)$, $A = (x_2,x_1)$, $O = (x_4,x_3)$, $X = (u_2,0)$, $Y = (x_6,x_5)$, $Z = (x_8,x_7)$, $H_1 = (x_{10},x_9)$.

The **nondegenerate** conditions: Points I, C and B are not collinear; $\angle CIB$ is not right; Points B, A and C are not collinear; Line BC is non-isotropic; Line AC is non-isotropic; Line AB is non-isotropic; Points Y, Z and X are not collinear.

Example 249 (A120-80: 24.03s, 1310). *If X_1 is the symmetric, with respect to the internal bisector of the angle A, of the point of contact of the side BC of the triangle*

ABC with the incircle, and A_1 is the midpoint of BC, show that the line A_1X_1 and its two analogues B_1Y_1, C_1Z_1 have a point in common. Is the proposition valid for an excircle?

Points B, C, I are arbitrarily chosen. Points A, X, Y, Z, D, E, F, X_1, Y_1, Z_1, A_1, B_1, C_1, O are constructed (in order) as follows: $\tan(CBI) = \tan(IBA)$; $\tan(BCI) = \tan(ICA)$; X is on line BC; $XI \perp BC$; Y is on line AC; $YI \perp AC$; Z is on line AB; $ZI \perp AB$; D is on line AI; $DX \perp AI$; E is on line BI; $EY \perp BI$; F is on line CI; $FZ \perp CI$; D is the midpoint of X and X_1; E is the midpoint of Y and Y_1; F is the midpoint of Z and Z_1; A_1 is the midpoint of B and C; B_1 is the midpoint of C and A; C_1 is the midpoint of A and B; O is on line B_1Y_1; O is on line A_1X_1. The **conclusion**: Points C_1, Z_1 and O are collinear.

$B = (0,0)$, $C = (u_1,0)$, $I = (u_2,u_3)$, $A = (x_2,x_1)$, $X = (u_2,0)$, $Y = (x_4,x_3)$, $Z = (x_6,x_5)$, $D = (x_8,x_7)$, $E = (x_{10},x_9)$, $F = (x_{12},x_{11})$, $X_1 = (x_{13},x_{14})$, $Y_1 = (x_{15},x_{16})$, $Z_1 = (x_{17},x_{18})$, $A_1 = (x_{19},0)$, $B_1 = (x_{20},x_{21})$, $C_1 = (x_{22},x_{23})$, $O = (x_{25},x_{24})$.

The **nondegenerate** conditions: Points I, C and B are not collinear; $\angle CIB$ is not right; Line BC is non-isotropic; Line AC is non-isotropic; Line AB is non-isotropic; Line AI is non-isotropic; Line BI is non-isotropic; Line CI is non-isotropic; Line A_1X_1 intersects line B_1Y_1.

Remark. Yes, it is valid for an excircle; our proof is for the four tritangent circles.

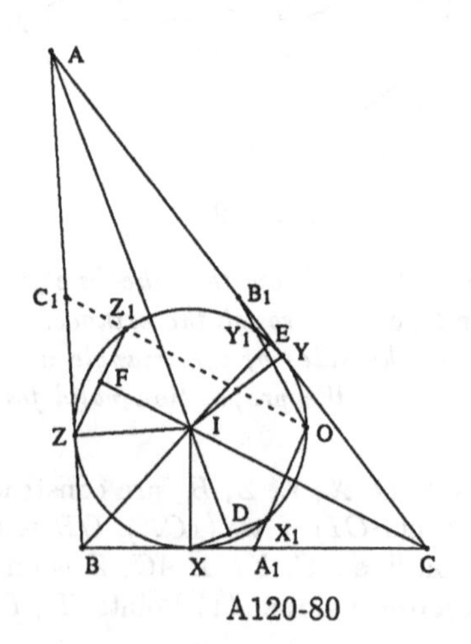

A120-80

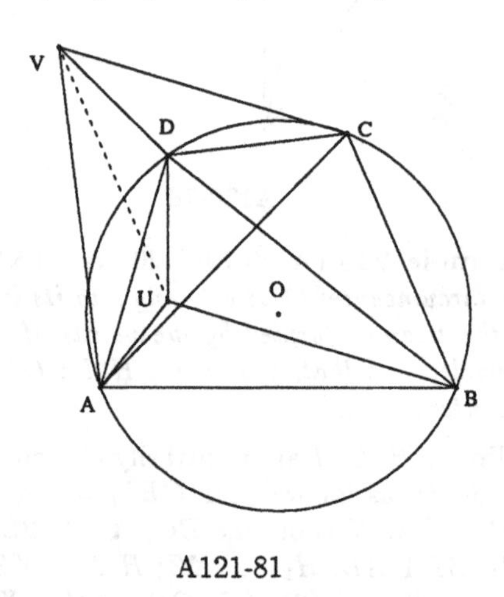

A121-81

Example 250 (A121-81: 8.12s, 47; 1.58s, 33). *A line AD through the vertex A meets the circumcircle of the triangle ABC in D. If U, V are the orthocenters of the triangle ABD, ACD, respectively, prove that UV is equal and parallel to BC.*

Points A, B, C are arbitrarily chosen. Points O, D, U, V are constructed (in order) as follows: $OA \equiv OC$; $OA \equiv OB$; $DO \equiv OA$; $UA \perp BD$; $UB \perp AD$; $VA \perp CD$; $VC \perp AD$. The **conclusion**: $VU \equiv BC$.

$A = (0,0)$, $B = (u_1,0)$, $C = (u_2,u_3)$, $O = (x_2,x_1)$, $D = (x_3,u_4)$, $U = (x_5,x_4)$, $V = (x_7,x_6)$.

The **nondegenerate** conditions: Points A, B and C are not collinear; Points A, D and B are not collinear; Points A, D and C are not collinear.

Example 251 (A121-82: 23.58s, 876; 2.72s, 83). *The internal bisectors of the angles B, C of the triangle ABC meet the line AX_a joining A to the point of contact of BC with the excircle relative to this side in the points L, M. Prove that $AL : AM = AB : AC$.*

Points B, C, I are arbitrarily chosen. Points A, I_A, X_A, L, M are constructed (in order) as follows: $\tan(CBI) = \tan(IBA)$; $\tan(BCI) = \tan(ICA)$; I_A is on line IA; $I_AB \perp BI$; X_A is on line BC; $X_AI_A \perp BC$; L is on line AX_A; L is on line BI; M is on line AX_A; M is on line CI. The **conclusion**: $AM \cdot AB = AL \cdot AC$.

$B = (0,0)$, $C = (u_1,0)$, $I = (u_2,u_3)$, $A = (x_2,x_1)$, $I_A = (x_4,x_3)$, $X_A = (x_4,0)$, $L = (x_6,x_5)$, $M = (x_8,x_7)$.

The **nondegenerate** conditions: Points I, C and B are not collinear; $\angle CIB$ is not right; Line BI is not perpendicular to line IA; Line BC is non-isotropic; Line BI intersects line AX_A; Line CI intersects line AX_A.

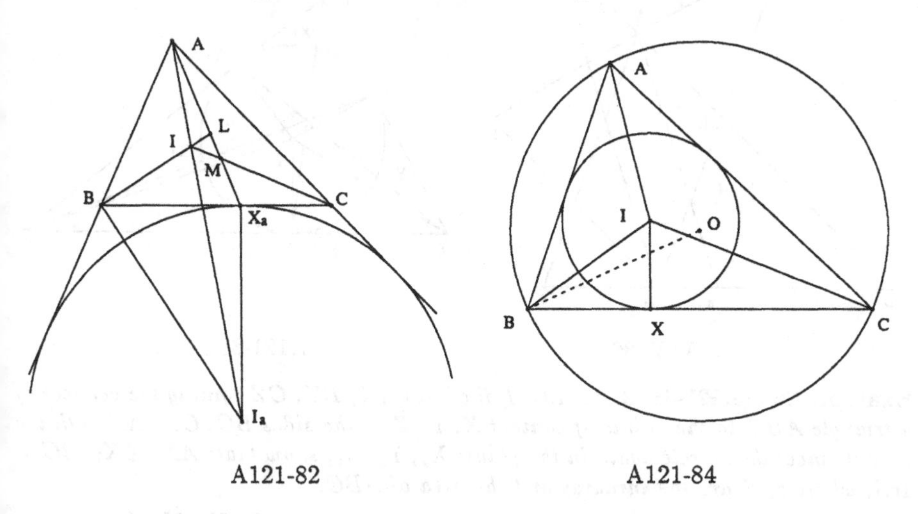

A121-82 A121-84

Example 252 (A121-84: 2.52s, 43). *Show that the product of the distances of the incenter of a triangle from the three vertices of the triangle is equal to $4Rr^2$. State and prove analogous formulas for the excenters.*

Points B, C, I are arbitrarily chosen. Points A, O, X are constructed (in order) as follows: $\tan(CBI) = \tan(IBA)$; $\tan(BCI) = \tan(ICA)$; $OB \equiv OA$; $OB \equiv OC$; X is on line BC; $XI \perp BC$. The **conclusion**: $16\ \overline{OB}^2\ \overline{IX}^2\ \overline{IX}^2 - \overline{IA}^2\ \overline{IB}^2\ \overline{IC}^2 = 0$.

$B = (0,0)$, $C = (u_1,0)$, $I = (u_2,u_3)$, $A = (x_2,x_1)$, $O = (x_4,x_3)$, $X = (u_2,0)$.

The **nondegenerate** conditions: Points I, C and B are not collinear; $\angle CIB$ is not right; Points B, C and A are not collinear; Line BC is non-isotropic.

Example 253 (A121-86: 2.03s, 11; 1.72s, 6). *A parallel to the median AA_1 of the* *triangle ABC meets BC, CA, AB in the points H, N, D. Prove that the symmetrics of H with respect to the midpoints of NC, BD are symmetrical with respect to the vertex A.*

Points B, C, A are arbitrarily chosen. Points H, A_1, N, D, K, L, H_1, H_2 are constructed (in order) as follows: H is on line BC; A_1 is the midpoint of B and C; N is on line CA; $NH \parallel AA_1$; D is on line HN; D is on line AB; K is the midpoint of N and C; L is the midpoint of B and D; K is the midpoint of H and H_1; L is the midpoint of H and H_2. The **conclusion**: A is the midpoint of H_1 and H_2.

$B = (0,0)$, $C = (u_1,0)$, $A = (u_2,u_3)$, $H = (u_4,0)$, $A_1 = (x_1,0)$, $N = (x_3,x_2)$, $D = (x_5,x_4)$, $K = (x_6,x_7)$, $L = (x_8,x_9)$, $H_1 = (x_{10},x_{11})$, $H_2 = (x_{12},x_{13})$.

The **nondegenerate** conditions: $B \neq C$; Points A, A_1 and C are not collinear; Line AB intersects line HN.

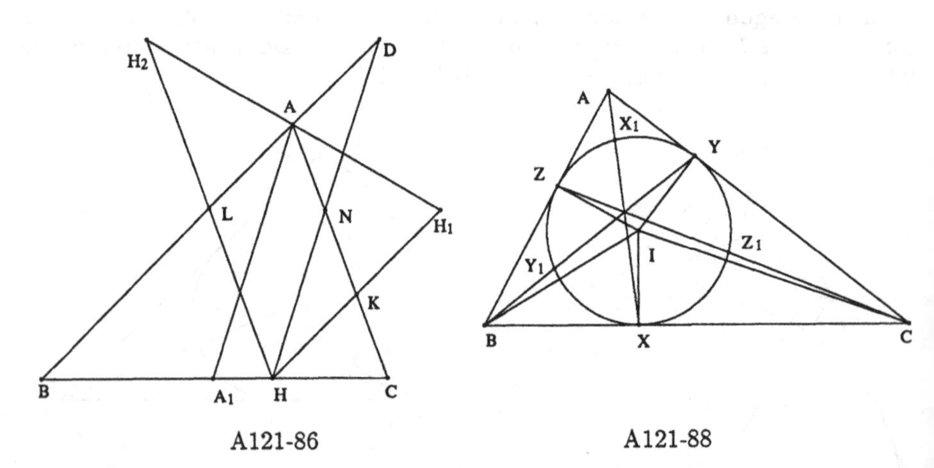

A121-86 A121-88

Example 254 (A121-88: 3.18s, 19). *If the lines AX, BY, CZ joining the vertices of a triangle ABC to the points of contact X, Y, Z of the sides BC, CA, AB with the incircle meet that circle again in the points X_1, Y_1, Z_1, show that: $AX \cdot XX_1 \cdot BC = 4rS$, where r, S are the inradius and the area of ABC.*

Points B, C, I are arbitrarily chosen. Points A, X, Y, Z, X_1, Y_1, Z_1 are constructed (in order) as follows: $\tan(CBI) = \tan(IBA)$; $\tan(BCI) = \tan(ICA)$; X is on line BC; $XI \perp BC$; Y is on line AC; $YI \perp AC$; Z is on line AB; $ZI \perp AB$; $X_1I \equiv IX$; X_1 is on line AX; $Y_1I \equiv IY$; Y_1 is on line BY; $Z_1I \equiv IZ$; Z_1 is on line CZ. The **conclusion**: $4\,\overline{IX}^2\ \nabla(ABC)\ \nabla(ABC) - \overline{XA}^2\ \overline{XX_1}^2\ \overline{BC}^2 = 0$.

$B = (0,0)$, $C = (u_1,0)$, $I = (u_2,u_3)$, $A = (x_2,x_1)$, $X = (u_2,0)$, $Y = (x_4,x_3)$, $Z = (x_6,x_5)$, $X_1 = (x_8,x_7)$, $Y_1 = (x_{10},x_9)$, $Z_1 = (x_{12},x_{11})$.

The **nondegenerate** conditions: Points I, C and B are not collinear; $\angle CIB$ is not right; Line BC is non-isotropic; Line AC is non-isotropic; Line AB is non-isotropic; Line AX is non-isotropic; Line BY is non-isotropic; Line CZ is non-isotropic. In addition, the following nondegenerate conditions, which come from reducibility and have been detected by our prover, should be also added: $Z_1 \neq Z$; $Y_1 \neq Y$; $X_1 \neq X$.

Example 255 (A121-89: 0.68s, 6). *The perpendiculars DP, DQ dropped from the foot D of the altitude AD of the triangle ABC upon the sides AB, AC meet the perpendiculars BP, CQ erected to BC at B, C in the points P, Q respectively. Prove that the line PQ passes through the orthocenter H of ABC.*

Points B, C, A are arbitrarily chosen. Points D, H, P, Q are constructed (in order) as follows: D is on line BC; $DA \perp BC$; $HB \perp AC$; H is on line AD; $PD \perp AB$; $PB \perp BC$; $QD \perp AC$; $QC \perp BC$. The **conclusion**: Points H, Q and P are collinear.

$B = (0,0)$, $C = (u_1, 0)$, $A = (u_2, u_3)$, $D = (u_2, 0)$, $H = (u_2, x_1)$, $P = (0, x_2)$, $Q = (u_1, x_3)$.

The **nondegenerate** conditions: Line BC is non-isotropic; Line AD is not perpendicular to line AC; Points B, C and A are not collinear; Points B, C and A are not collinear.

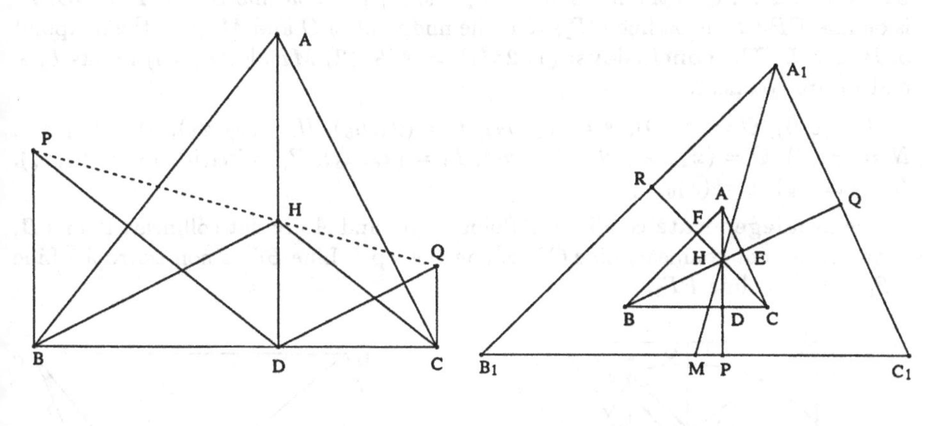

A121-89 A121-91

Example 256 (A121-91: 90.83s, 3469; 6.8s, 96). *The altitudes AHD, BHE, CHF of the triangle ABC are produced beyond D, E, F to the points P, Q, R by the lengths AH, BH, CH, respectively. The parallels through P, Q, R to the sides BC, CA, AB, form a triangle $A_1B_1C_1$. Show that H is the centroid of the triangle $A_1B_1C_1$.*

Points B, C, A are arbitrarily chosen. Points D, E, F, H, P, Q, R, B_1, A_1, C_1, M are constructed (in order) as follows: D is on line BC; $DA \perp BC$; E is on line AC; $EB \perp AC$; F is on line AB; $FC \perp AB$; H is on line BE; H is on line AD; P is on line AD; $\frac{\overline{AH}}{\overline{DP}} - 1 = 0$; Q is on line BE; $\frac{\overline{BH}}{\overline{EQ}} - 1 = 0$; R is on line CF; $\frac{\overline{CH}}{\overline{FR}} - 1 = 0$; $B_1R \parallel BA$; $B_1P \parallel CB$; A_1 is on line B_1R; $A_1Q \parallel AC$; C_1 is on line A_1Q; C_1 is on line B_1P; M is on line C_1B_1; M is on line HA_1. The **conclusion**: M is the midpoint of C_1 and B_1.

$B = (0,0)$, $C = (u_1, 0)$, $A = (u_2, u_3)$, $D = (u_2, 0)$, $E = (x_2, x_1)$, $F = (x_4, x_3)$, $H = (u_2, x_5)$, $P = (u_2, x_6)$, $Q = (x_8, x_7)$, $R = (x_{10}, x_9)$, $B_1 = (x_{11}, x_6)$, $A_1 = (x_{13}, x_{12})$, $C_1 = (x_{14}, x_6)$, $M = (x_{15}, x_6)$.

The **nondegenerate** conditions: Line BC is non-isotropic; Line AC is non-isotropic; Line AB is non-isotropic; Line AD intersects line BE; $-x_2 \neq 0$; $-x_4+u_1 \neq 0$; Points C, B and A are not collinear; Line AC intersects line B_1R; Line B_1P intersects line A_1Q; Line HA_1 intersects line C_1B_1.

Example 257 (A122-92: 10.73s, 13; 4.28s, 14). *The point H, O are the orthocenter and the circumcenter of the triangle ABC, and P, P_1 are two points symmetrical with respect to the mediator of BC. The perpendicular from P to BC meets BC in P_2 and OP_1 in P_3; M is the midpoint of HP. Prove that* (a) $MP_1 \parallel AP_3$; (b) $2MP_1 = AP_3$; (c) *the symmetric of P_2 with respect to the midpoint of OM lies on AP_1.*

Points B, C, A, P are arbitrarily chosen. Points H, M, N, O, Q, P_1, P_2, P_3, K, L are constructed (in order) as follows: $HB \perp AC$; $HA \perp BC$; M is the midpoint of H and P; N is the midpoint of B and C; $OB \equiv OA$; $ON \perp BC$; Q is on line ON; $QP \perp ON$; Q is the midpoint of P_1 and P; P_2 is on line BC; $P_2P \perp BC$; P_3 is on line PP_2; P_3 is on line OP_1; K is the midpoint of O and M; K is the midpoint of P_2 and L. The **conclusions**: (1) $2MP_2 = AP_3$; (2) $MP_2 \parallel AP_3$; (3) Points L, A and P_1 are collinear.

$B = (0,0)$, $C = (u_1,0)$, $A = (u_2,u_3)$, $P = (u_4,u_5)$, $H = (u_2,x_1)$, $M = (x_2,x_3)$, $N = (x_4,0)$, $O = (x_4,x_5)$, $Q = (x_4,u_5)$, $P_1 = (x_6,u_5)$, $P_2 = (u_4,0)$, $P_3 = (u_4,x_7)$, $K = (x_8,x_9)$, $L = (x_{10},x_{11})$.

The **nondegenerate** conditions: Points B, C and A are not collinear; Points B, C and A are not collinear; Line ON is non-isotropic; Line BC is non-isotropic; Line OP_1 intersects line PP_2.

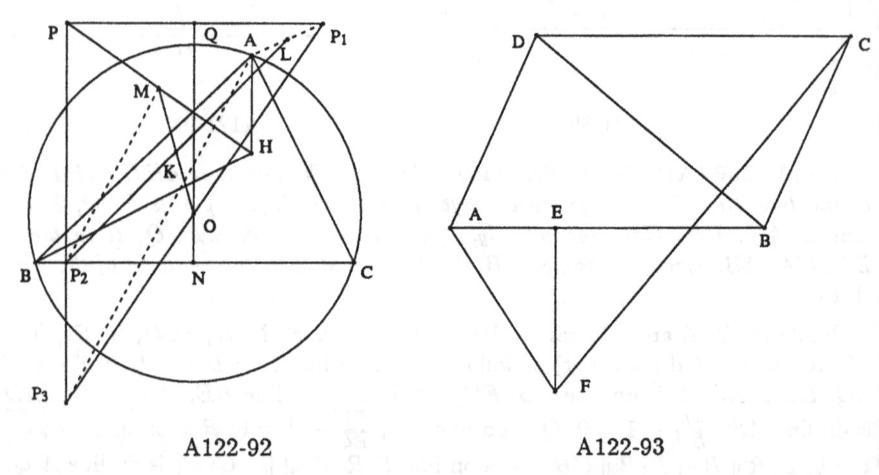

A122-92 A122-93

Example 258 (A122-93: 1.78s, 9). *The side AB of a parallelogram $ABCD$ is produced to E so that $BE = AD$. The perpendicular to ABE at E meets the perpendicular from C to the diagonal BD in F. Show that AF bisects the angle A.*

Points A, B, C are arbitrarily chosen. Points D, E, F are constructed (in order) as follows: $DA \parallel BC$; $DC \parallel BA$; $EB \equiv AD$; E is on line AB; $FC \perp BD$; $FE \perp AB$.

The **conclusion**: $\tan(BAF) = \tan(FAD)$.

$A = (0,0)$, $B = (u_1,0)$, $C = (u_2,u_3)$, $D = (x_1,u_3)$, $E = (x_2,0)$, $F = (x_2,x_3)$.

The **nondegenerate** conditions: Points B, A and C are not collinear; Line AB is non-isotropic; Points A, B and D are not collinear.

Example 259 (A122-96: 2.77s, 98). *If h, m, t are the altitude, the median, and the internal bisector issued from the same vertex of a triangle whose circumradius is R, show that $4R^2h^2(t^2-h^2) = t^4(m^2-h^2)$.*

Points B, C, I are arbitrarily chosen. Points A, D, A_1, O, T are constructed (in order) as follows: $\tan(CBI) = \tan(IBA)$; $\tan(BCI) = \tan(ICA)$; D is on line BC; $DA \perp BC$; A_1 is the midpoint of B and C; $OB \equiv OC$; $OA \equiv OB$; T is on line BC; T is on line IA. The **conclusion**: $\overline{OB}^2\ \overline{AD}^2\ 4\,(\overline{TA}^2 - \overline{AD}^2) - \overline{TA}^2\ \overline{TA}^2\ (\overline{AA_1}^2 - \overline{AD}^2) = 0$.

$B = (0,0)$, $C = (u_1,0)$, $I = (u_2,u_3)$, $A = (x_2,x_1)$, $D = (x_2,0)$, $A_1 = (x_3,0)$, $O = (x_5,x_4)$, $T = (x_6,0)$.

The **nondegenerate** conditions: Points I, C and B are not collinear; $\angle CIB$ is not right; Line BC is non-isotropic; Points A, B and C are not collinear; Line IA intersects line BC.

Remark. The formula remains valid if t is the external bisector.

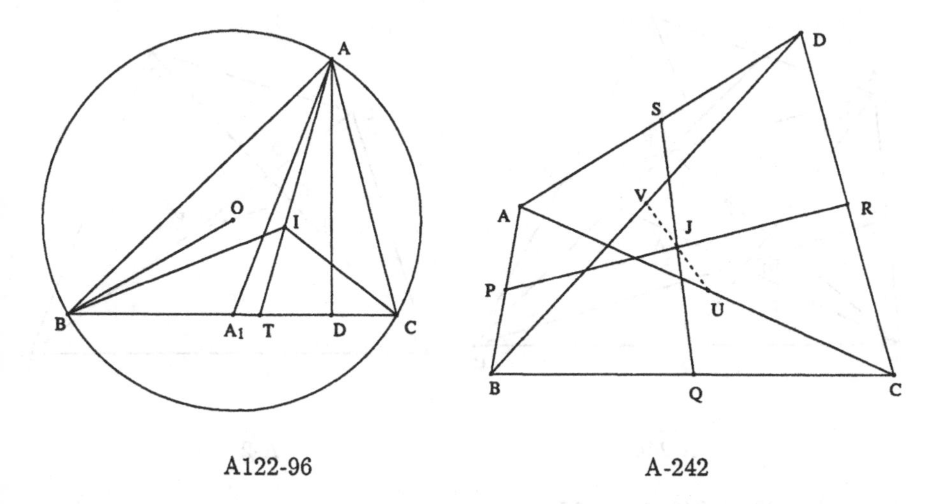

A122-96 A-242

Example 260 (A-242: 5.08s, 20; 2.32s, 20). *The lines joining the midpoints of the two pairs of opposite sides of a quadrilateral and the line joining the midpoints of the diagonals are concurrent and are bisected by their common point.*

Points B, C, D, A are arbitrarily chosen. Points P, Q, R, S, V, U, J are constructed (in order) as follows: P is the midpoint of A and B; Q is the midpoint of B and C; R is the midpoint of C and D; S is the midpoint of D and A; V is the midpoint of B and D; U is the midpoint of A and C; J is on line PR; J is on line SQ. The **conclusions**: (1) Points V, U and J are collinear; (2) J is the midpoint of V and U.

$B = (0,0)$, $C = (u_1,0)$, $D = (u_2,u_3)$, $A = (u_4,u_5)$, $P = (x_1,x_2)$, $Q = (x_3,0)$, $R = (x_4,x_5)$, $S = (x_6,x_7)$, $V = (x_8,x_9)$, $U = (x_{10},x_{11})$, $J = (x_{13},x_{12})$.

The **nondegenerate** conditions: Line SQ intersects line PR.

Example 261 (A-244: 2.3s, 16; 2.1s, 16). *The four lines obtained by joining each vertex of a quadrilateral to the centroid of the triangle determined by the remaining three vertices are concurrent.*

Points B, C, D, A are arbitrarily chosen. Points P, Q, R, S, A_1, B_1, D_1, J are constructed (in order) as follows: P is the midpoint of A and B; Q is the midpoint of B and C; R is the midpoint of C and D; S is the midpoint of D and A; A_1 is on line DQ; A_1 is on line BR; B_1 is on line CS; B_1 is on line AR; D_1 is on line AQ; D_1 is on line CP; J is on line DD_1; J is on line BB_1. The **conclusion**: Points A, A_1 and J are collinear.

$B = (0,0)$, $C = (u_1,0)$, $D = (u_2,u_3)$, $A = (u_4,u_5)$, $P = (x_1,x_2)$, $Q = (x_3,0)$, $R = (x_4,x_5)$, $S = (x_6,x_7)$, $A_1 = (x_9,x_8)$, $B_1 = (x_{11},x_{10})$, $D_1 = (x_{13},x_{12})$, $J = (x_{15},x_{14})$.

The **nondegenerate** conditions: Line BR intersects line DQ; Line AR intersects line CS; Line CP intersects line AQ; Line BB_1 intersects line DD_1.

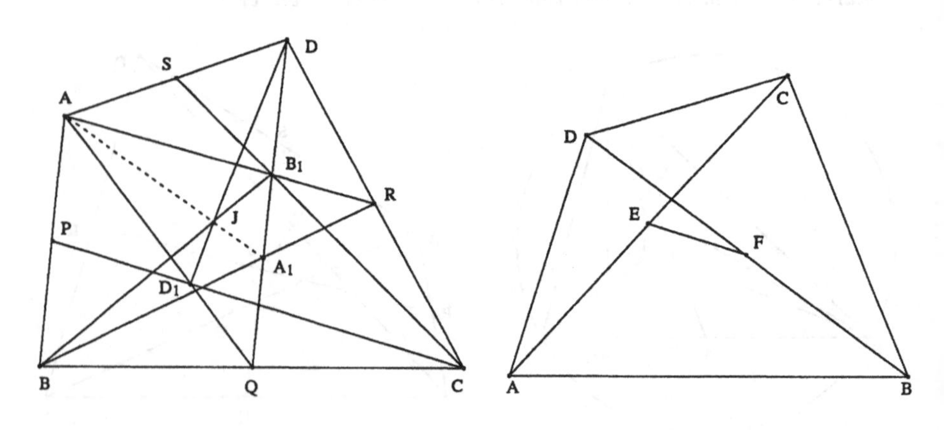

A-244 A-246

Example 262 (A-246: 1.12s, 15).

Points A, B, C, D are arbitrarily chosen. Points E, F are constructed (in order) as follows: E is the midpoint of A and C; F is the midpoint of B and D. The **conclusion**: $\overline{AB}^2 + \overline{CB}^2 + \overline{CD}^2 + \overline{DA}^2 - (\overline{AC}^2 + \overline{BD}^2 + 4\,\overline{EF}^2) = 0$.

$A = (0,0)$, $B = (u_1,0)$, $C = (u_2,u_3)$, $D = (u_4,u_5)$, $E = (x_1,x_2)$, $F = (x_3,x_4)$.

The **nondegenerate** conditions: none.

Example 263 (A-248: 0.83s, 17). *The sum of the squares of the diagonals of a* *quadrilateral is equal to twice the sum of the squares of the two lines joining the midpoints of the two pairs of opposite sides of the quadrilateral.*

Points B, C, D, A are arbitrarily chosen. Points P, Q, S, R are constructed (in order) as follows: P is the midpoint of A and B; Q is the midpoint of B and C; S is the midpoint of D and A; R is the midpoint of C and D. The **conclusion**: $\overline{AC}^2 + \overline{BD}^2 - 2\,(\overline{QS}^2 + \overline{PR}^2) = 0$.

$B = (0,0)$, $C = (u_1,0)$, $D = (u_2,u_3)$, $A = (u_4,u_5)$, $P = (x_1,x_2)$, $Q = (x_3,0)$, $S = (x_4,x_5)$, $R = (x_6,x_7)$.

The **nondegenerate** conditions: none.

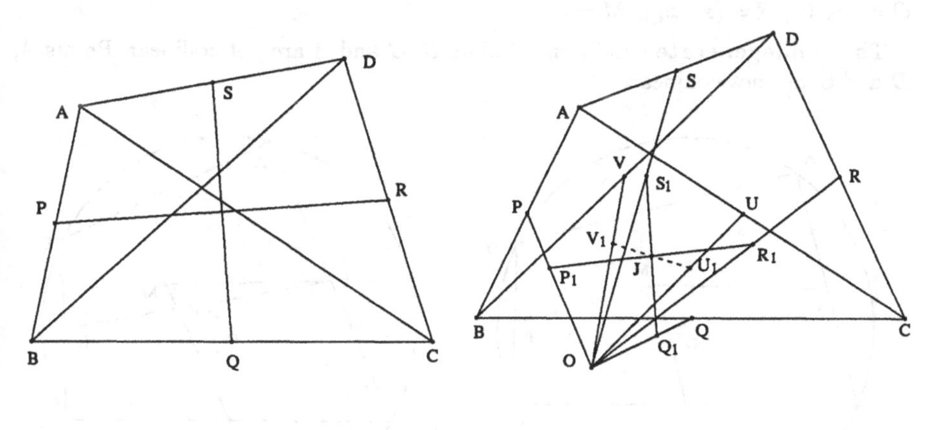

A-248 A127-2

Example 264 (A127-2: 47.23s, 1145; 19.03s, 210). *$ABCD$ is a quadrilateral, P, Q, R, S the midpoints of its sides taken in order, U, V, the midpoints of the diagonals, O any point; OP, OQ, OR, OS, OU, OV are divided in the same ratio in P_1, Q_1, R_1, S_1, U_1, V_1. Prove that P_1R_1, Q_1S_1, U_1V_1, are concurrent.*

Points B, C, D, A, O are arbitrarily chosen. Points P, Q, R, S, V, U, P_1, Q_1, R_1, S_1, V_1, U_1, J are constructed (in order) as follows: P is the midpoint of A and B; Q is the midpoint of B and C; R is the midpoint of C and D; S is the midpoint of D and A; V is the midpoint of B and D; U is the midpoint of A and C; P_1 is on line OP; $\frac{\overline{P_1P}}{\overline{OP}} - \frac{\overline{Q_1Q}}{\overline{OQ}} = 0$; Q_1 is on line OQ; $\frac{\overline{R_1R}}{\overline{OR}} - \frac{\overline{Q_1Q}}{\overline{OQ}} = 0$; R_1 is on line OR; $\frac{\overline{S_1S}}{\overline{OS}} - \frac{\overline{Q_1Q}}{\overline{OQ}} = 0$; S_1 is on line OS; $\frac{\overline{V_1V}}{\overline{OV}} - \frac{\overline{Q_1Q}}{\overline{OQ}} = 0$; V_1 is on line OV; $\frac{\overline{U_1U}}{\overline{OU}} - \frac{\overline{Q_1Q}}{\overline{OQ}} = 0$; U_1 is on line OU; J is on line Q_1S_1; J is on line P_1R_1. The **conclusion**: Points V_1, U_1 and J are collinear.

$B = (0,0)$, $C = (u_1,0)$, $D = (u_2,u_3)$, $A = (u_4,u_5)$, $O = (u_6,u_7)$, $P = (x_1,x_2)$, $Q = (x_3,0)$, $R = (x_4,x_5)$, $S = (x_6,x_7)$, $V = (x_8,x_9)$, $U = (x_{10},x_{11})$, $P_1 = (x_{12},u_8)$, $Q_1 = (x_{14},x_{13})$, $R_1 = (x_{16},x_{15})$, $S_1 = (x_{18},x_{17})$, $V_1 = (x_{20},x_{19})$, $U_1 = (x_{22},x_{21})$, $J = (x_{24},x_{23})$.

The **nondegenerate** conditions: $O \neq P$; $(x_1 - u_6)x_3 - u_6x_1 + u_6^2 \neq 0$; $-u_7 \neq 0$;

$(x_3-u_6)x_4-u_6x_3+u_6^2 \neq 0$; $-x_3+u_6 \neq 0$; $(x_3-u_6)x_6-u_6x_3+u_6^2 \neq 0$; $-x_3+u_6 \neq 0$; $(x_3-u_6)x_8-u_6x_3+u_6^2 \neq 0$; $-x_3+u_6 \neq 0$; $(x_3-u_6)x_{10}-u_6x_3+u_6^2 \neq 0$; $-x_3+u_6 \neq 0$; Line P_1R_1 intersects line Q_1S_1.

Example 265 (A-258: 9.32s, 17; 1.38s, 15). *The perpendiculars from the midpoints of the sides of a cyclic quadrilateral to the respectively opposite sides are concurrent.*

Points B, C, A are arbitrarily chosen. Points O, D, P, Q, R, M are constructed (in order) as follows: $OB \equiv OA$; $OB \equiv OC$; $DO \equiv OB$; P is the midpoint of A and B; Q is the midpoint of B and C; R is the midpoint of C and D; $MP \perp CD$; $MQ \perp AD$. The **conclusion**: $MR \perp AB$.

$B = (0,0)$, $C = (u_1,0)$, $A = (u_2,u_3)$, $O = (x_2,x_1)$, $D = (x_3,u_4)$, $P = (x_4,x_5)$, $Q = (x_6,0)$, $R = (x_7,x_8)$, $M = (x_{10},x_9)$.

The **nondegenerate** conditions: Points B, C and A are not collinear; Points A, D and C are not collinear.

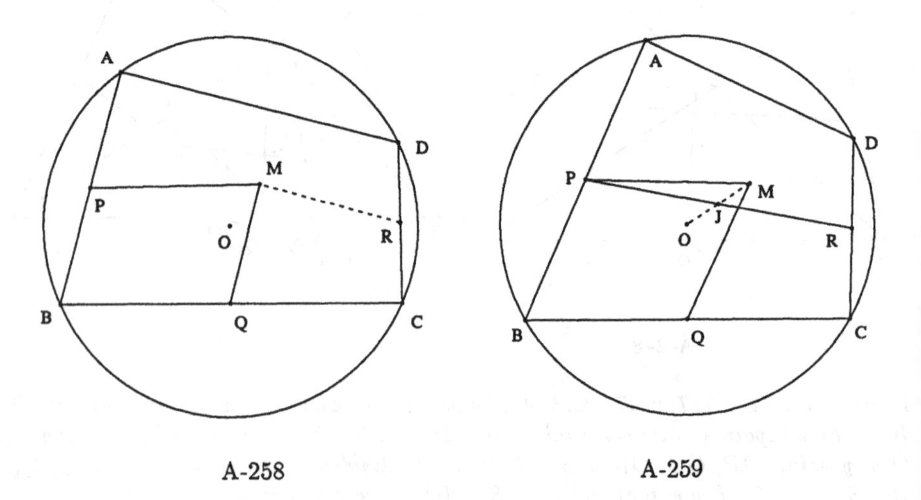

A-258 A-259

Example 266 (A-259: 2.0s, 61). *The perpendicular from the midpoint of each diagonal upon the other diagonal also passes through the anticenter of a cyclic quadrilateral.*

Points B, C, A are arbitrarily chosen. Points O, D, P, Q, R, J, M are constructed (in order) as follows: $OB \equiv OA$; $OB \equiv OC$; $DO \equiv OB$; P is the midpoint of A and B; Q is the midpoint of B and C; R is the midpoint of C and D; J is the midpoint of P and R; $MP \perp CD$; $MQ \perp AD$. The **conclusion**: Points O, J and M are collinear.

$B = (0,0)$, $C = (u_1,0)$, $A = (u_2,u_3)$, $O = (x_2,x_1)$, $D = (x_3,u_4)$, $P = (x_4,x_5)$, $Q = (x_6,0)$, $R = (x_7,x_8)$, $J = (x_9,x_{10})$, $M = (x_{12},x_{11})$.

The **nondegenerate** conditions: Points B, C and A are not collinear; Points A, D and C are not collinear.

Example 267 (A-261: 1.45s, 12). *The four lines obtained by joining each vertex* *of a cyclic quadrilateral to the orthocenter of the triangle formed by the remaining three vertices bisect each other.*

Points B, C, A are arbitrarily chosen. Points O, D, D_1, A_1, M are constructed (in order) as follows: $OB \equiv OA$; $OB \equiv OC$; $DO \equiv OB$; $D_1B \perp AC$; $D_1A \perp BC$; $A_1B \perp DC$; $A_1D \perp BC$; M is on line DD_1; M is on line AA_1. The **conclusion**: M is the midpoint of A and A_1.

$B = (0,0)$, $C = (u_1,0)$, $A = (u_2,u_3)$, $O = (x_2,x_1)$, $D = (x_3,u_4)$, $D_1 = (u_2,x_4)$, $A_1 = (x_3,x_5)$, $M = (x_7,x_6)$.

The **nondegenerate** conditions: Points B, C and A are not collinear; Points B, C and A are not collinear; Points B, C and D are not collinear; Line AA_1 intersects line DD_1.

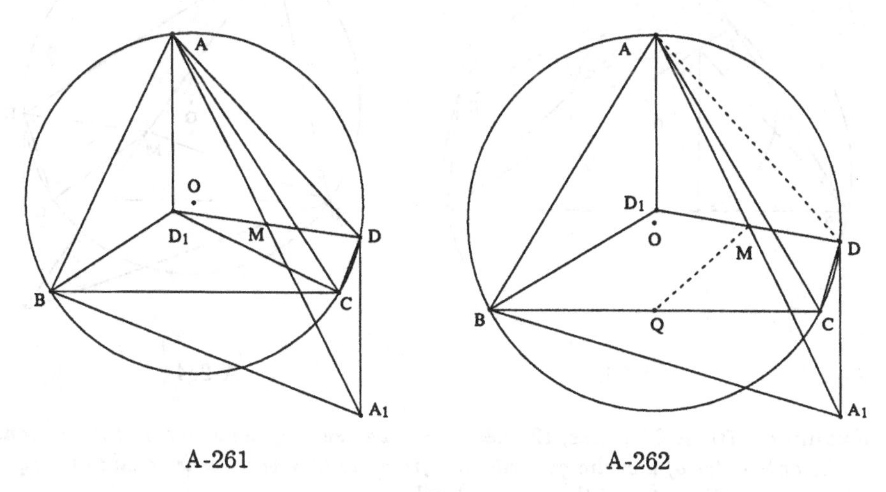

A-261 A-262

Example 268 (A-262: 1.37s, 34). *The point M in A-261 coincides with the anticenter of the quadrilateral.*

Points B, C, A are arbitrarily chosen. Points O, D, D_1, A_1, M, Q are constructed (in order) as follows: $OB \equiv OA$; $OB \equiv OC$; $DO \equiv OB$; $D_1B \perp AC$; $D_1A \perp BC$; $A_1B \perp DC$; $A_1D \perp BC$; M is on line DD_1; M is on line AA_1; Q is the midpoint of B and C. The **conclusion**: $MQ \perp AD$.

$B = (0,0)$, $C = (u_1,0)$, $A = (u_2,u_3)$, $O = (x_2,x_1)$, $D = (x_3,u_4)$, $D_1 = (u_2,x_4)$, $A_1 = (x_3,x_5)$, $M = (x_7,x_6)$, $Q = (x_8,0)$.

The **nondegenerate** conditions: Points B, C and A are not collinear; Points B, C and A are not collinear; Points B, C and D are not collinear; Line AA_1 intersects line DD_1.

Example 269 (A-263: 4.83s, 171). *The nine-point circles of the four triangles determined by the four vertices of a cyclic quadrilateral pass through the anticenter of the quadrilateral.*

Points B, C, A are arbitrarily chosen. Points O, D, H, G, E, F, N, M are constructed (in order) as follows: $OB \equiv OA$; $OB \equiv OC$; $DO \equiv OB$; $HB \perp AC$;

$HA \perp BC$; G is on line BC; G is on line AH; E is on line AC; E is on line BH; F is on line AB; F is on line CH; $NG \equiv NE$; $NG \equiv NF$; M is the midpoint of D and H. The **conclusion**: $NG \equiv NM$.

$B = (0,0)$, $C = (u_1, 0)$, $A = (u_2, u_3)$, $O = (x_2, x_1)$, $D = (x_3, u_4)$, $H = (u_2, x_4)$, $G = (u_2, 0)$, $E = (x_6, x_5)$, $F = (x_8, x_7)$, $N = (x_{10}, x_9)$, $M = (x_{11}, x_{12})$.

The **nondegenerate** conditions: Points B, C and A are not collinear; Points B, C and A are not collinear; Line AH intersects line BC; Line BH intersects line AC; Line CH intersects line AB; Points G, F and E are not collinear.

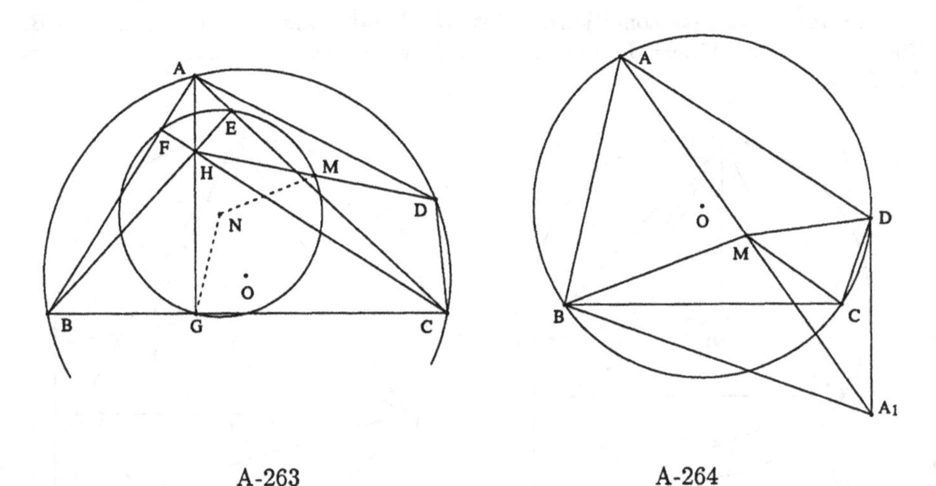

A-263 A-264

Example 270 (A-264: 2.2s, 12; 0.85s, 18). *The sum of the squares of the distances of the anticenter of a cyclic quadrilateral from the four vertices is equal to the square of the circumdiameter of the quadrilateral.*

Points B, C, A are arbitrarily chosen. Points O, D, A_1, M are constructed (in order) as follows: $OB \equiv OA$; $OB \equiv OC$; $DO \equiv OB$; $A_1B \perp DC$; $A_1D \perp BC$; M is the midpoint of A and A_1. The **conclusion**: $\overline{MA}^2 + \overline{MB}^2 + \overline{MC}^2 + \overline{MD}^2 - 4\ \overline{OB}^2 = 0$.

$B = (0,0)$, $C = (u_1, 0)$, $A = (u_2, u_3)$, $O = (x_2, x_1)$, $D = (x_3, u_4)$, $A_1 = (x_3, x_4)$, $M = (x_5, x_6)$.

The **nondegenerate** conditions: Points B, C and A are not collinear; Points B, C and D are not collinear.

Example 271 (A134-1: 7.78s, 453). *Show that in a cyclic quadrilateral the distances of the point of intersection of the diagonals from two opposite sides are proportional to these sides.*

Points B, C, A are arbitrarily chosen. Points O, D, I, Q, S are constructed (in order) as follows: $OB \equiv OA$; $OB \equiv OC$; $DO \equiv OB$; I is on line BD; I is on line AC; Q is on line BC; $QI \perp BC$; S is on line AD; $SI \perp AD$. The **conclusion**: $IS \cdot BC = IQ \cdot AD$.

$B = (0,0)$, $C = (u_1,0)$, $A = (u_2,u_3)$, $O = (x_2,x_1)$, $D = (x_3,u_4)$, $I = (x_5,x_4)$, $Q = (x_5,0)$, $S = (x_7,x_6)$.

The **nondegenerate** conditions: Points B, C and A are not collinear; Line AC intersects line BD; Line BC is non-isotropic; Line AD is non-isotropic.

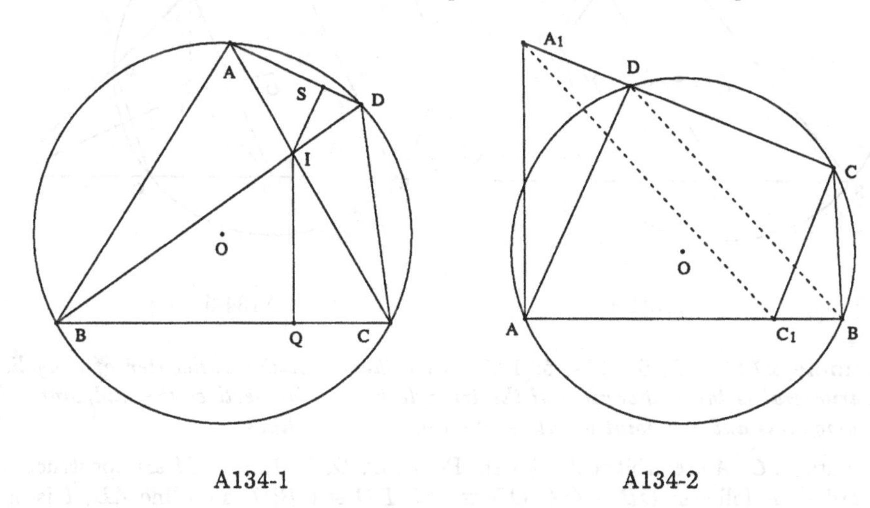

A134-1 A134-2

Example 272 (A134-2: 0.93s, 12). *In the cyclic quadrilateral $ABCD$ the perpendicular to AB at A meets CD in A_1, and the perpendicular to CD at C meets AB in C_1. Show that the line A_1C_1 is parallel to the diagonal BD.*

Points A, B, C are arbitrarily chosen. Points O, D, A_1, C_1 are constructed (in order) as follows: $OB \equiv OA$; $OA \equiv OC$; $DO \equiv OB$; A_1 is on line CD; $A_1A \perp AB$; C_1 is on line AB; $C_1C \perp CD$. The **conclusion**: $A_1C_1 \parallel BD$.

$A = (0,0)$, $B = (u_1,0)$, $C = (u_2,u_3)$, $O = (x_2,x_1)$, $D = (x_3,u_4)$, $A_1 = (0,x_4)$, $C_1 = (x_5,0)$.

The **nondegenerate** conditions: Points A, C and B are not collinear; Line AB is not perpendicular to line CD; Line CD is not perpendicular to line AB.

Example 273 (A134-4: 1.47s, 13). *Show that the perpendicular from the point of intersection of two opposite sides, produced, of a cyclic quadrilateral upon the line joining the midpoints of the two sides considered passes through the anticenter of the quadrilateral.*

Points B, C, A are arbitrarily chosen. Points O, D, I, Q, S, J, M are constructed (in order) as follows: $OB \equiv OA$; $OB \equiv OC$; $DO \equiv OB$; I is on line AD; I is on line BC; Q is the midpoint of B and C; S is the midpoint of A and D; J is the midpoint of S and Q; J is the midpoint of O and M. The **conclusion**: $IM \perp SQ$.

$B = (0,0)$, $C = (u_1,0)$, $A = (u_2,u_3)$, $O = (x_2,x_1)$, $D = (x_3,u_4)$, $I = (x_4,0)$, $Q = (x_5,0)$, $S = (x_6,x_7)$, $J = (x_8,x_9)$, $M = (x_{10},x_{11})$.

The **nondegenerate** conditions: Points B, C and A are not collinear; Line BC intersects line AD.

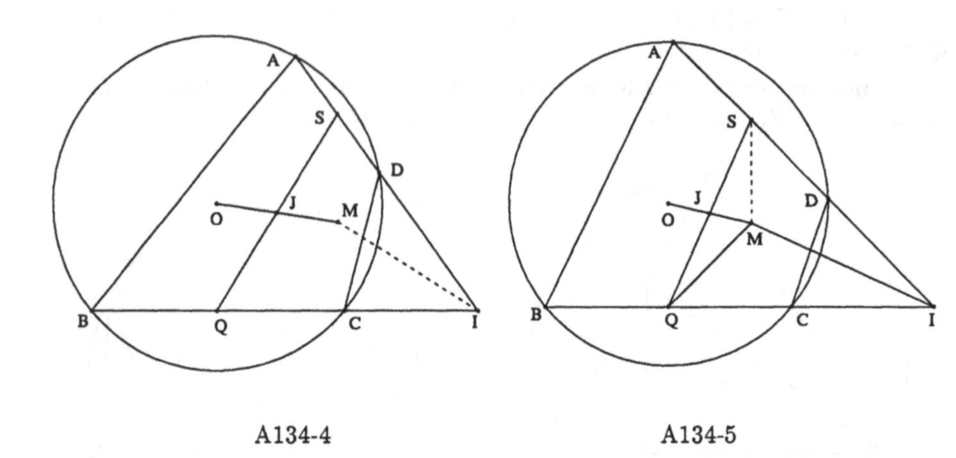

A134-4 A134-5

Example 274 (A134-5: 2.7s, 5; 1.27s, 6). *Show that the anticenter of a cyclic quadrilateral is the orthocenter of the triangle having for vertices the midpoints of the diagonals and the point of intersection of those two lines.*

Points B, C, A are arbitrarily chosen. Points O, D, I, Q, S, J, M are constructed (in order) as follows: $OB \equiv OA$; $OB \equiv OC$; $DO \equiv OB$; I is on line AD; I is on line BC; Q is the midpoint of B and C; S is the midpoint of A and D; J is the midpoint of S and Q; J is the midpoint of O and M. The **conclusion**: $MS \perp BC$.

$B = (0,0)$, $C = (u_1, 0)$, $A = (u_2, u_3)$, $O = (x_2, x_1)$, $D = (x_3, u_4)$, $I = (x_4, 0)$, $Q = (x_5, 0)$, $S = (x_6, x_7)$, $J = (x_8, x_9)$, $M = (x_{10}, x_{11})$.

The **nondegenerate** conditions: Points B, C and A are not collinear; Line BC intersects line AD.

Example 275 (A134-6: 1.58s, 14). *Show that the anticenter of a cyclic quadrilateral is collinear with the two symmetrics of the circumcenter of the quadrilateral with respect to a pair of opposite sides.*

Points B, C, A are arbitrarily chosen. Points O, D, Q, S, J, M, Q_1, S_1 are constructed (in order) as follows: $OB \equiv OA$; $OB \equiv OC$; $DO \equiv OB$; Q is the midpoint of B and C; S is the midpoint of A and D; J is the midpoint of S and Q; J is the midpoint of O and M; Q is the midpoint of O and Q_1; S is the midpoint of O and S_1. The **conclusion**: Points Q_1, S_1 and M are collinear.

$B = (0,0)$, $C = (u_1, 0)$, $A = (u_2, u_3)$, $O = (x_2, x_1)$, $D = (x_3, u_4)$, $Q = (x_4, 0)$, $S = (x_5, x_6)$, $J = (x_7, x_8)$, $M = (x_9, x_{10})$, $Q_1 = (x_{11}, x_{12})$, $S_1 = (x_{13}, x_{14})$.

The **nondegenerate** conditions: Points B, C and A are not collinear.

Example 276 (A135-7: 1.17s, 25). *If H_a, H_b, H_c, H_d are the orthocenters of the four triangles determined by the vertices of the cyclic quadrilateral $ABCD$, show that the vertices of $ABCD$ are the orthocenters of the four triangles determined by the points H_a, H_b, H_c, H_d.*

Points B, C, A are arbitrarily chosen. Points O, D, H_D, H_A, H_C are constructed

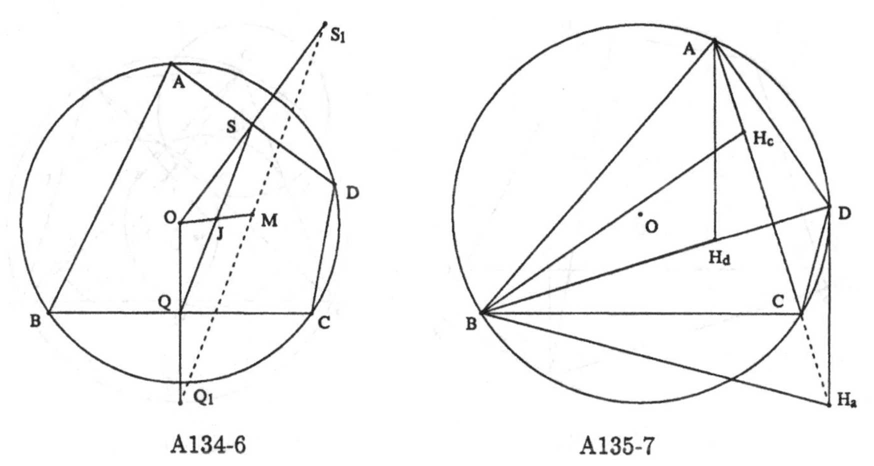

A134-6 A135-7

(in order) as follows: $OB \equiv OA$; $OB \equiv OC$; $DO \equiv OB$; $H_DB \perp AC$; $H_DA \perp BC$; $H_AB \perp DC$; $H_AD \perp BC$; $H_CA \perp BD$; $H_CB \perp AD$. The **conclusion**: $H_DB \perp H_CH_A$.

$B = (0,0)$, $C = (u_1,0)$, $A = (u_2,u_3)$, $O = (x_2,x_1)$, $D = (x_3,u_4)$, $H_D = (u_2,x_4)$, $H_A = (x_3,x_5)$, $H_C = (x_7,x_6)$.

The **nondegenerate** conditions: Points B, C and A are not collinear; Points B, C and A are not collinear; Points B, C and D are not collinear; Points A, D and B are not collinear.

Example 277 (A135-8: 187.55s, 11318). *Show that the product of the distances of two opposite sides of a cyclic quadrilateral from a point on the circumcircles is equal to the product of the distances of the other two sides from the same point.*

Points B, C, A are arbitrarily chosen. Points O, D, E, P, Q, R, S are constructed (in order) as follows: $OB \equiv OA$; $OB \equiv OC$; $DO \equiv OB$; $EO \equiv OB$; P is on line AB; $PE \perp AB$; Q is on line BC; $QE \perp BC$; R is on line CD; $RE \perp CD$; S is on line AD; $SE \perp AD$. The **conclusion**: $ES \cdot EQ = EP \cdot ER$.

$B = (0,0)$, $C = (u_1,0)$, $A = (u_2,u_3)$, $O = (x_2,x_1)$, $D = (x_3,u_4)$, $E = (x_4,u_5)$, $P = (x_6,x_5)$, $Q = (x_4,0)$, $R = (x_8,x_7)$, $S = (x_{10},x_9)$.

The **nondegenerate** conditions: Points B, C and A are not collinear; Line AB is non-isotropic; Line BC is non-isotropic; Line CD is non-isotropic; Line AD is non-isotropic.

Example 278 (A135-9: 12.93s, 268; 9.93s, 49). *Show that the four lines obtained by joining each vertex of a cyclic quadrilateral to the nine-point center of the triangle formed by the remaining three vertices, are concurrent.*

Points B, C, A are arbitrarily chosen. Points O, D, P, Q, R, S, V, U, N_D, N_A, N_C, I are constructed (in order) as follows: $OB \equiv OA$; $OB \equiv OC$; $DO \equiv OB$; P is the midpoint of A and B; Q is the midpoint of B and C; R is the midpoint of C and D; S is the midpoint of D and A; V is the midpoint of B and D; U is the midpoint of A and C; $N_DQ \equiv N_DU$; $N_DQ \equiv N_DP$; $N_AQ \equiv N_AV$; $N_AQ \equiv N_AR$;

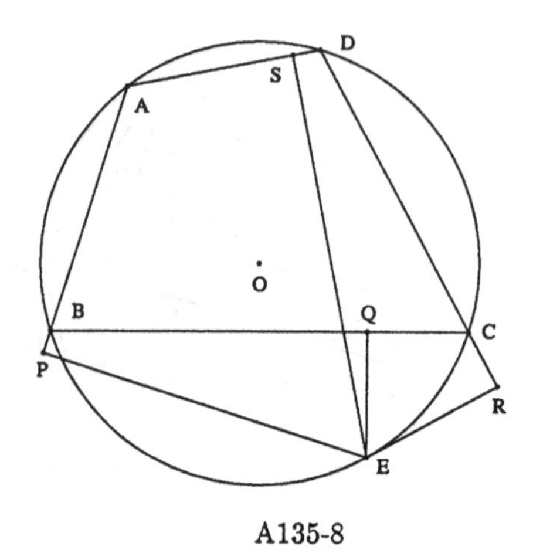

A135-8

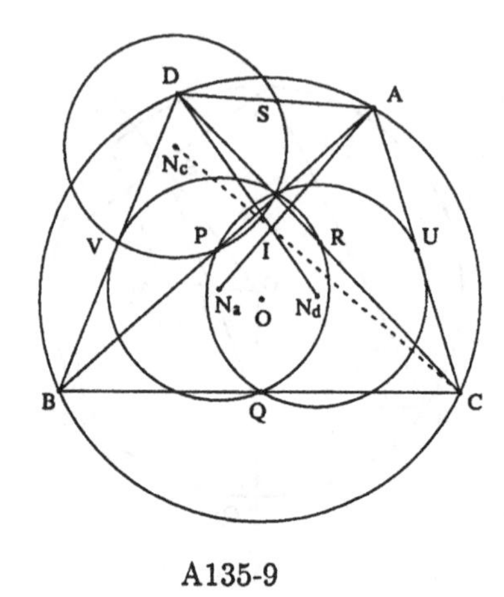

A135-9

$N_CP \equiv N_CS$; $N_CP \equiv N_CV$; I is on line DN_D; I is on line AN_A. The **conclusion**: Points I, C and N_C are collinear.

$B = (0,0)$, $C = (u_1,0)$, $A = (u_2,u_3)$, $O = (x_2,x_1)$, $D = (x_3,u_4)$, $P = (x_4,x_5)$, $Q = (x_6,0)$, $R = (x_7,x_8)$, $S = (x_9,x_{10})$, $V = (x_{11},x_{12})$, $U = (x_{13},x_{14})$, $N_D = (x_{16},x_{15})$, $N_A = (x_{18},x_{17})$, $N_C = (x_{20},x_{19})$, $I = (x_{22},x_{21})$.

The **nondegenerate** conditions: Points B, C and A are not collinear; Points Q, P and U are not collinear; Points Q, R and V are not collinear; Points P, V and S are not collinear; Line AN_A intersects line DN_D.

Example 279 (A-273: 1.15s, 12). *In an orthodiagonal quadrilateral the two lines joining the midpoints of the pairs of opposite sides are equal.*

Points A, C, B are arbitrarily chosen. Points D, P, Q, R, S are constructed (in order) as follows: $DB \perp CA$; P is the midpoint of A and B; Q is the midpoint of B and C; R is the midpoint of C and D; S is the midpoint of D and A. The **conclusion**: $SQ \equiv PR$.

$A = (0,0)$, $C = (u_1,0)$, $B = (u_2,u_3)$, $D = (u_2,u_4)$, $P = (x_1,x_2)$, $Q = (x_3,x_4)$, $R = (x_5,x_6)$, $S = (x_7,x_8)$.

The **nondegenerate** conditions: $C \neq A$.

Example 280 (A-274: 1.28s, 16). *In an orthodiagonal quadrilateral the midpoints of the sides lie on a circle having for center the centroid of the quadrilateral.*

Points A, C, B are arbitrarily chosen. Points D, P, Q, R, S, O are constructed (in order) as follows: $DB \perp CA$; P is the midpoint of A and B; Q is the midpoint of B and C; R is the midpoint of C and D; S is the midpoint of D and A; O is on line QS; O is on line PR. The **conclusion**: $OS \equiv OR$.

$A = (0,0)$, $C = (u_1,0)$, $B = (u_2,u_3)$, $D = (u_2,u_4)$, $P = (x_1,x_2)$, $Q = (x_3,x_4)$, $R = (x_5,x_6)$, $S = (x_7,x_8)$, $O = (x_{10},x_9)$.

The **nondegenerate** conditions: $C \neq A$; Line PR intersects line QS.

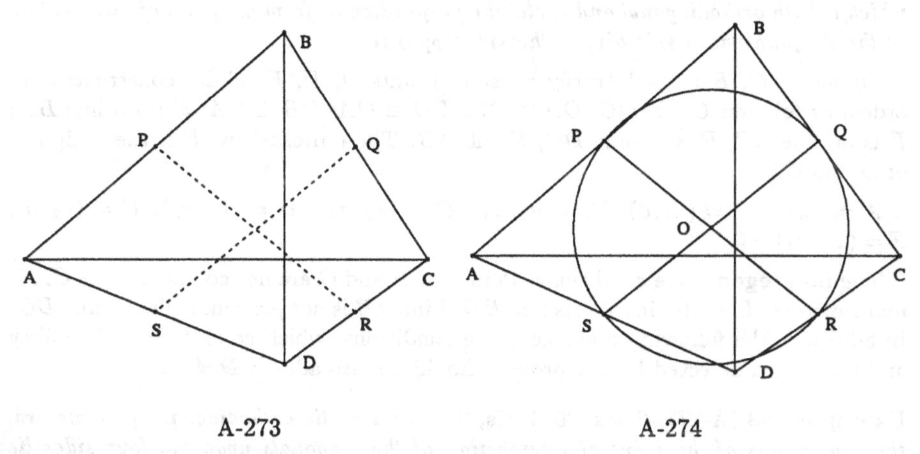

A-273 A-274

Example 281 (A-275: 1.93s, 7). *If an orthodiagonal quadrilateral is cyclic, the anticenter coincides with the point of intersection of its diagonals.*

Points M, A are arbitrarily chosen. Points C, B, O, D, P, R, J are constructed (in order) as follows: C is on line AM; $BM \perp MA$; $OA \equiv OB$; $OA \equiv OC$; D is on line BM; $DO \equiv OA$; P is the midpoint of A and B; R is the midpoint of C and D; J is the midpoint of O and M. The **conclusion**: Points P, J and R are collinear.

$M = (0,0)$, $A = (u_1, 0)$, $C = (u_2, 0)$, $B = (0, u_3)$, $O = (x_2, x_1)$, $D = (0, x_3)$, $P = (x_4, x_5)$, $R = (x_6, x_7)$, $J = (x_8, x_9)$.

The **nondegenerate** conditions: $A \neq M$; $M \neq A$; Points A, C and B are not collinear; Line BM is non-isotropic. In addition, the following nondegenerate conditions, which come from reducibility and have been detected by our prover, should be also added: $D \neq B$.

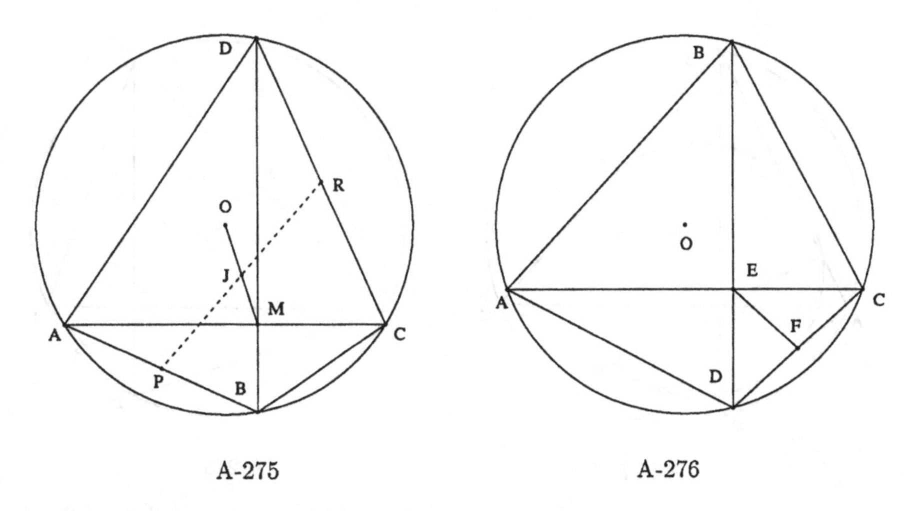

A-275 A-276

Example 282 (A-276: 1.03s, 8). (*Theorem of Brahmagupta*) *In a quadrilateral* *which is both orthodiagonal and cyclic the perpendicular from the point of intersection of the diagonals to a side bisects the side opposite.*

Points A, C, B are arbitrarily chosen. Points O, D, E, F are constructed (in order) as follows: $OA \equiv OC$; $OA \equiv OB$; $DO \equiv OA$; $DB \perp CA$; E is on line BD; E is on line AC; F is on line DC; $FE \perp AB$. The **conclusion**: F is the midpoint of D and C.

$A = (0,0)$, $C = (u_1,0)$, $B = (u_2,u_3)$, $O = (x_2,x_1)$, $D = (u_2,x_3)$, $E = (u_2,0)$, $F = (x_5,x_4)$.

The **nondegenerate** conditions: Points A, B and C are not collinear; Line CA is non-isotropic; Line AC intersects line BD; Line AB is not perpendicular to line DC. In addition, the following nondegenerate conditions, which come from reducibility and have been detected by our prover, should be also added: $D \neq B$.

Example 283 (A-277: 2.08s, 96; 1.92s, 96). *In a cyclic orthodiagonal quadrilateral the projections of the point of intersection of the diagonals upon the four sides lie on the circle passing through the midpoints of the sides.*

Points A, C, B are arbitrarily chosen. Points O, D, E, F, P, Q, R are constructed (in order) as follows: $OA \equiv OC$; $OA \equiv OB$; $DB \perp CA$; $DO \equiv OA$; E is on line BD; E is on line AC; F is on line AB; $FE \perp AB$; P is the midpoint of A and B; Q is the midpoint of B and C; R is the midpoint of C and D. The **conclusion**: Points P, Q, R and F are on the same circle.

$A = (0,0)$, $C = (u_1,0)$, $B = (u_2,u_3)$, $O = (x_2,x_1)$, $D = (u_2,x_3)$, $E = (u_2,0)$, $F = (x_5,x_4)$, $P = (x_6,x_7)$, $Q = (x_8,x_9)$, $R = (x_{10},x_{11})$.

The **nondegenerate** conditions: Points A, B and C are not collinear; Line CA is non-isotropic; Line AC intersects line BD; Line AB is non-isotropic. In addition, the following nondegenerate conditions, which come from reducibility and have been detected by our prover, should be also added: $D \neq B$.

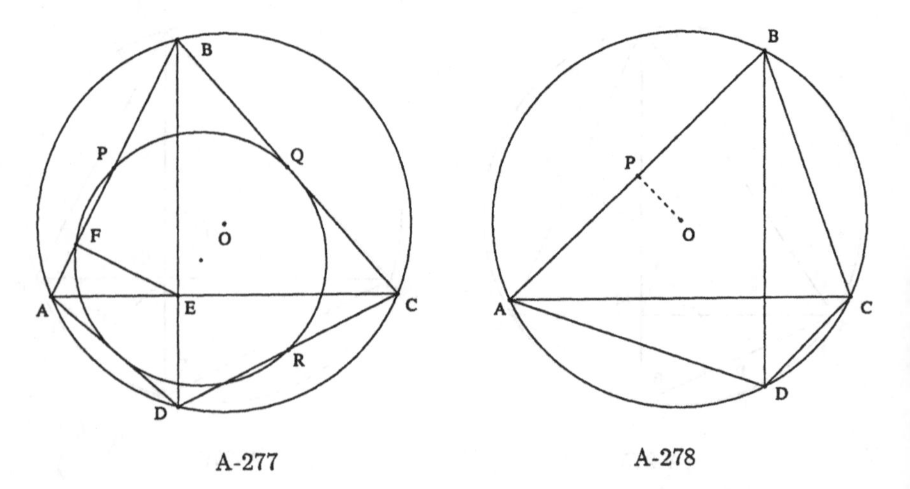

A-277 A-278

Example 284 (A-278: 0.75s, 10). *In a cyclic orthodiagonal quadrilateral the dis-*

tance of a side from the circumcenter of the quadrilateral is equal to half the opposite side.

Points A, C, B are arbitrarily chosen. Points O, D, P are constructed (in order) as follows: $OA \equiv OC$; $OA \equiv OB$; $DB \perp CA$; $DO \equiv OA$; P is the midpoint of A and B. The **conclusion**: $2OP = CD$.

$A = (0,0)$, $C = (u_1, 0)$, $B = (u_2, u_3)$, $O = (x_2, x_1)$, $D = (u_2, x_3)$, $P = (x_4, x_5)$.

The **nondegenerate** conditions: Points A, B and C are not collinear; Line CA is non-isotropic. In addition, the following nondegenerate conditions, which come from reducibility and have been detected by our prover, should be also added: $D \neq B$.

Example 285 (A-279: 3.15s, 8; 0.77s, 9). *If a quadrilateral is both cyclic and orthodiagonal, the sum of the squares of two opposite sides is equal to the square of the circumdiameter of the quadrilateral.*

Points A, C, B are arbitrarily chosen. Points O, D are constructed (in order) as follows: $OA \equiv OC$; $OA \equiv OB$; $DB \perp CA$; $DO \equiv OA$. The **conclusion**: $\overline{AB}^2 + \overline{DC}^2 - 4\,\overline{OA}^2 = 0$.

$A = (0,0)$, $C = (u_1, 0)$, $B = (u_2, u_3)$, $O = (x_2, x_1)$, $D = (u_2, x_3)$.

The **nondegenerate** conditions: Points A, B and C are not collinear; Line CA is non-isotropic. In addition, the following nondegenerate conditions, which come from reducibility and have been detected by our prover, should be also added: $D \neq B$.

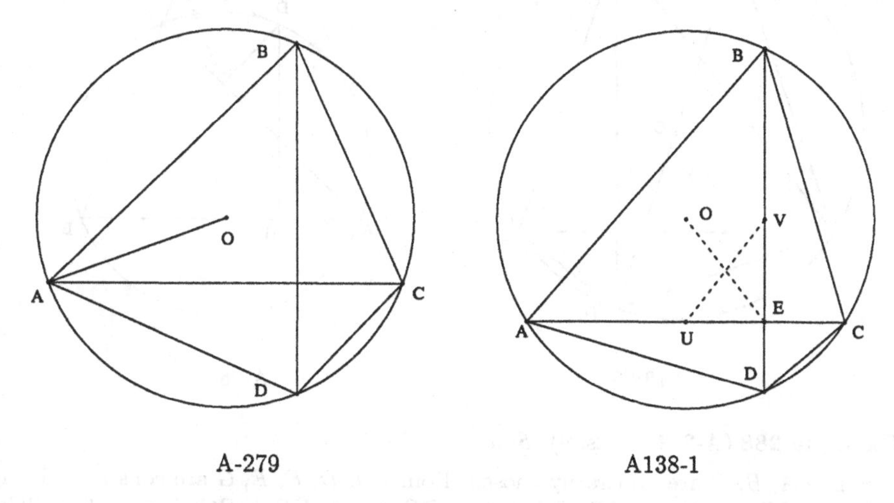

A-279 A138-1

Example 286 (A138-1: 0.98s, 6). *Show that the line joining the midpoints of the diagonals of a cyclic orthodiagonal quadrilateral is equal to the distance of the point of intersection of the diagonals from the circumcenter of the quadrilateral.*

Points A, C, B are arbitrarily chosen. Points O, D, E, U, V are constructed (in order) as follows: $OA \equiv OC$; $OA \equiv OB$; $DO \equiv OA$; $DB \perp CA$; E is on line BD; E is on line AC; U is the midpoint of A and C; V is the midpoint of B and D. The **conclusion**: $UV \equiv OE$.

$A = (0,0)$, $C = (u_1,0)$, $B = (u_2,u_3)$, $O = (x_2,x_1)$, $D = (u_2,x_3)$, $E = (u_2,0)$, $U = (x_4,0)$, $V = (u_2,x_5)$.

The **nondegenerate** conditions: Points A, B and C are not collinear; Line CA is non-isotropic; Line AC intersects line BD. In addition, the following nondegenerate conditions, which come from reducibility and have been detected by our prover, should be also added: $D \neq B$.

Example 287 (A138-2: 1.52s, 16). *If the diagonals of a cyclic quadrilateral ABCD are orthogonal, and E is the diametric opposite of D on its circumcircle, show that* $AE = CB$.

Points A, C, B are arbitrarily chosen. Points O, D, E are constructed (in order) as follows: $OA \equiv OC$; $OA \equiv OB$; $DO \equiv OA$; $DB \perp CA$; E is on line OD; $EO \equiv OA$. The **conclusion**: $AE \equiv CB$.

$A = (0,0)$, $C = (u_1,0)$, $B = (u_2,u_3)$, $O = (x_2,x_1)$, $D = (u_2,x_3)$, $E = (x_5,x_4)$.

The **nondegenerate** conditions: Points A, B and C are not collinear; Line CA is non-isotropic; Line OD is non-isotropic. In addition, the following nondegenerate conditions, which come from reducibility and have been detected by our prover, should be also added: $E \neq D$; $D \neq B$.

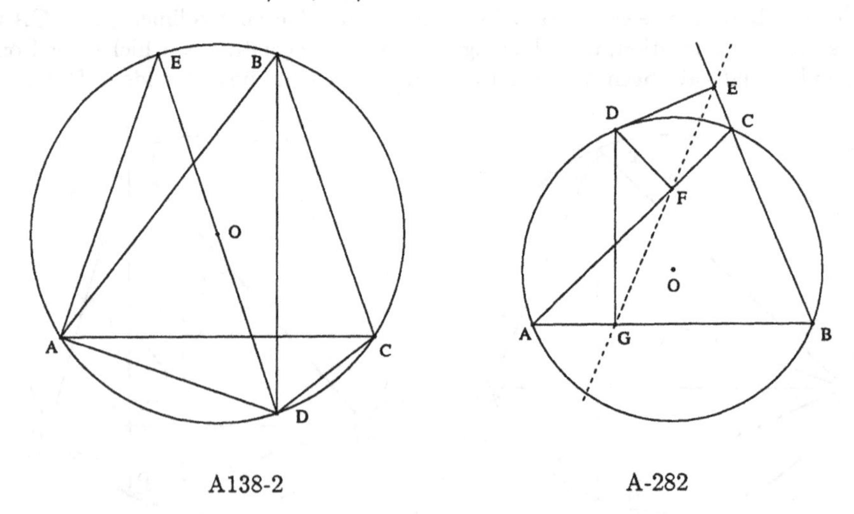

A138-2 A-282

Example 288 (A-282: 1.22s, 9). *Simson's Theorem.*

Points A, B, C are arbitrarily chosen. Points O, D, E, F, G are constructed (in order) as follows: $OA \equiv OB$; $OA \equiv OC$; $DO \equiv OA$; $ED \perp BC$; E is on line BC; $FD \perp AC$; F is on line AC; $GD \perp AB$; G is on line AB. The **conclusion**: Points E, F and G are collinear.

$A = (0,0)$, $B = (u_1,0)$, $C = (u_2,u_3)$, $O = (x_2,x_1)$, $D = (x_3,u_4)$, $E = (x_5,x_4)$, $F = (x_7,x_6)$, $G = (x_3,0)$.

The **nondegenerate** conditions: Points A, C and B are not collinear; Line BC is non-isotropic; Line AC is non-isotropic; Line AB is non-isotropic.

Example 289 (A-284: 3.53s, 55). *If three chords drawn through a point of a circle are taken for diameters of three circles, these circles intersect, in pairs, in three new points, which are collinear.*

Points A, B, C are arbitrarily chosen. Points O, D, M_1, M_2, M_3, E, F, G are constructed (in order) as follows: $OA \equiv OB$; $OA \equiv OC$; $DO \equiv OA$; M_1 is the midpoint of A and D; M_2 is the midpoint of B and D; M_3 is the midpoint of C and D; $EM_2 \equiv M_2B$; $EM_1 \equiv M_1A$; $FM_3 \equiv M_3C$; $FM_1 \equiv M_1A$; $GM_3 \equiv M_3C$; $GM_2 \equiv M_2B$. The **conclusion**: Points E, F and G are collinear.

$A = (0,0)$, $B = (u_1,0)$, $C = (u_2,u_3)$, $O = (x_2,x_1)$, $D = (x_3,u_4)$, $M_1 = (x_4,x_5)$, $M_2 = (x_6,x_7)$, $M_3 = (x_8,x_9)$, $E = (x_{11},x_{10})$, $F = (x_{13},x_{12})$, $G = (x_{15},x_{14})$.

The **nondegenerate** conditions: Points A, C and B are not collinear; Line M_1M_2 is non-isotropic; Line M_1M_3 is non-isotropic; Line M_2M_3 is non-isotropic. In addition, the following nondegenerate conditions, which come from reducibility and have been detected by our prover, should be also added: $G \neq D$; $F \neq D$; $E \neq D$.

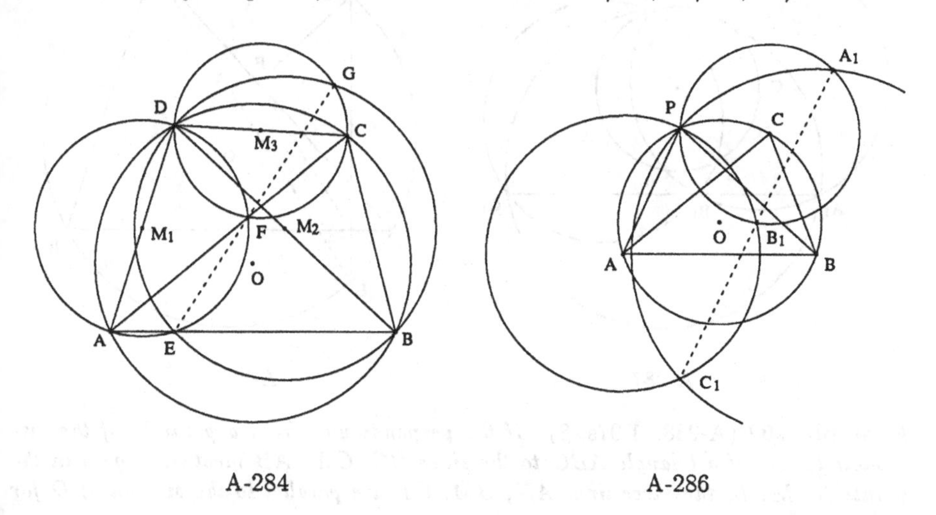

A-284 A-286

Example 290 (A-286: 3.13s, 17; 1.75s, 33). *If three circles pass through the same point of the circumcircle of the triangle of their centers, these circles intersect, in pairs, in three collinear points.*

Points A, B, C are arbitrarily chosen. Points O, P, A_1, B_1, C_1 are constructed (in order) as follows: $OA \equiv OC$; $OA \equiv OB$; $PO \equiv OA$; $A_1B \equiv BP$; $A_1C \equiv CP$; $B_1C \equiv CP$; $B_1A \equiv AP$; $C_1B \equiv BP$; $C_1A \equiv AP$. The **conclusion**: Points A_1, B_1 and C_1 are collinear.

$A = (0,0)$, $B = (u_1,0)$, $C = (u_2,u_3)$, $O = (x_2,x_1)$, $P = (x_3,u_4)$, $A_1 = (x_5,x_4)$, $B_1 = (x_7,x_6)$, $C_1 = (x_9,x_8)$.

The **nondegenerate** conditions: Points A, B and C are not collinear; Line CB is non-isotropic; Line AC is non-isotropic; Line AB is non-isotropic. In addition, the following nondegenerate conditions, which come from reducibility and have been detected by our prover, should be also added: $C_1 \neq P$; $B_1 \neq P$; $A_1 \neq P$.

Example 291 (A-287: 2.12s, 92; 1.48s, 76). *If three circles having a point in common intersect in pairs in three collinear points, their common point is concyclic with their centers.*

Points A_1, B_1, P are arbitrarily chosen. Points C_1, A, B, C are constructed (in order) as follows: C_1 is on line B_1A_1; $AC_1 \equiv AP$; $AC_1 \equiv AB_1$; $BA_1 \equiv BP$; $BA_1 \equiv BC_1$; $CP \equiv CA_1$; $CA_1 \equiv CB_1$. The **conclusion**: Points A, B, C and P are on the same circle.

$A_1 = (0,0)$, $B_1 = (u_1,0)$, $P = (u_2,u_3)$, $C_1 = (u_4,0)$, $A = (x_2,x_1)$, $B = (x_4,x_3)$, $C = (x_6,x_5)$.

The **nondegenerate** conditions: $B_1 \neq A_1$; Points C_1, B_1 and P are not collinear; Points A_1, C_1 and P are not collinear; Points A_1, B_1 and P are not collinear.

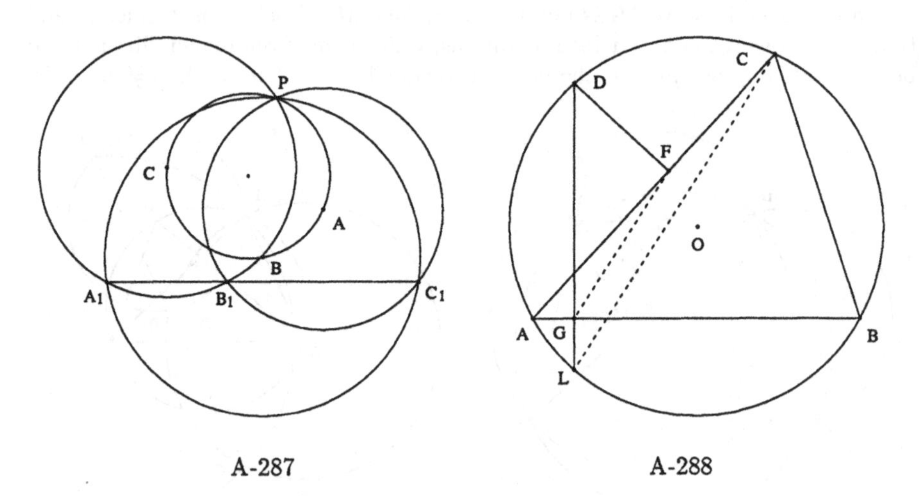

A-287 A-288

Example 292 (A-288: 1.07s, 8). *If the perpendicular from a point D of the circumcircle (O) of a triangle ABC to the sides BC, CA,, AB meet (O) again in the points N, M, L, the three lines AN, BM, CL are parallel to the simson of D for ABC.*

Points A, B, C are arbitrarily chosen. Points O, D, F, G, L are constructed (in order) as follows: $OA \equiv OC$; $OA \equiv OB$; $DO \equiv OA$; $FD \perp CA$; F is on line AC; $GD \perp BA$; G is on line AB; $LO \equiv OA$; L is on line DG. The **conclusion**: $CL \parallel FG$.

$A = (0,0)$, $B = (u_1,0)$, $C = (u_2,u_3)$, $O = (x_2,x_1)$, $D = (x_3,u_4)$, $F = (x_5,x_4)$, $G = (x_3,0)$, $L = (x_3,x_6)$.

The **nondegenerate** conditions: Points A, B and C are not collinear; Line AC is non-isotropic; Line AB is non-isotropic; Line DG is non-isotropic. In addition, the following nondegenerate conditions, which come from reducibility and have been detected by our prover, should be also added: $L \neq D$.

Example 293 (A-290: 1.6s, 15). *The Simson line bisects the line joining its pole to the orthocenter of the triangle.*

Points A, B, C are arbitrarily chosen. Points O, D, F, G, H, N are constructed (in order) as follows: $OA \equiv OC$; $OA \equiv OB$; $DO \equiv OA$; F is on line AC; $FD \perp AC$; G is on line AB; $GD \perp AB$; $HB \perp AC$; $HC \perp AB$; N is on line GF; N is on line DH. The **conclusion**: N is the midpoint of D and H.

$A = (0,0)$, $B = (u_1,0)$, $C = (u_2,u_3)$, $O = (x_2,x_1)$, $D = (x_3,u_4)$, $F = (x_5,x_4)$, $G = (x_3,0)$, $H = (u_2,x_6)$, $N = (x_8,x_7)$.

The **nondegenerate** conditions: Points A, B and C are not collinear; Line AC is non-isotropic; Line AB is non-isotropic; Points A, B and C are not collinear; Line DH intersects line GF.

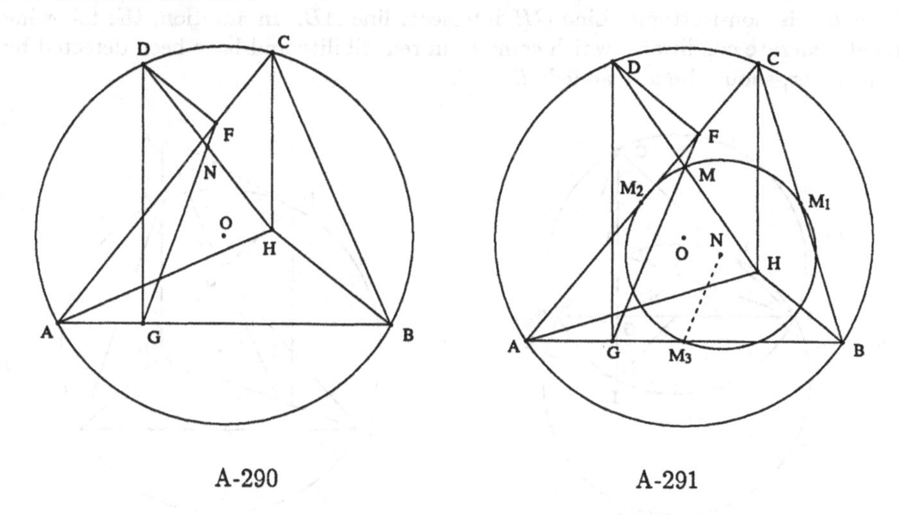

A-290 A-291

Example 294 (A-291: 4.53s, 119; 2.28s, 37). *The midpoint of DH lies on the nine-point circle of ABC.*

Points A, B, C are arbitrarily chosen. Points O, D, F, G, H, M, M_3, M_2, M_1, N are constructed (in order) as follows: $OA \equiv OC$; $OA \equiv OB$; $DO \equiv OA$; F is on line AC; $FD \perp AC$; G is on line AB; $GD \perp AB$; $HB \perp AC$; $HC \perp AB$; M is on line GF; M is on line DH; M_3 is the midpoint of A and B; M_2 is the midpoint of A and C; M_1 is the midpoint of B and C; $NM_3 \equiv NM_2$; $NM_3 \equiv NM_1$. The **conclusion**: $NM_3 \equiv NM$.

$A = (0,0)$, $B = (u_1,0)$, $C = (u_2,u_3)$, $O = (x_2,x_1)$, $D = (x_3,u_4)$, $F = (x_5,x_4)$, $G = (x_3,0)$, $H = (u_2,x_6)$, $M = (x_8,x_7)$, $M_3 = (x_9,0)$, $M_2 = (x_{10},x_{11})$, $M_1 = (x_{12},x_{13})$, $N = (x_{15},x_{14})$.

The **nondegenerate** conditions: Points A, B and C are not collinear; Line AC is non-isotropic; Line AB is non-isotropic; Points A, B and C are not collinear; Line DH intersects line GF; Points M_3, M_1 and M_2 are not collinear.

Example 295 (A145-2: 2.98s, 57; 2.95s, 100). *Show that the simson of the point where an altitude cuts the circumcircle again passes through the foot of the altitude and is antiparallel to the corresponding side of the triangle with respect to the other two sides.*

Points A, B, C are arbitrarily chosen. Points O, H, D, E, F, G are constructed (in order) as follows: $OA \equiv OC$; $OA \equiv OB$; $HC \perp AB$; $HA \perp BC$; D is on line CH; $DO \equiv OA$; E is on line AC; $ED \perp AC$; F is on line BC; $FD \perp BC$; G is on line AB; G is on line CH. The **conclusions**: (1) Points E, G and F are collinear; (2) Points E, F, A and B are on the same circle.

$A = (0,0)$, $B = (u_1,0)$, $C = (u_2,u_3)$, $O = (x_2,x_1)$, $H = (u_2,x_3)$, $D = (u_2,x_4)$, $E = (x_6,x_5)$, $F = (x_8,x_7)$, $G = (u_2,0)$.

The **nondegenerate** conditions: Points A, B and C are not collinear; Points B, C and A are not collinear; Line CH is non-isotropic; Line AC is non-isotropic; Line BC is non-isotropic; Line CH intersects line AB. In addition, the following nondegenerate conditions, which come from reducibility and have been detected by our prover, should be also added: $D \neq C$.

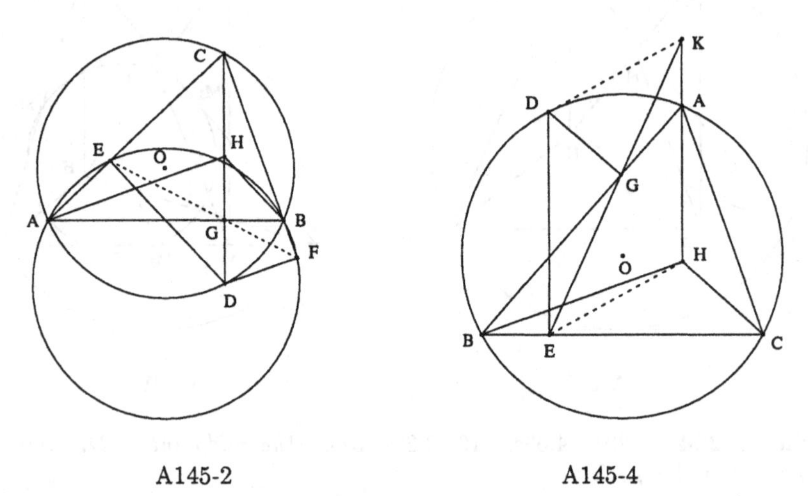

A145-2 A145-4

Example 296 (A145-4: 1.1s, 15). *If the Simson line $D(ABC)$ meets BC in E and the altitude from A in K, show that the line DK is parallel to EH, where H is the orthocenter of ABC.*

Points B, C, A are arbitrarily chosen. Points O, D, E, G, H, K are constructed (in order) as follows: $OB \equiv OA$; $OB \equiv OC$; $DO \equiv OB$; $ED \perp CB$; E is on line BC; $GD \perp BA$; G is on line AB; $HB \perp AC$; $HA \perp BC$; K is on line EG; K is on line AH. The **conclusion**: $DK \parallel EH$.

$B = (0,0)$, $C = (u_1,0)$, $A = (u_2,u_3)$, $O = (x_2,x_1)$, $D = (x_3,u_4)$, $E = (x_3,0)$, $G = (x_5,x_4)$, $H = (u_2,x_6)$, $K = (u_2,x_7)$.

The **nondegenerate** conditions: Points B, C and A are not collinear; Line BC is non-isotropic; Line AB is non-isotropic; Points B, C and A are not collinear; Line AH intersects line EG.

Example 297 (A145-5: 1.22s, 10). *Let D be a point on the circumcircle of triangle ABC. If line DA is parallel to BC, show that the Simson line $D(ABC)$ is parallel to the circumradius OA.*

Points A, B, C are arbitrarily chosen. Points O, D, F, G are constructed (in order) as follows: $OA \equiv OC$; $OA \equiv OB$; $DA \parallel BC$; $DO \equiv OA$; $FD \perp CA$; F is on line AC; $GD \perp BA$; G is on line AB. The **conclusion**: $OA \parallel GF$.

$A = (0,0)$, $B = (u_1,0)$, $C = (u_2,u_3)$, $O = (x_2,x_1)$, $D = (x_4,x_3)$, $F = (x_6,x_5)$, $G = (x_4,0)$.

The **nondegenerate** conditions: Points A, B and C are not collinear; Line BC is non-isotropic; Line AC is non-isotropic; Line AB is non-isotropic. In addition, the following nondegenerate conditions, which come from reducibility and have been detected by our prover, should be also added: $D \neq A$.

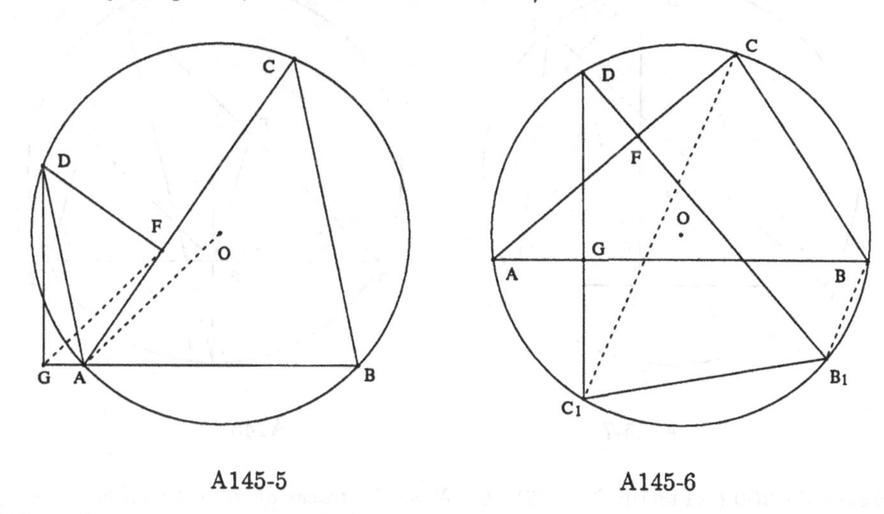

A145-5 A145-6

Example 298 (A145-6: 4.27s, 160). *If the perpendiculars dropped from a point D of the circumcircle (O) of the triangle ABC upon the sides, meet (O) in the points A_1, B_1, C_1, show that the two triangle ABC, $A_1B_1C_1$ are congruent and symmetrical with respect to an axis.*

Points A, B, C are arbitrarily chosen. Points O, D, F, G, C_1, B_1 are constructed (in order) as follows: $OA \equiv OC$; $OA \equiv OB$; $DO \equiv OA$; $FD \perp CA$; F is on line AC; $GD \perp BA$; G is on line AB; $C_1O \equiv OA$; C_1 is on line DG; $B_1O \equiv OA$; B_1 is on line DF. The **conclusion**: $C_1C \parallel B_1B$.

$A = (0,0)$, $B = (u_1,0)$, $C = (u_2,u_3)$, $O = (x_2,x_1)$, $D = (x_3,u_4)$, $F = (x_5,x_4)$, $G = (x_3,0)$, $C_1 = (x_3,x_6)$, $B_1 = (x_8,x_7)$.

The **nondegenerate** conditions: Points A, B and C are not collinear; Line AC is non-isotropic; Line AB is non-isotropic; Line DG is non-isotropic; Line DF is non-isotropic. In addition, the following nondegenerate conditions, which come from reducibility and have been detected by our prover, should be also added: $B_1 \neq D$; $C_1 \neq D$.

Example 299 (A145-7: 1.73s, 52). *If E, F, G are the feet of the perpendiculars from a point D of the circumcircle of a triangle ABC upon its sides BC, CA, AB, prove that the triangle DFG, DAC are similar.*

Points A, B, C are arbitrarily chosen. Points O, D, F, G are constructed (in order) as follows: $OA \equiv OC$; $OA \equiv OB$; $DO \equiv OA$; F is on line AC; $FD \perp AC$; G is on line AB; $GD \perp AB$. The **conclusion**: $DB \cdot DF = DC \cdot DG$.

$A = (0,0)$, $B = (u_1,0)$, $C = (u_2,u_3)$, $O = (x_2,x_1)$, $D = (x_3,u_4)$, $F = (x_5,x_4)$, $G = (x_3,0)$.

The **nondegenerate** conditions: Points A, B and C are not collinear; Line AC is non-isotropic; Line AB is non-isotropic.

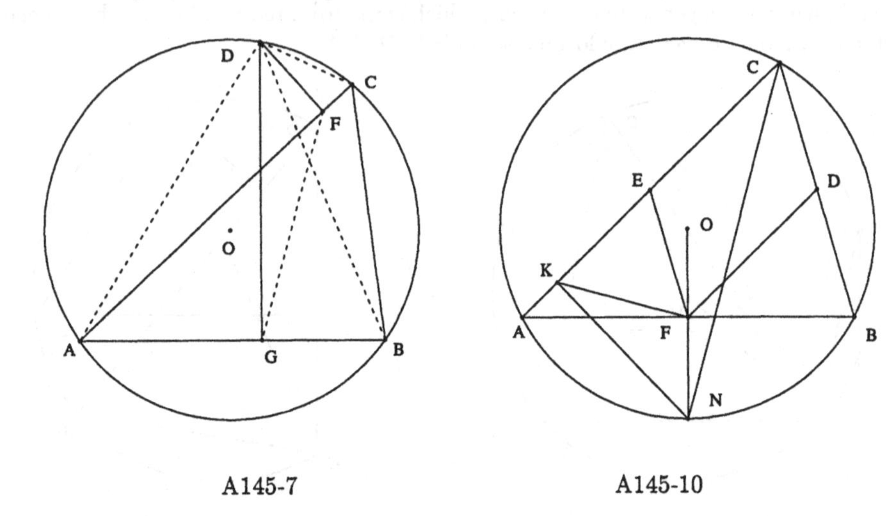

A145-7 A145-10

Example 300 (A145-10: 2.1s, 26). *Let N be the traces of the internal bisectors of the triangle ABC on the circumscribed circle (O). Show that the Simson line of N is the external bisector of the medial triangle of ABC.*

Points A, B, C are arbitrarily chosen. Points O, D, E, F, N, K are constructed (in order) as follows: $OA \equiv OC$; $OA \equiv OB$; D is the midpoint of B and C; E is the midpoint of A and C; F is the midpoint of B and A; $NO \equiv OA$; N is on line OF; K is on line AC; $KN \perp AC$. The **conclusion**: $\tan(EFK) = \tan(KFD)$.

$A = (0,0)$, $B = (u_1,0)$, $C = (u_2,u_3)$, $O = (x_2,x_1)$, $D = (x_3,x_4)$, $E = (x_5,x_6)$, $F = (x_7,0)$, $N = (x_9,x_8)$, $K = (x_{11},x_{10})$.

The **nondegenerate** conditions: Points A, B and C are not collinear; Line OF is non-isotropic; Line AC is non-isotropic.

Example 301 (A145-11: 1.42s, 9). *Show that the symmetrics, with respect to the sides of a triangle, of a point on its circumcircle lie on a line passing through the orthocenter of the triangle.*

Points A, B, C are arbitrarily chosen. Points O, D, F, G, G_1, F_1, H are constructed (in order) as follows: $OA \equiv OC$; $OA \equiv OB$; $DO \equiv OA$; $FD \perp CA$; F is on line AC; $GD \perp BA$; G is on line AB; G is the midpoint of D and G_1; F is the midpoint of D and F_1; $HA \perp BC$; $HC \perp AB$. The **conclusion**: Points F_1, G_1 and H are collinear.

$A = (0,0)$, $B = (u_1,0)$, $C = (u_2,u_3)$, $O = (x_2,x_1)$, $D = (x_3,u_4)$, $F = (x_5,x_4)$, $G = (x_3,0)$, $G_1 = (x_3,x_6)$, $F_1 = (x_7,x_8)$, $H = (u_2,x_9)$.

The **nondegenerate** conditions: Points A, B and C are not collinear; Line AC is non-isotropic; Line AB is non-isotropic; Points A, B and C are not collinear.

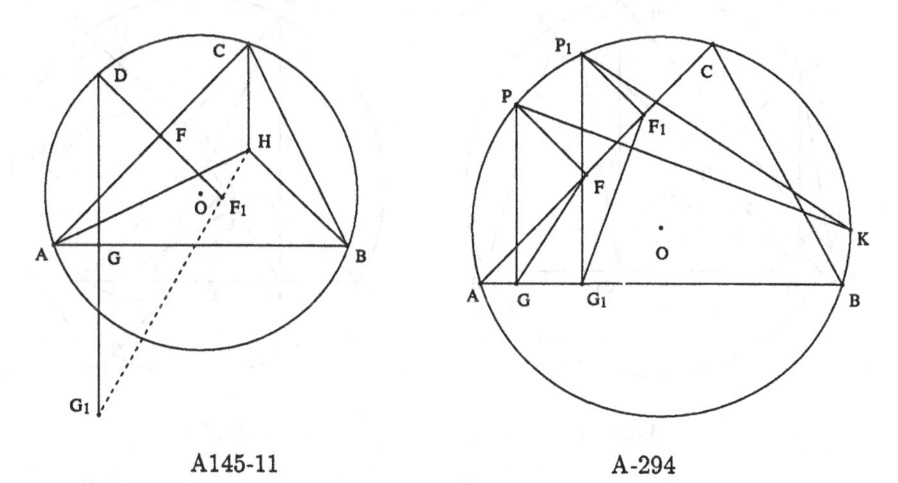

A145-11 A-294

Example 302 (A-294: 48.3s, 1402). *The perpendiculars through the points P, P_1 to the Simson lines $P_1(ABC)$, $P(ABC)$, respectively, intersect on the circumcircle of the triangle ABC.*

Points A, B, C are arbitrarily chosen. Points O, P, P_1, G, G_1, F, F_1, K are constructed (in order) as follows: $OA \equiv OC$; $OA \equiv OB$; $PO \equiv OA$; $P_1O \equiv OA$; G is on line AB; $GP \perp AB$; G_1 is on line AB; $G_1P_1 \perp AB$; F is on line AC; $FP \perp AC$; F_1 is on line AC; $F_1P_1 \perp AC$; $KP_1 \perp FG$; $KP \perp F_1G_1$. The **conclusion**: $OA \equiv OK$.

$A = (0,0)$, $B = (u_1,0)$, $C = (u_2,u_3)$, $O = (x_2,x_1)$, $P = (x_3,u_4)$, $P_1 = (x_4,u_5)$, $G = (x_3,0)$, $G_1 = (x_4,0)$, $F = (x_6,x_5)$, $F_1 = (x_8,x_7)$, $K = (x_{10},x_9)$.

The **nondegenerate** conditions: Points A, B and C are not collinear; Line AB is non-isotropic; Line AB is non-isotropic; Line AC is non-isotropic; Line AC is non-isotropic; Line F_1G_1 intersects line FG.

Example 303 (A-294-a: 14.52s, 214). *The parallels through the points P, P_1 to the Simson lines $P_1(ABC)$, $P(ABC)$, respectively, intersect on the circumcircle of the triangle ABC.*

Points A, B, C are arbitrarily chosen. Points O, P, P_1, G, G_1, F, F_1, K are constructed (in order) as follows: $OA \equiv OC$; $OA \equiv OB$; $PO \equiv OA$; $P_1O \equiv OA$; G is on line AB; $GP \perp AB$; G_1 is on line AB; $G_1P_1 \perp AB$; F is on line AC; $FP \perp AC$; F_1 is on line AC; $F_1P_1 \perp AC$; $KP_1 \parallel FG$; $KP \parallel F_1G_1$. The **conclusion**: $OA \equiv OK$.

$A = (0,0)$, $B = (u_1,0)$, $C = (u_2,u_3)$, $O = (x_2,x_1)$, $P = (x_3,u_4)$, $P_1 = (x_4,u_5)$, $G = (x_3,0)$, $G_1 = (x_4,0)$, $F = (x_6,x_5)$, $F_1 = (x_8,x_7)$, $K = (x_{10},x_9)$.

The **nondegenerate** conditions: Points A, B and C are not collinear; Line AB is non-isotropic; Line AB is non-isotropic; Line AC is non-isotropic; Line AC is non-isotropic; Line F_1G_1 intersects line FG.

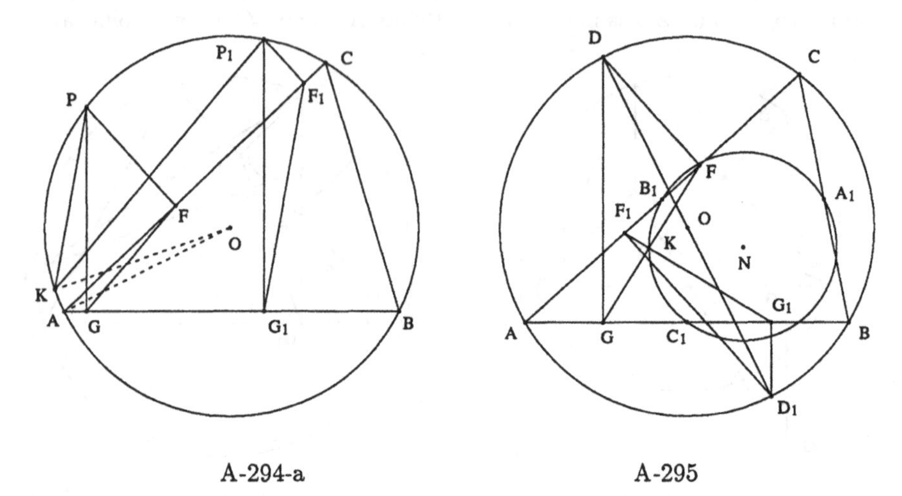

A-294-a A-295

Example 304 (A-295: 10.68s, 303). *The Simson lines of two diametrically opposite points are perpendicular at a point of the 9-point circle.*

Points A, B, C are arbitrarily chosen. Points O, D, D_1, F, G, F_1, G_1, K, A_1, B_1, C_1, N are constructed (in order) as follows: $OA \equiv OC$; $OA \equiv OB$; $DO \equiv OA$; O is the midpoint of D and D_1; $FD \perp CA$; F is on line AC; $GD \perp BA$; G is on line AB; $F_1D_1 \perp CA$; F_1 is on line AC; $G_1D_1 \perp BA$; G_1 is on line AB; K is on line G_1F_1; K is on line GF; A_1 is the midpoint of C and B; B_1 is the midpoint of A and C; C_1 is the midpoint of A and B; $NC_1 \equiv NB_1$; $NC_1 \equiv NA_1$. The **conclusions**: (1) $NC_1 \equiv NK$; (2) $FG \perp F_1G_1$.

$A = (0,0)$, $B = (u_1,0)$, $C = (u_2,u_3)$, $O = (x_2,x_1)$, $D = (x_3,u_4)$, $D_1 = (x_4,x_5)$, $F = (x_7,x_6)$, $G = (x_3,0)$, $F_1 = (x_9,x_8)$, $G_1 = (x_4,0)$, $K = (x_{11},x_{10})$, $A_1 = (x_{12},x_{13})$, $B_1 = (x_{14},x_{15})$, $C_1 = (x_{16},0)$, $N = (x_{18},x_{17})$.

The **nondegenerate** conditions: Points A, B and C are not collinear; Line AC is non-isotropic; Line AB is non-isotropic; Line AC is non-isotropic; Line AB is non-isotropic; Line GF intersects line G_1F_1; Points C_1, A_1 and B_1 are not collinear.

Example 305 (A-293: 47.45s, 1199). *The angle formed by the Simson lines of two points for the same triangle is measured by half the arc between the two points.*

Points A, B, C are arbitrarily chosen. Points O, P, P_1, M, G, G_1, F, F_1, K are constructed (in order) as follows: $OA \equiv OC$; $OA \equiv OB$; $PO \equiv OA$; $P_1O \equiv OA$; M is the midpoint of P_1 and P; G is on line AB; $GP \perp AB$; G_1 is on line AB; $G_1P_1 \perp AB$; F is on line AC; $FP \perp AC$; F_1 is on line AC; $F_1P_1 \perp AC$; K is on line F_1G_1; K is on line FG. The **conclusion**: $\tan(POM) = \tan(F_1KF)$.

$A = (0,0)$, $B = (u_1,0)$, $C = (u_2,u_3)$, $O = (x_2,x_1)$, $P = (x_3,u_4)$, $P_1 = (x_4,u_5)$, $M = (x_5,x_6)$, $G = (x_3,0)$, $G_1 = (x_4,0)$, $F = (x_8,x_7)$, $F_1 = (x_{10},x_9)$, $K =$

(x_{12}, x_{11}).

The **nondegenerate** conditions: Points A, B and C are not collinear; Line AB is non-isotropic; Line AB is non-isotropic; Line AC is non-isotropic; Line AC is non-isotropic; Line FG intersects line F_1G_1.

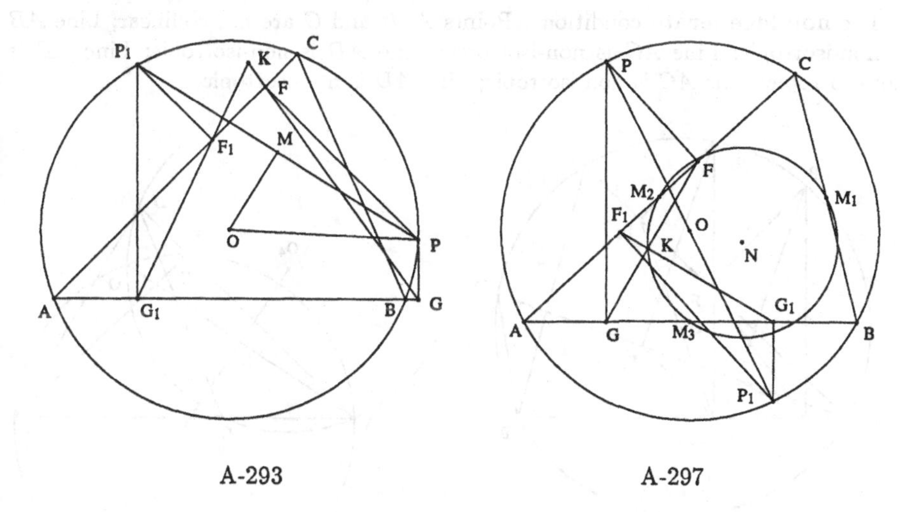

A-293 A-297

Example 306 (A-297: 12.82s, 283). *the point of intersection of the pairs of Simson lines corresponding to pairs of diametrically opposite points on the circumcircle of a triangle is on the nine-point circle of the triangle.*

Points A, B, C are arbitrarily chosen. Points O, M_1, M_2, M_3, N, P, P_1, G, G_1, F, F_1, K are constructed (in order) as follows: $OA \equiv OC$; $OA \equiv OB$; M_1 is the midpoint of B and C; M_2 is the midpoint of A and C; M_3 is the midpoint of A and B; $NM_3 \equiv NM_2$; $NM_3 \equiv NM_1$; $PO \equiv OA$; O is the midpoint of P and P_1; G is on line AB; $GP \perp AB$; G_1 is on line AB; $G_1P_1 \perp AB$; F is on line AC; $FP \perp AC$; F_1 is on line AC; $F_1P_1 \perp AC$; K is on line F_1G_1; K is on line FG. The **conclusion**: $NM_3 \equiv NK$.

$A = (0,0)$, $B = (u_1,0)$, $C = (u_2,u_3)$, $O = (x_2,x_1)$, $M_1 = (x_3,x_4)$, $M_2 = (x_5,x_6)$, $M_3 = (x_7,0)$, $N = (x_9,x_8)$, $P = (x_{10},u_4)$, $P_1 = (x_{11},x_{12})$, $G = (x_{10},0)$, $G_1 = (x_{11},0)$, $F = (x_{14},x_{13})$, $F_1 = (x_{16},x_{15})$, $K = (x_{18},x_{17})$.

The **nondegenerate** conditions: Points A, B and C are not collinear; Points M_3, M_1 and M_2 are not collinear; Line AB is non-isotropic; Line AB is non-isotropic; Line AC is non-isotropic; Line AC is non-isotropic; Line FG intersects line F_1G_1.

Example 307 (A-298: 10.07s, 302). *The angle formed by the two Simson lines of the same points with respect to two triangles inscribed in the same circle is the same for all positions of the point on the circle.*

Points A, B, C are arbitrarily chosen. Points O, D, P, P_1, G, F, F_1, G_1, F_2, F_3 are constructed (in order) as follows: $OA \equiv OC$; $OA \equiv OB$; $DO \equiv OA$; $PO \equiv OA$; $P_1O \equiv OA$; G is on line AB; $GP \perp AB$; F is on line AC; $FP \perp AC$; F_1 is on line AD; $F_1P \perp AD$; G_1 is on line AB; $G_1P_1 \perp AB$; F_2 is on line AC; $F_2P_1 \perp AC$; F_3

is on line AD; $F_3P_1 \perp AD$. The **conclusion**: $\tan(FGF_1) = \tan(F_2G_1F_3)$.

$A = (0,0)$, $B = (u_1,0)$, $C = (u_2,u_3)$, $O = (x_2,x_1)$, $D = (x_3,u_4)$, $P = (x_4,u_5)$, $P_1 = (x_5,u_6)$, $G = (x_4,0)$, $F = (x_7,x_6)$, $F_1 = (x_9,x_8)$, $G_1 = (x_5,0)$, $F_2 = (x_{11},x_{10})$, $F_3 = (x_{13},x_{12})$.

The **nondegenerate** conditions: Points A, B and C are not collinear; Line AB is non-isotropic; Line AC is non-isotropic; Line AD is non-isotropic; Line AB is non-isotropic; Line AC is non-isotropic; Line AD is non-isotropic.

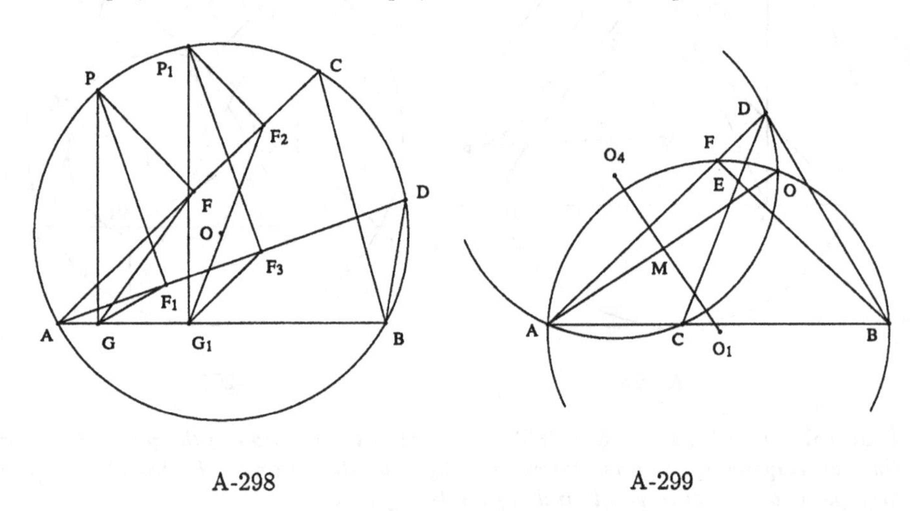

A-298 A-299

Example 308 (A-299: 49.1s, 1779). *The four circumcircles of the four triangles determined by four given straight lines, taken three at a time, have a point in common (the Miquel point).*

Points A, B, D are arbitrarily chosen. Points C, E, F, O_1, O_4, M, O are constructed (in order) as follows: C is on line AB; E is on line CD; F is on line DA; F is on line EB; $O_1A \equiv O_1F$; $O_1A \equiv O_1B$; $O_4A \equiv O_4D$; $O_4A \equiv O_4C$; $MA \perp O_1O_4$; M is on line O_1O_4; M is the midpoint of A and O. The **conclusion**: Points B, C, E and O are on the same circle.

$A = (0,0)$, $B = (u_1,0)$, $D = (u_2,u_3)$, $C = (u_4,0)$, $E = (x_1,u_5)$, $F = (x_3,x_2)$, $O_1 = (x_5,x_4)$, $O_4 = (x_7,x_6)$, $M = (x_9,x_8)$, $O = (x_{10},x_{11})$.

The **nondegenerate** conditions: $A \neq B$; $C \neq D$; Line EB intersects line DA; Points A, B and F are not collinear; Points A, C and D are not collinear; Line O_1O_4 is non-isotropic.

Remark. The point O is called the Miquel point of the four lines p, q, r, s.

Example 309 (A-302: 871.48s, 7133; 63.83s, 2654). *The center of the four circles (pqr), (qrs), (rsp), (spq) and the Miquel point lie on the same circle.*

Points A, B, D are arbitrarily chosen. Points C, E, F, O_1, O_2, O_3, O_4 are constructed (in order) as follows: C is on line AB; E is on line CD; F is on line DA; F is on line EB; $O_1A \equiv O_1F$; $O_1A \equiv O_1B$; $O_2B \equiv O_2E$; $O_2B \equiv O_2C$;

$O_3D \equiv O_3E$; $O_3D \equiv O_3F$; $O_4A \equiv O_4D$; $O_4A \equiv O_4C$. The **conclusion**: Points O_1, O_2, O_3 and O_4 are on the same circle.

$A = (0,0)$, $B = (u_1, 0)$, $D = (u_2, u_3)$, $C = (u_4, 0)$, $E = (x_1, u_5)$, $F = (x_3, x_2)$, $O_1 = (x_5, x_4)$, $O_2 = (x_7, x_6)$, $O_3 = (x_9, x_8)$, $O_4 = (x_{11}, x_{10})$.

The **nondegenerate** conditions: $A \neq B$; $C \neq D$; Line EB intersects line DA; Points A, B and F are not collinear; Points B, C and E are not collinear; Points D, F and E are not collinear; Points A, C and D are not collinear.

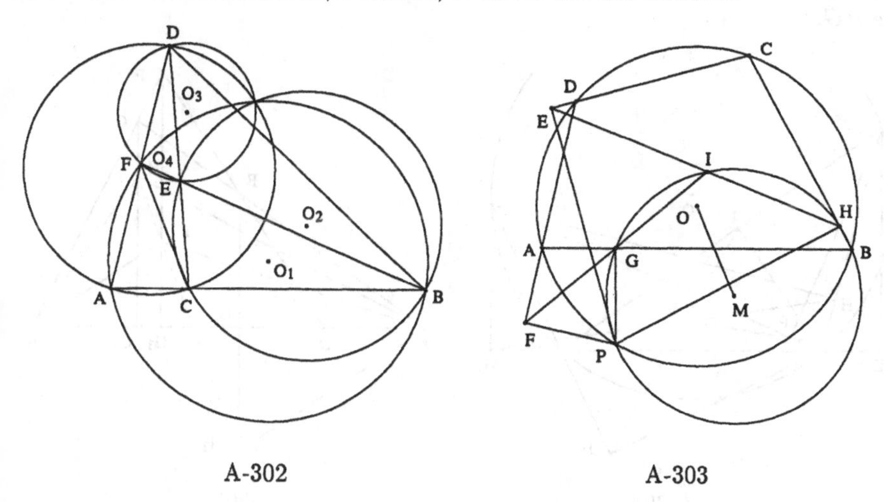

A-302 A-303

Example 310 (A-303: 56.4s, 2251). *The four Simson lines of a point of a circle for the four triangles determined by the vertices of a quadrilateral inscribed in that circle, admit the point considered for their Miquel point.*

Points A, B, C are arbitrarily chosen. Points O, D, P, E, F, G, H, I, M are constructed (in order) as follows: $OA \equiv OC$; $OA \equiv OB$; $DO \equiv OA$; $PO \equiv OA$; E is on line CD; $EP \perp CD$; F is on line DA; $FP \perp DA$; G is on line AB; $GP \perp AB$; H is on line BC; $HP \perp BC$; I is on line EH; I is on line GF; $MH \equiv MI$; $MG \equiv MI$. The **conclusion**: $MG \equiv MP$.

$A = (0,0)$, $B = (u_1, 0)$, $C = (u_2, u_3)$, $O = (x_2, x_1)$, $D = (x_3, u_4)$, $P = (x_4, u_5)$, $E = (x_6, x_5)$, $F = (x_8, x_7)$, $G = (x_4, 0)$, $H = (x_{10}, x_9)$, $I = (x_{12}, x_{11})$, $M = (x_{14}, x_{13})$.

The **nondegenerate** conditions: Points A, B and C are not collinear; Line CD is non-isotropic; Line DA is non-isotropic; Line AB is non-isotropic; Line BC is non-isotropic; Line GF intersects line EH; Points G, I and H are not collinear.

Example 311 (A-304: 12.0s, 118; 11.13s, 122). *The four Simson lines of four points of a circle, each taken for the triangle formed by the remaining three points, are concurrent.*

Points A, B, C are arbitrarily chosen. Points O, D, E, F, G, H, I, J, K are constructed (in order) as follows: $OA \equiv OC$; $OA \equiv OB$; $DO \equiv OA$; E is on line AB; $ED \perp AB$; F is on line AC; $FD \perp AC$; G is on line AB; $GC \perp AB$; H is on line AD; $HC \perp AD$; I is on line DB; $IA \perp DB$; J is on line BC; $JA \perp BC$; K is on line HG; K is on line EF. The **conclusion**: Points I, J and K are collinear.

$A = (0,0)$, $B = (u_1,0)$, $C = (u_2,u_3)$, $O = (x_2,x_1)$, $D = (x_3,u_4)$, $E = (x_3,0)$, $F = (x_5,x_4)$, $G = (u_2,0)$, $H = (x_7,x_6)$, $I = (x_9,x_8)$, $J = (x_{11},x_{10})$, $K = (x_{13},x_{12})$.

The **nondegenerate** conditions: Points A, B and C are not collinear; Line AB is non-isotropic; Line AC is non-isotropic; Line AB is non-isotropic; Line AD is non-isotropic; Line DB is non-isotropic; Line BC is non-isotropic; Line EF intersects line HG.

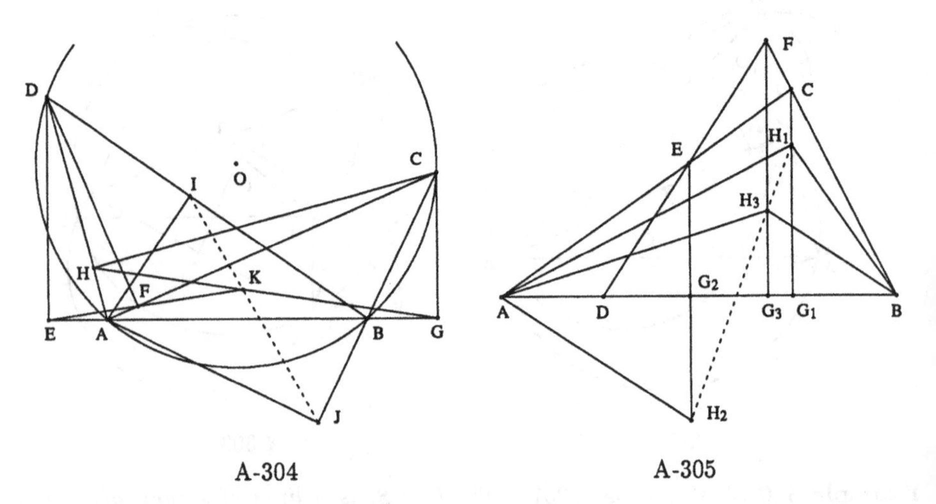

A-304 A-305

Example 312 (A-305: 1.47s, 20). *The orthocenters of the four triangles formed by four lines taken three at a time are collinear.*

Points A, B, C are arbitrarily chosen. Points D, E, F, G_1, H_1, G_2, H_2, G_3, H_3 are constructed (in order) as follows: D is on line AB; E is on line AC; F is on line ED; F is on line CB; G_1 is on line AB; $G_1C \perp AB$; H_1 is on line CG_1; $H_1A \perp BC$; G_2 is on line AB; $G_2E \perp AB$; H_2 is on line EG_2; $H_2A \perp ED$; G_3 is on line AB; $G_3F \perp AB$; H_3 is on line FG_3; $H_3B \perp ED$. The **conclusion**: Points H_1, H_2 and H_3 are collinear.

$A = (0,0)$, $B = (u_1,0)$, $C = (u_2,u_3)$, $D = (u_4,0)$, $E = (x_1,u_5)$, $F = (x_3,x_2)$, $G_1 = (u_2,0)$, $H_1 = (u_2,x_4)$, $G_2 = (x_1,0)$, $H_2 = (x_1,x_5)$, $G_3 = (x_3,0)$, $H_3 = (x_3,x_6)$.

The **nondegenerate** conditions: $A \neq B$; $A \neq C$; Line CB intersects line ED; Line AB is non-isotropic; Line BC is not perpendicular to line CG_1; Line AB is non-isotropic; Line ED is not perpendicular to line EG_2; Line AB is non-isotropic; Line ED is not perpendicular to line FG_3.

Example 313 (A149-1: 4.28s, 38; 2.73s, 39). *If two triangles are inscribed in the same circle and are symmetrical with respect to the center of that circle, show that the two simsons of any point of the circle for these triangles are rectangular.*

Points A, B, C are arbitrarily chosen. Points O, D, A_1, B_1, C_1, G, F, G_1, F_1

are constructed (in order) as follows: $OA \equiv OC$; $OA \equiv OB$; $DO \equiv OA$; O is the midpoint of A and A_1; O is the midpoint of B and B_1; O is the midpoint of C and C_1; G is on line AB; $GD \perp AB$; F is on line AC; $FD \perp AC$; G_1 is on line A_1B_1; $G_1D \perp A_1B_1$; F_1 is on line A_1C_1; $F_1D \perp A_1C_1$. The **conclusion:** $GF \perp G_1F_1$.

$A = (0,0)$, $B = (u_1,0)$, $C = (u_2,u_3)$, $O = (x_2,x_1)$, $D = (x_3,u_4)$, $A_1 = (x_4,x_5)$, $B_1 = (x_6,x_7)$, $C_1 = (x_8,x_9)$, $G = (x_3,0)$, $F = (x_{11},x_{10})$, $G_1 = (x_{13},x_{12})$, $F_1 = (x_{15},x_{14})$.

The **nondegenerate** conditions: Points A, B and C are not collinear; Line AB is non-isotropic; Line AC is non-isotropic; Line A_1B_1 is non-isotropic; Line A_1C_1 is non-isotropic.

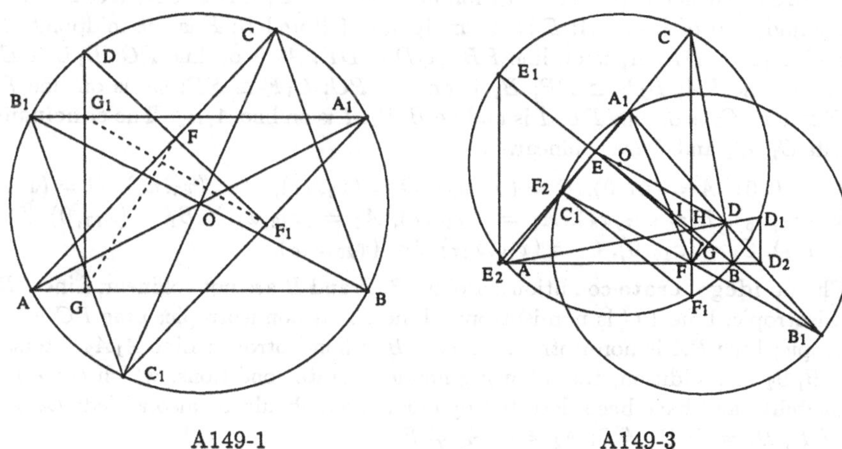

A149-1 A149-3

Example 314 (A149-3: 4960.25s, 23267; 54.33s, 446). *Show that the Simson lines of the three points where the altitudes of a triangle cut the circumcircle again form a triangle homothetic to the orthic triangle, and its circumcenter coincides with orthocenter of the orthic triangle.*

Points A, B, C are arbitrarily chosen. Points O, D, E, F, H, G, D_1, E_1, F_1, D_2, E_2, F_2, A_1, B_1, C_1, I are constructed (in order) as follows: $OA \equiv OC$; $OA \equiv OB$; $DA \perp CB$; D is on line BC; $EB \perp CA$; E is on line AC; $FC \perp BA$; F is on line AB; H is on line EB; H is on line AD; $GF \perp ED$; $GE \perp DF$; D is the midpoint of H and D_1; E is the midpoint of H and E_1; F is the midpoint of H and F_1; D_2 is on line AB; $D_2D_1 \perp AB$; E_2 is on line AB; $E_2E_1 \perp AB$; F_2 is on line AC; $F_2F_1 \perp AC$; A_1 is on line EE_2; A_1 is on line DD_2; B_1 is on line DD_2; B_1 is on line FF_2; C_1 is on line EE_2; C_1 is on line FF_2; I is on line A_1F; I is on line B_1E. The **conclusions**: (1) $\frac{\overline{IB_1}}{\overline{IE}} - \frac{\overline{IA_1}}{\overline{IF}} = 0$; (2) $GA_1 \equiv GB_1$.

$A = (0,0)$, $B = (u_1,0)$, $C = (u_2,u_3)$, $O = (x_2,x_1)$, $D = (x_4,x_3)$, $E = (x_6,x_5)$, $F = (u_2,0)$, $H = (x_8,x_7)$, $G = (x_{10},x_9)$, $D_1 = (x_{11},x_{12})$, $E_1 = (x_{13},x_{14})$, $F_1 = (x_{15},x_{16})$, $D_2 = (x_{11},0)$, $E_2 = (x_{13},0)$, $F_2 = (x_{18},x_{17})$, $A_1 = (x_{20},x_{19})$, $B_1 = (x_{22},x_{21})$, $C_1 = (x_{24},x_{23})$, $I = (x_{26},x_{25})$.

The **nondegenerate** conditions: Points A, B and C are not collinear; Line BC is non-isotropic; Line AC is non-isotropic; Line AB is non-isotropic; Line AD intersects line EB; Points D, F and E are not collinear; Line AB is non-isotropic; Line AB is non-isotropic; Line AC is non-isotropic; Line DD_2 intersects line EE_2; Line FF_2 intersects line DD_2; Line FF_2 intersects line EE_2; Line B_1E intersects line A_1F.

Example 315 (A149-4: 3.9s, 52). *The chords PA, PB, PC of a given circle are the diameters of three circles (PA), (PB), (PC). The circle (PA) meets the lines PB, PC in the points A_1, A_2; the circle (PB) meets PA, PC in B_1, B_2; the circle (PC) meets PA, PB in C_1, C_2. Show that the three lines A_1A_2, B_1B_2, C_1C_2 are concurrent.*

Points P, A, B are arbitrarily chosen. Points O, C, D, E, F, A_1, A_2, B_1, B_2, C_1, C_2, I are constructed (in order) as follows: $OP \equiv OB$; $OP \equiv OA$; $CO \equiv OP$; D is the midpoint of P and A; E is the midpoint of P and B; F is the midpoint of P and C; $A_1D \equiv DP$; A_1 is on line PB; $A_2D \equiv DP$; A_2 is on line PC; $B_1E \equiv EP$; B_1 is on line PA; $B_2E \equiv EP$; B_2 is on line PC; $C_1F \equiv FP$; C_1 is on line PA; $C_2F \equiv FP$; C_2 is on line PB; I is on line B_1B_2; I is on line A_1A_2. The **conclusion**: Points C_1, C_2 and I are collinear.

$P = (0,0)$, $A = (u_1,0)$, $B = (u_2,u_3)$, $O = (x_2,x_1)$, $C = (x_3,u_4)$, $D = (x_4,0)$, $E = (x_5,x_6)$, $F = (x_7,x_8)$, $A_1 = (x_{10},x_9)$, $A_2 = (x_{12},x_{11})$, $B_1 = (x_{13},0)$, $B_2 = (x_{15},x_{14})$, $C_1 = (x_{16},0)$, $C_2 = (x_{18},x_{17})$, $I = (x_{20},x_{19})$.

The **nondegenerate** conditions: Points P, A and B are not collinear; Line PB is non-isotropic; Line PC is non-isotropic; Line PA is non-isotropic; Line PC is non-isotropic; Line PA is non-isotropic; Line PB is non-isotropic; Line A_1A_2 intersects line B_1B_2. In addition, the following nondegenerate conditions, which come from reducibility and have been detected by our prover, should be also added: $C_2 \neq P$; $C_1 \neq P$; $B_2 \neq P$; $B_1 \neq P$; $A_2 \neq P$; $A_1 \neq P$.

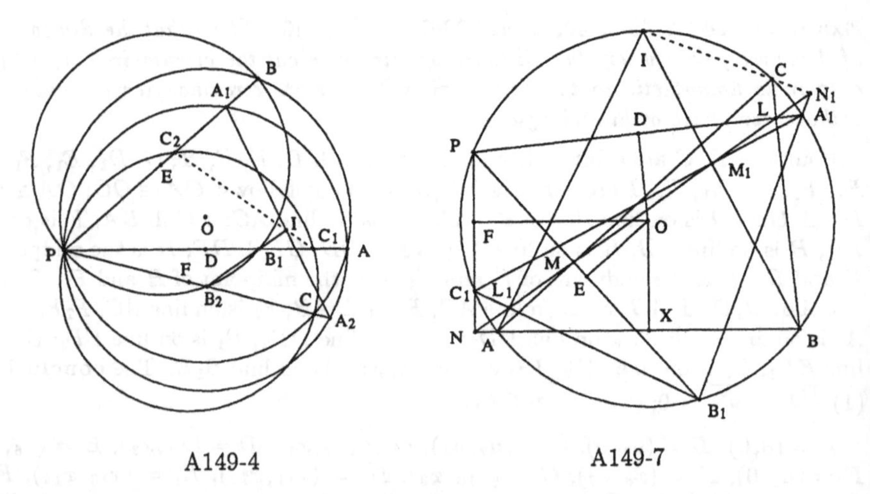

A149-4 A149-7

Example 316 (A149-7: 20058.6s, 49980). *The perpendiculars dropped upon the sides BC, CA, AB of the triangle from a point P on its circumcircle meet these sides in L, M, N and the circle in A_1, B_1, C_1. The Simson line LMN meets B_1C_1,*

C_1A_1, A_1B_1 in L_1, M_1, N_1. Prove that the lines AL_1, BM_1, CN_1 are concurrent.

Points A, B are arbitrarily chosen. Points X, O, C, P, L, M, N, D, E, F, A_1, B_1, C_1, L_1, M_1, N_1, I are constructed (in order) as follows: X is the midpoint of A and B; $OX \perp AB$; $CO \equiv OA$; $PO \equiv OC$; L is on line BC; $LP \perp BC$; M is on line AC; $MP \perp AC$; N is on line AB; $NP \perp AB$; D is on line PL; $DO \perp PL$; E is on line PM; $EO \perp PM$; F is on line PN; $FO \perp PN$; D is the midpoint of P and A_1; E is the midpoint of P and B_1; F is the midpoint of P and C_1; L_1 is on line B_1C_1; L_1 is on line MN; M_1 is on line C_1A_1; M_1 is on line MN; N_1 is on line A_1B_1; N_1 is on line MN; I is on line BM_1; I is on line AL_1. The **conclusion**: Points C, N_1 and I are collinear.

$A = (0,0)$, $B = (u_1,0)$, $X = (x_1,0)$, $O = (x_1,u_2)$, $C = (x_2,u_3)$, $P = (x_3,u_4)$, $L = (x_5,x_4)$, $M = (x_7,x_6)$, $N = (x_3,0)$, $D = (x_9,x_8)$, $E = (x_{11},x_{10})$, $F = (x_3,u_2)$, $A_1 = (x_{12},x_{13})$, $B_1 = (x_{14},x_{15})$, $C_1 = (x_3,x_{16})$, $L_1 = (x_{18},x_{17})$, $M_1 = (x_{20},x_{19})$, $N_1 = (x_{22},x_{21})$, $I = (x_{24},x_{23})$.

The **nondegenerate** conditions: $A \neq B$; Line BC is non-isotropic; Line AC is non-isotropic; Line AB is non-isotropic; Line PL is non-isotropic; Line PM is non-isotropic; Line PN is non-isotropic; Line MN intersects line B_1C_1; Line MN intersects line C_1A_1; Line MN intersects line A_1B_1; Line AL_1 intersects line BM_1.

Remark. There are several problems which could beyond the space and time limits of our computers. This is almost one among them. We proved it to show the the expense: it took about 6 hours on a Symbolics 3600.

Example 317 (A150-11: 1.67s, 28). *The circumradius OP of the triangle ABC meets the sides of the triangle in the points A_1, B_1, C_1. Show that the projections A_2, B_2, C_2 of the points A_1, B_1, C_1 upon the lines AP, BP, CP lie on the Simson line of P for ABC.*

Points A, B, C are arbitrarily chosen. Points O, P, A_1, A_2, G, F are constructed (in order) as follows: $OA \equiv OC$; $OA \equiv OB$; $PO \equiv OA$; A_1 is on line BC; A_1 is on line PO; A_2 is on line PA; $A_2A_1 \perp AP$; G is on line AB; $GP \perp AB$; F is on line AC; $FP \perp AC$. The **conclusion**: Points G, F and A_2 are collinear.

$A = (0,0)$, $B = (u_1,0)$, $C = (u_2,u_3)$, $O = (x_2,x_1)$, $P = (x_3,u_4)$, $A_1 = (x_5,x_4)$, $A_2 = (x_7,x_6)$, $G = (x_3,0)$, $F = (x_9,x_8)$.

The **nondegenerate** conditions: Points A, B and C are not collinear; Line PO intersects line BC; Line AP is non-isotropic; Line AB is non-isotropic; Line AC is non-isotropic.

Example 318 (A150-12: 76.38s, 1255; 29.07s, 342). *Let L, M, N be the projections of the point P of the circumcircle of the triangle ABC upon the sides BC, CA, AB, and let the Simson line LMN meet the altitudes AD, BE, CF in the points L_1, M_1, N_1. Show that (a) the segments LM, L_1M_1 are equal to the projection of the sides upon the Simson line; (b) the projection of the segments AL_1, BM_1, CN_1 upon LMN are equal.*

Points A, B, C are arbitrarily chosen. Points H, O, P, L, M, N, L_1, M_1, N_1, A_1, B_1 are constructed (in order) as follows: $HC \perp AB$; $HA \perp BC$; $OA \equiv OC$;

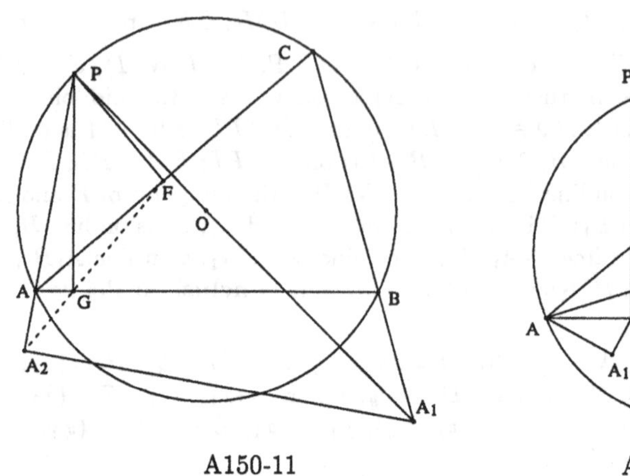

A150-11

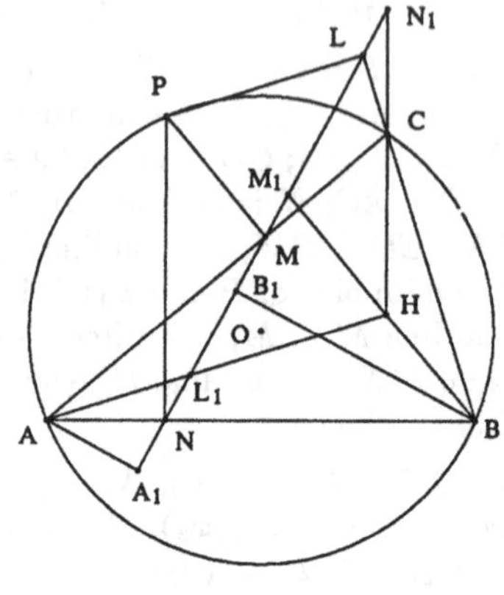

A150-12

$OA \equiv OB$; $PO \equiv OC$; L is on line BC; $LP \perp BC$; M is on line AC; $MP \perp AC$; N is on line AB; $NP \perp AB$; L_1 is on line AH; L_1 is on line NM; M_1 is on line BH; M_1 is on line NM; N_1 is on line CH; N_1 is on line NM; A_1 is on line MN; $A_1A \perp MN$; B_1 is on line MN; $B_1B \perp MN$. The **conclusions**: (1) $\frac{\overline{L_1M_1}}{\overline{ML}} - 1 = 0$; (2) $\frac{\overline{A_1B_1}}{\overline{ML}} - 1 = 0$; (3) $\frac{\overline{A_1L_1}}{\overline{B_1M_1}} - 1 = 0$.

$A = (0,0)$, $B = (u_1,0)$, $C = (u_2,u_3)$, $H = (u_2,x_1)$, $O = (x_3,x_2)$, $P = (x_4,u_4)$, $L = (x_6,x_5)$, $M = (x_8,x_7)$, $N = (x_4,0)$, $L_1 = (x_{10},x_9)$, $M_1 = (x_{12},x_{11})$, $N_1 = (u_2,x_{13})$, $A_1 = (x_{15},x_{14})$, $B_1 = (x_{17},x_{16})$.

The **nondegenerate** conditions: Points B, C and A are not collinear; Points A, B and C are not collinear; Line BC is non-isotropic; Line AC is non-isotropic; Line AB is non-isotropic; Line NM intersects line AH; Line NM intersects line BH; Line NM intersects line CH; Line MN is non-isotropic; Line MN is non-isotropic.

Example 319 (A153-3: 6.97s, 15; 5.58s, 144). *If H, O, G, I are the orthocenter, the centroid, the circumcenter, and the incenter of a triangle, show that: $HI^2 + 2OI = 3(IG^2 + 2OG^2)$.*

Points B, C, I are arbitrarily chosen. Points A, H, O, A_1, C_1, G are constructed (in order) as follows: $\tan(CBI) = \tan(IBA)$; $\tan(BCI) = \tan(ICA)$; $HC \perp AB$; $HA \perp BC$; $OB \equiv OC$; $OB \equiv OA$; A_1 is the midpoint of B and C; C_1 is the midpoint of A and B; G is on line C_1C; G is on line A_1A. The **conclusion**: $\overline{HI}^2 + 2\,\overline{OI}^2 - 3\,(\overline{GI}^2 + 2\,\overline{OG}^2) = 0$.

$B = (0,0)$, $C = (u_1,0)$, $I = (u_2,u_3)$, $A = (x_2,x_1)$, $H = (x_2,x_3)$, $O = (x_5,x_4)$, $A_1 = (x_6,0)$, $C_1 = (x_7,x_8)$, $G = (x_{10},x_9)$.

The **nondegenerate** conditions: Points I, C and B are not collinear; $\angle CIB$ is not right; Points B, C and A are not collinear; Points B, A and C are not collinear; Line A_1A intersects line C_1C.

Example 320 (A-310: 0.58s, 8). (*Menelaus' Theorem.*) *The six segments determined by a transversal on the sides of a triangle are such that the product of three*

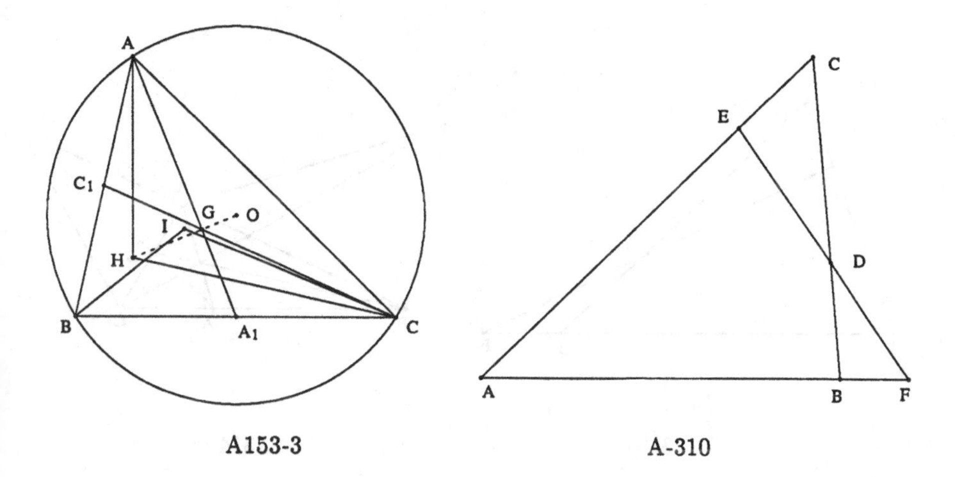

A153-3

A-310

nonconsecutive segments is equal to the product of the remaining three.

Points A, B, C are arbitrarily chosen. Points D, E, F are constructed (in order) as follows: D is on line BC; E is on line AC; $\frac{\overline{AF}}{\overline{FB}}\frac{\overline{BD}}{\overline{DC}}\frac{\overline{CE}}{\overline{EA}}+1=0$; F is on line AB. The **conclusion**: Points F, E and D are collinear.

$A=(0,0)$, $B=(u_1,0)$, $C=(u_2,u_3)$, $D=(x_1,u_4)$, $E=(x_2,u_5)$, $F=(x_3,0)$.

The **nondegenerate** conditions: $B\neq C$; $A\neq C$; $(-u_2+u_1)x_2+u_2x_1-u_1u_2\neq 0$.

Example 321 (A-313: 2.68s, 8; 1.3s, 10). (*The Converse of Menelaus' Theorem.*) *If three points are taken, one on each side of a triangle, so that these points divide the sides into six segments such that the products of the segments in each of the two set of nonconsecutive segments are equal in magnitude and opposite in sign, the three points are collinear.*

Points A, B, C are arbitrarily chosen. Points D, E, F are constructed (in order) as follows: D is on line BC; E is on line AC; F is on line ED; F is on line AB. The **conclusion**: $\frac{\overline{AF}}{\overline{FB}}\frac{\overline{BD}}{\overline{DC}}\frac{\overline{CE}}{\overline{EA}}+1=0$.

$A=(0,0)$, $B=(u_1,0)$, $C=(u_2,u_3)$, $D=(x_1,u_4)$, $E=(x_2,u_5)$, $F=(x_3,0)$.

The **nondegenerate** conditions: $B\neq C$; $A\neq C$; Line AB intersects line ED.

Remark. Here we use the ratios of directed segments to express the main hypothesis. If we use the lengths of the segments, then problem will become reducible and can be confirmed only using decomposition.

Example 322 (A-314: 1.4s, 21). *(A) The external bisectors of the angles of a triangle meet the opposite sides in three collinear points. (B) Two internal bisectors and the external bisector of the third angle meet the opposite sides in three collinear points.*

Points B, C, I are arbitrarily chosen. Points A, A_1, B_1, C_1 are constructed (in order) as follows: $\tan(CBI)=\tan(IBA)$; $\tan(BCI)=\tan(ICA)$; A_1 is on line BC; $A_1A\perp IA$; B_1 is on line AC; $B_1B\perp IB$; C_1 is on line AB; $C_1C\perp IC$. The

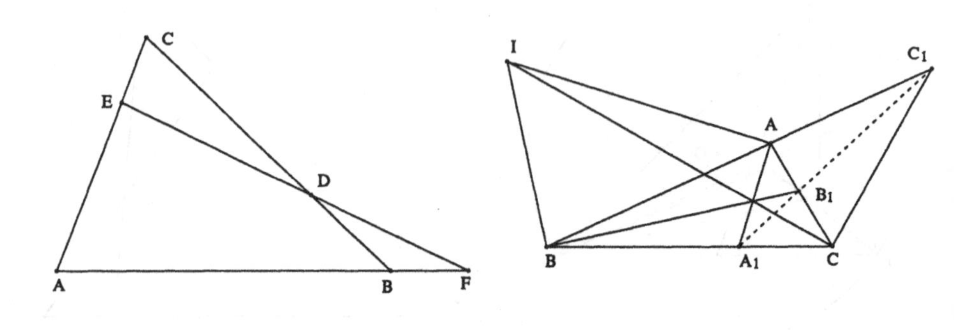

A-313 A-314

conclusion: Points A_1, B_1 and C_1 are collinear.

$B = (0,0)$, $C = (u_1,0)$, $I = (u_2,u_3)$, $A = (x_2,x_1)$, $A_1 = (x_3,0)$, $B_1 = (x_5,x_4)$, $C_1 = (x_7,x_6)$.

The **nondegenerate** conditions: Points I, C and B are not collinear; $\angle CIB$ is not right; Line IA is not perpendicular to line BC; Line IB is not perpendicular to line AC; Line IC is not perpendicular to line AB.

Remark. Our mechanical proof is for both (A) and (B).

Example 323 (A-316: 1.62s, 20). *The sides of the orthic triangle meet the sides of the given triangle in three collinear points.*

Points B, C, A are arbitrarily chosen. Points D, E, F, A_1, B_1, C_1 are constructed (in order) as follows: D is on line BC; $DA \perp BC$; E is on line AC; $EB \perp AC$; F is on line AB; $FC \perp AB$; A_1 is on line BC; A_1 is on line EF; B_1 is on line AC; B_1 is on line DF; C_1 is on line AB; C_1 is on line DE. The **conclusion**: Points C_1, A_1 and B_1 are collinear.

$B = (0,0)$, $C = (u_1,0)$, $A = (u_2,u_3)$, $D = (u_2,0)$, $E = (x_2,x_1)$, $F = (x_4,x_3)$, $A_1 = (x_5,0)$, $B_1 = (x_7,x_6)$, $C_1 = (x_9,x_8)$.

The **nondegenerate** conditions: Line BC is non-isotropic; Line AC is non-isotropic; Line AB is non-isotropic; Line EF intersects line BC; Line DF intersects line AC; Line DE intersects line AB.

Example 324 (A-321: 1.78s, 11). *The isotomic points of three collinear points are collinear.*

Points A, B, C are arbitrarily chosen. Points D, E, F, M_1, M_2, M_3, D_1, E_1, F_1 are constructed (in order) as follows: D is on line BC; E is on line AC; F is on line DE; F is on line AB; M_1 is the midpoint of B and C; M_2 is the midpoint of C and A; M_3 is the midpoint of A and B; M_1 is the midpoint of D and D_1; M_2 is the midpoint of E and E_1; M_3 is the midpoint of F and F_1. The **conclusion**: Points F_1, D_1 and E_1 are collinear.

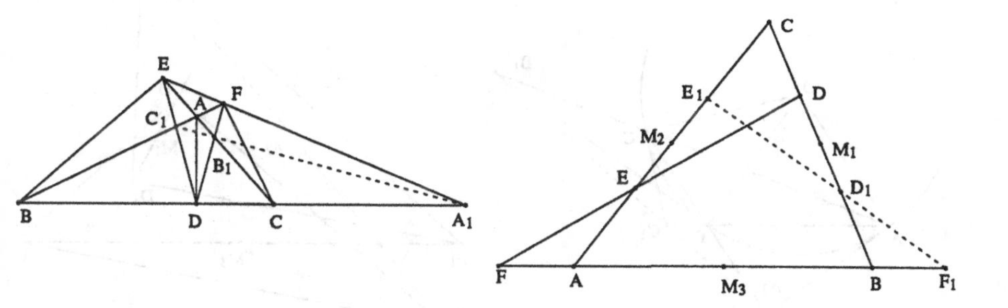

A-316 A-321

$A=(0,0)$, $B=(u_1,0)$, $C=(u_2,u_3)$, $D=(x_1,u_4)$, $E=(x_2,u_5)$, $F=(x_3,0)$, $M_1=(x_4,x_5)$, $M_2=(x_6,x_7)$, $M_3=(x_8,0)$, $D_1=(x_9,x_{10})$, $E_1=(x_{11},x_{12})$, $F_1=(x_{13},0)$.

The **nondegenerate** conditions: $B\neq C$; $A\neq C$; Line AB intersects line DE.

Example 325 (A-318: 1.1s, 13). *The lines tangent to the circumcircle of a triangle at the vertices meet opposite sides in three collinear points. (The Lemoine axis of the given triangle.)*

Points A, B, C are arbitrarily chosen. Points O, A_1, B_1, C_1 are constructed (in order) as follows: $OA\equiv OC$; $OA\equiv OB$; A_1 is on line BC; $A_1A\perp OA$; B_1 is on line AC; $B_1B\perp OB$; C_1 is on line AB; $C_1C\perp OC$. The **conclusion**: Points B_1, C_1 and A_1 are collinear.

$A=(0,0)$, $B=(u_1,0)$, $C=(u_2,u_3)$, $O=(x_2,x_1)$, $A_1=(x_4,x_3)$, $B_1=(x_6,x_5)$, $C_1=(x_7,0)$.

The **nondegenerate** conditions: Points A, B and C are not collinear; Line OA is not perpendicular to line BC; Line OB is not perpendicular to line AC; Line OC is not perpendicular to line AB.

Example 326 (A-324: 1.27s, 12). *If the pairs of points D, D_1; E, E_1; F, F_1 are isotomic on the sides BC, CA, AB of the triangle ABC, the two triangles DEF, $D_1E_1F_1$ are equivalent.*

Points A, B, C are arbitrarily chosen. Points D, E, F, M_1, M_2, M_3, D_1, E_1, F_1 are constructed (in order) as follows: D is on line BC; E is on line AC; F is on line AB; M_1 is the midpoint of B and C; M_2 is the midpoint of C and A; M_3 is the midpoint of A and B; M_1 is the midpoint of D and D_1; M_2 is the midpoint of E and E_1; M_3 is the midpoint of F and F_1. The **conclusion**: $\nabla(D_1E_1F_1)-\nabla(DEF)=0$.

$A=(0,0)$, $B=(u_1,0)$, $C=(u_2,u_3)$, $D=(x_1,u_4)$, $E=(x_2,u_5)$, $F=(u_6,0)$, $M_1=(x_3,x_4)$, $M_2=(x_5,x_6)$, $M_3=(x_7,0)$, $D_1=(x_8,x_9)$, $E_1=(x_{10},x_{11})$, $F_1=(x_{12},0)$.

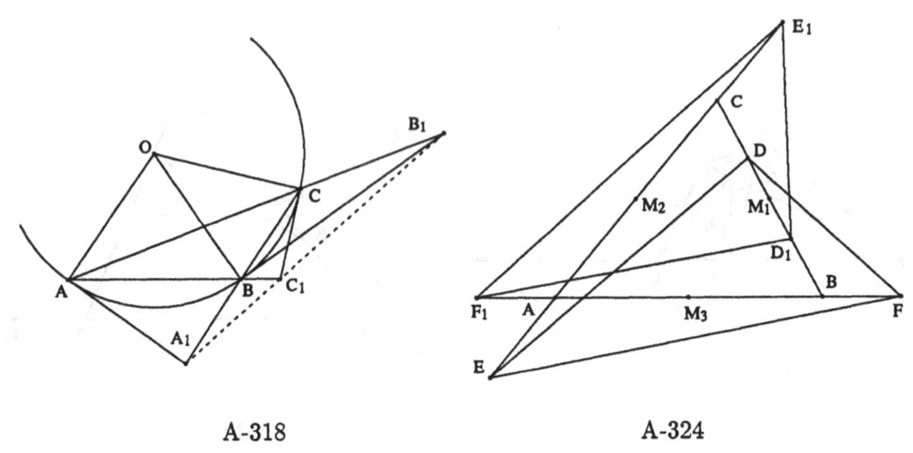

A-318 A-324

The **nondegenerate** conditions: $B \neq C$; $A \neq C$; $A \neq B$.

Example 327 (A158-1: 0.82s, 4). *The median AA_1 of the triangle ABC meets the side B_1A_1 of the medial triangle $A_1B_1C_1$ in P, and CP meets AB in Q. Show that $AB = 3AQ$.*

Points B, C, A are arbitrarily chosen. Points A_1, C_1, P, Q are constructed (in order) as follows: A_1 is the midpoint of B and C; C_1 is the midpoint of A and B; P is on line AA_1; $PC_1 \parallel BC$; Q is on line AB; Q is on line PC. The **conclusion**: Points Q, A, B are collinear, and $\overline{QA}/\overline{AB} = 1/(-3)$.

$B = (0,0)$, $C = (u_1, 0)$, $A = (u_2, u_3)$, $A_1 = (x_1, 0)$, $C_1 = (x_2, x_3)$, $P = (x_4, x_3)$, $Q = (x_6, x_5)$.

The **nondegenerate** conditions: Line BC intersects line AA_1; Line PC intersects line AB.

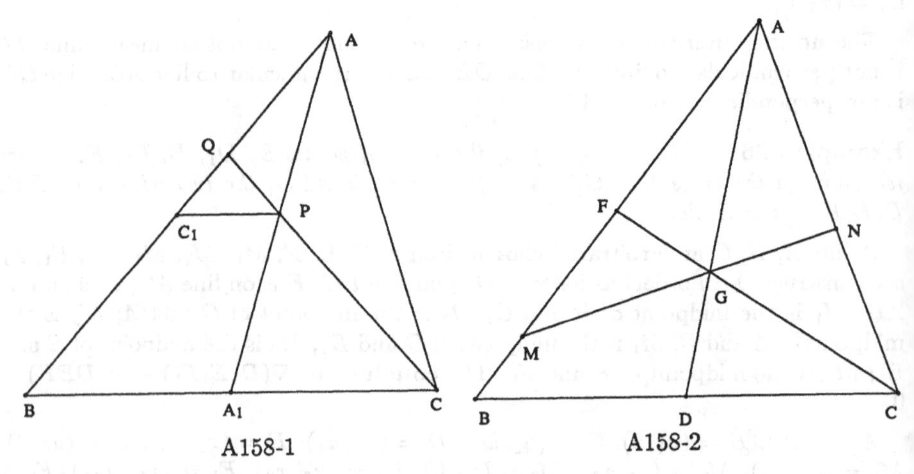

A158-1 A158-2

Example 328 (A158-2: 1.33s, 11). *Show that, if a line through the centroid G of* *the triangle ABC meets AB in M and AC in N, we have, both in magnitude and in sign, $AN \cdot MB + AM \cdot NC = AM \cdot AN$.*

Points B, C, A are arbitrarily chosen. Points D, F, G, M, N are constructed (in order) as follows: D is the midpoint of B and C; F is the midpoint of A and B; G is on line CF; G is on line AD; M is on line AB; N is on line GM; N is on line AC. The **conclusion**: $\frac{\overline{MB}}{\overline{AM}} + \frac{\overline{NC}}{\overline{AN}} - 1 = 0$.

$B = (0,0)$, $C = (u_1,0)$, $A = (u_2,u_3)$, $D = (x_1,0)$, $F = (x_2,x_3)$, $G = (x_5,x_4)$, $M = (x_6,u_4)$, $N = (x_8,x_7)$.

The **nondegenerate** conditions: Line AD intersects line CF; $A \neq B$; Line AC intersects line GM.

Example 329 (A158-4: 14.13s, 395; 12.53s, 200). *Prove that the triangle formed by the points of contact of the sides of a given triangle with the excircles corresponding to these sides is equivalent to the triangle formed by the points of contact of the sides of the triangle with the inscribed circle.*

Points B, C, I are arbitrarily chosen. Points A, I_A, X, Y, Z, X_A, I_B, I_C, Y_B, Z_C are constructed (in order) as follows: $\tan(BCI) = \tan(ICA)$; $\tan(CBI) = \tan(IBA)$; I_A is on line AI; $I_AB \perp BI$; $XI \perp BC$; X is on line BC; Y is on line AC; $YI \perp AC$; Z is on line AB; $ZI \perp AB$; X_A is on line BC; $X_AI_A \perp BC$; I_B is on line BI; $I_BC \perp CI$; I_C is on line CI; $I_CB \perp BI$; Y_B is on line AC; $Y_BI_B \perp AC$; Z_C is on line BA; $Z_CI_C \perp BA$. The **conclusion**: $\nabla(XYZ) - \nabla(X_AY_BZ_C) = 0$.

$B = (0,0)$, $C = (u_1,0)$, $I = (u_2,u_3)$, $A = (x_2,x_1)$, $I_A = (x_4,x_3)$, $X = (u_2,0)$, $Y = (x_6,x_5)$, $Z = (x_8,x_7)$, $X_A = (x_4,0)$, $I_B = (x_{10},x_9)$, $I_C = (x_{12},x_{11})$, $Y_B = (x_{14},x_{13})$, $Z_C = (x_{16},x_{15})$.

The **nondegenerate** conditions: Points I, B and C are not collinear; $\angle BIC$ is not right; Line BI is not perpendicular to line AI; Line BC is non-isotropic; Line AC is non-isotropic; Line AB is non-isotropic; Line BC is non-isotropic; Line CI is not perpendicular to line BI; Line BI is not perpendicular to line CI; Line AC is non-isotropic; Line BA is non-isotropic.

Remark. Actually, our proof is for the four pairs of such equivalent triangles.

Example 330 (A158-5: 2.45s, 34). *The sides BC, CA, AB of a triangle ABC are met by two transversal PQR, $P_1Q_1R_1$ in the pairs of points P, P_1; Q, Q_1; R, R_1. Show that the points $X = BC \cap QR_1$, $Y = CA \cap RP_1$, $Z = AB \cap PQ_1$ are collinear.*

Points B, C, A are arbitrarily chosen. Points P, Q, R, P_1, Q_1, R_1, X, Y, Z are constructed (in order) as follows: P is on line BC; Q is on line CA; R is on line PQ; R is on line AB; P_1 is on line BC; Q_1 is on line CA; R_1 is on line P_1Q_1; R_1 is on line AB; X is on line QR_1; X is on line BC; Y is on line RP_1; Y is on line CA; Z is on line PQ_1; Z is on line AB. The **conclusion**: Points X, Y and Z are collinear.

$B = (0,0)$, $C = (u_1,0)$, $A = (u_2,u_3)$, $P = (u_4,0)$, $Q = (x_1,u_5)$, $R = (x_3,x_2)$, $P_1 = (u_6,0)$, $Q_1 = (x_4,u_7)$, $R_1 = (x_6,x_5)$, $X = (x_7,0)$, $Y = (x_9,x_8)$, $Z = (x_{11},x_{10})$.

The **nondegenerate** conditions: $B \neq C$; $C \neq A$; Line AB intersects line PQ; $B \neq C$; $C \neq A$; Line AB intersects line P_1Q_1; Line BC intersects line QR_1; Line

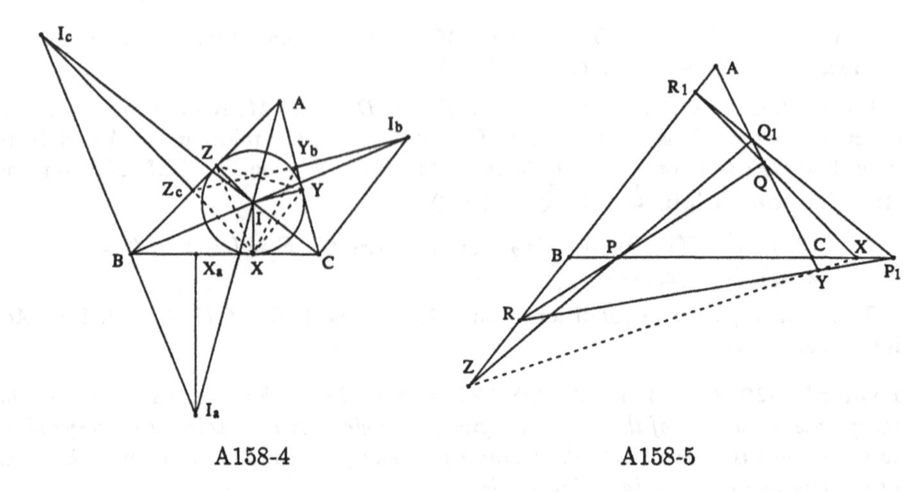

A158-4 A158-5

CA intersects line RP_1; Line AB intersects line PQ_1.

Example 331 (A158-6: 7.28s, 405). *Show that the projections of a point of the circumcircle of a cyclic quadrilateral upon the sides divide the sides into eight segments such that the product of four nonconsecutive segments is equal to the product of the remaining four.*

Points B, C, A are arbitrarily chosen. Points O, D, P, A_1, B_1, C_1, D_1 are constructed (in order) as follows: $OB \equiv OC$; $OB \equiv OA$; $DO \equiv OB$; $PO \equiv OB$; A_1 is on line AB; $A_1P \perp AB$; B_1 is on line BC; $B_1P \perp BC$; C_1 is on line CD; $C_1P \perp CD$; D_1 is on line DA; $D_1P \perp DA$. The **conclusion**: $1 - \frac{\overline{AA_1}}{\overline{A_1B}} \frac{\overline{BB_1}}{\overline{B_1C}} \frac{\overline{CC_1}}{\overline{C_1D}} \frac{\overline{DD_1}}{\overline{D_1A}} = 0$.

$B = (0,0)$, $C = (u_1,0)$, $A = (u_2,u_3)$, $O = (x_2,x_1)$, $D = (x_3,u_4)$, $P = (x_4,u_5)$, $A_1 = (x_6,x_5)$, $B_1 = (x_4,0)$, $C_1 = (x_8,x_7)$, $D_1 = (x_{10},x_9)$.

The **nondegenerate** conditions: Points B, A and C are not collinear; Line AB is non-isotropic; Line BC is non-isotropic; Line CD is non-isotropic; Line DA is non-isotropic.

Example 332 (A158-7: 3.28s, 86). *A circle whose center is equidistant from the vertices A, B of the triangle ABC cuts the sides BC, AC in the pairs of points P, P_1; Q, Q_1. Show that the lines PQ, P_1Q_1 meet AB in two isotomic points.*

Points A, B, C are arbitrarily chosen. Points M, O, P, D, P_1, Q, E, Q_1, X, Y are constructed (in order) as follows: M is the midpoint of A and B; $OM \perp AB$; P is on line AC; D is on line AC; $DO \perp AC$; D is the midpoint of P and P_1; $QO \equiv OP$; Q is on line BC; E is on line BC; $EO \perp BC$; E is the midpoint of Q and Q_1; X is on line PQ; X is on line AB; Y is on line P_1Q_1; Y is on line AB. The **conclusion**: M is the midpoint of Y and X.

$A = (0,0)$, $B = (u_1,0)$, $C = (u_2,u_3)$, $M = (x_1,0)$, $O = (x_1,u_4)$, $P = (x_2,u_5)$, $D = (x_4,x_3)$, $P_1 = (x_5,x_6)$, $Q = (x_8,x_7)$, $E = (x_{10},x_9)$, $Q_1 = (x_{11},x_{12})$, $X = (x_{13},0)$, $Y = (x_{14},0)$.

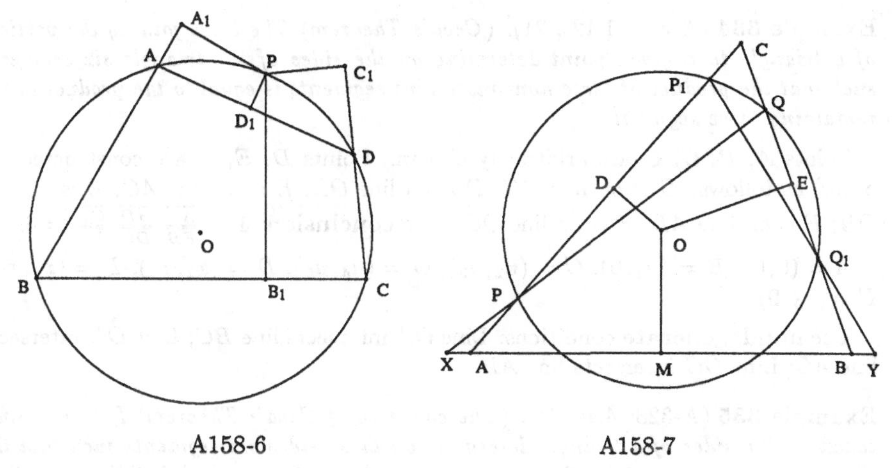

A158-6 A158-7

The **nondegenerate** conditions: $A \neq B$; $A \neq C$; Line AC is non-isotropic; Line BC is non-isotropic; Line BC is non-isotropic; Line AB intersects line PQ; Line AB intersects line P_1Q_1.

Example 333 (A158-8: 1.87s, 60). *Two equal segments AE, AF are taken on the sides AB, AC of the triangle ABC. Show that median issued from A divides EF in the ratio of the sides AC, AB.*

Points A, B, C are arbitrarily chosen. Points E, F, A_1, D are constructed (in order) as follows: E is on line AB; $FA \equiv AE$; F is on line AC; A_1 is the midpoint of B and C; D is on line EF; D is on line AA_1. The **conclusion**: $ED \cdot AB = AC \cdot FD$.

$A = (0,0)$, $B = (u_1,0)$, $C = (u_2,u_3)$, $E = (u_4,0)$, $F = (x_2,x_1)$, $A_1 = (x_3,x_4)$, $D = (x_6,x_5)$.

The **nondegenerate** conditions: $A \neq B$; Line AC is non-isotropic; Line AA_1 intersects line EF.

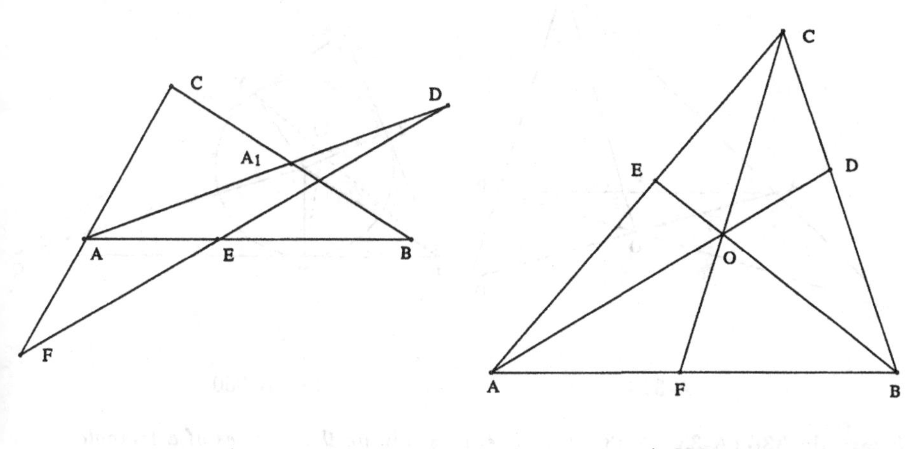

A158-8 A-326

Example 334 (A-326: 1.12s, 21). (*Ceva's Theorem*) *The lines joining the vertices of a triangle to a given point determine on the sides of the triangle six segments such that the product of three nonconsecutive segments is equal to the product of the remaining three segments.*

Points A, B, C, O are arbitrarily chosen. Points D, E, F are constructed (in order) as follows: D is on line BC; D is on line OA; E is on line AC; E is on line OB; F is on line AB; F is on line OC. The **conclusion**: $1 - \frac{\overline{AF}}{\overline{FB}} \frac{\overline{BD}}{\overline{DC}} \frac{\overline{CE}}{\overline{EA}} = 0$.

$A = (0,0)$, $B = (u_1, 0)$, $C = (u_2, u_3)$, $O = (u_4, u_5)$, $D = (x_2, x_1)$, $E = (x_4, x_3)$, $F = (x_5, 0)$.

The **nondegenerate** conditions: Line OA intersects line BC; Line OB intersects line AC; Line OC intersects line AB.

Example 335 (A-328: 4.4s, 21). (*The converse of Ceva's Theorem*) *If three points taken on the sides of a triangle determine on these sides six segments such that the products of the segments in the two nonconsecutive sets are equal, both in magnitude and in sign, the lines joining these points to the respectively opposite vertices are concurrent.*

Points A, B, C, O are arbitrarily chosen. Points D, E, F are constructed (in order) as follows: D is on line BC; D is on line OA; E is on line AC; E is on line OB; $1 - \frac{\overline{AF}}{\overline{FB}} \frac{\overline{BD}}{\overline{DC}} \frac{\overline{CE}}{\overline{EA}} = 0$; F is on line AB. The **conclusion**: Points O, F and C are collinear.

$A = (0,0)$, $B = (u_1, 0)$, $C = (u_2, u_3)$, $O = (u_4, u_5)$, $D = (x_2, x_1)$, $E = (x_4, x_3)$, $F = (x_5, 0)$.

The **nondegenerate** conditions: Line OA intersects line BC; Line OB intersects line AC; $(2x_2 - u_2 - u_1)x_4 - u_2 x_2 + u_1 u_2 \neq 0$.

Remark. See the remark of A-313.

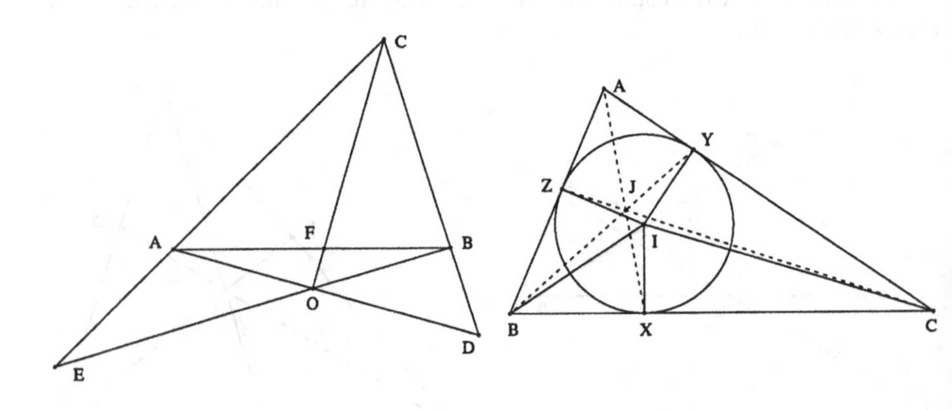

A-328 A-330

Example 336 (A-330: 2.43s, 21). *The lines joining the vertices of a triangle to the* *points of contact of the opposite sides with the inscribed circle are concurrent (The Gergonne Point).*

Points B, C, I are arbitrarily chosen. Points A, X, Y, Z, J are constructed (in order) as follows: $\tan(BCI) = \tan(ICA)$; $\tan(CBI) = \tan(IBA)$; $XI \perp BC$; X is on line BC; Y is on line AC; $YI \perp AC$; Z is on line AB; $ZI \perp AB$; J is on line BY; J is on line AX. The **conclusion**: Points J, C and Z are collinear.

$B = (0,0)$, $C = (u_1,0)$, $I = (u_2,u_3)$, $A = (x_2,x_1)$, $X = (u_2,0)$, $Y = (x_4,x_3)$, $Z = (x_6,x_5)$, $J = (x_8,x_7)$.

The **nondegenerate** conditions: Points I, B and C are not collinear; $\angle BIC$ is not right; Line BC is non-isotropic; Line AC is non-isotropic; Line AB is non-isotropic; Line AX intersects line BY.

Remark. Actually there are four Gergonne points: in our proof, article 332 in the book is a repetition of A-330.

Example 337 (A-333: 4.12s, 81). *The lines joining the vertices of a triangle to the points of contact of the opposite sides with the excircles relative to those sides are concurrent (The Nagel Point).*

Points B, C, I are arbitrarily chosen. Points A, I_A, I_B, I_C, X_A, Y_B, Z_C, J are constructed (in order) as follows: $\tan(BCI) = \tan(ICA)$; $\tan(CBI) = \tan(IBA)$; I_A is on line AI; $I_AB \perp BI$; I_B is on line I_AC; I_B is on line BI; I_C is on line CI; I_C is on line I_AB; $X_AI_A \perp BC$; X_A is on line BC; Y_B is on line AC; $Y_BI_B \perp AC$; Z_C is on line AB; $Z_CI_C \perp AB$; J is on line BY_B; J is on line AX_A. The **conclusion**: Points J, C and Z_C are collinear.

$B = (0,0)$, $C = (u_1,0)$, $I = (u_2,u_3)$, $A = (x_2,x_1)$, $I_A = (x_4,x_3)$, $I_B = (x_6,x_5)$, $I_C = (x_8,x_7)$, $X_A = (x_4,0)$, $Y_B = (x_{10},x_9)$, $Z_C = (x_{12},x_{11})$, $J = (x_{14},x_{13})$.

The **nondegenerate** conditions: Points I, B and C are not collinear; $\angle BIC$ is not right; Line BI is not perpendicular to line AI; Line BI intersects line I_AC; Line I_AB intersects line CI; Line BC is non-isotropic; Line AC is non-isotropic; Line AB is non-isotropic; Line AX_A intersects line BY_B.

Remark. There should be four Nagel points for a triangle. The subsequent examples related to the Nagel points are valid for the four Nagel points.

Example 338 (A161-2: 0.85s, 5). *A parallel to the side BC of the triangle ABC meets AB, AC in B_1, C_1. Prove that the lines BC_1, B_1C meet on the median from A.*

Points B, C, A are arbitrarily chosen. Points M, E, F, I are constructed (in order) as follows: M is the midpoint of B and C; E is on line AB; F is on line AC; $FE \parallel BC$; I is on line FB; I is on line EC. The **conclusion**: Points I, A and M are collinear.

$B = (0,0)$, $C = (u_1,0)$, $A = (u_2,u_3)$, $M = (x_1,0)$, $E = (x_2,u_4)$, $F = (x_3,u_4)$, $I = (x_5,x_4)$.

The **nondegenerate** conditions: $A \neq B$; Points B, C and A are not collinear; Line EC intersects line FB.

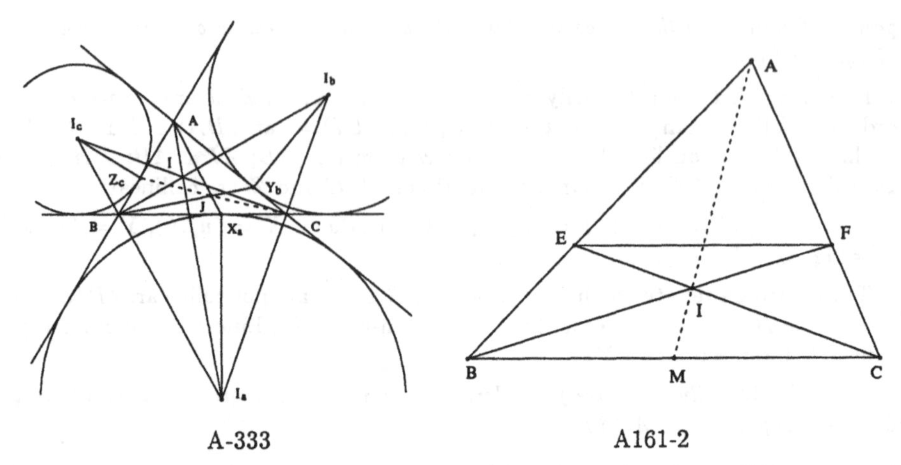

A-333 A161-2

Example 339 (A161-4: 2.37s, 54). *With a point M of the side BC of the triangle ABC as center circles are drawn passing through B and C, respectively, meeting AB, AC again in N, P. For what position of M will the lines AM, BP, CN be concurrent? (The midpoints of the traces on BC of the mediators of AB, AC.)*

Points B, C, A are arbitrarily chosen. Points M_1, M_2, M, N, P, O are constructed (in order) as follows: $M_1A \equiv M_1B$; M_1 is on line BC; $M_2A \equiv M_2C$; M_2 is on line BC; M is the midpoint of M_1 and M_2; N is on line AB; $NM \equiv MB$; P is on line AC; $PM \equiv MC$; O is on line BP; O is on line AM. The **conclusion**: Points O, C and N are collinear.

$B = (0,0)$, $C = (u_1,0)$, $A = (u_2,u_3)$, $M_1 = (x_1,0)$, $M_2 = (x_2,0)$, $M = (x_3,0)$, $N = (x_5,x_4)$, $P = (x_7,x_6)$, $O = (x_9,x_8)$.

The **nondegenerate** conditions: Line BC is not perpendicular to line AB; Line BC is not perpendicular to line AC; Line AB is non-isotropic; Line AC is non-isotropic; Line AM intersects line BP. In addition, the following nondegenerate conditions, which come from reducibility and have been detected by our prover, should be also added: $P \neq C$; $N \neq B$.

Example 340 (A-335: 2.75s, 27). *If the three lines joining three points marked on the sides of a triangle to the respectively opposite vertices are concurrent, the same is true of the isotomics of the given points.*

Points A, B, C, O are arbitrarily chosen. Points D, E, F, M_1, M_2, M_3, D_1, E_1, F_1, H are constructed (in order) as follows: D is on line AO; D is on line CB; E is on line BO; E is on line AC; F is on line CO; F is on line AB; M_1 is the midpoint of B and C; M_2 is the midpoint of A and C; M_3 is the midpoint of A and B; M_1 is the midpoint of D and D_1; M_2 is the midpoint of E and E_1; M_3 is the midpoint of F and F_1; H is on line BE_1; H is on line AD_1. The **conclusion**: Points H, C and F_1 are collinear.

$A = (0,0)$, $B = (u_1,0)$, $C = (u_2,u_3)$, $O = (u_4,u_5)$, $D = (x_2,x_1)$, $E = (x_4,x_3)$, $F = (x_5,0)$, $M_1 = (x_6,x_7)$, $M_2 = (x_8,x_9)$, $M_3 = (x_{10},0)$, $D_1 = (x_{11},x_{12})$, $E_1 = (x_{13},x_{14})$, $F_1 = (x_{15},0)$, $H = (x_{17},x_{16})$.

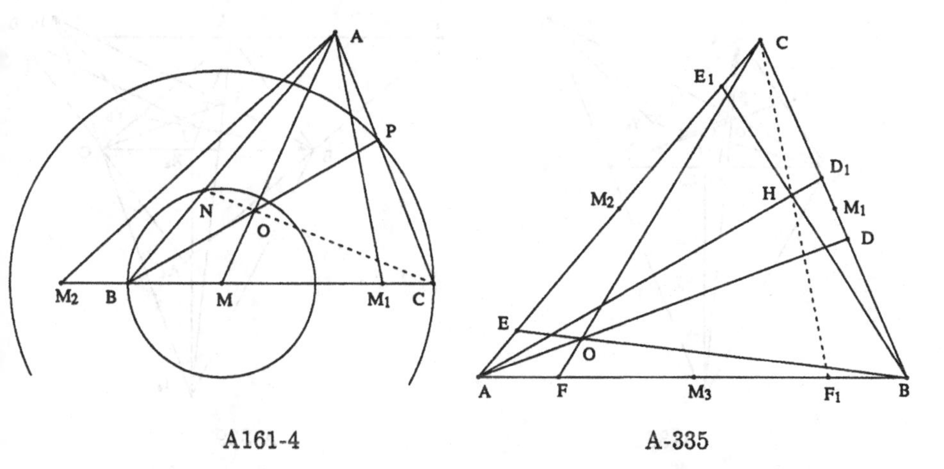

A161-4 A-335

The **nondegenerate** conditions: Line CB intersects line AO; Line AC intersects line BO; Line AB intersects line CO; Line AD_1 intersects line BE_1.

Example 341 (A-338: 4.25s, 96). *The Nagel point M of ABC and I, its incenter, are collinear with the centroid G of ABC and $2\,IG = GM$.*

Points B, C, I are arbitrarily chosen. Points A, I_A, I_B, X_A, Y_B, J, C_1, A_1, G are constructed (in order) as follows: $\tan(BCI) = \tan(ICA)$; $\tan(CBI) = \tan(IBA)$; I_A is on line AI; $I_AB \perp BI$; I_B is on line I_AC; I_B is on line BI; $X_AI_A \perp BC$; X_A is on line BC; Y_B is on line AC; $Y_BI_B \perp AC$; J is on line BY_B; J is on line AX_A; C_1 is the midpoint of B and A; A_1 is the midpoint of B and C; G is on line CC_1; G is on line AA_1. The **conclusion**: Points I, G, J are collinear, and $\overline{IG}/\overline{GJ} = 1/2$.

$B = (0,0)$, $C = (u_1,0)$, $I = (u_2,u_3)$, $A = (x_2,x_1)$, $I_A = (x_4,x_3)$, $I_B = (x_6,x_5)$, $X_A = (x_4,0)$, $Y_B = (x_8,x_7)$, $J = (x_{10},x_9)$, $C_1 = (x_{11},x_{12})$, $A_1 = (x_{13},0)$, $G = (x_{15},x_{14})$.

The **nondegenerate** conditions: Points I, B and C are not collinear; $\angle BIC$ is not right; Line BI is not perpendicular to line AI; Line BI intersects line I_AC; Line BC is non-isotropic; Line AC is non-isotropic; Line AX_A intersects line BY_B; Line AA_1 intersects line CC_1.

Example 342 (A-339: 27.38s, 828; 16.45s, 528). *The Nagel point of a triangle is the incenter of the anticomplementary triangle.*

Points B, C, I are arbitrarily chosen. Points A, I_A, I_B, X_A, Y_B, J, C_1, B_1, A_1 are constructed (in order) as follows: $\tan(BCI) = \tan(ICA)$; $\tan(CBI) = \tan(IBA)$; I_A is on line AI; $I_AB \perp BI$; I_B is on line I_AC; I_B is on line BI; $X_AI_A \perp BC$; X_A is on line BC; Y_B is on line AC; $Y_BI_B \perp AC$; J is on line BY_B; J is on line AX_A; $C_1B \parallel AC$; $C_1A \parallel BC$; $B_1C \parallel AB$; B_1 is on line C_1A; A_1 is on line B_1C; A_1 is on line BC_1. The **conclusion**: $\tan(A_1B_1J) = \tan(JB_1C_1)$.

$B = (0,0)$, $C = (u_1,0)$, $I = (u_2,u_3)$, $A = (x_2,x_1)$, $I_A = (x_4,x_3)$, $I_B = (x_6,x_5)$, $X_A = (x_4,0)$, $Y_B = (x_8,x_7)$, $J = (x_{10},x_9)$, $C_1 = (x_{11},x_1)$, $B_1 = (x_{12},x_1)$, $A_1 = (x_{14},x_{13})$.

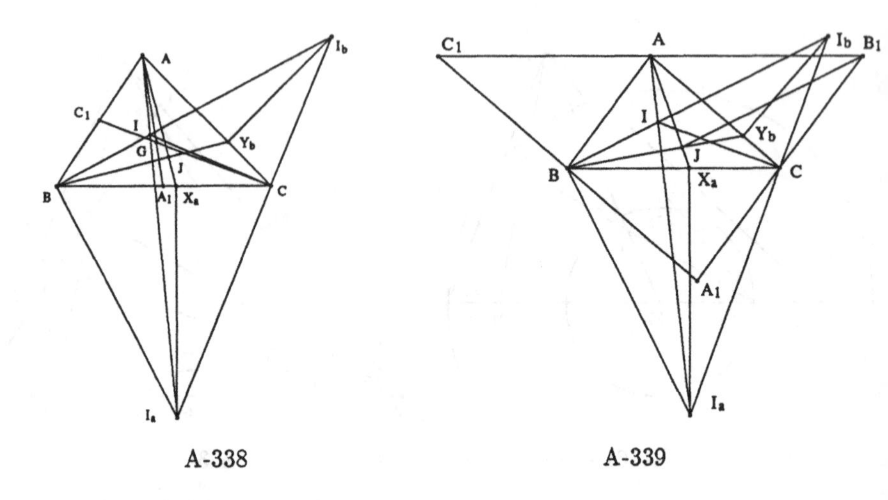

A-338 A-339

The **nondegenerate** conditions: Points I, B and C are not collinear; $\angle BIC$ is not right; Line BI is not perpendicular to line AI; Line BI intersects line I_AC; Line BC is non-isotropic; Line AC is non-isotropic; Line AX_A intersects line BY_B; Points B, C and A are not collinear; Points C_1, A and B are not collinear; Line BC_1 intersects line B_1C.

Example 343 (A-340: 4.45s, 19; 3.6s, 15). *If LMN is the cevian triangle of the point S for the triangle ABC, we have* $SL/AL + SM/BM + SN/CN = 1$.

Points B, C, A, S are arbitrarily chosen. Points L, N, M are constructed (in order) as follows: L is on line BC; L is on line AS; N is on line BA; N is on line CS; M is on line AC; M is on line BS. The **conclusion**: $1 - (\frac{\overline{SL}}{\overline{AL}} + \frac{\overline{SM}}{\overline{BM}} + \frac{\overline{SN}}{\overline{CN}}) = 0$.

$B = (0,0)$, $C = (u_1,0)$, $A = (u_2,u_3)$, $S = (u_4,u_5)$, $L = (x_1,0)$, $N = (x_3,x_2)$, $M = (x_5,x_4)$.

The **nondegenerate** conditions: Line AS intersects line BC; Line CS intersects line BA; Line BS intersects line AC.

Example 344 (A-341: 1.98s, 19).

Points B, C, A, S are arbitrarily chosen. Points L, N, M are constructed (in order) as follows: L is on line BC; L is on line AS; N is on line BA; N is on line CS; M is on line AC; M is on line BS. The **conclusion**: $2 - (\frac{\overline{AS}}{\overline{AL}} + \frac{\overline{BS}}{\overline{BM}} + \frac{\overline{CS}}{\overline{CN}}) = 0$.

$B = (0,0)$, $C = (u_1,0)$, $A = (u_2,u_3)$, $S = (u_4,u_5)$, $L = (x_1,0)$, $N = (x_3,x_2)$, $M = (x_5,x_4)$.

The **nondegenerate** conditions: Line AS intersects line BC; Line CS intersects line BA; Line BS intersects line AC.

Example 345 (A-342: 1.18s, 14). *If LMN is the cevian triangle of the point S for the triangle ABC, we have* $AS/SL = AM/MC + AN/NB$.

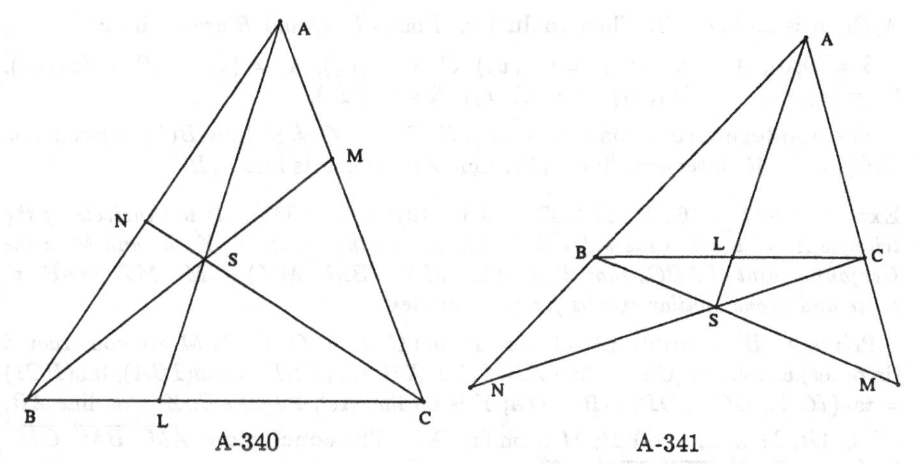

A-340 A-341

Points B, C, A, S are arbitrarily chosen. Points L, N, M are constructed (in order) as follows: L is on line BC; L is on line AS; N is on line BA; N is on line CS; M is on line AC; M is on line BS. The **conclusion:** $\frac{\overline{AM}}{\overline{MC}} + \frac{\overline{AN}}{\overline{NB}} - \frac{\overline{AS}}{\overline{SL}} = 0$.

$B = (0,0)$, $C = (u_1, 0)$, $A = (u_2, u_3)$, $S = (u_4, u_5)$, $L = (x_1, 0)$, $N = (x_3, x_2)$, $M = (x_5, x_4)$.

The **nondegenerate** conditions: Line AS intersects line BC; Line CS intersects line BA; Line BS intersects line AC.

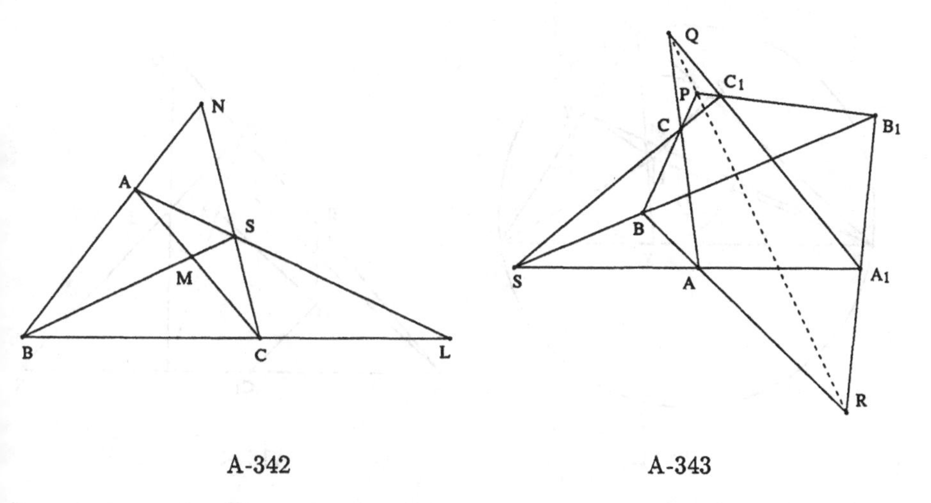

A-342 A-343

Example 346 (A-343: 3.72s, 402; 2.28s, 66). *Desargues' Theorem. Given two triangle ABC, $A_1B_1C_1$, if the three lines AA_1, BB_1, CC_1 meet in a point, S, the three points $P = BC \cap B_1C_1$, $Q = CA \cap C_1A_1$, $R = AB \cap A_1B_1$ lie on a line.*

Points S, A, B, C are arbitrarily chosen. Points A_1, B_1, C_1, P, Q, R are constructed (in order) as follows: A_1 is on line AS; B_1 is on line BS; C_1 is on line CS; P is on line B_1C_1; P is on line BC; Q is on line A_1C_1; Q is on line AC; R is on line

A_1B_1; R is on line AB. The **conclusion**: Points P, Q and R are collinear.

$S = (0,0)$, $A = (u_1, 0)$, $B = (u_2, u_3)$, $C = (u_4, u_5)$, $A_1 = (u_6, 0)$, $B_1 = (x_1, u_7)$, $C_1 = (x_2, u_8)$, $P = (x_4, x_3)$, $Q = (x_6, x_5)$, $R = (x_8, x_7)$.

The **nondegenerate** conditions: $A \neq S$; $B \neq S$; $C \neq S$; Line BC intersects line B_1C_1; Line AC intersects line A_1C_1; Line AB intersects line A_1B_1.

Example 347 (A164-1: 2352.97s, 14031; 1012.18s, 18363). *If the incircle of the triangle ABC touches the sides BC, CA, AB in the points X, Y, Z, and M is the Gergonne point of ABC, show that $(AM : MX)\cdot(BM : MY)\cdot(CM : MZ) = 4R : r$. State and prove similar results for the excircles.*

Points X, B are arbitrarily chosen. Points C, I, A, O, Y, Z, M are constructed (in order) as follows: C is on line XB; $IX \perp XB$; $\tan(CBI) = \tan(IBA)$; $\tan(BCI) = \tan(ICA)$; $OC \equiv OB$; $OB \equiv OA$; Y is on line AC; $YI \perp AC$; Z is on line AB; $ZI \perp AB$; M is on line YB; M is on line XA. The **conclusion**: $\overline{AM}^2\ \overline{BM}^2\ \overline{CM}^2\ \overline{IX}^2 - 16\ \overline{XM}^2\ \overline{YM}^2\ \overline{ZM}^2\ \overline{OB}^2 = 0$.

$X = (0,0)$, $B = (u_1, 0)$, $C = (u_2, 0)$, $I = (0, u_3)$, $A = (x_2, x_1)$, $O = (x_4, x_3)$, $Y = (x_6, x_5)$, $Z = (x_8, x_7)$, $M = (x_{10}, x_9)$.

The **nondegenerate** conditions: $X \neq B$; $X \neq B$; Points I, C and B are not collinear; $\angle CIB$ is not right; Points B, A and C are not collinear; Line AC is non-isotropic; Line AB is non-isotropic; Line XA intersects line YB.

Remark. Our proof is four the four Gergonne points.

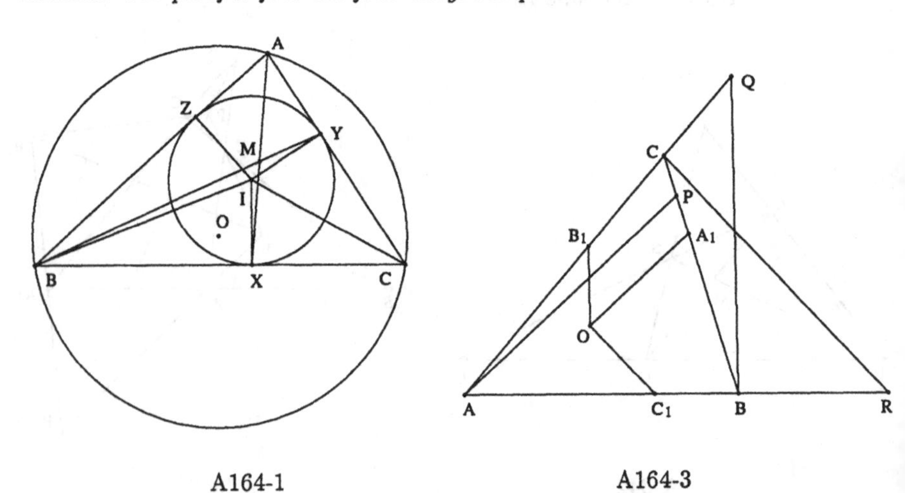

A164-1 A164-3

Example 348 (A164-3: 2.82s, 64; 2.02s, 70). *The lines AP, BQ, CR through the vertices of a triangle ABC parallel, respectively, to the lines OA_1, OB_1, OC_1 joining any point O to the points A_1, B_1, C_1 marked in any manner whatever, on the sides of BC, CA, AB meet these sides in the points P, Q, R. Show that $OA_1/AP + OB_1/BQ + OC_1/CR = 1$.*

Points A, B, C, O are arbitrarily chosen. Points A_1, B_1, C_1, P, Q, R are constructed (in order) as follows: A_1 is on line BC; B_1 is on line CA; C_1 is on line BA; P is on line BC; $PA \parallel A_1O$; Q is on line AC; $QB \parallel B_1O$; R is on line AB; $RC \parallel C_1O$. The **conclusion**: $1 - (\frac{\overline{OA_1}}{\overline{AP}} + \frac{\overline{OB_1}}{\overline{BQ}} + \frac{\overline{OC_1}}{\overline{CR}}) = 0$.

$A = (0,0)$, $B = (u_1,0)$, $C = (u_2,u_3)$, $O = (u_4,u_5)$, $A_1 = (x_1,u_6)$, $B_1 = (x_2,u_7)$, $C_1 = (u_8,0)$, $P = (x_4,x_3)$, $Q = (x_6,x_5)$, $R = (x_7,0)$.

The **nondegenerate** conditions: $B \neq C$; $C \neq A$; $B \neq A$; Line A_1O intersects line BC; Line B_1O intersects line AC; Line C_1O intersects line AB.

Example 349 (A164-4: 4.62s, 40; 3.68s, 43). *If the altitudes AD, BE, CF of the triangle ABC meet the circumcircle of ABC again in P, Q, R, show that we have $(AP : AD) + (BQ : BE) + (CR : CF) = 4$.*

Points A, B, C are arbitrarily chosen. Points O, D, E, F, H, P, Q, R are constructed (in order) as follows: $OA \equiv OC$; $OA \equiv OB$; D is on line BC; $DA \perp BC$; E is on line AC; $EB \perp AC$; F is on line AB; $FC \perp AB$; H is on line AD; H is on line CF; P is on line AD; $PO \equiv OA$; Q is on line BE; $QO \equiv OA$; R is on line CF; $RO \equiv OA$. The **conclusion**: $4 - (\frac{\overline{AP}}{\overline{AD}} + \frac{\overline{BQ}}{\overline{BE}} + \frac{\overline{CR}}{\overline{CF}}) = 0$.

$A = (0,0)$, $B = (u_1,0)$, $C = (u_2,u_3)$, $O = (x_2,x_1)$, $D = (x_4,x_3)$, $E = (x_6,x_5)$, $F = (u_2,0)$, $H = (u_2,x_7)$, $P = (x_9,x_8)$, $Q = (x_{11},x_{10})$, $R = (u_2,x_{12})$.

The **nondegenerate** conditions: Points A, B and C are not collinear; Line BC is non-isotropic; Line AC is non-isotropic; Line AB is non-isotropic; Line CF intersects line AD; Line AD is non-isotropic; Line BE is non-isotropic; Line CF is non-isotropic. In addition, the following nondegenerate conditions, which come from reducibility and have been detected by our prover, should be also added: $R \neq C$; $Q \neq B$; $P \neq A$.

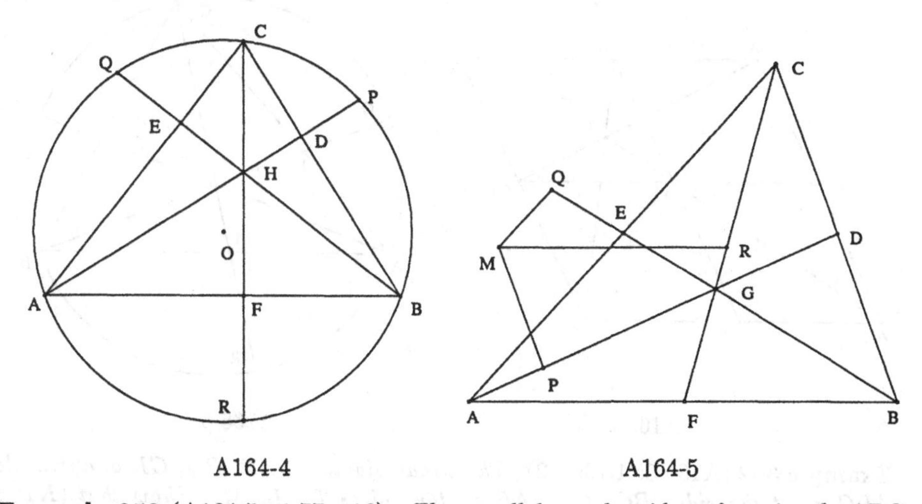

A164-4 A164-5

Example 350 (A164-5: 1.77s, 16). *The parallels to the sides of a triangle ABC through the same point, M, meet the respective medians in the points P, Q, R. Prove that we have, both in magnitude and in sign, $(GP : GA) + (GQ : GB) + (GR : GC) = 0$.*

Points A, B, C, M are arbitrarily chosen. Points D, E, F, G, P, Q, R are constructed (in order) as follows: D is the midpoint of B and C; E is the midpoint of C and A; F is the midpoint of A and B; G is on line AD; G is on line CF; P is on line AD; $PM \parallel BC$; Q is on line BE; $QM \parallel AC$; R is on line CF; $RM \parallel AB$. The **conclusion**: $\frac{\overline{GP}}{\overline{GA}} + \frac{\overline{GQ}}{\overline{GB}} + \frac{\overline{GR}}{\overline{GC}} = 0$.

$A = (0,0)$, $B = (u_1, 0)$, $C = (u_2, u_3)$, $M = (u_4, u_5)$, $D = (x_1, x_2)$, $E = (x_3, x_4)$, $F = (x_5, 0)$, $G = (x_7, x_6)$, $P = (x_9, x_8)$, $Q = (x_{11}, x_{10})$, $R = (x_{12}, u_5)$.

The **nondegenerate** conditions: Line CF intersects line AD; Line BC intersects line AD; Line AC intersects line BE; Line AB intersects line CF.

Example 351 (A165-5: 1.2s, 11). *Three parallel lines are cut by three parallel transversals in the points A, B, C; A_1, B_1, C_1; A_2, B_2, C_2. Show that B_2C, C_1A_2, AB_1 are concurrent.*

Points A, B, A_1 are arbitrarily chosen. Points C, B_1, C_1, A_2, B_2, C_2, I are constructed (in order) as follows: C is on line AB; $B_1A_1 \parallel AB$; $B_1B \parallel AA_1$; C_1 is on line A_1B_1; $C_1C \parallel AA_1$; A_2 is on line AA_1; $B_2A_2 \parallel AB$; B_2 is on line BB_1; C_2 is on line A_2B_2; C_2 is on line CC_1; I is on line C_1A_2; I is on line B_2C. The **conclusion**: Points I, A and B_1 are collinear.

$A = (0,0)$, $B = (u_1, 0)$, $A_1 = (u_2, u_3)$, $C = (u_4, 0)$, $B_1 = (x_1, u_3)$, $C_1 = (x_2, u_3)$, $A_2 = (x_3, u_5)$, $B_2 = (x_4, u_5)$, $C_2 = (x_5, u_5)$, $I = (x_7, x_6)$.

The **nondegenerate** conditions: $A \neq B$; Points A, A_1 and B are not collinear; Points A, A_1 and B_1 are not collinear; $A \neq A_1$; Points B, B_1 and A are not collinear; Line CC_1 intersects line A_2B_2; Line B_2C intersects line C_1A_2.

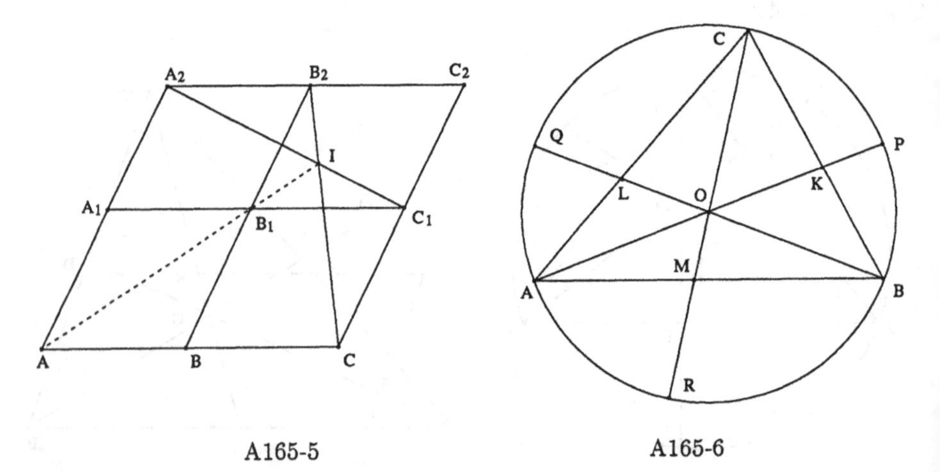

A165-5 A165-6

Example 352 (A165-6: 1.73s, 12). *The circumdiameters AP, BQ, CR of a triangle ABC meet the sides BC, CA, AB in the points K, L, M. Show that $(KP : AK) + (LQ : BL) + (MR : AM) = 1$.*

Points A, B, C are arbitrarily chosen. Points O, P, Q, R, K, L, M are constructed (in order) as follows: $OA \equiv OC$; $OA \equiv OB$; O is the midpoint of A and P; O is the midpoint of B and Q; O is the midpoint of C and R; K is on line BC; K is on line AO; L is on line CA; L is on line BO; M is on line AB; M is on line CO. The **conclusion**: $1-(\frac{\overline{KP}}{\overline{AK}}+\frac{\overline{LQ}}{\overline{BL}}+\frac{\overline{MR}}{\overline{CM}})=0$.

$A=(0,0)$, $B=(u_1,0)$, $C=(u_2,u_3)$, $O=(x_2,x_1)$, $P=(x_3,x_4)$, $Q=(x_5,x_6)$, $R=(x_7,x_8)$, $K=(x_{10},x_9)$, $L=(x_{12},x_{11})$, $M=(x_{13},0)$.

The **nondegenerate** conditions: Points A, B and C are not collinear; Line AO intersects line BC; Line BO intersects line CA; Line CO intersects line AB.

Example 353 (A165-7: 48.42s, 1189; 18.85s, 54). *Show that the line joining the incenter of the triangle ABC to the midpoint of the segment joining A to the Nagel point of ABC is bisected by the median issued from A.*

Points B, C, I are arbitrarily chosen. Points A, I_A, I_B, X_A, Y_B, N, A_1, S, P are constructed (in order) as follows: $\tan(CBI)=\tan(IBA)$; $\tan(BCI)=\tan(ICA)$; $I_AB \perp IB$; I_A is on line AI; I_B is on line BI; I_B is on line CI_A; X_A is on line BC; $X_AI_A \perp BC$; Y_B is on line CA; $Y_BI_B \perp AC$; N is on line AX_A; N is on line BY_B; A_1 is the midpoint of B and C; S is the midpoint of N and A; P is on line A_1A; P is on line IS. The **conclusion**: P is the midpoint of I and S.

$B=(0,0)$, $C=(u_1,0)$, $I=(u_2,u_3)$, $A=(x_2,x_1)$, $I_A=(x_4,x_3)$, $I_B=(x_6,x_5)$, $X_A=(x_4,0)$, $Y_B=(x_8,x_7)$, $N=(x_{10},x_9)$, $A_1=(x_{11},0)$, $S=(x_{12},x_{13})$, $P=(x_{15},x_{14})$.

The **nondegenerate** conditions: Points I, C and B are not collinear; $\angle CIB$ is not right; Line AI is not perpendicular to line IB; Line CI_A intersects line BI; Line BC is non-isotropic; Line AC is non-isotropic; Line BY_B intersects line AX_A; Line IS intersects line A_1A.

Remark. Our proof is four the four Nagel points.

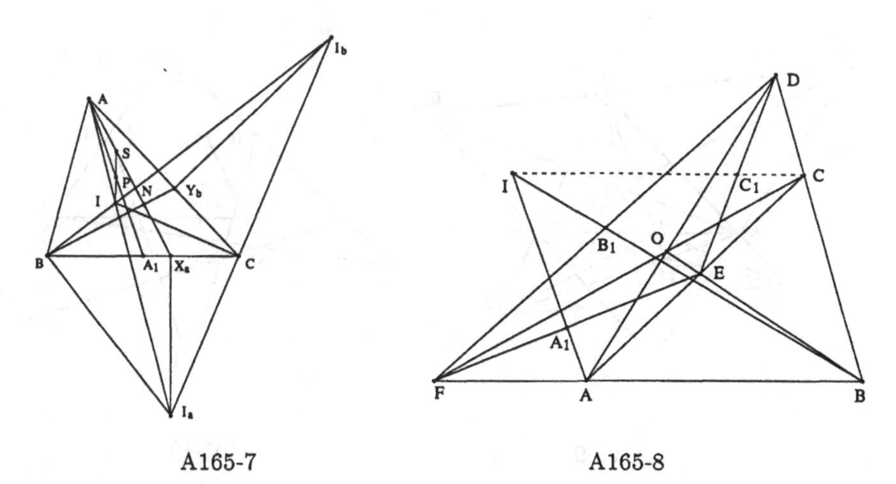

A165-7 A165-8

Example 354 (A165-8: 10.83s, 46). *Through the vertices of a triangle ABC lines are drawn intersecting in O and meeting the opposite sides in D, E, F. Prove that the lines joining A, B, C to the midpoints of EF, FD, DE are concurrent.*

Points A, B, C, O are arbitrarily chosen. Points D, E, F, A_1, B_1, C_1, I are constructed (in order) as follows: D is on line BC; D is on line AO; E is on line AC; E is on line BO; F is on line AB; F is on line CO; A_1 is the midpoint of E and F; B_1 is the midpoint of F and D; C_1 is the midpoint of D and E; I is on line BB_1; I is on line AA_1. The **conclusion**: Points I, C and C_1 are collinear.

$A = (0,0)$, $B = (u_1,0)$, $C = (u_2,u_3)$, $O = (u_4,u_5)$, $D = (x_2,x_1)$, $E = (x_4,x_3)$, $F = (x_5,0)$, $A_1 = (x_6,x_7)$, $B_1 = (x_8,x_9)$, $C_1 = (x_{10},x_{11})$, $I = (x_{13},x_{12})$.

The **nondegenerate** conditions: Line AO intersects line BC; Line BO intersects line AC; Line CO intersects line AB; Line AA_1 intersects line BB_1.

Example 355 (A165-9: 35.53s, 1422; 15.53s, 130). *If (Q) is the cevian triangle of a point M for the triangle (P), show that the triangle formed by the parallels through the vertices of (P) to the corresponding sides of (Q) is perspective to (P).*

Points A, B, C, M are arbitrarily chosen. Points A_1, B_1, C_1, C_2, A_2, B_2, P are constructed (in order) as follows: A_1 is on line BC; A_1 is on line AM; B_1 is on line AC; B_1 is on line BM; C_1 is on line AB; C_1 is on line CM; $C_2B \parallel A_1C_1$; $C_2A \parallel B_1C_1$; $A_2C \parallel A_1B_1$; A_2 is on line C_2B; B_2 is on line C_2A; B_2 is on line A_2C; P is on line BB_2; P is on line AA_2. The **conclusion**: Points C, C_2 and P are collinear.

$A = (0,0)$, $B = (u_1,0)$, $C = (u_2,u_3)$, $M = (u_4,u_5)$, $A_1 = (x_2,x_1)$, $B_1 = (x_4,x_3)$, $C_1 = (x_5,0)$, $C_2 = (x_7,x_6)$, $A_2 = (x_9,x_8)$, $B_2 = (x_{11},x_{10})$, $P = (x_{13},x_{12})$.

The **nondegenerate** conditions: Line AM intersects line BC; Line BM intersects line AC; Line CM intersects line AB; Points B_1, C_1 and A_1 are not collinear; Line C_2B intersects line A_1B_1; Line A_2C intersects line C_2A; Line AA_2 intersects line BB_2.

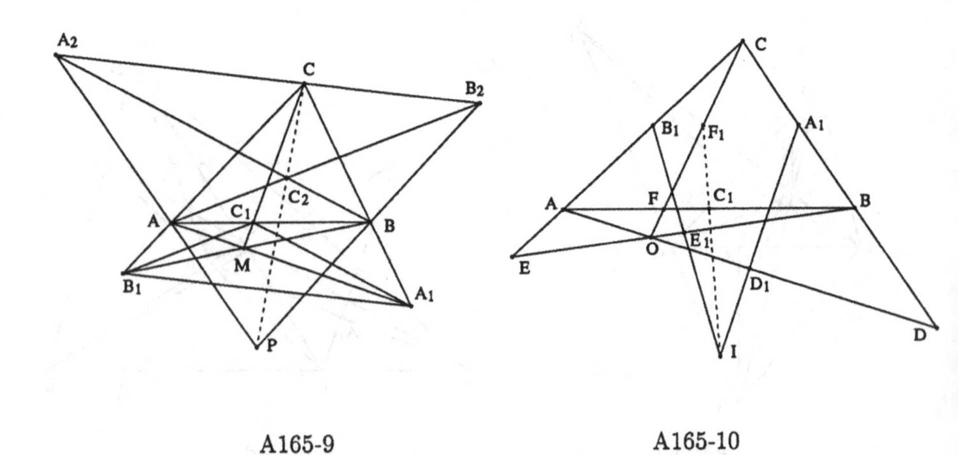

A165-9 A165-10

Example 356 (A165-10: 4.15s, 27). *Prove that the lines joining the midpoints of three concurrent cevians to the midpoints of the corresponding sides of the given triangle are concurrent.*

Points A, B, C, O are arbitrarily chosen. Points D, E, F, D_1, E_1, F_1, A_1, B_1, C_1, I are constructed (in order) as follows: D is on line BC; D is on line AO; E is on line AC; E is on line BO; F is on line AB; F is on line CO; D_1 is the midpoint of A and D; E_1 is the midpoint of B and E; F_1 is the midpoint of C and F; A_1 is the midpoint of B and C; B_1 is the midpoint of C and A; C_1 is the midpoint of A and B; I is on line B_1E_1; I is on line A_1D_1. The **conclusion**: Points I, C_1 and F_1 are collinear.

$A = (0,0)$, $B = (u_1,0)$, $C = (u_2,u_3)$, $O = (u_4,u_5)$, $D = (x_2,x_1)$, $E = (x_4,x_3)$, $F = (x_5,0)$, $D_1 = (x_6,x_7)$, $E_1 = (x_8,x_9)$, $F_1 = (x_{10},x_{11})$, $A_1 = (x_{12},x_{13})$, $B_1 = (x_{14},x_{15})$, $C_1 = (x_{16},0)$, $I = (x_{18},x_{17})$.

The **nondegenerate** conditions: Line AO intersects line BC; Line BO intersects line AC; Line CO intersects line AB; Line A_1D_1 intersects line B_1E_1.

Example 357 (A168-2: 2.97s, 12). *Given the parallelogram $MDOM_1$, the vertex O is joined to the midpoint C of MM_1. If the internal and external bisectors of the angle COD meet MD in A and B, show that $MD^2 = MA \cdot MB$.*

Points O, M_1, M are arbitrarily chosen. Points D, C, A, B are constructed (in order) as follows: $DO \parallel M_1M$; $DM \parallel M_1O$; C is the midpoint of M_1 and M; A is on line MD; $\tan(COA) = \tan(AOD)$; B is on line MD; $BO \perp AO$. The **conclusion**: $MD \cdot MD = MA \cdot MB$.

$O = (0,0)$, $M_1 = (u_1,0)$, $M = (u_2,u_3)$, $D = (x_1,u_3)$, $C = (x_2,x_3)$, $A = (x_4,u_3)$, $B = (x_5,u_3)$.

The **nondegenerate** conditions: Points M_1, O and M are not collinear; $x_1x_3 + u_3x_2 \neq 0$; Line AO is not perpendicular to line MD.

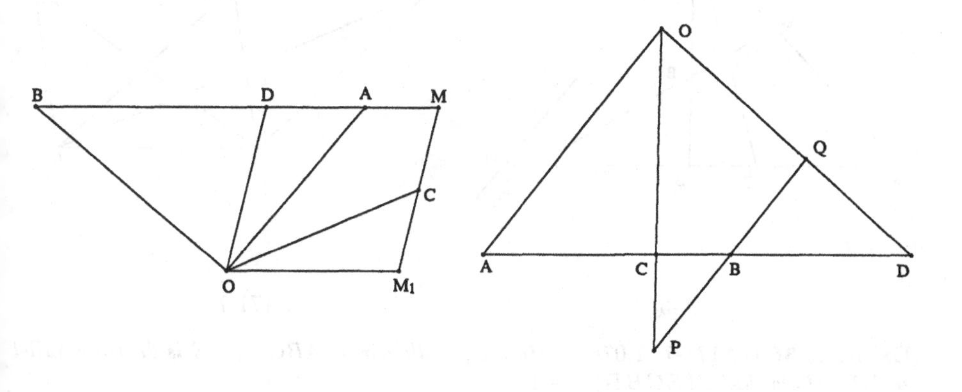

A168-2 A-351

Example 358 (A-351: 3.13s, 9; 2.8s, 7). *Given $(ABCD) = -1$ and a point O outside the line AB, if a parallel through B to OA meets OC, OD in P, Q, we then have $PB = BQ$.*

Points A, B, O are arbitrarily chosen. Points C, D, P, Q are constructed (in order) as follows: C is on line AB; D is on line AB; Points A, B, C and D form a harmonic set; P is on line OC; $PB \parallel OA$; Q is on line OD; Q is on line BP. The **conclusion**: B is the midpoint of P and Q.

$A = (0,0)$, $B = (u_1,0)$, $O = (u_2,u_3)$, $C = (u_4,0)$, $D = (x_1,0)$, $P = (x_3,x_2)$, $Q = (x_5,x_4)$.

The **nondegenerate** conditions: $A \neq B$; $A \neq B$; C is not the midpoint of A and B. Points O, A and C are not collinear; Line BP intersects line OD.

Example 359 (A-353: 2.68s, 160; 2.52s, 67). *Given $(ABCD) = -1$ and a point O outside the line AB, any transversal cuts the four lines OA, OB, OC, OD in four harmonic points.*

Points A, B, O are arbitrarily chosen. Points C, D, A_1, B_1, C_1, D_1 are constructed (in order) as follows: C is on line AB; D is on line AB; Points A, B, C and D form a harmonic set; A_1 is on line OA; B_1 is on line OB; C_1 is on line A_1B_1; C_1 is on line OC; D_1 is on line A_1B_1; D_1 is on line OD. The **conclusion**: Points A_1, B_1, C_1 and D_1 form a harmonic set.

$A = (0,0)$, $B = (u_1,0)$, $O = (u_2,u_3)$, $C = (u_4,0)$, $D = (x_1,0)$, $A_1 = (x_2,u_5)$, $B_1 = (x_3,u_6)$, $C_1 = (x_5,x_4)$, $D_1 = (x_7,x_6)$.

The **nondegenerate** conditions: $A \neq B$; $A \neq B$; C is not the midpoint of A and B. $O \neq A$; $O \neq B$; Line OC intersects line A_1B_1; Line OD intersects line A_1B_1.

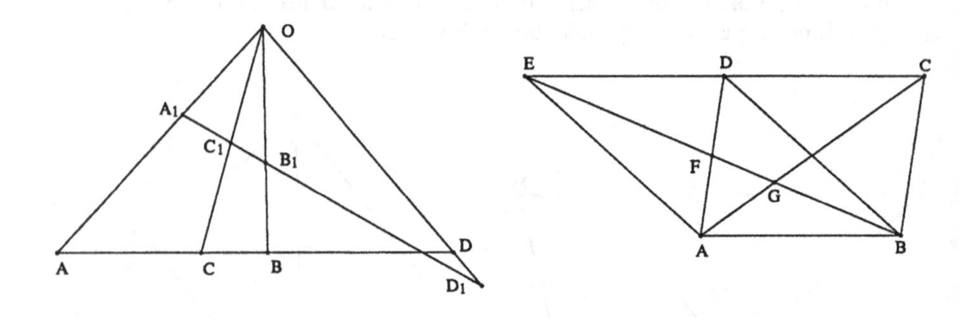

A-353 A171-7

Example 360 (A171-7: 1.07s, 6). *In the parallelogram $ABCD$, AE is drawn parallel to BD; show that $A(ECBD) = -1$.*

Points A, B, C are arbitrarily chosen. Points D, E, F, G are constructed (in order) as follows: $DA \parallel BC$; $DC \parallel BA$; E is on line DC; $EA \parallel BD$; F is on line BE; F is on line AD; G is on line EB; G is on line AC. The **conclusion**: Points E, G, F and B form a harmonic set.

$A = (0,0)$, $B = (u_1,0)$, $C = (u_2,u_3)$, $D = (x_1,u_3)$, $E = (x_2,u_3)$, $F = (x_4,x_3)$, $G = (x_6,x_5)$.

The **nondegenerate** conditions: Points B, A and C are not collinear; Points B, D and C are not collinear; Line AD intersects line BE; Line AC intersects line EB.

Example 361 (A171-8: 1.15s, 6). *If A_1, B_1, C_1 are the midpoint of the sides of the triangle ABC, prove that A_1A is the harmonic conjugate of A_1C with respect to A_1B_1, A_1C_1.*

Points B, C, A are arbitrarily chosen. Points A_1, B_1, C_1, G, D are constructed (in order) as follows: A_1 is the midpoint of C and B; B_1 is the midpoint of A and C; C_1 is the midpoint of B and A; G is on line C_1C; G is on line AA_1; D is on line C_1C; D is on line A_1B_1. The **conclusion**: Points C_1, D, G and C form a harmonic set.

$B = (0,0)$, $C = (u_1,0)$, $A = (u_2,u_3)$, $A_1 = (x_1,0)$, $B_1 = (x_2,x_3)$, $C_1 = (x_4,x_5)$, $G = (x_7,x_6)$, $D = (x_9,x_8)$.

The **nondegenerate** conditions: Line AA_1 intersects line C_1C; Line A_1B_1 intersects line C_1C.

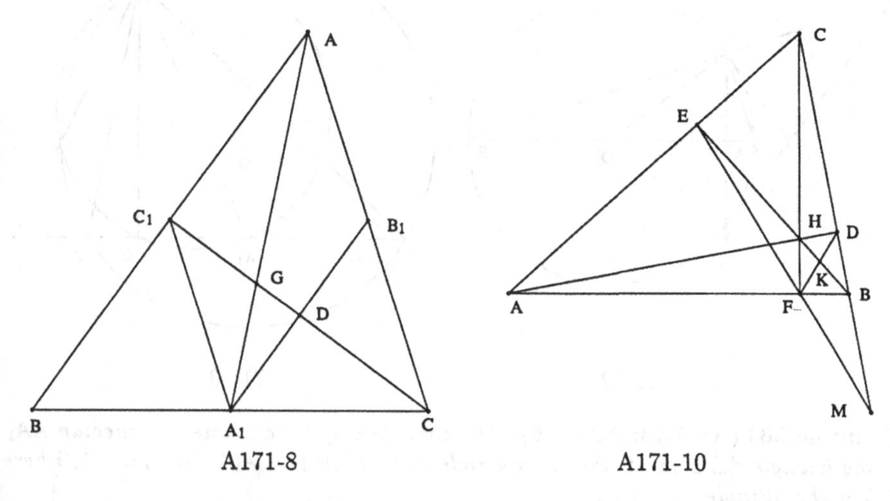

A171-8 A171-10

Example 362 (A171-10: 3.53s, 16). *With the usual notations for the triangle ABC, if DF meets BE in K, show that $(BHKE) = -1$; also if EF meets BC in M, show that $(BCDM) = -1$.*

Points A, B, C are arbitrarily chosen. Points D, E, F, H, K, M are constructed (in order) as follows: D is on line BC; $DA \perp BC$; E is on line AC; $EB \perp AC$; F is on line AB; $FC \perp AB$; H is on line AD; H is on line CF; K is on line DF; K is on line BE; M is on line BC; M is on line EF. The **conclusions**: (1) Points B, C, D and M form a harmonic set; (2) Points B, H, K and E form a harmonic set.

$A = (0,0)$, $B = (u_1,0)$, $C = (u_2,u_3)$, $D = (x_2,x_1)$, $E = (x_4,x_3)$, $F = (u_2,0)$, $H = (u_2,x_5)$, $K = (x_7,x_6)$, $M = (x_9,x_8)$.

The **nondegenerate** conditions: Line BC is non-isotropic; Line AC is non-isotropic; Line AB is non-isotropic; Line CF intersects line AD; Line BE intersects line DF; Line EF intersects line BC.

Example 363 (A171-12: 1.42s, 12). *The tangent to a circle at the point C meets the diameter AB, produced, in T; Prove that the other tangent from T to the circle is divided harmonically by CA, CB, CT and its point of contact.*

Points A, O are arbitrarily chosen. Points B, C, E, D, T, A_1, B_1 are constructed (in order) as follows: O is the midpoint of A and B; $CO \equiv OA$; E is on line AO; $EC \perp AO$; E is the midpoint of C and D; $TC \perp CO$; T is on line AO; A_1 is on line TD; A_1 is on line CA; B_1 is on line TD; B_1 is on line CB. The **conclusion**: Points T, D, A_1 and B_1 form a harmonic set.

$A = (0,0)$, $O = (u_1,0)$, $B = (x_1,0)$, $C = (x_2,u_2)$, $E = (x_2,0)$, $D = (x_2,x_3)$, $T = (x_4,0)$, $A_1 = (x_6,x_5)$, $B_1 = (x_8,x_7)$.

The **nondegenerate** conditions: Line AO is non-isotropic; Line AO is not perpendicular to line CO; Line CA intersects line TD; Line CB intersects line TD.

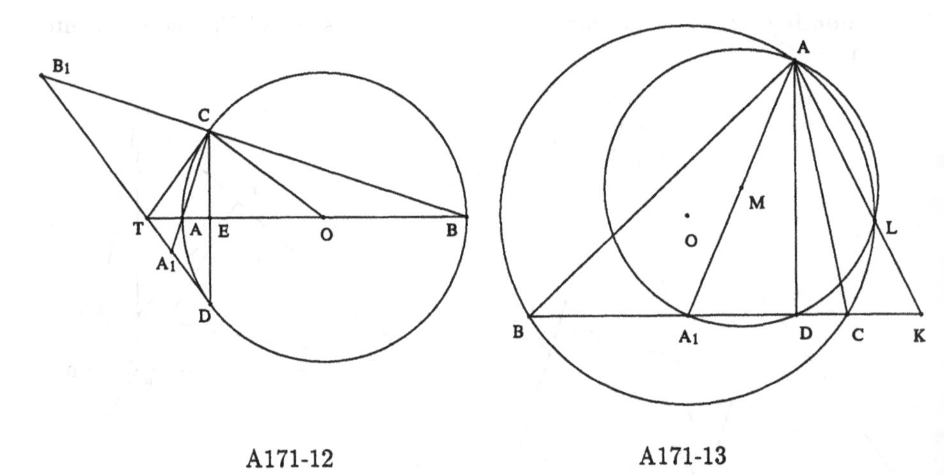

A171-12 A171-13

Example 364 (A171-13: 2.92s, 36). *The circle having for diameter the median AA_1 of the triangle ABC meets the circumcircle in L: show that $A(LDBC) = -1$, where AD is the altitude.*

Points B, C, A are arbitrarily chosen. Points A_1, D, O, M, L, K are constructed (in order) as follows: A_1 is the midpoint of B and C; D is on line BC; $DA \perp BC$; $OB \equiv OA$; $OB \equiv OC$; M is the midpoint of A_1 and A; $LM \equiv MA$; $LO \equiv OB$; K is on line BC; K is on line AL. The **conclusion**: Points K, D, B and C form a harmonic set.

$B = (0,0)$, $C = (u_1,0)$, $A = (u_2,u_3)$, $A_1 = (x_1,0)$, $D = (u_2,0)$, $O = (x_3,x_2)$, $M = (x_4,x_5)$, $L = (x_7,x_6)$, $K = (x_8,0)$.

The **nondegenerate** conditions: Line BC is non-isotropic; Points B, C and A are not collinear; Line OM is non-isotropic; Line AL intersects line BC. In addition, the following nondegenerate conditions, which come from reducibility and have been detected by our prover, should be also added: $L \neq A$.

Example 365 (A171-4: 1.47s, 12). *The sides AB, AC intercept the segments DE,*

FG on the parallels to the side BC through the tritangent centers I and I_a. *Show that:* $2/BC = 1/DE + 1/FG$.

Points B, C, I are arbitrarily chosen. Points A, I_A, D, E, F, G are constructed (in order) as follows: $\tan(ACI) = \tan(ICB)$; $\tan(ABI) = \tan(IBC)$; I_A is on line AI; $I_AB \perp IB$; D is on line AB; $DI \parallel BC$; E is on line AC; E is on line DI; F is on line AB; $FI_A \parallel BC$; G is on line AC; G is on line FI_A. The **conclusion**: $2/\overline{BC} - (1/\overline{DE} + 1/\overline{FG}) = 0$.

$B = (0,0)$, $C = (u_1, 0)$, $I = (u_2, u_3)$, $A = (x_2, x_1)$, $I_A = (x_4, x_3)$, $D = (x_5, u_3)$, $E = (x_6, u_3)$, $F = (x_7, x_3)$, $G = (x_8, x_3)$.

The **nondegenerate** conditions: Points B, I and C are not collinear; $\angle BIC$ is not right; Line IB is not perpendicular to line AI; Points B, C and A are not collinear; Line DI intersects line AC; Points B, C and A are not collinear; Line FI_A intersects line AC.

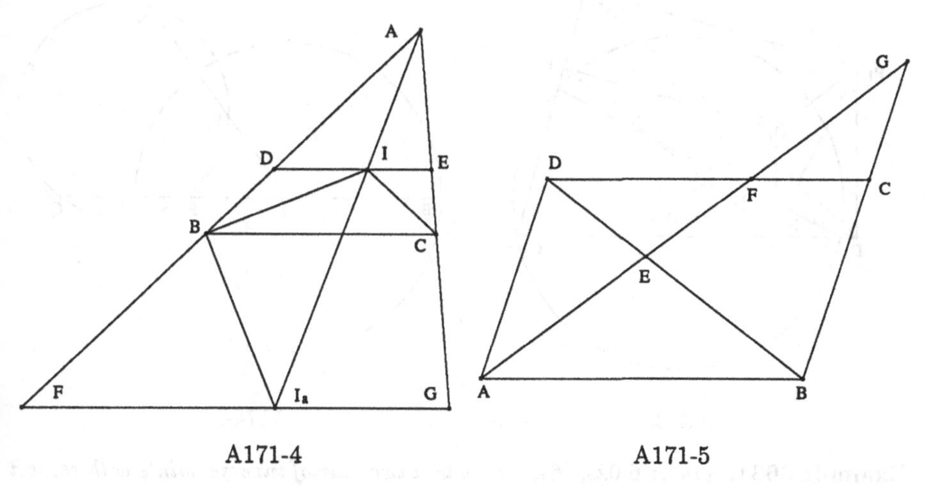

A171-4 A171-5

Example 366 (A171-5: 0.77s, 7). *A secant through the vertex A of the parallelogram ABCD meets the diagonal BD and the sides BC, CD in the points E, F, G. Show that:* $1/AE = 1/AF + 1/AG$.

Points A, B, C are arbitrarily chosen. Points D, E, F, G are constructed (in order) as follows: $DA \parallel CB$; $DC \parallel BA$; E is on line BD; F is on line EA; F is on line DC; G is on line FA; G is on line BC. The **conclusion**: $1/\overline{AE} - (1/\overline{AF} + 1/\overline{AG}) = 0$.

$A = (0,0)$, $B = (u_1, 0)$, $C = (u_2, u_3)$, $D = (x_1, u_3)$, $E = (x_2, u_4)$, $F = (x_3, u_3)$, $G = (x_5, x_4)$.

The **nondegenerate** conditions: Points B, A and C are not collinear; $B \neq D$; Line DC intersects line EA; Line BC intersects line FA.

Example 367 (A-362: 57.02s, 4070). *The pedal triangles, for a given triangle, of two points inverse for the circumcircle of the given triangle are similar.*

Points B, C, A are arbitrarily chosen. Points O, H, P, P_1, E, F, D, F_1, D_1, E_1

are constructed (in order) as follows: $OB \equiv OC$; $OA \equiv OB$; $HO \equiv OB$; P is on line OH; P_1 is on line OH; P and P_1 are inversive wrpt circle (O, OH); E is on line AC; $EP \perp AC$; F is on line BA; $FP \perp BA$; D is on line BC; $DP \perp BC$; F_1 is on line BA; $F_1P_1 \perp BA$; D_1 is on line BC; $D_1P_1 \perp BC$; E_1 is on line AC; $E_1P_1 \perp AC$. The **conclusion**: $DF \cdot D_1E_1 = D_1F_1 \cdot DE$.

$B = (0,0)$, $C = (u_1,0)$, $A = (u_2,u_3)$, $O = (x_2,x_1)$, $H = (x_3,u_4)$, $P = (x_4,u_5)$, $P_1 = (x_6,x_5)$, $E = (x_8,x_7)$, $F = (x_{10},x_9)$, $D = (x_4,0)$, $F_1 = (x_{12},x_{11})$, $D_1 = (x_6,0)$, $E_1 = (x_{14},x_{13})$.

The **nondegenerate** conditions: Points A, B and C are not collinear; $O \neq H$; $O \neq H$; $P \neq O$; Line AC is non-isotropic; Line BA is non-isotropic; Line BC is non-isotropic; Line BA is non-isotropic; Line BC is non-isotropic; Line AC is non-isotropic.

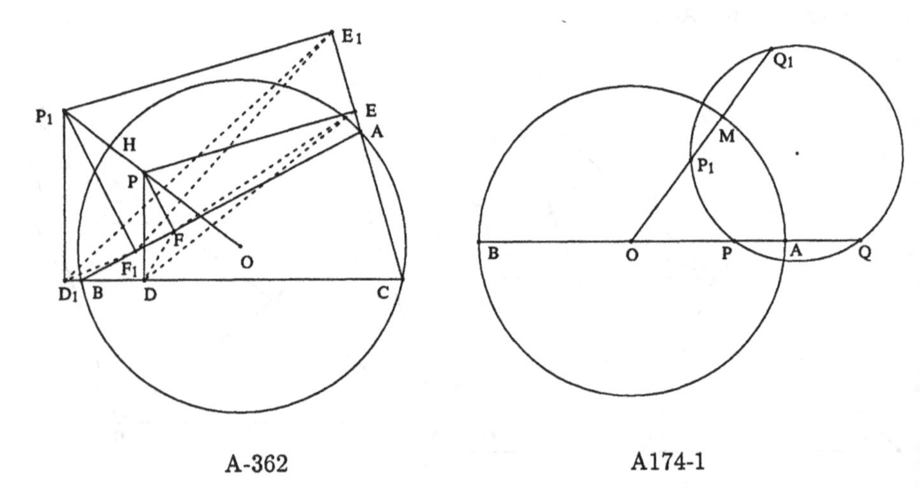

A-362 A174-1

Example 368 (A174-1: 6.02s, 16). *Prove that two pair of inverse points with respect to the same circle are cyclic, or collinear.*

Points O, A are arbitrarily chosen. Points M, B, P, Q, P_1, Q_1 are constructed (in order) as follows: $MO \equiv OA$; O is the midpoint of A and B; P is on line OA; P and Q are inversive wrpt circle (O, OA); Q is on line OB; P_1 is on line OM; P_1 and Q_1 are inversive wrpt circle (O, OM); Q_1 is on line OM. The **conclusion**: Points P, Q, P_1 and Q_1 are on the same circle.

$O = (0,0)$, $A = (u_1,0)$, $M = (x_1,u_2)$, $B = (x_2,0)$, $P = (u_3,0)$, $Q = (x_3,0)$, $P_1 = (x_4,u_4)$, $Q_1 = (x_6,x_5)$.

The **nondegenerate** conditions: $O \neq A$; $-u_3 \neq 0$; $O \neq M$; $O \neq M$; $P_1 \neq O$.

Example 369 (A174-3: 1.33s, 3). *If the circle (B) passes through the center A of the circle (A), and a diameter of (A) meets the common chord of the two circles in F and the circle (B) again in G, show that the points F, G are inverse for the circle (A).*

Points A, B are arbitrarily chosen. Points E, I, D, F, G are constructed (in order) as follows: $EB \equiv BA$; $IE \perp AB$; I is on line AB; $DA \equiv AE$; F is on line EI; F is on line AD; $GB \equiv BA$; G is on line AD. The **conclusion**: G and F are inversive wrpt circle (A, AD).

$A = (0,0)$, $B = (u_1,0)$, $E = (x_1,u_2)$, $I = (x_1,0)$, $D = (x_2,u_3)$, $F = (x_1,x_3)$, $G = (x_5,x_4)$.

The **nondegenerate** conditions: Line AB is non-isotropic; Line AD intersects line EI; Line AD is non-isotropic. In addition, the following nondegenerate conditions, which come from reducibility and have been detected by our prover, should be also added: $G \neq A$.

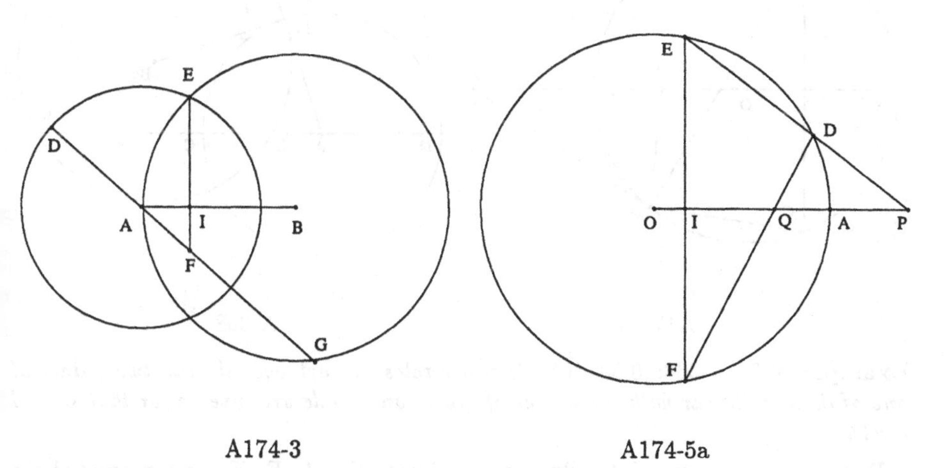

A174-3 A174-5a

Example 370 (A174-5a: 1.4s, 4). *Show that the two lines joining any point of a circle to the ends of a given chord meet the diameter perpendicular to that chord in two inverse points.*

Points O, A are arbitrarily chosen. Points E, I, F, D, P, Q are constructed (in order) as follows: $EO \equiv OA$; $IE \perp AO$; I is on line OA; I is the midpoint of E and F; $DO \equiv OA$; P is on line OA; P is on line DE; Q is on line OA; Q is on line DF. The **conclusion**: P and Q are inversive wrpt circle (O, OA).

$O = (0,0)$, $A = (u_1,0)$, $E = (x_1,u_2)$, $I = (x_1,0)$, $F = (x_1,x_2)$, $D = (x_3,u_3)$, $P = (x_4,0)$, $Q = (x_5,0)$.

The **nondegenerate** conditions: Line OA is non-isotropic; Line DE intersects line OA; Line DF intersects line OA.

Example 371 (A174-6: 0.97s, 36). *If P, Q are two inverse points with respect to a circle and CD is a chord perpendicular to the diameter containing P, Q, the angels which the segments CP, DQ subtend at any point M of the circle are either equal or supplementary.*

Points A, B are arbitrarily chosen. Points O, P, Q, D, X, C, M are constructed (in order) as follows: O is the midpoint of A and B; P is on line AB; Q is on line AB; P and Q are inversive wrpt circle (O, OA); $DO \equiv OA$; X is on line

AB; $XD \perp AB$; X is the midpoint of D and C; $MO \equiv OA$. The **conclusion**: $\tan(CMP) = \tan(QMD)$.

$A = (0,0)$, $B = (u_1,0)$, $O = (x_1,0)$, $P = (u_2,0)$, $Q = (x_2,0)$, $D = (x_3,u_3)$, $X = (x_3,0)$, $C = (x_3,x_4)$, $M = (x_5,u_4)$.

The **nondegenerate** conditions: $A \neq B$; $x_1 - u_2 \neq 0$; Line AB is non-isotropic.

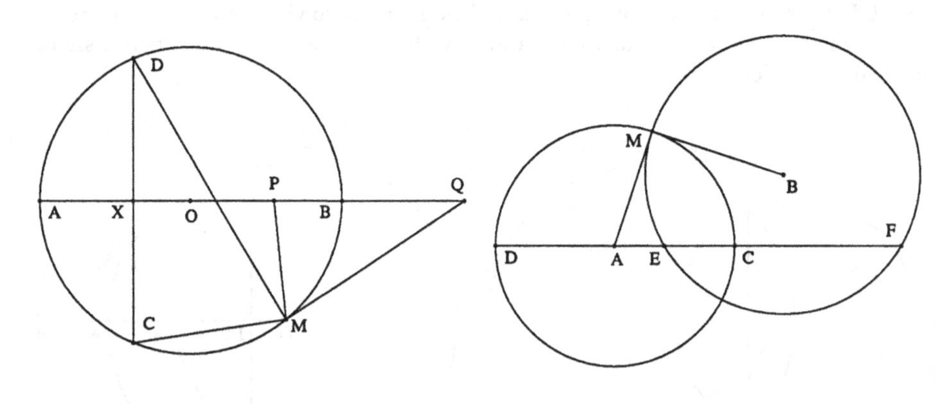

A174-6 A-368

Example 372 (A-368: 0.88s, 6). *If two circles are orthogonal, any two points of one of them collinear with the center of the second circle are inverse for that second circle.*

Points A, C, B are arbitrarily chosen. Points D, M, E, F are constructed (in order) as follows: A is the midpoint of C and D; $MA \equiv AC$; M is on the circle with diagonal AB; E is on line AC; $EB \equiv BM$; F is on line AD; $FB \equiv BM$. The **conclusion**: E and F are inversive wrpt circle (A, AC).

$A = (0,0)$, $C = (u_1,0)$, $B = (u_2,u_3)$, $D = (x_1,0)$, $M = (x_3,x_2)$, $E = (x_4,0)$, $F = (x_5,0)$.

The **nondegenerate** conditions: The line joining the midpoint of A and B and the point A is non-isotropic; Line AC is non-isotropic; Line AD is non-isotropic. In addition, the following nondegenerate conditions, which come from reducibility and have been detected by our prover, should be also added: $F \neq E$.

Example 373 (A-372: 1.15s, 17). *The two lines joining the points of intersection of two orthogonal circles to a point on one of the circles meet the other circle in two diametrically opposite points.*

Points Q, P are arbitrarily chosen. Points E, I, F, A, C, D are constructed (in order) as follows: E is on the circle with diagonal PQ; I is on line PQ; $IE \perp PQ$; I is the midpoint of E and F; $AP \equiv PE$; $CQ \equiv QE$; C is on line AE; Q is the midpoint of C and D. The **conclusion**: Points A, D and F are collinear.

$Q = (0,0)$, $P = (u_1,0)$, $E = (x_1,u_2)$, $I = (x_1,0)$, $F = (x_1,x_2)$, $A = (x_3,u_3)$, $C = (x_5,x_4)$, $D = (x_6,x_7)$.

The **nondegenerate** conditions: Line PQ is non-isotropic; Line AE is non-isotropic. In addition, the following nondegenerate conditions, which come from reducibility and have been detected by our prover, should be also added: $C \neq E$.

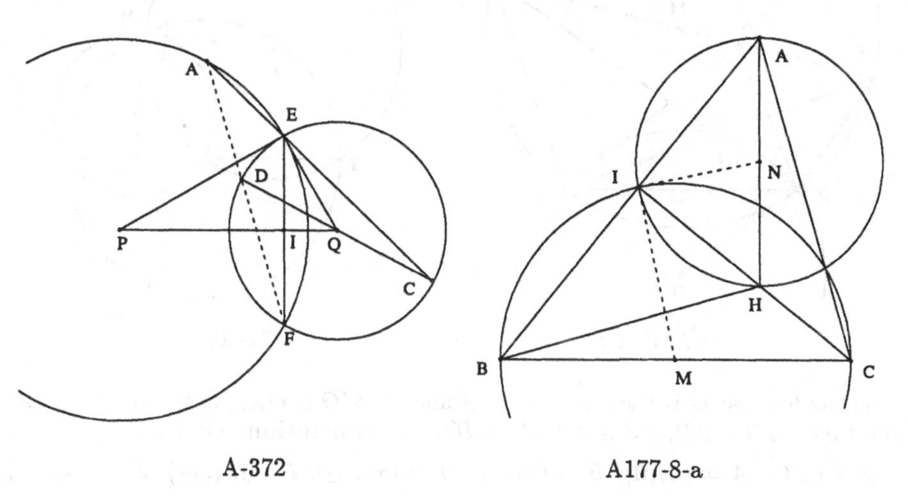

A-372 A177-8-a

Example 374 (A177-8-a: 1.63s, 28). *Show that in a triangle ABC the circles on AH and BC as diameters are orthogonal.*

Points B, C, A are arbitrarily chosen. Points H, M, N, I are constructed (in order) as follows: $HB \perp AC$; $HA \perp BC$; M is the midpoint of B and C; N is the midpoint of A and H; $IN \equiv NA$; $IM \equiv MB$. The **conclusion**: $MI \perp NI$.

$B = (0,0)$, $C = (u_1, 0)$, $A = (u_2, u_3)$, $H = (u_2, x_1)$, $M = (x_2, 0)$, $N = (u_2, x_3)$, $I = (x_5, x_4)$.

The **nondegenerate** conditions: Points B, C and A are not collinear; Line MN is non-isotropic.

Example 375 (A177-8-b: 2.38s, 13; 1.77s, 12). *The circle IBC is orthogonal to the circle on I_bI_c as diameter.*

Points B, C, I are arbitrarily chosen. Points A, I_B, I_C, O, M are constructed (in order) as follows: $\tan(CBI) = \tan(IBA)$; $\tan(BCI) = \tan(ICA)$; I_B is on line BI; $I_BC \perp IC$; I_C is on line CI; $I_CB \perp IB$; $OB \equiv OC$; $OB \equiv OI$; M is the midpoint of I_B and I_C. The **conclusion**: $MB \perp OB$.

$B = (0,0)$, $C = (u_1, 0)$, $I = (u_2, u_3)$, $A = (x_2, x_1)$, $I_B = (x_4, x_3)$, $I_C = (x_6, x_5)$, $O = (x_8, x_7)$, $M = (x_9, x_{10})$.

The **nondegenerate** conditions: Points I, C and B are not collinear; $\angle CIB$ is not right; Line IC is not perpendicular to line BI; Line IB is not perpendicular to line CI; Points B, I and C are not collinear.

Example 376 (A177-10: 1.33s, 18). *Show that if AB is a diameter and M any point of a circle, center O, the two circles AMO, BMO are orthogonal.*

Points O, A are arbitrarily chosen. Points B, M, I, J, P are constructed (in

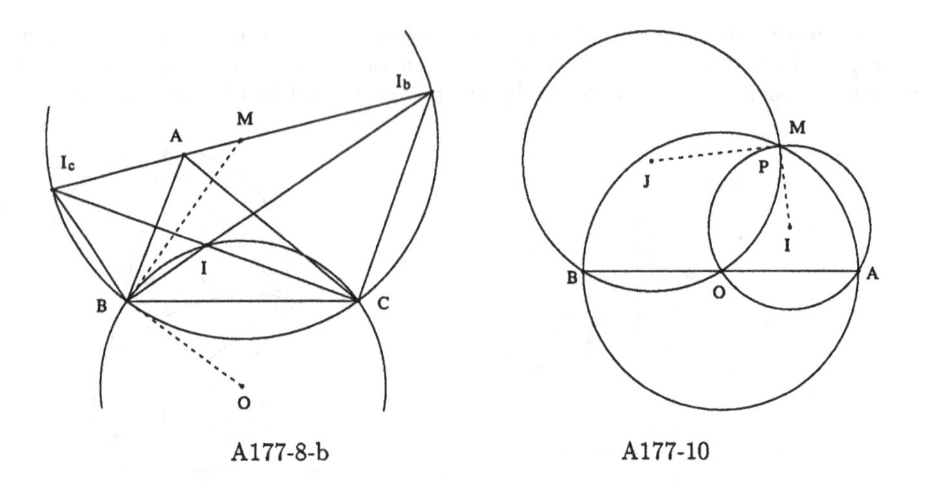

A177-8-b A177-10

order) as follows: O is the midpoint of A and B; $MO \equiv OA$; $IO \equiv IM$; $IO \equiv IA$; $JO \equiv JM$; $JO \equiv JB$; $PJ \equiv JO$; $PI \equiv IO$. The **conclusion:** $IP \perp JP$.

$O = (0,0)$, $A = (u_1, 0)$, $B = (x_1, 0)$, $M = (x_2, u_2)$, $I = (x_4, x_3)$, $J = (x_6, x_5)$, $P = (x_8, x_7)$.

The **nondegenerate** conditions: Points O, A and M are not collinear; Points O, B and M are not collinear; Line IJ is non-isotropic. In addition, the following nondegenerate conditions, which come from reducibility and have been detected by our prover, should be also added: $P \neq M$.

Example 377 (A177-12a: 2.55s, 50). *Show that given two perpendicular diameters of two orthogonal circles, the lines joining an end of one of these diameters to the ends of the other pass through the points common to the two circles.*

Points A, B are arbitrarily chosen. Points P, D, E, F, G, I are constructed (in order) as follows: P is on the circle with diagonal AB; $DA \equiv AP$; A is the midpoint of E and D; $FB \perp AD$; $FB \equiv BP$; B is the midpoint of F and G; I is on line EG; I is on line DF. The **conclusion:** $BP \equiv BI$.

$A = (0,0)$, $B = (u_1, 0)$, $P = (x_1, u_2)$, $D = (x_2, u_3)$, $E = (x_3, x_4)$, $F = (x_6, x_5)$, $G = (x_7, x_8)$, $I = (x_{10}, x_9)$.

The **nondegenerate** conditions: Line AD is non-isotropic; Line DF intersects line EG.

Example 378 (A-386: 1.1s, 5). *If two circles are orthogonal, any two diametrically opposite points of one circle are conjugate with respect to the other circle.*

Points O, O_1 are arbitrarily chosen. Points D, E, A, F, G are constructed (in order) as follows: D is on the circle with diagonal OO_1; $EO_1 \equiv O_1D$; $AO \equiv OD$; A is on line OE; O_1 is the midpoint of E and F; G is on line OE; $GF \perp OE$. The **conclusion:** G and E are inversive wrpt circle (O, OA).

$O = (0,0)$, $O_1 = (u_1, 0)$, $D = (x_1, u_2)$, $E = (x_2, u_3)$, $A = (x_4, x_3)$, $F = (x_5, x_6)$, $G = (x_8, x_7)$.

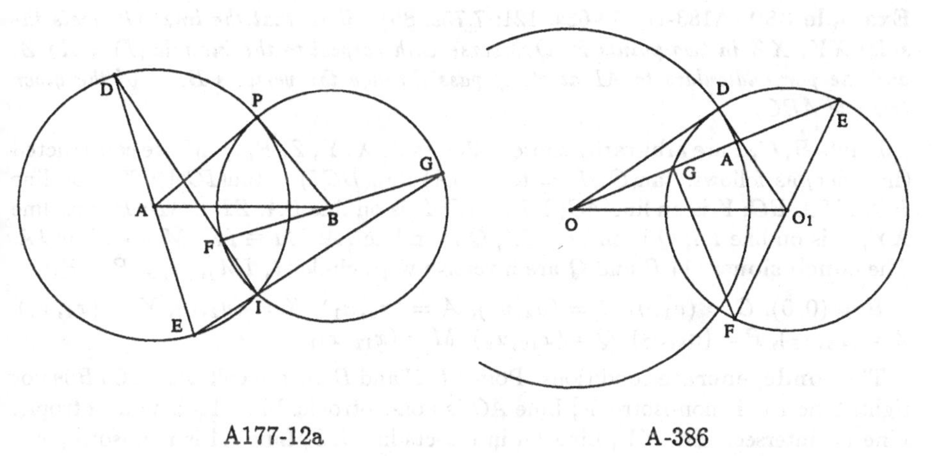

A177-12a A-386

The **nondegenerate** conditions: Line OE is non-isotropic; Line OE is non-isotropic.

Example 379 (A182-3-a: 2.22s, 57). *TP, TQ are the tangents at the extremities of a chord PQ of a circle. The tangent at any point R of the circle meets PQ in S; prove that TR is the polar of S.*

Points P, A are arbitrarily chosen. Points O, Q, R, T, S, D, S_1 are constructed (in order) as follows: $OP \perp PA$; $QO \equiv OP$; $RO \equiv OP$; T is on line AP; $TQ \perp QO$; S is on line PQ; $SR \perp RO$; D is on line OS; $DO \equiv OP$; S_1 is on line TR; S_1 is on line OS. The **conclusion**: S and S_1 are inversive wrpt circle (O, OD).

$P = (0,0)$, $A = (u_1,0)$, $O = (0,u_2)$, $Q = (x_1,u_3)$, $R = (x_2,u_4)$, $T = (x_3,0)$, $S = (x_5,x_4)$, $D = (x_7,x_6)$, $S_1 = (x_9,x_8)$.

The **nondegenerate** conditions: $P \neq A$; Line QO is not perpendicular to line AP; Line RO is not perpendicular to line PQ; Line OS is non-isotropic; Line OS intersects line TR.

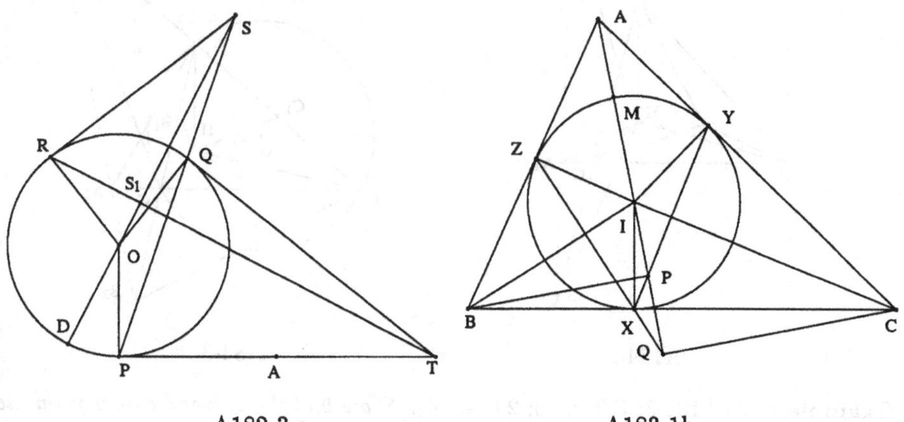

A182-3-a A183-1b

Example 380 (A183-1b: 14.63s, 121; 7.75s, 85). *Show that the line AI meets the sides XY, XZ in two points P, Q inverse with respect to the incircle $(I) = XYZ$, and the perpendiculars to AI at P, Q pass through the vertices B, C of the given triangle ABC.*

Points B, C, I are arbitrarily chosen. Points A, X, Y, Z, P, Q, M are constructed (in order) as follows: $\tan(CBI) = \tan(IBA)$; $\tan(BCI) = \tan(ICA)$; X is on line BC; $XI \perp BC$; Y is on line AC; $YI \perp AC$; Z is on line BA; $ZI \perp AB$; P is on line XY; P is on line IA; Q is on line XZ; Q is on line IA; $MI \equiv IX$; M is on line IA. The **conclusions**: (1) P and Q are inversive wrpt circle (I, IM); (2) $BP \perp PA$.

$B = (0,0)$, $C = (u_1,0)$, $I = (u_2,u_3)$, $A = (x_2,x_1)$, $X = (u_2,0)$, $Y = (x_4,x_3)$, $Z = (x_6,x_5)$, $P = (x_8,x_7)$, $Q = (x_{10},x_9)$, $M = (x_{12},x_{11})$.

The **nondegenerate** conditions: Points I, C and B are not collinear; $\angle CIB$ is not right; Line BC is non-isotropic; Line AC is non-isotropic; Line AB is non-isotropic; Line IA intersects line XY; Line IA intersects line XZ; Line IA is non-isotropic.

Example 381 (A184-2: 1.4s, 12). *Show that the perpendiculars dropped from the orthocenter of a triangle upon the lines joining the vertices to to a given points meet the respectively opposite sides of the triangle in three collinear points.*

Points B, C, A, O are arbitrarily chosen. Points H, A_1, B_1, C_1 are constructed (in order) as follows: $HB \perp AC$; $HA \perp BC$; A_1 is on line BC; $A_1H \perp OA$; B_1 is on line AC; $B_1H \perp OB$; C_1 is on line AB; $C_1H \perp OC$. The **conclusion**: Points A_1, B_1 and C_1 are collinear.

$B = (0,0)$, $C = (u_1,0)$, $A = (u_2,u_3)$, $O = (u_4,u_5)$, $H = (u_2,x_1)$, $A_1 = (x_2,0)$, $B_1 = (x_4,x_3)$, $C_1 = (x_6,x_5)$.

The **nondegenerate** conditions: Points B, C and A are not collinear; Line OA is not perpendicular to line BC; Line OB is not perpendicular to line AC; Line OC is not perpendicular to line AB.

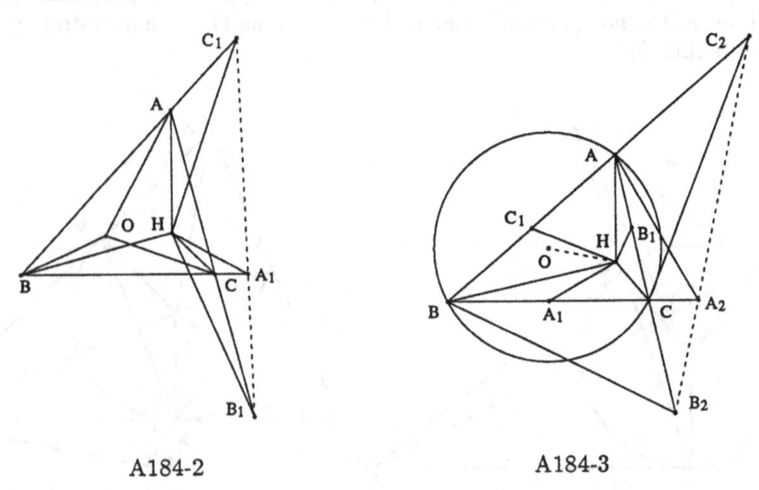

A184-2 A184-3

Example 382 (A184-3: 3.28s, 16; 2.82s, 24). *Show that the perpendiculars from the* *vertices of a triangle to the lines joining the midpoints of the respectively opposite sides to the orthocenter of the triangle meet these sides in three points of a straight line perpendicular to the Euler line of the triangle.*

Points B, C, A are arbitrarily chosen. Points O, H, A_1, A_2, B_1, B_2, C_1, C_2 are constructed (in order) as follows: $OB \equiv OC$; $OB \equiv OA$; $HB \perp AC$; $HA \perp BC$; A_1 is the midpoint of B and C; A_2 is on line BC; $A_2A \perp A_1H$; B_1 is the midpoint of C and A; B_2 is on line AC; $B_2B \perp HB_1$; C_1 is the midpoint of A and B; C_2 is on line AB; $C_2C \perp C_1H$. The **conclusions**: (1) Points A_2, B_2 and C_2 are collinear; (2) $OH \perp A_2C_2$.

$B = (0,0)$, $C = (u_1,0)$, $A = (u_2,u_3)$, $O = (x_2,x_1)$, $H = (u_2,x_3)$, $A_1 = (x_4,0)$, $A_2 = (x_5,0)$, $B_1 = (x_6,x_7)$, $B_2 = (x_9,x_8)$, $C_1 = (x_{10},x_{11})$, $C_2 = (x_{13},x_{12})$.

The **nondegenerate** conditions: Points B, A and C are not collinear; Points B, C and A are not collinear; Line A_1H is not perpendicular to line BC; Line HB_1 is not perpendicular to line AC; Line C_1H is not perpendicular to line AB.

Example 383 (A184-4: 3.27s, 24). *Show that the perpendiculars to the internal bisectors of a triangle at the incenter meet the respective sides in three points lying on a line perpendicular to the line joining the incenter to the circumcenter of the triangle.*

Points B, C, I are arbitrarily chosen. Points A, O, X, Y, Z are constructed (in order) as follows: $\tan(CBI) = \tan(IBA)$; $\tan(BCI) = \tan(ICA)$; $OB \equiv OA$; $OB \equiv OC$; X is on line BC; $XI \perp IA$; Y is on line AC; $YI \perp IB$; Z is on line AB; $ZI \perp IC$. The **conclusions**: (1) Points X, Y and Z are collinear; (2) $OI \perp XZ$.

$B = (0,0)$, $C = (u_1,0)$, $I = (u_2,u_3)$, $A = (x_2,x_1)$, $O = (x_4,x_3)$, $X = (x_5,0)$, $Y = (x_7,x_6)$, $Z = (x_9,x_8)$.

The **nondegenerate** conditions: Points I, C and B are not collinear; $\angle CIB$ is not right; Points B, C and A are not collinear; Line IA is not perpendicular to line BC; Line IB is not perpendicular to line AC; Line IC is not perpendicular to line AB.

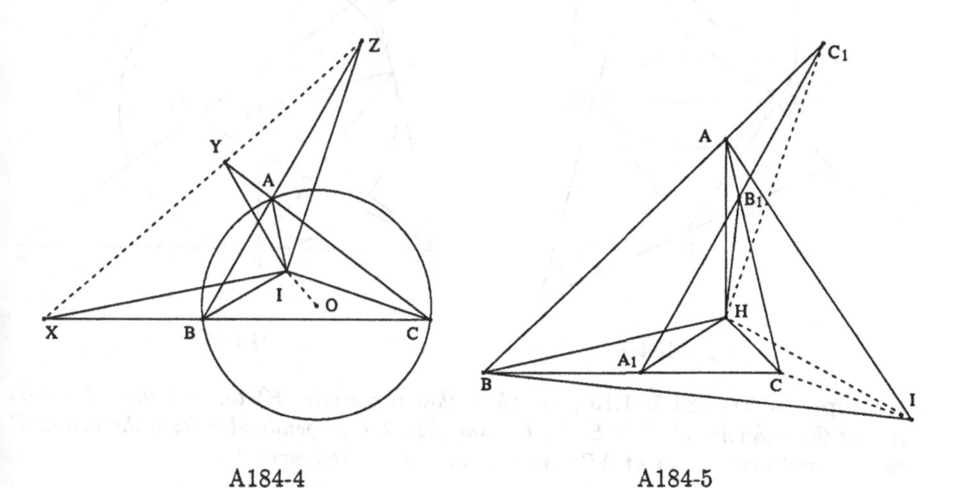

A184-4 A184-5

Example 384 (A184-5: 1.68s, 44). *If A_1, B_1, C_1 are the traces of a transversal on the sides BC, CA, AB of a triangle ABC whose orthocenter is H, prove that the perpendiculars from A, B, C upon the lines HA_1, HB_1, HC_1, respectively, are concurrent in a point on the perpendicular from H to $A_1B_1C_1$.*

Points B, C, A are arbitrarily chosen. Points H, A_1, B_1, C_1, I are constructed (in order) as follows: $HB \perp AC$; $HA \perp BC$; A_1 is on line BC; B_1 is on line AC; C_1 is on line B_1A_1; C_1 is on line AB; $IB \perp HB_1$; $IA \perp HA_1$. The **conclusions:** (1) $IC \perp HC_1$; (2) $IH \perp A_1B_1$.

$B = (0,0)$, $C = (u_1,0)$, $A = (u_2,u_3)$, $H = (u_2,x_1)$, $A_1 = (u_4,0)$, $B_1 = (x_2,u_5)$, $C_1 = (x_4,x_3)$, $I = (x_6,x_5)$.

The **nondegenerate** conditions: Points B, C and A are not collinear; $B \neq C$; $A \neq C$; Line AB intersects line B_1A_1; Points H, A_1 and B_1 are not collinear.

Example 385 (A184-7: 2.05s, 47). *Through the point of intersection of the tangents DB, DC to the circumcircle (O) of the triangle ABC a parallel is drawn to the line touching (O) at A. If this parallel meets AB, AC in E, F, show that D bisects EF.*

Points B, C, A are arbitrarily chosen. Points O, D, F, E are constructed (in order) as follows: $OB \equiv OC$; $OA \equiv OB$; $DC \perp OC$; $DB \perp OB$; F is on line AB; $FD \perp OA$; E is on line AC; E is on line FD. The **conclusion**: D is the midpoint of E and F.

$B = (0,0)$, $C = (u_1,0)$, $A = (u_2,u_3)$, $O = (x_2,x_1)$, $D = (x_4,x_3)$, $F = (x_6,x_5)$, $E = (x_8,x_7)$.

The **nondegenerate** conditions: Points A, B and C are not collinear; Points O, B and C are not collinear; Line OA is not perpendicular to line AB; Line FD intersects line AC.

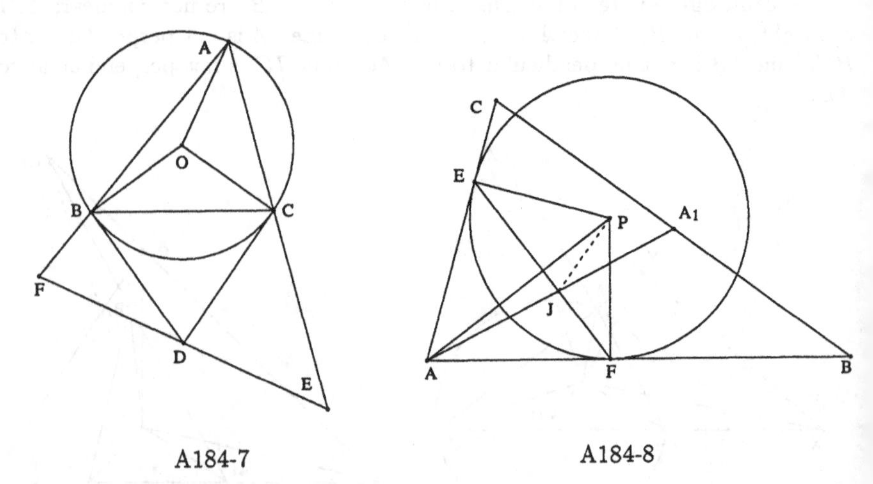

A184-7 A184-8

Example 386 (A184-8: 1.13s, 9). *Show that if a circle (P) touches the sides AB, AC of the triangle ABC in E, F, the line EF, the perpendicular from the center P to BC, and the median of ABC issued from A are concurrent.*

Points A, B, P are arbitrarily chosen. Points C, F, E, A_1, J are constructed (in order) as follows: $\tan(BAP) = \tan(PAC)$; $FP \perp AB$; F is on line AB; E is on line AC; $EP \perp CA$; A_1 is the midpoint of B and C; J is on line EF; J is on line AA_1. The **conclusion**: $JP \perp BC$.

$A = (0,0)$, $B = (u_1,0)$, $P = (u_2,u_3)$, $C = (x_1,u_4)$, $F = (u_2,0)$, $E = (x_3,x_2)$, $A_1 = (x_4,x_5)$, $J = (x_7,x_6)$.

The **nondegenerate** conditions: $A \neq B$, $A \neq P$; Line AB is non-isotropic; Line CA is non-isotropic; Line AA_1 intersects line EF.

Example 387 (A189-8: 0.97s, 6). *The side BC of the triangle ABC touches the incircle (I) in X and the excircle (I_a) relative to BC in X_a. Show that the line AX_a passes through the diametric opposite X_1 of X on (I). State a similar proposition about diametric opposite of X_a on (I_a).*

Points B, C, I are arbitrarily chosen. Points A, I_A, X, X_A, X_1 are constructed (in order) as follows: $\tan(BCI) = \tan(ICA)$; $\tan(CBI) = \tan(IBA)$; I_A is on line AI; $I_AB \perp BI$; X is on line BC; $XI \perp BC$; X_A is on line BC; $X_AI_A \perp BC$; I is the midpoint of X and X_1. The **conclusion**: Points A, X_A and X_1 are collinear.

$B = (0,0)$, $C = (u_1,0)$, $I = (u_2,u_3)$, $A = (x_2,x_1)$, $I_A = (x_4,x_3)$, $X = (u_2,0)$, $X_A = (x_4,0)$, $X_1 = (u_2,x_5)$.

The **nondegenerate** conditions: Points I, B and C are not collinear; $\angle BIC$ is not right; Line BI is not perpendicular to line AI; Line BC is non-isotropic; Line BC is non-isotropic.

Remark. Our specification and proof are for both statements.

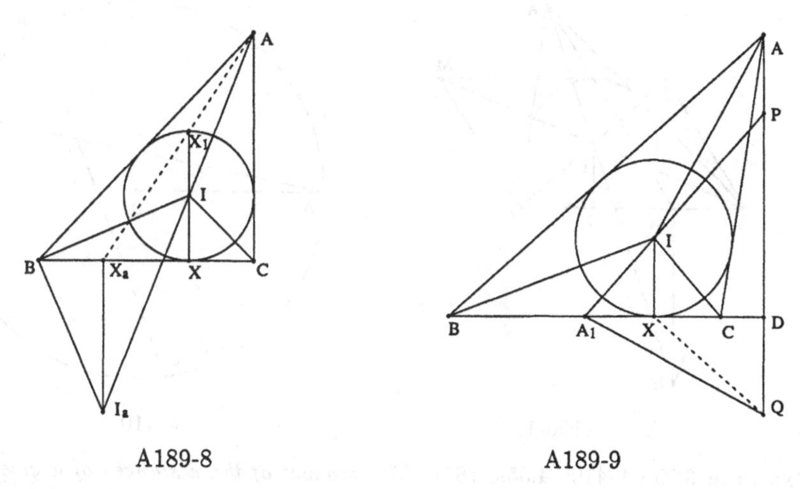

A189-8 A189-9

Example 388 (A189-9: 1.58s, 11). *With the notations of A189-8, show that if the line A_1I meets the altitude AD of ABC in P, then AP is equal to the inradius of ABC. State and prove a similar proposition for the excircle.*

Points B, C, I are arbitrarily chosen. Points A, X, D, A_1, P, Q are constructed (in order) as follows: $\tan(BCI) = \tan(ICA)$; $\tan(CBI) = \tan(IBA)$; X is on line

BC; $XI \perp BC$; $DA \perp BC$; D is on line BC; A_1 is the midpoint of B and C; P is on line AD; P is on line IA_1; Q is on line AD; $QA_1 \perp AI$. The **conclusions**: (1) $QX \perp A_1I$; (2) $AP \equiv IX$.

$B = (0,0)$, $C = (u_1,0)$, $I = (u_2,u_3)$, $A = (x_2,x_1)$, $X = (u_2,0)$, $D = (x_2,0)$, $A_1 = (x_3,0)$, $P = (x_2,x_4)$, $Q = (x_2,x_5)$.

The **nondegenerate** conditions: Points I, B and C are not collinear; $\angle BIC$ is not right; Line BC is non-isotropic; Line BC is non-isotropic; Line IA_1 intersects line AD; Line AI is not perpendicular to line AD.

Remark. Our specification and proof are for both statements.

Example 389 (A190-12: 1.53s, 10). *With the notations of A189-8, if the parallels to AX_a through B, C meet the bisectors CI, BI in L, M show that the line LM is parallel to BC.*

Points B, C, I are arbitrarily chosen. Points A, I_A, X, X_A, L, M are constructed (in order) as follows: $\tan(BCI) = \tan(ICA)$; $\tan(CBI) = \tan(IBA)$; I_A is on line AI; $I_AB \perp BI$; X is on line BC; $XI \perp BC$; X_A is on line BC; $X_AI_A \perp BC$; L is on line CI; $LB \parallel AX_A$; M is on line BI; $MC \parallel AX_A$. The **conclusion**: $LM \parallel BC$.

$B = (0,0)$, $C = (u_1,0)$, $I = (u_2,u_3)$, $A = (x_2,x_1)$, $I_A = (x_4,x_3)$, $X = (u_2,0)$, $X_A = (x_4,0)$, $L = (x_6,x_5)$, $M = (x_8,x_7)$.

The **nondegenerate** conditions: Points I, B and C are not collinear; $\angle BIC$ is not right; Line BI is not perpendicular to line AI; Line BC is non-isotropic; Line BC is non-isotropic; Line AX_A intersects line CI; Line AX_A intersects line BI.

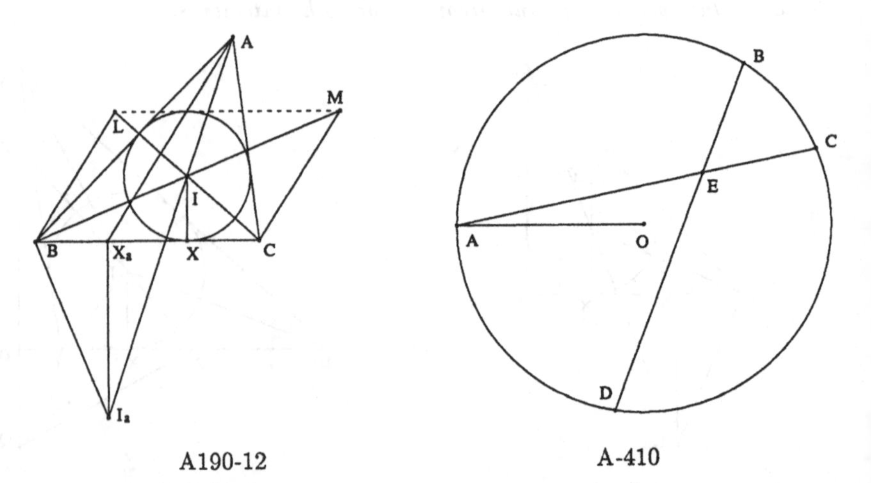

A190-12 A-410

Example 390 (A-410: 4.55s, 183). *The product of the distances of a given point, from any two points which are collinear with the given point and lie on the circle, is a constant. This constant is called the power of the point with respect to the circle.*

Points A, O are arbitrarily chosen. Points B, C, D, E are constructed (in order) as follows: $BO \equiv OA$; $CO \equiv OA$; $DO \equiv OA$; E is on line BD; E is on line AC. The **conclusion**: $AE \cdot CE = BE \cdot DE$.

$A = (0,0)$, $O = (u_1, 0)$, $B = (x_1, u_2)$, $C = (x_2, u_3)$, $D = (x_3, u_4)$, $E = (x_5, x_4)$.

The **nondegenerate** conditions: Line AC intersects line BD.

Remark. This is an extreme example for the Greobner basis method. It took more than 3 hours to reduce the conclusion polynomial to zero. Compare with Wu's method.

Example 391 (A193-7: 3.5s, 83). *Show that the power of the orthocenter of a triangle with respect to the circumcircle is equal to four times the power of the same point with respect to the nine-point circle of the triangle.*

Points B, C, A are arbitrarily chosen. Points D, E, F, H, O, N are constructed (in order) as follows: D is on line BC; $DA \perp BC$; E is on line AC; $EB \perp AC$; F is on line AB; $FC \perp AB$; H is on line BE; H is on line AD; $OC \equiv OB$; $OA \equiv OB$; $NF \equiv ND$; $NE \equiv ND$. The **conclusion**: $\overline{HO}^2 - \overline{OB}^2 - 4(\overline{HN}^2 - \overline{ND}^2) = 0$.

$B = (0,0)$, $C = (u_1, 0)$, $A = (u_2, u_3)$, $D = (u_2, 0)$, $E = (x_2, x_1)$, $F = (x_4, x_3)$, $H = (u_2, x_5)$, $O = (x_7, x_6)$, $N = (x_9, x_8)$.

The **nondegenerate** conditions: Line BC is non-isotropic; Line AC is non-isotropic; Line AB is non-isotropic; Line AD intersects line BE; Points A, B and C are not collinear; Points E, D and F are not collinear.

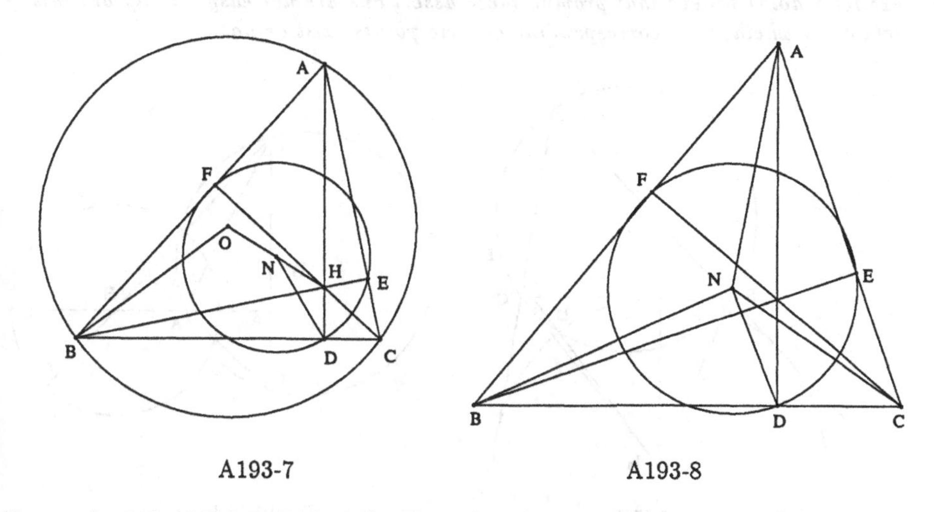

A193-7 A193-8

Example 392 (A193-8: 2.18s, 58). *Show that the sum of the powers of the vertices of a triangle for the nine-point circle of the triangle is equal to one fourth the sum of the squares of the sides of the triangle.*

Points B, C, A are arbitrarily chosen. Points D, E, F, N are constructed (in order) as follows: D is on line BC; $DA \perp BC$; E is on line AC; $EB \perp AC$; F is on line AB; $FC \perp AB$; $NF \equiv ND$; $NE \equiv ND$. The **conclusion**: $4(\overline{NA}^2 - \overline{ND}^2 + \overline{NB}^2 - \overline{ND}^2 + \overline{NC}^2 - \overline{ND}^2) - (\overline{AB}^2 + \overline{BC}^2 + \overline{CA}^2) = 0$.

$B = (0,0)$, $C = (u_1, 0)$, $A = (u_2, u_3)$, $D = (u_2, 0)$, $E = (x_2, x_1)$, $F = (x_4, x_3)$, $N = (x_6, x_5)$.

The **nondegenerate** conditions: Line BC is non-isotropic; Line AC is non-isotropic; Line AB is non-isotropic; Points E, D and F are not collinear.

Example 393 (A195-3: 8639.73s, 22090; 389.52s, 2979). *The escribed circle (I_a) of a triangle ABC meets the circumcircle (O) of ABC in D, and I_aD meets (O) in E. Show that I_aE is equal to the circumdiameter of ABC.*

Points B, C, I are arbitrarily chosen. Points A, X, O, D, E are constructed (in order) as follows: $\tan(CBI) = \tan(IBA)$; $\tan(BCI) = \tan(ICA)$; X is on line BC; $XI \perp BC$; $OB \equiv OA$; $OB \equiv OC$; $DI \equiv IX$; $DO \equiv OB$; E is on line ID; $EO \equiv OB$. The **conclusion**: $4\ \overline{OB}^2 - \overline{IE}^2 = 0$.

$B = (0,0)$, $C = (u_1, 0)$, $I = (u_2, u_3)$, $A = (x_2, x_1)$, $X = (u_2, 0)$, $O = (x_4, x_3)$, $D = (x_6, x_5)$, $E = (x_8, x_7)$.

The **nondegenerate** conditions: Points I, C and B are not collinear; $\angle CIB$ is not right; Line BC is non-isotropic; Points B, C and A are not collinear; Line OI is non-isotropic; Line ID is non-isotropic. In addition, the following nondegenerate conditions, which come from reducibility and have been detected by our prover, should be also added: $E \neq D$.

Remark. In Euclidean geometry, incircle does not intersect with circumcircle. But excircles do. I believe that proving these assertions are not easy. In algebra this is related to whether the corresponding generic points exist or not.

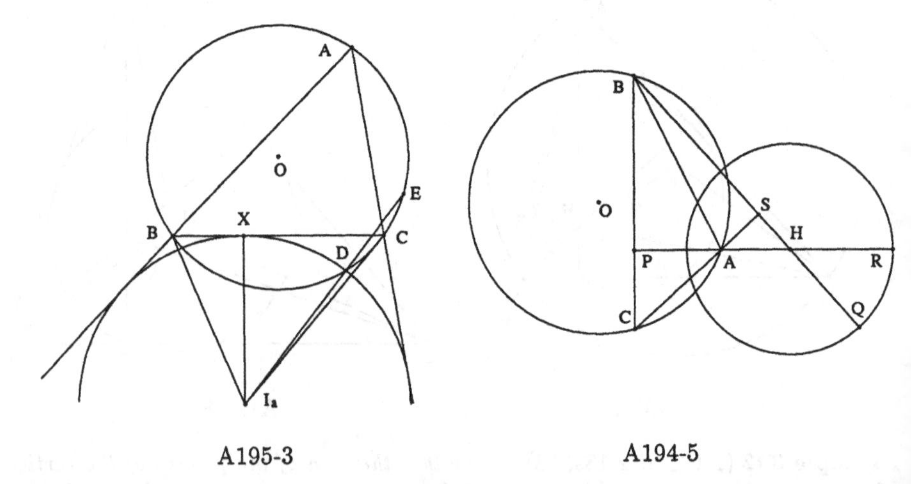

A195-3 A194-5

Example 394 (A194-5: 1.27s, 11). *Show that the square of the radius of the polar circle of a triangle is equal to half the power of the orthocenter of the triangle with respect to the circumcircle of the triangle.*

Points H, R are arbitrarily chosen. Points A, P, B, Q, S, C, O are constructed (in order) as follows: A is on line HR; A and P are inversive wrpt circle (H, HR); P is on line HR; $BP \perp HR$; $QH \equiv HR$; Q is on line HB; B and S are inversive wrpt circle (H, HQ); S is on line HB; C is on line PB; C is on line SA; $OA \equiv OC$; $OA \equiv OB$. The **conclusion**: $2\ \overline{HP}^2 - (\overline{HO}^2 - \overline{OA}^2) = 0$.

$H = (0,0)$, $R = (u_1,0)$, $A = (u_2,0)$, $P = (x_1,0)$, $B = (x_1,u_3)$, $Q = (x_3,x_2)$, $S = (x_5,x_4)$, $C = (x_1,x_6)$, $O = (x_8,x_7)$.

The **nondegenerate** conditions: $H \neq R$; $H \neq R$; $A \neq H$; $H \neq R$; Line HB is non-isotropic; $-u_3 \neq 0$; $u_3 \neq 0$; Line SA intersects line PB; Points A, B and C are not collinear.

Example 395 (A194-7: 7.27s, 342). *Two unequal circles are tangent internally at A. The tangent to the smaller circle at a point B meets the larger circle in C, D. Show that AB bisects the angle CAD.*

Points A, O_1 are arbitrarily chosen. Points O_2, B, C, D are constructed (in order) as follows: O_2 is on line AO_1; $BO_1 \equiv O_1A$; $CO_2 \equiv O_2A$; $CB \perp BO_1$; $DO_2 \equiv O_2A$; D is on line BC. The **conclusion**: $\tan(CAB) = \tan(BAD)$.

$A = (0,0)$, $O_1 = (u_1,0)$, $O_2 = (u_2,0)$, $B = (x_1,u_3)$, $C = (x_3,x_2)$, $D = (x_5,x_4)$.

The **nondegenerate** conditions: $A \neq O_1$; Line BO_1 is non-isotropic; Line BC is non-isotropic. In addition, the following nondegenerate conditions, which come from reducibility and have been detected by our prover, should be also added: $D \neq C$.

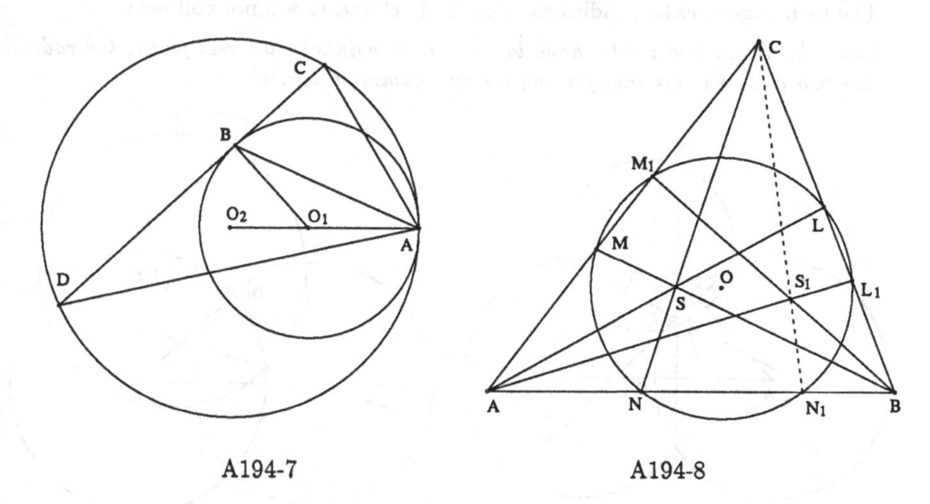

A194-7 A194-8

Example 396 (A194-8: 2371.9s, 20916). *If the lines joining the vertices A, B, C of a triangle ABC to a point S meet the respectively opposite sides in L, M, N, and the circle (LMN) meets these sides again in L_1, M_1, N_1, show that the lines AL_1, BM_1, CN_1 are concurrent.*

Points A, B, C, S are arbitrarily chosen. Points L, M, N, O, L_1, M_1, N_1, S_1 are constructed (in order) as follows: L is on line BC; L is on line SA; M is on line AC; M is on line SB; N is on line AB; N is on line SC; $ON \equiv OM$; $ON \equiv OL$; L_1 is on line BC; $L_1O \equiv OL$; M_1 is on line AC; $M_1O \equiv OL$; N_1 is on line AB; $N_1O \equiv OL$; S_1 is on line BM_1; S_1 is on line AL_1. The **conclusion**: Points C, N_1 and S_1 are collinear.

$A = (0,0)$, $B = (u_1,0)$, $C = (u_2,u_3)$, $S = (u_4,u_5)$, $L = (x_2,x_1)$, $M = (x_4,x_3)$, $N = (x_5,0)$, $O = (x_7,x_6)$, $L_1 = (x_9,x_8)$, $M_1 = (x_{11},x_{10})$, $N_1 = (x_{12},0)$, $S_1 =$

(x_{14}, x_{13}).

The **nondegenerate** conditions: Line SA intersects line BC; Line SB intersects line AC; Line SC intersects line AB; Points N, L and M are not collinear; Line BC is non-isotropic; Line AC is non-isotropic; Line AB is non-isotropic; Line AL_1 intersects line BM_1. In addition, the following nondegenerate conditions, which come from reducibility and have been detected by our prover, should be also added: $N_1 \neq N$; $M_1 \neq M$; $L_1 \neq L$.

Example 397 (A-425: 0.83s, 12). *The radical axes of three circles, with noncollinear centers, taken in pairs are concurrent. (The radical center of the three circles.)*

Points A, B, C, D, E, F are arbitrarily chosen. Points O are constructed (in order) as follows: Point O is on the radical axises of circle (A, AD) and circle (C, CF); Point O is on the radical axises of circle (A, AD) and circle (B, BE). The **conclusion**: Point O is on the radical axises of circle (B, BE) and circle (C, CF).

$A = (0,0)$, $B = (u_1, 0)$, $C = (u_2, u_3)$, $D = (u_4, u_5)$, $E = (u_6, u_7)$, $F = (u_8, u_9)$, $O = (x_2, x_1)$.

The **nondegenerate** conditions: Points A, B and C are not collinear.

Remark. When two circles have two common points in the real plane, the radical of the two circles is the line joining the two common points.

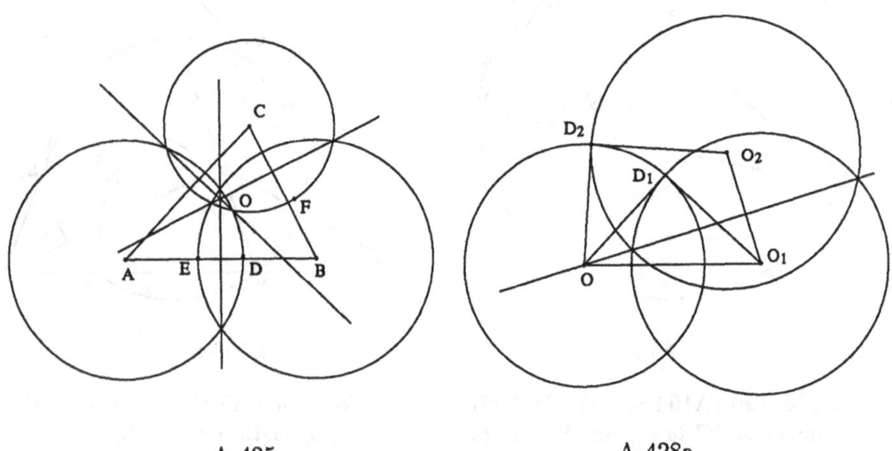

A-425 A-428a

Example 398 (A-428a: 0.77s, 10). *A circle orthogonal to two given circles has its center on the radical axis of the two circles.*

Points O, O_1, O_2 are arbitrarily chosen. Points D_1, D_2 are constructed (in order) as follows: D_1 is on the circle with diagonal OO_1; D_2 is on the circle with diagonal OO_2; $D_2O \equiv OD_1$. The **conclusion**: Point O is on the radical axises of circle (O_1, O_1D_1) and circle (O_2, O_2D_2).

$O = (0,0)$, $O_1 = (u_1, 0)$, $O_2 = (u_2, u_3)$, $D_1 = (x_1, u_4)$, $D_2 = (x_3, x_2)$.

The **nondegenerate** conditions: The line joining the midpoint of O and O_2 and the point O is non-isotropic.

Example 399 (A-429: 3.47s, 94). *The polars, with respect to two given circles, of a point on their radical axis, intersect on the radical axis.*

Points O, M, A are arbitrarily chosen. Points P, B, Q, S, T are constructed (in order) as follows: P is on line OA; B is on line MP; Point P is on the radical axises of circle (O, OA) and circle (M, MB); P and Q are inversive wrpt circle (O, OA); Q is on line OA; P and S are inversive wrpt circle (M, MB); S is on line MB; $TQ \perp AP$; $TS \perp BP$. The **conclusion**: Point T is on the radical axises of circle (O, OA) and circle (M, MB).

$O = (0,0)$, $M = (u_1,0)$, $A = (u_2,u_3)$, $P = (x_1,u_4)$, $B = (x_3,x_2)$, $Q = (x_5,x_4)$, $S = (x_7,x_6)$, $T = (x_9,x_8)$.

The **nondegenerate** conditions: $O \neq A$; Line OM is not perpendicular to line MP; $O \neq A$; $P \neq O$; $M \neq B$; $P \neq M$; Points B, P and A are not collinear.

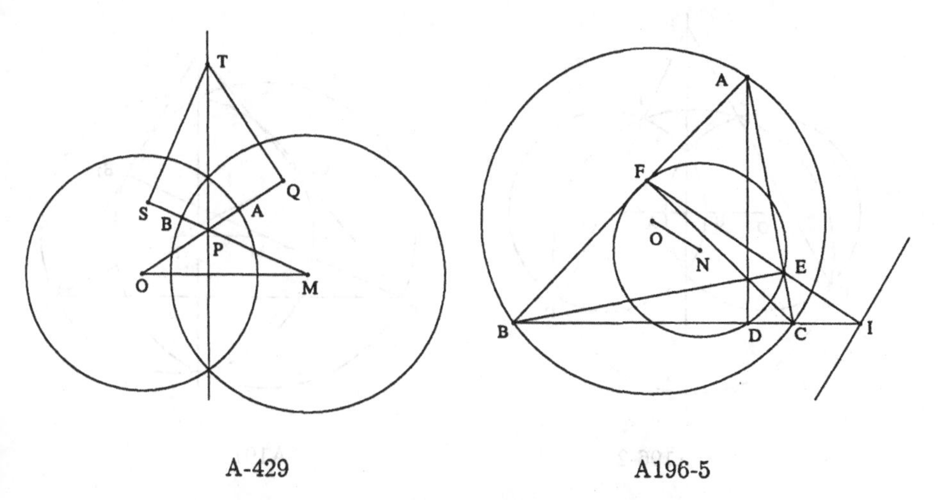

A-429 A196-5

Example 400 (A196-5: 2.73s, 38). *The orthic axis of a triangle is the radical axis of the circumcircle and the nine-point circle of the triangle.*

Points B, C, A are arbitrarily chosen. Points D, E, F, O, I, N are constructed (in order) as follows: D is on line BC; $DA \perp BC$; E is on line AC; $EB \perp AC$; F is on line AB; $FC \perp AB$; $OB \equiv OA$; $OB \equiv OC$; I is on line EF; I is on line BC; $ND \equiv NF$; $ND \equiv NE$. The **conclusion**: Point I is on the radical axises of circle (N, ND) and circle (O, OB).

$B = (0,0)$, $C = (u_1,0)$, $A = (u_2,u_3)$, $D = (u_2,0)$, $E = (x_2,x_1)$, $F = (x_4,x_3)$, $O = (x_6,x_5)$, $I = (x_7,0)$, $N = (x_9,x_8)$.

The **nondegenerate** conditions: Line BC is non-isotropic; Line AC is non-isotropic; Line AB is non-isotropic; Points B, C and A are not collinear; Line BC intersects line EF; Points D, E and F are not collinear.

Example 401 (A196-2: 1533.7s, 22637). *If from a point on the radical axis of two* *circles secants are drawn for each of the circles, show that the four points determined on the two circles are cyclic.*

Points O, A are arbitrarily chosen. Points H, B, P, C, D, E, F are constructed (in order) as follows: H is on line OA; B is on line OH; Point P is on the radical axises of circle (O, OA) and circle (H, HB); $CO \equiv OA$; $DH \equiv HB$; E is on line PC; $EO \equiv OA$; F is on line PD; $FH \equiv HB$. The **conclusion**: Points C, E, D and F are on the same circle.

$O = (0,0)$, $A = (u_1, 0)$, $H = (u_2, 0)$, $B = (u_3, 0)$, $P = (x_1, u_4)$, $C = (x_2, u_5)$, $D = (x_3, u_6)$, $E = (x_5, x_4)$, $F = (x_7, x_6)$.

The **nondegenerate** conditions: $O \neq A$; $O \neq H$; $O \neq H$; Line PC is non-isotropic; Line PD is non-isotropic. In addition, the following nondegenerate conditions, which come from reducibility and have been detected by our prover, should be also added: $F \neq D$; $E \neq C$.

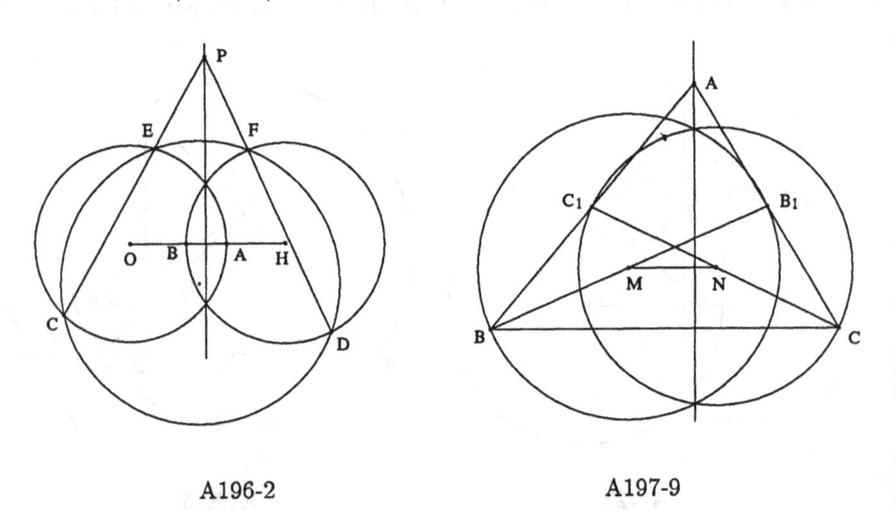

A196-2 A197-9

Example 402 (A197-9: 0.87s, 6). *Show that an altitude of a triangle is the radical axis of the two circles having for diameters the medians issued from the other two vertices.*

Points B, C, A are arbitrarily chosen. Points B_1, C_1, M, N are constructed (in order) as follows: B_1 is the midpoint of A and C; C_1 is the midpoint of A and B; M is the midpoint of B and B_1; N is the midpoint of C and C_1. The **conclusion**: Point A is on the radical axises of circle (M, MB) and circle (N, NC).

$B = (0,0)$, $C = (u_1, 0)$, $A = (u_2, u_3)$, $B_1 = (x_1, x_2)$, $C_1 = (x_3, x_4)$, $M = (x_5, x_6)$, $N = (x_7, x_8)$.

The **nondegenerate** conditions: none.

Example 403 (A197-12: 19.27s, 677; 13.28s, 244). *Show that the four tritangent circles of a triangle taken in pairs have for their radical axes the bisectors of the angles of the medial triangle of the given triangle.*

Points B, C, I are arbitrarily chosen. Points A, I_A, X, X_A, P, A_1, B_1, C_1 are constructed (in order) as follows: $\tan(BCI) = \tan(ICA)$; $\tan(CBI) = \tan(IBA)$; I_A is on line AI; $I_AB \perp BI$; X is on line BC; $XI \perp BC$; X_A is on line BC; $X_AI_A \perp BC$; Point P is on the radical axises of circle (I_A, I_AX_A) and circle (I, IX); A_1 is the midpoint of B and C; B_1 is the midpoint of C and A; C_1 is the midpoint of B and A. The **conclusions**: (1) $\tan(B_1A_1P) = \tan(PA_1C_1)$; (2) Point A_1 is on the radical axises of circle (I_A, I_AX_A) and circle (I, IX).

$B = (0,0)$, $C = (u_1,0)$, $I = (u_2,u_3)$, $A = (x_2,x_1)$, $I_A = (x_4,x_3)$, $X = (u_2,0)$, $X_A = (x_4,0)$, $P = (x_5,u_4)$, $A_1 = (x_6,0)$, $B_1 = (x_7,x_8)$, $C_1 = (x_9,x_{10})$.

The **nondegenerate** conditions: Points I, B and C are not collinear; $\angle BIC$ is not right; Line BI is not perpendicular to line AI; Line BC is non-isotropic; Line BC is non-isotropic; $I_A \neq I$.

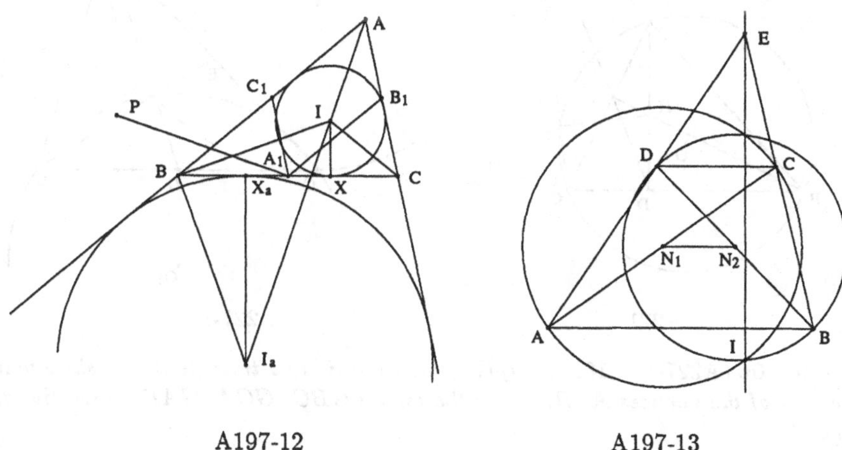

A197-12 A197-13

Example 404 (A197-13: 0.93s, 7). *Show that the radical axis of the two circles having for diameters the diagonals AC, BD of a trapezoid $ABCD$ passes through the point of intersection E of the nonparallel sides BC, AD.*

Points A, B, C are arbitrarily chosen. Points D, E, N_1, N_2 are constructed (in order) as follows: $DC \parallel BA$; E is on line AD; E is on line BC; N_1 is the midpoint of A and C; N_2 is the midpoint of B and D. The **conclusion**: Point E is on the radical axises of circle (N_1, N_1A) and circle (N_2, N_2B).

$A = (0,0)$, $B = (u_1,0)$, $C = (u_2,u_3)$, $D = (u_4,u_3)$, $E = (x_2,x_1)$, $N_1 = (x_3,x_4)$, $N_2 = (x_5,x_6)$.

The **nondegenerate** conditions: $B \neq A$; Line BC intersects line AD.

Example 405 (A221-1: 2.8s, 69). *Show that the foot of the perpendicular from the orthocenter of a triangle upon the line joining a vertex to the point of intersection of the opposite side with the corresponding side of the orthic triangle lies on the circumcircle of the triangle.*

Points B, C, A are arbitrarily chosen. Points D, E, H, X, S, O are constructed (in order) as follows: D is on line BC; $DA \perp BC$; E is on line AC; $EB \perp AC$; H is on line AD; H is on line BE; X is on line ED; X is on line AB; S is on line CX; $SH \perp CX$; $OA \equiv OC$; $OA \equiv OB$. The **conclusion**: $OS \equiv OA$.

$B = (0,0)$, $C = (u_1,0)$, $A = (u_2,u_3)$, $D = (u_2,0)$, $E = (x_2,x_1)$, $H = (u_2,x_3)$, $X = (x_5,x_4)$, $S = (x_7,x_6)$, $O = (x_9,x_8)$.

The **nondegenerate** conditions: Line BC is non-isotropic; Line AC is non-isotropic; Line BE intersects line AD; Line AB intersects line ED; Line CX is non-isotropic; Points A, B and C are not collinear.

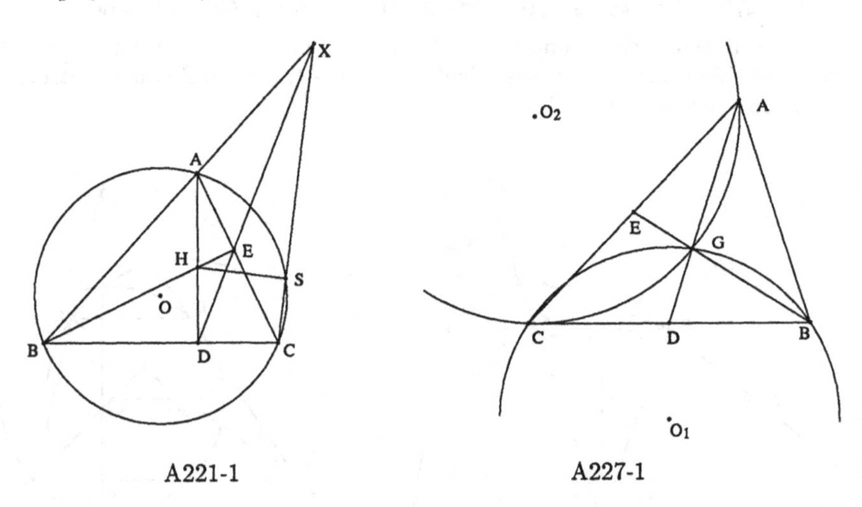

A221-1 A227-1

Example 406 (A227-1: 1.18s, 9). *If G is the centroid of a triangle ABC, show that the powers of the vertices A, B, C for the circles GBC, GCA, GAB, respectively, are equal.*

Points C, B, A are arbitrarily chosen. Points D, E, G, O_1, O_2 are constructed (in order) as follows: D is the midpoint of B and C; E is the midpoint of A and C; G is on line BE; G is on line AD; $O_1C \equiv O_1G$; $O_1B \equiv O_1C$; $O_2C \equiv O_2G$; $O_2A \equiv O_2C$. The **conclusion**: $\overline{AO_1}^2 - \overline{O_1C}^2 - (\overline{BO_2}^2 - \overline{O_2C}^2) = 0$.

$C = (0,0)$, $B = (u_1,0)$, $A = (u_2,u_3)$, $D = (x_1,0)$, $E = (x_2,x_3)$, $G = (x_5,x_4)$, $O_1 = (x_7,x_6)$, $O_2 = (x_9,x_8)$.

The **nondegenerate** conditions: Line AD intersects line BE; Points B, C and G are not collinear; Points A, C and G are not collinear.

Example 407 (A227-3: 1.63s, 34). *ABC is a triangle, A_1 the midpoint of BC, P the pole of BC for the circumcircle; through the midpoint of A_1P a line is drawn parallel to BC meeting AB, AC, produced, in Q, R; show that A, P, Q, R are cyclic.*

Points B, C are arbitrarily chosen. Points A_1, O, A, D, P, N, Q, R are constructed (in order) as follows: A_1 is the midpoint of B and C; $OA_1 \perp BC$; $AO \equiv OB$; $DO \equiv OB$; D is on line OA_1; P is on line OA_1; P and A_1 are inversive wrpt circle (O, OD); N is the midpoint of A_1 and P; $QN \parallel BC$; Q is on line AB; R is on line

NQ; R is on line AC. The **conclusion**: Points A, P, Q and R are on the same circle.

$B = (0,0)$, $C = (u_1,0)$, $A_1 = (x_1,0)$, $O = (x_1,u_2)$, $A = (x_2,u_3)$, $D = (x_1,x_3)$, $P = (x_1,x_4)$, $N = (x_1,x_5)$, $Q = (x_6,x_5)$, $R = (x_7,x_5)$.

The **nondegenerate** conditions: $B \neq C$; Line OA_1 is non-isotropic; $u_2 \neq 0$; Points A, B and C are not collinear; Line AC intersects line NQ.

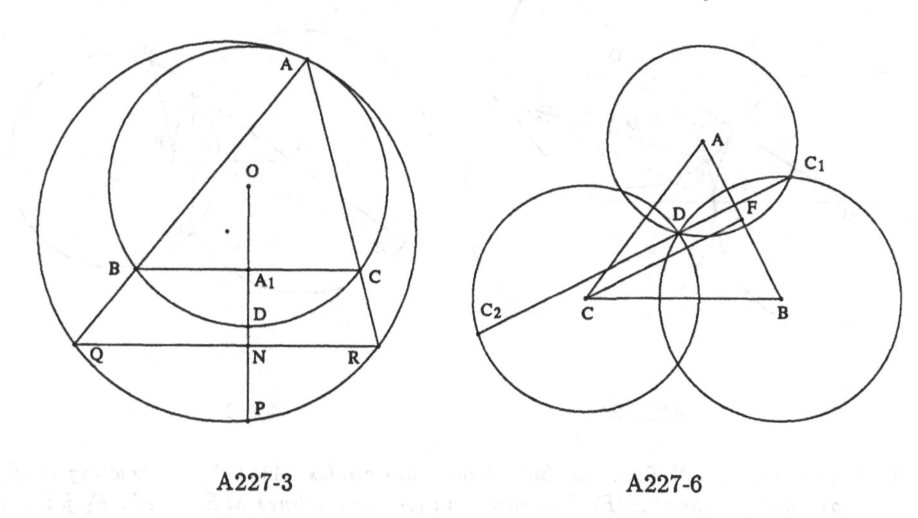

A227-3 A227-6

Example 408 (A227-6: 9.53s, 450). *Three circles, centers A, B, C, have a point D in common and intersect two-by-two in the points A_1, B_1, C_1. The common chord DC_1 of the first two circles meets the third in C_2. Let A_2, B_2 be the analogous points on the other two circles. Prove that the segments A_1A_2, B_1B_2, C_1C_2 are twice as long as the altitudes of the triangle ABC.*

Points C, B, A, D are arbitrarily chosen. Points F, C_1, C_2 are constructed (in order) as follows: F is on line BA; $FC \perp AB$; $C_1B \equiv BD$; $C_1A \equiv AD$; $C_2C \equiv CD$; C_2 is on line DC_1. The **conclusions**: (1) $\frac{\overline{C_2C_1}}{\overline{CF}} - 2 = 0$; (2) $DC_1 \parallel CF$.

$C = (0,0)$, $B = (u_1,0)$, $A = (u_2,u_3)$, $D = (u_4,u_5)$, $F = (x_2,x_1)$, $C_1 = (x_4,x_3)$, $C_2 = (x_6,x_5)$.

The **nondegenerate** conditions: Line AB is non-isotropic; Line AB is non-isotropic; Line DC_1 is non-isotropic. In addition, the following nondegenerate conditions, which come from reducibility and have been detected by our prover, should be also added: $C_2 \neq D$; $C_1 \neq D$.

Example 409 (A229-24: 41.83s, 1267). *If two circles intersect in A, B and touch a third circle in C, D, show that $AC : AD = BC : BD$.*

Points O, C are arbitrarily chosen. Points D, P, Q, A, M, B are constructed (in order) as follows: $DO \equiv OC$; P is on line OC; Q is on line OD; $AQ \equiv QD$; $AP \equiv PC$; $MA \perp PQ$; M is on line PQ; M is the midpoint of A and B. The **conclusion**: $AC \cdot BD = AD \cdot BC$.

$O = (0,0)$, $C = (u_1, 0)$, $D = (x_1, u_2)$, $P = (u_3, 0)$, $Q = (x_2, u_4)$, $A = (x_4, x_3)$, $M = (x_6, x_5)$, $B = (x_7, x_8)$.

The **nondegenerate** conditions: $O \neq C$; $O \neq D$; Line PQ is non-isotropic; Line PQ is non-isotropic.

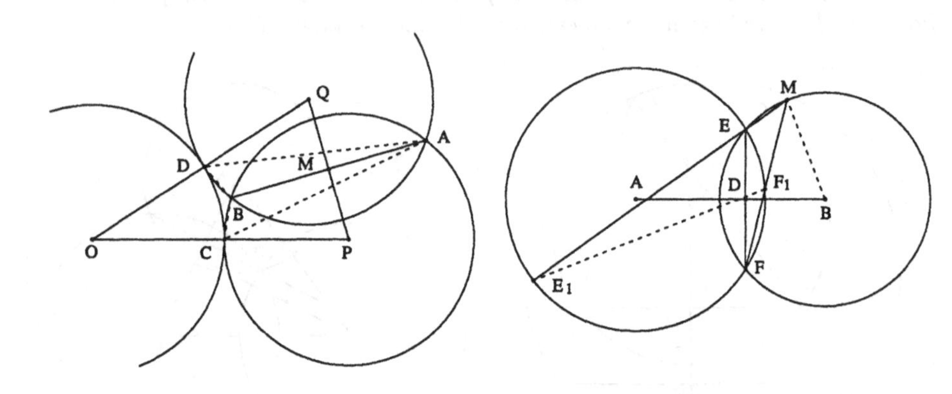

A229-24 A242-9

Example 410 (A242-9: 1.65s, 36). *Given two circles (A), (B) intersecting in E, F, show that the chord E_1F_1 determined in (A) by the lines MEE_1, MFF_1 joining E, F to any point M of (B) is perpendicular to MB.*

Points D, A are arbitrarily chosen. Points B, E, F, M, E_1, F_1 are constructed (in order) as follows: B is on line AD; $ED \perp DA$; D is the midpoint of E and F; $MB \equiv BE$; $E_1A \equiv AE$; E_1 is on line ME; $F_1A \equiv AF$; F_1 is on line MF. The **conclusion**: $MB \perp E_1F_1$.

$D = (0,0)$, $A = (u_1, 0)$, $B = (u_2, 0)$, $E = (0, u_3)$, $F = (0, x_1)$, $M = (x_2, u_4)$, $E_1 = (x_4, x_3)$, $F_1 = (x_6, x_5)$.

The **nondegenerate** conditions: $A \neq D$; $D \neq A$; Line ME is non-isotropic; Line MF is non-isotropic. In addition, the following nondegenerate conditions, which come from reducibility and have been detected by our prover, should be also added: $F_1 \neq F$; $E_1 \neq E$.

Example 411 (A-547-a: 2.75s, 60).

Points A, B, C, P are arbitrarily chosen. Points A_1, B_1, C_1, A_2, B_2, C_2 are constructed (in order) as follows: A_1 is on line BC; A_1 is on line AP; B_1 is on line AC; B_1 is on line BP; C_1 is on line AB; C_1 is on line CP; A_2 is on line BC; A_2 is on line B_1C_1; B_2 is on line AC; B_2 is on line A_1C_1; C_2 is on line AB; C_2 is on line A_1B_1. The **conclusion**: Points C_2, B_2 and A_2 are collinear.

$A = (0,0)$, $B = (u_1, 0)$, $C = (u_2, u_3)$, $P = (u_4, u_5)$, $A_1 = (x_2, x_1)$, $B_1 = (x_4, x_3)$, $C_1 = (x_5, 0)$, $A_2 = (x_7, x_6)$, $B_2 = (x_9, x_8)$, $C_2 = (x_{10}, 0)$.

The **nondegenerate** conditions: Line AP intersects line BC; Line BP intersects line AC; Line CP intersects line AB; Line B_1C_1 intersects line BC; Line A_1C_1 intersects line AC; Line A_1B_1 intersects line AB.

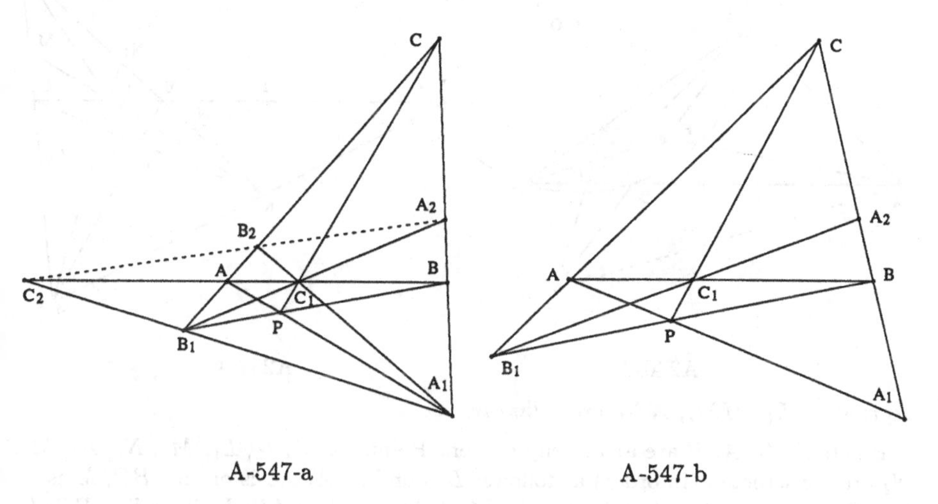

A-547-a A-547-b

Example 412 (A-547-b: 4.1s, 30).

Points A, B, C, P are arbitrarily chosen. Points A_1, B_1, C_1, A_2 are constructed (in order) as follows: A_1 is on line BC; A_1 is on line AP; B_1 is on line AC; B_1 is on line BP; C_1 is on line AB; C_1 is on line CP; A_2 is on line BC; A_2 is on line B_1C_1. The **conclusion**: Points B, C, A_1 and A_2 form a harmonic set.

$A = (0,0)$, $B = (u_1,0)$, $C = (u_2,u_3)$, $P = (u_4,u_5)$, $A_1 = (x_2,x_1)$, $B_1 = (x_4,x_3)$, $C_1 = (x_5,0)$, $A_2 = (x_7,x_6)$.

The **nondegenerate** conditions: Line AP intersects line BC; Line BP intersects line AC; Line CP intersects line AB; Line B_1C_1 intersects line BC.

Example 413 (A246-2: 4.05s, 56).

Points B, C, I are arbitrarily chosen. Points A, E, F, D, X, Z, O are constructed (in order) as follows: $\tan(CBI) = \tan(IBA)$; $\tan(BCI) = \tan(ICA)$; E is on line AC; E is on line BI; F is on line BA; F is on line CI; D is on line BC; D is on line AI; X is on line BC; X is on line EF; Z is on line AB; Z is on line DE; $OB \equiv OA$; $OB \equiv OC$. The **conclusions**: (1) $XZ \perp OI$; (2) $XA \perp AI$.

$B = (0,0)$, $C = (u_1,0)$, $I = (u_2,u_3)$, $A = (x_2,x_1)$, $E = (x_4,x_3)$, $F = (x_6,x_5)$, $D = (x_7,0)$, $X = (x_8,0)$, $Z = (x_{10},x_9)$, $O = (x_{12},x_{11})$.

The **nondegenerate** conditions: Points I, C and B are not collinear; $\angle CIB$ is not right; Line BI intersects line AC; Line CI intersects line BA; Line AI intersects line BC; Line EF intersects line BC; Line DE intersects line AB; Points B, C and A are not collinear.

Example 414 (A247-4: 4.02s, 46). *If L, M, N are the traces of the lines AP, BP, CP on the sides BC, CA, AB of the triangle ABC, and L_1, M_1, N_1 the traces, on the same sides, of the trilinear polar of P for ABC, show that the midpoints of the*

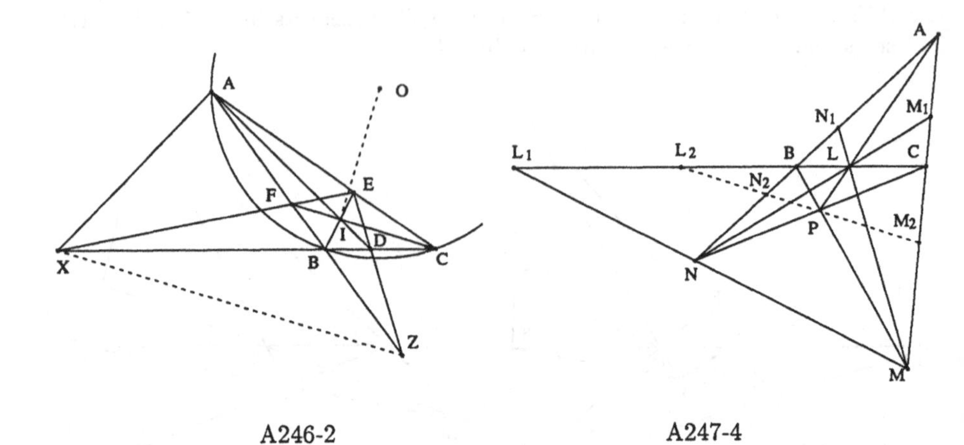

A246-2 A247-4

segments LL_1, MM_1, NN_1 are collinear.

Points B, C, A, P are arbitrarily chosen. Points L, M, N, L_1, M_1, N_1, L_2, M_2, N_2 are constructed (in order) as follows: L is on line AP; L is on line BC; M is on line BP; M is on line AC; N is on line CP; N is on line AB; L_1 is on line BC; L_1 is on line MN; M_1 is on line AC; M_1 is on line LN; N_1 is on line AB; N_1 is on line LM; L_2 is the midpoint of L and L_1; M_2 is the midpoint of M and M_1; N_2 is the midpoint of N and N_1. The **conclusion**: Points L_2, M_2 and N_2 are collinear.

$B = (0,0)$, $C = (u_1, 0)$, $A = (u_2, u_3)$, $P = (u_4, u_5)$, $L = (x_1, 0)$, $M = (x_3, x_2)$, $N = (x_5, x_4)$, $L_1 = (x_6, 0)$, $M_1 = (x_8, x_7)$, $N_1 = (x_{10}, x_9)$, $L_2 = (x_{11}, 0)$, $M_2 = (x_{12}, x_{13})$, $N_2 = (x_{14}, x_{15})$.

The **nondegenerate** conditions: Line BC intersects line AP; Line AC intersects line BP; Line AB intersects line CP; Line MN intersects line BC; Line LN intersects line AC; Line LM intersects line AB.

Example 415 (A247-5: 2.57s, 75). *If A_1 is the point of intersection of the side BC of the triangle ABC with the trilinear polar p of a point P on the circumcircle of ABC, show that the circle APA_1 passes through the midpoint of BC.*

Points B, C, A are arbitrarily chosen. Points O, P, M, N, A_1, G are constructed (in order) as follows: $OB \equiv OC$; $OB \equiv OA$; $PO \equiv OB$; M is on line BP; M is on line AC; N is on line CP; N is on line AB; A_1 is on line BC; A_1 is on line MN; G is the midpoint of B and C. The **conclusion**: Points A, P, A_1 and G are on the same circle.

$B = (0,0)$, $C = (u_1, 0)$, $A = (u_2, u_3)$, $O = (x_2, x_1)$, $P = (x_3, u_4)$, $M = (x_5, x_4)$, $N = (x_7, x_6)$, $A_1 = (x_8, 0)$, $G = (x_9, 0)$.

The **nondegenerate** conditions: Points B, A and C are not collinear; Line AC intersects line BP; Line AB intersects line CP; Line MN intersects line BC.

Example 416 (A247-7: 6.18s, 90). *Let L, M, N be the feet of the cevians AP, BP, CP of the triangle ABC and let P_1 be a point on the trilinear polar of P for ABC. If the lines AP_1 BP_1, CP_1 meet MN, NL, LM in X, Y, Z, show that the triangle*

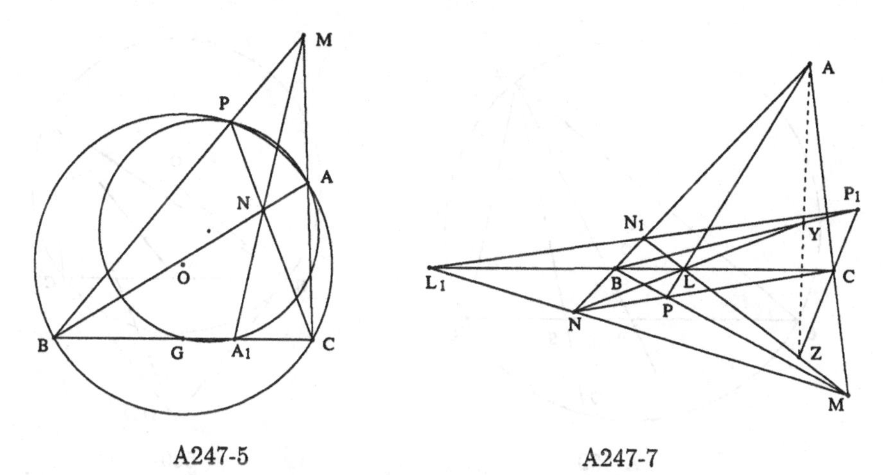

A247-5 A247-7

XYZ is circumscribed about the triangle ABC.

Points B, C, A, P are arbitrarily chosen. Points L, M, N, L_1, N_1, P_1, Y, Z are constructed (in order) as follows: L is on line AP; L is on line BC; M is on line BP; M is on line AC; N is on line CP; N is on line AB; L_1 is on line BC; L_1 is on line MN; N_1 is on line AB; N_1 is on line LM; P_1 is on line N_1L_1; Y is on line NL; Y is on line BP_1; Z is on line LM; Z is on line CP_1. The **conclusion**: Points Y, Z and A are collinear.

$B = (0,0)$, $C = (u_1,0)$, $A = (u_2,u_3)$, $P = (u_4,u_5)$, $L = (x_1,0)$, $M = (x_3,x_2)$, $N = (x_5,x_4)$, $L_1 = (x_6,0)$, $N_1 = (x_8,x_7)$, $P_1 = (x_9,u_6)$, $Y = (x_{11},x_{10})$, $Z = (x_{13},x_{12})$.

The **nondegenerate** conditions: Line BC intersects line AP; Line AC intersects line BP; Line AB intersects line CP; Line MN intersects line BC; Line LM intersects line AB; $N_1 \neq L_1$; Line BP_1 intersects line NL; Line CP_1 intersects line LM.

Example 417 (A-559: 2.88s, 30). *The traces, on the circumcircle of a triangle, of a median and the corresponding symmedian determine a line parallel to the side of the triangle opposite the vertex considered.*

Points B, C, A are arbitrarily chosen. Points O, A_1, S, Q, P are constructed (in order) as follows: $OA \equiv OB$; $OC \equiv OB$; A_1 is the midpoint of B and C; S is on line BC; $\tan(BAS) = \tan(A_1AC)$; $QO \equiv OB$; Q is on line AS; $PO \equiv OB$; P is on line AA_1. The **conclusion**: $PQ \parallel BC$.

$B = (0,0)$, $C = (u_1,0)$, $A = (u_2,u_3)$, $O = (x_2,x_1)$, $A_1 = (x_3,0)$, $S = (x_4,0)$, $Q = (x_6,x_5)$, $P = (x_8,x_7)$.

The **nondegenerate** conditions: Points C, B and A are not collinear; $((2u_2 - u_1)u_3)x_3 - u_3^3 - u_2^2u_3 \neq 0$; Line AS is non-isotropic; Line AA_1 is non-isotropic. In addition, the following nondegenerate conditions, which come from reducibility and have been detected by our prover, should be also added: $P \neq A$; $Q \neq A$.

Example 418 (A-560: 0.95s, 14). *The symmedian issued from a vertex of triangle*

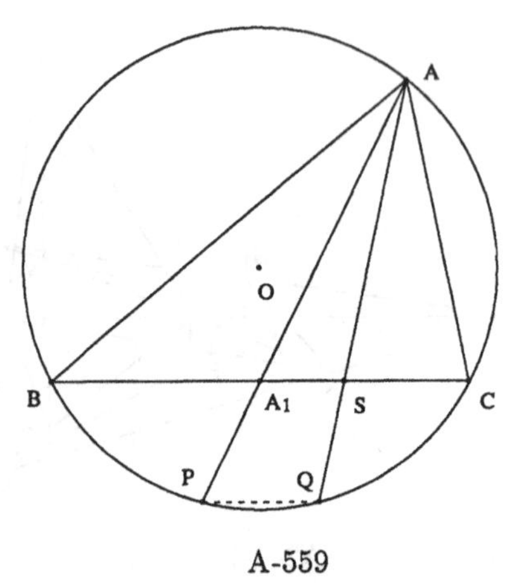

A-559

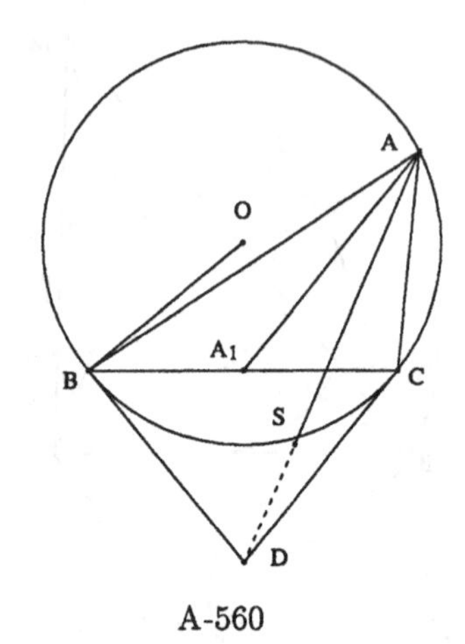

A-560

passes through the point of intersection of the tangents to the circumcircle at the other two vertices of the triangle.

Points B, C, A are arbitrarily chosen. Points O, A_1, S, D are constructed (in order) as follows: $OA \equiv OB$; $OC \equiv OB$; A_1 is the midpoint of B and C; $\tan(BAS) = \tan(A_1AC)$; $DB \perp OB$; $DB \equiv DC$. The **conclusion**: Points A, S and D are collinear.

$B = (0,0)$, $C = (u_1,0)$, $A = (u_2,u_3)$, $O = (x_2,x_1)$, $A_1 = (x_3,0)$, $S = (x_4,u_4)$, $D = (x_6,x_5)$.

The **nondegenerate** conditions: Points C, B and A are not collinear; $A \neq B$, $A \neq A_1$, $A \neq C$; Points B, C and O are not collinear.

Example 419 (A-562: 1.02s, 9). *A symmedian of a triangle is the harmonic conjugate of the tangent to the circumcircle at the vertex considered with respect to the two sides passing through that vertex.*

Points B, C, A are arbitrarily chosen. Points O, A_1, S, T_A are constructed (in order) as follows: $OA \equiv OB$; $OC \equiv OB$; A_1 is the midpoint of B and C; S is on line BC; $\tan(BAS) = \tan(A_1AC)$; T_A is on line BC; $T_AA \perp AO$. The **conclusion**: Points S, T_A, B and C form a harmonic set.

$B = (0,0)$, $C = (u_1,0)$, $A = (u_2,u_3)$, $O = (x_2,x_1)$, $A_1 = (x_3,0)$, $S = (x_4,0)$, $T_A = (x_5,0)$.

The **nondegenerate** conditions: Points C, B and A are not collinear; $((2u_2 - u_1)u_3)x_3 - u_3^3 - u_2^2u_3 \neq 0$; Line AO is not perpendicular to line BC.

Example 420 (A-564: 2.78s, 117). *A symmedian of a triangle bisects any antiparallel to the side of the triangle relative to the symmedian considered.*

Points B, C, A are arbitrarily chosen. Points A_1, D, O, E, S are constructed (in order) as follows: A_1 is the midpoint of B and C; D is on line AB; $OD \equiv OB$; $OC \equiv OB$; $EO \equiv OB$; E is on line AC; S is on line DE; $\tan(BAS) = \tan(A_1AC)$. The **conclusion**: S is the midpoint of D and E.

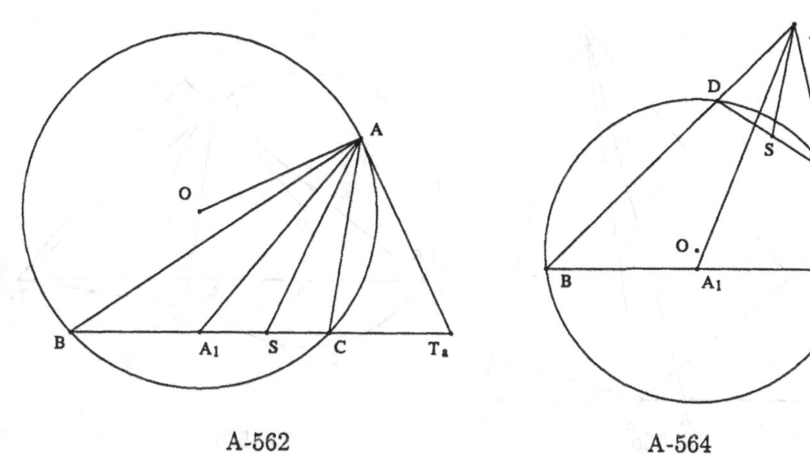

A-562 A-564

$B = (0,0)$, $C = (u_1,0)$, $A = (u_2,u_3)$, $A_1 = (x_1,0)$, $D = (x_2,u_4)$, $O = (x_4,x_3)$, $E = (x_6,x_5)$, $S = (x_8,x_7)$.

The **nondegenerate** conditions: $A \neq B$; Points C, B and D are not collinear; Line AC is non-isotropic; $(((-2u_2+u_1)u_3)x_1+u_3^3+u_2^2u_3)x_6+((-u_3^2+u_2^2-u_1u_2)x_1+(-u_2+u_1)u_3^2-u_2^3+u_1u_2^2)x_5+(((2u_2-u_1)u_3)x_1-u_3^3-u_2^2u_3)x_2+((u_3^2-u_2^2+u_1u_2)u_4)x_1+((u_2-u_1)u_3^2+u_2^3-u_1u_2^2)u_4 \neq 0$. In addition, the following nondegenerate conditions, which come from reducibility and have been detected by our prover, should be also added: $E \neq C$.

Example 421 (A-565: 2.12s, 78). *The distances from a point on a symmedian of a triangle to the two including sides are proportional to these sides.*

Points B, C, A are arbitrarily chosen. Points A_1, S, X, Y are constructed (in order) as follows: A_1 is the midpoint of B and C; S is on line BC; $\tan(BAS) = \tan(A_1AC)$; X is on line AC; $XS \perp AC$; Y is on line AB; $YS \perp AB$. The **conclusion**: $SY \cdot AC = SX \cdot AB$.

$B = (0,0)$, $C = (u_1,0)$, $A = (u_2,u_3)$, $A_1 = (x_1,0)$, $S = (x_2,0)$, $X = (x_4,x_3)$, $Y = (x_6,x_5)$.

The **nondegenerate** conditions: $((2u_2 - u_1)u_3)x_1 - u_3^3 - u_2^2u_3 \neq 0$; Line AC is non-isotropic; Line AB is non-isotropic.

Example 422 (A-566: 2.53s, 79). *The preceding proposition (A-565) is valid for the points of the external symmedian of the triangle.*

Points B, C, A are arbitrarily chosen. Points O, A_1, T_A, X, Y are constructed (in order) as follows: $OA \equiv OB$; $OC \equiv OB$; A_1 is the midpoint of B and C; T_A is on line BC; $T_AA \perp AO$; X is on line AC; $XT_A \perp AC$; Y is on line AB; $YT_A \perp AB$. The **conclusion**: $T_AY \cdot AC = T_AX \cdot AB$.

$B = (0,0)$, $C = (u_1,0)$, $A = (u_2,u_3)$, $O = (x_2,x_1)$, $A_1 = (x_3,0)$, $T_A = (x_4,0)$, $X = (x_6,x_5)$, $Y = (x_8,x_7)$.

The **nondegenerate** conditions: Points C, B and A are not collinear; Line AO

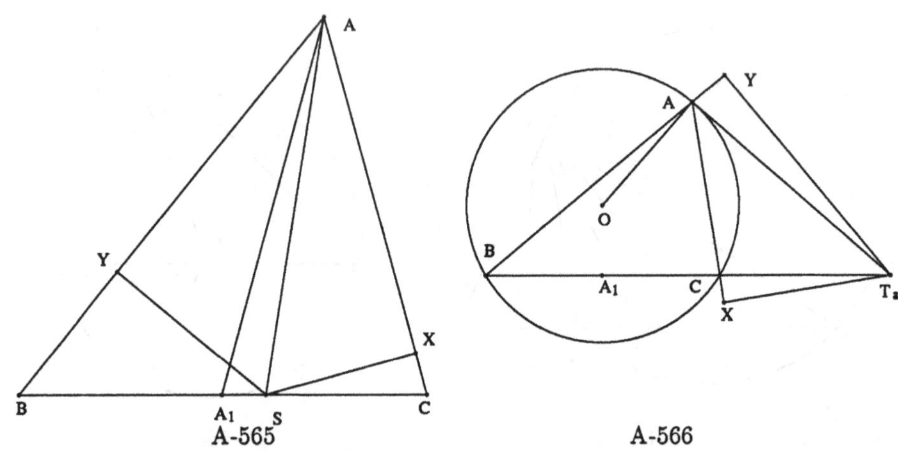

A-565 A-566

is not perpendicular to line BC; Line AC is non-isotropic; Line AB is non-isotropic.

Example 423 (A-568-a: 1.38s, 27). *If from a point on the symmedian perpendiculars are drawn to the including sides of the triangle, the line joining the feet of these perpendiculars is perpendicular to the corresponding median of the triangle.*

Points B, C, A are arbitrarily chosen. Points A_1, S, N, G, H are constructed (in order) as follows: A_1 is the midpoint of B and C; S is on line BC; $\tan(BAS)$ = $\tan(A_1AC)$; N is on line AA_1; G is on line AB; $GN \perp AB$; H is on line AC; $HN \perp AC$. The **conclusion**: $GH \perp AS$.

$B = (0,0)$, $C = (u_1,0)$, $A = (u_2,u_3)$, $A_1 = (x_1,0)$, $S = (x_2,0)$, $N = (x_3,u_4)$, $G = (x_5,x_4)$, $H = (x_7,x_6)$.

The **nondegenerate** conditions: $((2u_2 - u_1)u_3)x_1 - u_3^3 - u_2^2u_3 \neq 0$; $A \neq A_1$; Line AB is non-isotropic; Line AC is non-isotropic.

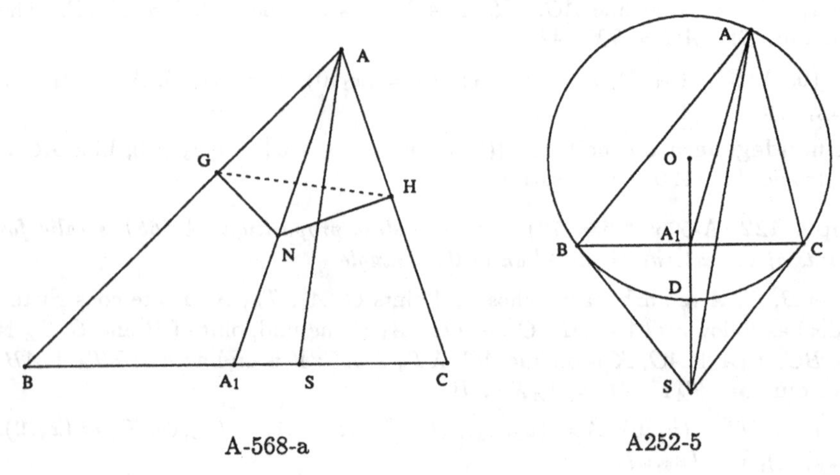

A-568-a A252-5

Example 424 (A252-5: 0.95s, 13). *Show that a side of a triangle and the corre-*

sponding symmedian are conjugate with respect to the circumcircle of the triangle.

Points B, C, A are arbitrarily chosen. Points A_1, O, D, S are constructed (in order) as follows: A_1 is the midpoint of B and C; $OB \equiv OC$; $OA \equiv OB$; $DO \perp BC$; $DO \equiv OB$; S is on line OD; A_1 and S are inversive wrpt circle (O, OD). The **conclusion**: $\tan(BAS) = \tan(A_1AC)$.

$B = (0,0)$, $C = (u_1,0)$, $A = (u_2,u_3)$, $A_1 = (x_1,0)$, $O = (x_3,x_2)$, $D = (x_3,x_4)$, $S = (x_3,x_5)$.

The **nondegenerate** conditions: Points A, B and C are not collinear; Line BC is non-isotropic; $O \neq D$; $A_1 \neq O$.

Example 425 (A252-7b: 2.17s, 43). *If the symmedian issued from the vertex A of the triangle ABC meets the circumcircle in D, and P, R, Q are the projections of D upon BC, CA, AB, show that $PQ = PR$.*

Points B, C, A are arbitrarily chosen. Points O, A_1, S, D, P, Q, R are constructed (in order) as follows: $OA \equiv OB$; $OC \equiv OB$; A_1 is the midpoint of B and C; S is on line BC; $\tan(BAS) = \tan(A_1AC)$; $DO \equiv OB$; D is on line AS; P is on line BC; $PD \perp BC$; Q is on line AB; $QD \perp AB$; R is on line CA; $RD \perp CA$. The **conclusion**: P is the midpoint of R and Q.

$B = (0,0)$, $C = (u_1,0)$, $A = (u_2,u_3)$, $O = (x_2,x_1)$, $A_1 = (x_3,0)$, $S = (x_4,0)$, $D = (x_6,x_5)$, $P = (x_6,0)$, $Q = (x_8,x_7)$, $R = (x_{10},x_9)$.

The **nondegenerate** conditions: Points C, B and A are not collinear; $((2u_2 - u_1)u_3)x_3 - u_3^3 - u_2^2u_3 \neq 0$; Line AS is non-isotropic; Line BC is non-isotropic; Line AB is non-isotropic; Line CA is non-isotropic. In addition, the following nondegenerate conditions, which come from reducibility and have been detected by our prover, should be also added: $D \neq A$.

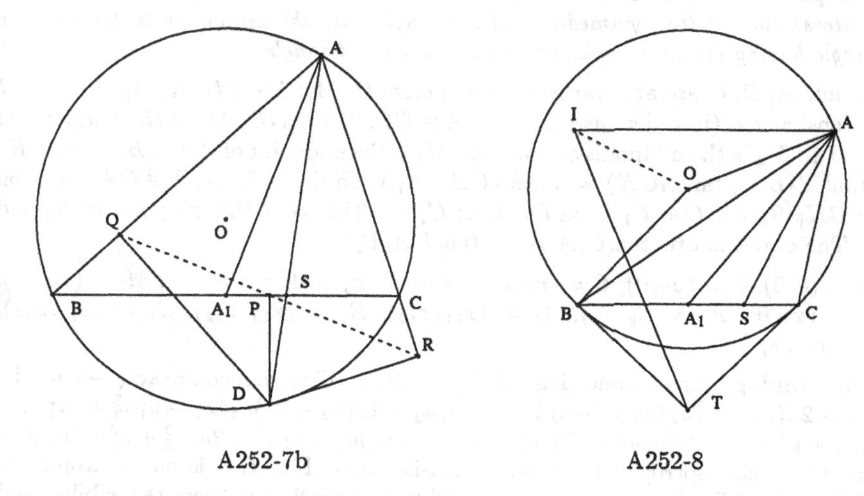

A252-7b A252-8

Example 426 (A252-8: 1.35s, 12). *Show that the parallel to the side BC through the vertex A of ABC, the perpendicular from the circumcenter O upon symmedian issued from A, and the perpendicular upon AO from the point of intersection T of* *the tangents to the circumcircle at B and C are three concurrent lines.*

Points B, C, A are arbitrarily chosen. Points O, A_1, S, T, I are constructed (in order) as follows: $OA \equiv OB$; $OC \equiv OB$; A_1 is the midpoint of B and C; S is on line BC; $\tan(BAS) = \tan(A_1AC)$; $TB \perp OB$; $TB \equiv TC$; $IA \parallel BC$; $IT \perp AO$. The **conclusion**: $OI \perp AS$.

$B = (0,0)$, $C = (u_1,0)$, $A = (u_2,u_3)$, $O = (x_2,x_1)$, $A_1 = (x_3,0)$, $S = (x_4,0)$, $T = (x_6,x_5)$, $I = (x_7,u_3)$.

The **nondegenerate** conditions: Points C, B and A are not collinear; $((2u_2 - u_1)u_3)x_3 - u_3^3 - u_2^2u_3 \neq 0$; Points B, C and O are not collinear; Line AO is not perpendicular to line BC.

Example 427 (A252-9: 1.17s, 6). *If B, B_1 are the inverse points of A, A_1 in an inversion of center O, show that the median and the symmedian of the triangle AOA_1 issued from O are respectively the symmedian and the median of the triangle BO_1B_1.*

Points O, R are arbitrarily chosen. Points A, B, R_1, A_1, B_1, M, N are constructed (in order) as follows: A is on line OR; A and B are inversive wrpt circle (O, OR); B is on line OR; $R_1O \equiv OR$; A_1 is on line OR_1; A_1 and B_1 are inversive wrpt circle (O, OR_1); B_1 is on line OR_1; M is the midpoint of A and A_1; N is the midpoint of B_1 and B. The **conclusion**: $\tan(AON) = \tan(MOA_1)$.

$O = (0,0)$, $R = (u_1,0)$, $A = (u_2,0)$, $B = (x_1,0)$, $R_1 = (x_2,u_3)$, $A_1 = (x_3,u_4)$, $B_1 = (x_5,x_4)$, $M = (x_6,x_7)$, $N = (x_8,x_9)$.

The **nondegenerate** conditions: $O \neq R$; $O \neq R$; $A \neq O$; $O \neq R_1$; $O \neq R_1$; $A_1 \neq O$.

Example 428 (A252-10: 1419.87s, 9844; 78.97s, 2343). *Show that the three points of intersection of the symmedians of a triangle with the circumcircle determine a triangle having the same symmedians as the given triangle.*

Points A, B, C are arbitrarily chosen. Points O, M_1, M_2, M_3, K, A_1, B_1, C_1, N are constructed (in order) as follows: $OA \equiv OC$; $OA \equiv OB$; M_1 is the midpoint of B and C; M_2 is the midpoint of A and C; M_3 is the midpoint of A and B; $\tan(CBK) = \tan(M_2BA)$; $\tan(ACK) = \tan(M_3CB)$; A_1 is on line AK; $A_1O \equiv OA$; B_1 is on line BK; $B_1O \equiv OA$; C_1 is on line CK; $C_1O \equiv OA$; N is the midpoint of B_1 and C_1. The **conclusion**: $\tan(C_1A_1N) = \tan(KA_1B_1)$.

$A = (0,0)$, $B = (u_1,0)$, $C = (u_2,u_3)$, $O = (x_2,x_1)$, $M_1 = (x_3,x_4)$, $M_2 = (x_5,x_6)$, $M_3 = (x_7,0)$, $K = (x_9,x_8)$, $A_1 = (x_{11},x_{10})$, $B_1 = (x_{13},x_{12})$, $C_1 = (x_{15},x_{14})$, $N = (x_{16},x_{17})$.

The **nondegenerate** conditions: Points A, B and C are not collinear; $((-u_1u_2u_3^2 - u_1u_2^3 + 2u_1^2u_2^2 - u_1^3u_2)x_6 + (-u_1u_3^3 + (-u_1u_2^2 + 2u_1^2u_2 - u_1^3)u_3)x_5 + u_1^2u_3^3 + (u_1^2u_2^2 - 2u_1^3u_2 + u_1^4)u_3)x_7 + (u_1u_3^4 + (2u_1u_2^2 - 2u_1^2u_2 + u_1^3)u_3^2 + u_1u_2^4 - 2u_1^2u_2^3 + u_1^3u_2^2)x_6 \neq 0$; Line AK is non-isotropic; Line BK is non-isotropic; Line CK is non-isotropic. In addition, the following nondegenerate conditions, which come from reducibility and have been detected by our prover, should be also added: $C_1 \neq C$; $B_1 \neq B$; $A_1 \neq A$.

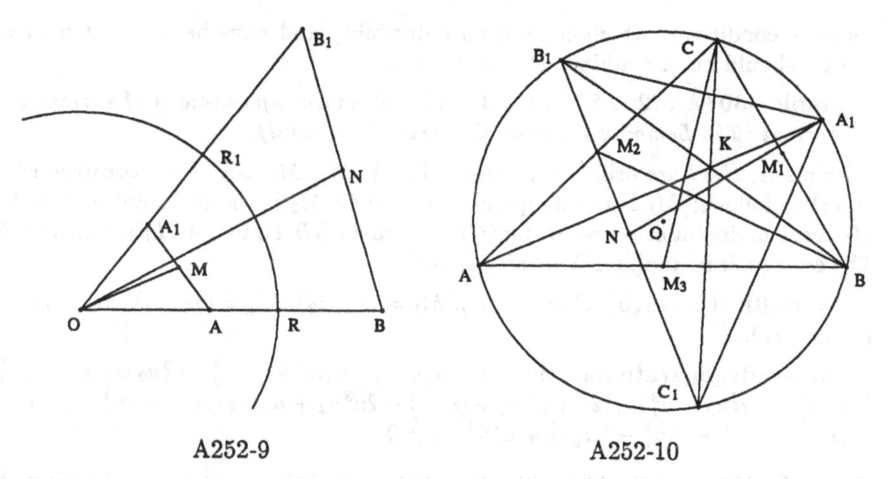

A252-9

A252-10

Example 429 (A252-11-a: 8.12s, 70). *The median and the symmedian of a triangle ABC issued from A meet the circumcircle in P, Q. Show that the Simson lines of P, Q are respectively perpendicular to AP, AQ.*

Points B, C, A are arbitrarily chosen. Points O, A_1, S, Q, P, D, G are constructed (in order) as follows: $OA \equiv OB$; $OC \equiv OB$; A_1 is the midpoint of B and C; S is on line BC; $\tan(BAS) = \tan(A_1AC)$; $QO \equiv OB$; Q is on line AS; $PO \equiv OB$; P is on line AA_1; $DQ \perp BC$; D is on line BC; $GQ \perp AB$; G is on line AB. The **conclusion**: $DG \perp AP$.

$B = (0,0)$, $C = (u_1,0)$, $A = (u_2,u_3)$, $O = (x_2,x_1)$, $A_1 = (x_3,0)$, $S = (x_4,0)$, $Q = (x_6,x_5)$, $P = (x_8,x_7)$, $D = (x_6,0)$, $G = (x_{10},x_9)$.

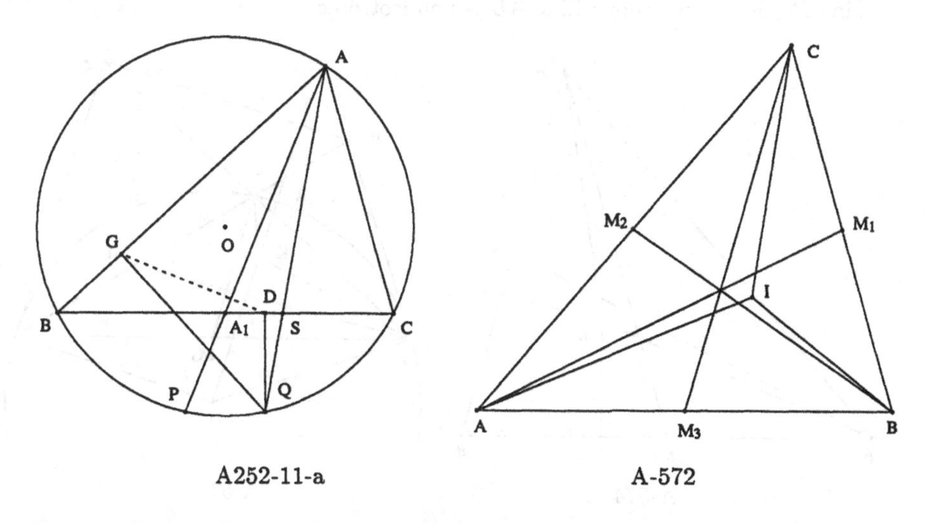

A252-11-a

A-572

The **nondegenerate** conditions: Points C, B and A are not collinear; $((2u_2 - u_1)u_3)x_3 - u_3^3 - u_2^2u_3 \neq 0$; Line AS is non-isotropic; Line AA_1 is non-isotropic; Line BC is non-isotropic; Line AB is non-isotropic. In addition, the following nondegenerate conditions, which come from reducibility and have been detected by our prover, should be also added: $P \neq A$; $Q \neq A$.

Example 430 (A-572: 3.87s, 33; 3.33s, 33). *The three symmedians of a triangle are concurrent (The Lemoine Point or the symmedian point).*

Points A, B, C are arbitrarily chosen. Points M_1, M_2, M_3, I are constructed (in order) as follows: M_1 is the midpoint of B and C; M_2 is the midpoint of A and C; M_3 is the midpoint of A and B; $\tan(CBI) = \tan(M_2BA)$; $\tan(ACI) = \tan(M_3CB)$. The **conclusion**: $\tan(BAI) = \tan(M_1AC)$.

$A = (0,0)$, $B = (u_1, 0)$, $C = (u_2, u_3)$, $M_1 = (x_1, x_2)$, $M_2 = (x_3, x_4)$, $M_3 = (x_5, 0)$, $I = (x_7, x_6)$.

The **nondegenerate** conditions: $((-u_1u_2u_3^2 - u_1u_2^3 + 2u_1^2u_2^2 - u_1^3u_2)x_4 + (-u_1u_3^3 + (-u_1u_2^2 + 2u_1^2u_2 - u_1^3)u_3)x_3 + u_1^2u_3^3 + (u_1^2u_2^2 - 2u_1^3u_2 + u_1^4)u_3)x_5 + (u_1u_3^4 + (2u_1u_2^2 - 2u_1^2u_2 + u_1^3)u_3^2 + u_1u_2^4 - 2u_1^2u_2^3 + u_1^3u_2^2)x_4 \neq 0$.

Example 431 (A-574: 2.65s, 39). *The distances of the Lemoine point K from the sides of the triangle are proportional to the respective sides.*

Points A, B, C are arbitrarily chosen. Points M_1, M_3, K, D, F are constructed (in order) as follows: M_1 is the midpoint of B and C; M_3 is the midpoint of A and B; $\tan(BAK) = \tan(M_1AC)$; $\tan(ACK) = \tan(M_3CB)$; D is on line BC; $DK \perp BC$; F is on line AB; $FK \perp AB$. The **conclusion**: $KF \cdot BC = KD \cdot AB$.

$A = (0,0)$, $B = (u_1, 0)$, $C = (u_2, u_3)$, $M_1 = (x_1, x_2)$, $M_3 = (x_3, 0)$, $K = (x_5, x_4)$, $D = (x_7, x_6)$, $F = (x_5, 0)$.

The **nondegenerate** conditions: $(((u_1u_2 - u_1^2)u_3^2 + u_1u_2^3 - u_1^2u_2^2)x_2 + (u_1u_3^3 + u_1u_2^2u_3)x_1)x_3 + (-u_1u_3^4 + (-2u_1u_2^2 + u_1^2u_2)u_3^2 - u_1u_2^4 + u_1^2u_2^3)x_2 + (-u_1^2u_3^3 - u_1^2u_2^2u_3)x_1 \neq 0$; Line BC is non-isotropic; Line AB is non-isotropic.

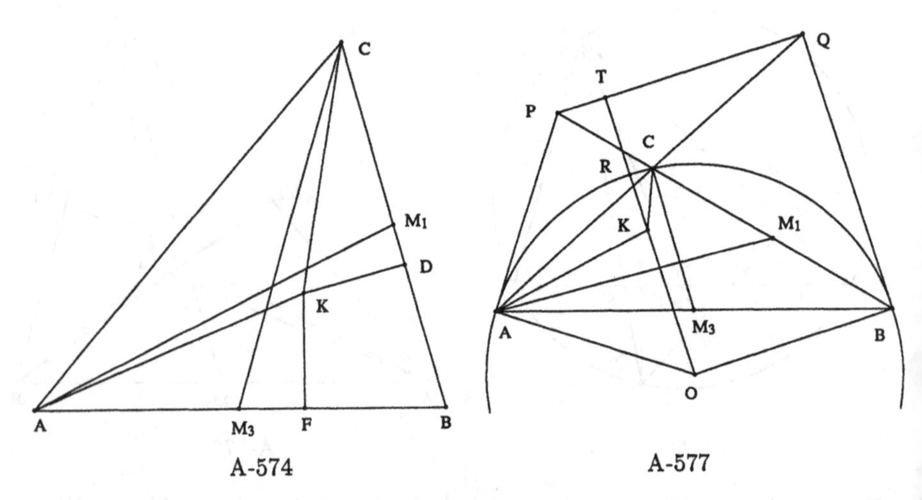

A-574 A-577

Example 432 (A-577: 11.97s, 476). *The Lemoine point of a triangle is the pole of the Lemoine axis with respect to the circumcircle of the triangle.*

Points A, B, C are arbitrarily chosen. Points O, P, Q, M_3, M_1, K, R, T are constructed (in order) as follows: $OA \equiv OC$; $OA \equiv OB$; P is on line BC; $PA \perp OA$; Q is on line AC; $QB \perp OB$; M_3 is the midpoint of A and B; M_1 is the midpoint of B and C; $\tan(BAK) = \tan(M_1AC)$; $\tan(ACK) = \tan(M_3CB)$; $RO \equiv OA$; R is on line OK; T is on line PQ; T is on line OK. The **conclusions**: (1) $KO \perp PQ$; (2) K and T are inversive wrpt circle (O, OR).

$A = (0,0)$, $B = (u_1,0)$, $C = (u_2,u_3)$, $O = (x_2,x_1)$, $P = (x_4,x_3)$, $Q = (x_6,x_5)$, $M_3 = (x_7,0)$, $M_1 = (x_8,x_9)$, $K = (x_{11},x_{10})$, $R = (x_{13},x_{12})$, $T = (x_{15},x_{14})$.

The **nondegenerate** conditions: Points A, B and C are not collinear; Line OA is not perpendicular to line BC; Line OB is not perpendicular to line AC; $(((-u_1u_2+u_1^2)u_3^2 - u_1u_2^3 + u_1^2u_2^2)x_7 + u_1u_3^4 + (2u_1u_2^2 - u_1^2u_2)u_3^2 + u_1u_2^4 - u_1^2u_2^3)x_9 + ((-u_1u_3^3 - u_1u_2^2u_3)x_7 + u_1^2u_3^3 + u_1^2u_2^2u_3)x_8 \neq 0$; Line OK is non-isotropic; Line OK intersects line PQ.

Example 433 (A-579: 2.98s, 33). *The Brocard diameter of a triangle is perpendicular to the Lemoine axis.*

Points A, B, C are arbitrarily chosen. Points O, M_1, M_2, M_3, I, B_1, C_1 are constructed (in order) as follows: $OA \equiv OC$; $OA \equiv OB$; M_1 is the midpoint of B and C; M_2 is the midpoint of A and C; M_3 is the midpoint of A and B; $\tan(CBI) = \tan(M_2BA)$; $\tan(ACI) = \tan(M_3CB)$; B_1 is on line AC; $B_1B \perp OB$; $C_1C \perp CO$; C_1 is on line AB. The **conclusion**: $OI \perp C_1B_1$.

$A = (0,0)$, $B = (u_1,0)$, $C = (u_2,u_3)$, $O = (x_2,x_1)$, $M_1 = (x_3,x_4)$, $M_2 = (x_5,x_6)$, $M_3 = (x_7,0)$, $I = (x_9,x_8)$, $B_1 = (x_{11},x_{10})$, $C_1 = (x_{12},0)$.

The **nondegenerate** conditions: Points A, B and C are not collinear; $((-u_1u_2u_3^2 - u_1u_2^3 + 2u_1^2u_2^2 - u_1^3u_2)x_6 + (-u_1u_3^3 + (-u_1u_2^2 + 2u_1^2u_2 - u_1^3)u_3)x_5 + u_1^2u_3^3 + (u_1^2u_2^2 - 2u_1^3u_2 + u_1^4)u_3)x_7 + (u_1u_3^4 + (2u_1u_2^2 - 2u_1^2u_2 + u_1^3)u_3^2 + u_1u_2^4 - 2u_1^2u_2^3 + u_1^3u_2^2)x_6 \neq 0$; Line OB is not perpendicular to line AC; Line AB is not perpendicular to line CO.

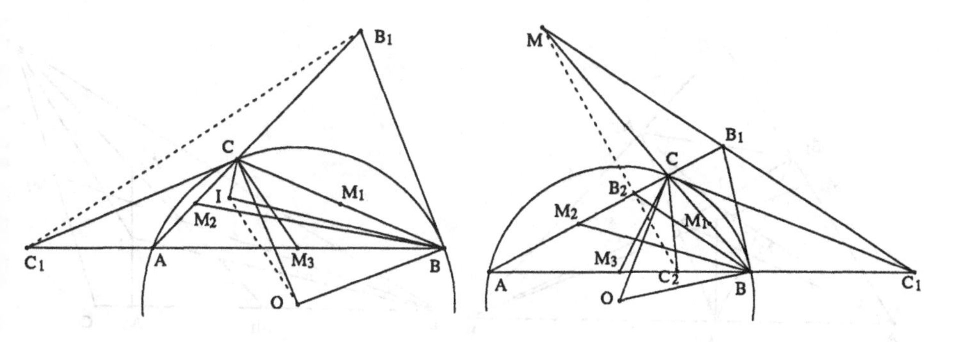

A-579 A-581

Example 434 (A-581: 2.43s, 32). *The Lemoine point of a triangle is the trilinear* *pole, for the triangle, of the Lemoine axis of the triangle.*

Points A, B, C are arbitrarily chosen. Points O, M_1, M_2, M_3, C_2, B_2, B_1, C_1, M are constructed (in order) as follows: $OA \equiv OC$; $OA \equiv OB$; M_1 is the midpoint of B and C; M_2 is the midpoint of A and C; M_3 is the midpoint of A and B; C_2 is on line AB; $\tan(ACC_2) = \tan(M_3CB)$; B_2 is on line AC; $\tan(CBB_2) = \tan(M_2BA)$; B_1 is on line AC; $B_1B \perp OB$; $C_1C \perp CO$; C_1 is on line AB; M is on line B_1C_1; M is on line BC. The **conclusion**: Points M, B_2 and C_2 are collinear.

$A = (0,0)$, $B = (u_1,0)$, $C = (u_2,u_3)$, $O = (x_2,x_1)$, $M_1 = (x_3,x_4)$, $M_2 = (x_5,x_6)$, $M_3 = (x_7,0)$, $C_2 = (x_8,0)$, $B_2 = (x_{10},x_9)$, $B_1 = (x_{12},x_{11})$, $C_1 = (x_{13},0)$, $M = (x_{15},x_{14})$.

The **nondegenerate** conditions: Points A, B and C are not collinear; $((2u_2 - u_1)u_3)x_7 - u_3^3 - u_2^2u_3 \neq 0$; $(u_1u_3^2 + u_1u_2^2 - u_1^2u_2)x_6 - u_1^2u_3x_5 + u_1^3u_3 \neq 0$; Line OB is not perpendicular to line AC; Line AB is not perpendicular to line CO; Line BC intersects line B_1C_1.

Example 435 (A-583: 2.6s, 22). *The pedal triangle of the Lemoine point has this point for its centroid.*

Points A, B, C are arbitrarily chosen. Points M_1, M_2, I, B_1, C_1, A_1, M are constructed (in order) as follows: M_1 is the midpoint of B and C; M_2 is the midpoint of A and C; $\tan(BAI) = \tan(M_1AC)$; $\tan(CBI) = \tan(M_2BA)$; B_1 is on line CA; $B_1I \perp CA$; C_1 is on line AB; $C_1I \perp AB$; A_1 is on line BC; $A_1I \perp BC$; M is the midpoint of C_1 and B_1. The **conclusion**: Points A_1, M and I are collinear.

$A = (0,0)$, $B = (u_1,0)$, $C = (u_2,u_3)$, $M_1 = (x_1,x_2)$, $M_2 = (x_3,x_4)$, $I = (x_6,x_5)$, $B_1 = (x_8,x_7)$, $C_1 = (x_6,0)$, $A_1 = (x_{10},x_9)$, $M = (x_{11},x_{12})$.

The **nondegenerate** conditions: $(-u_1^3u_3x_2+(u_1^2u_3^2+u_1^2u_2^2-u_1^3u_2)x_1)x_4+((-u_1^2u_3^2-u_1^2u_2^2+u_1^3u_2)x_2-u_1^3u_3x_1)x_3+(u_1^3u_3^2+u_1^3u_2^2-u_1^4u_2)x_2+u_1^4u_3x_1 \neq 0$; Line CA is non-isotropic; Line AB is non-isotropic; Line BC is non-isotropic.

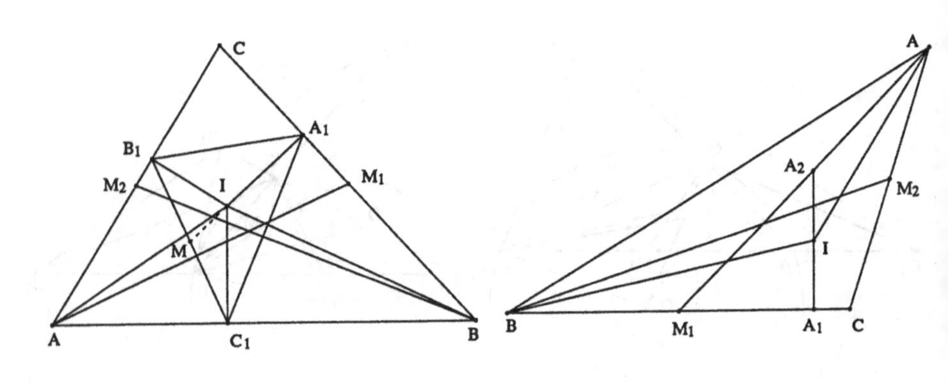

A-583 A-584

Example 436 (A-584: 1.27s, 21). *The symmetrics, with respect to the Lemoine* *point, of the projection of this point upon a side of the triangle lies on the median of the triangle relative to the side considered.*

Points B, C, A are arbitrarily chosen. Points M_1, M_2, I, A_1, A_2 are constructed (in order) as follows: M_1 is the midpoint of B and C; M_2 is the midpoint of A and C; $\tan(BAI) = \tan(M_1AC)$; $\tan(CBI) = \tan(M_2BA)$; A_1 is on line BC; $A_1I \perp BC$; I is the midpoint of A_1 and A_2. The **conclusion**: Points A, M_1 and A_2 are collinear.

$B = (0,0)$, $C = (u_1,0)$, $A = (u_2,u_3)$, $M_1 = (x_1,0)$, $M_2 = (x_2,x_3)$, $I = (x_5,x_4)$, $A_1 = (x_5,0)$, $A_2 = (x_5,x_6)$.

The **nondegenerate** conditions: $(((-u_1u_2 + u_1^2)u_3^2 - u_1u_2^3 + u_1^2u_2^2)x_1 + u_1u_3^4 + (2u_1u_2^2 - u_1^2u_2)u_3^2 + u_1u_2^4 - u_1^2u_2^3)x_3 + ((-u_1u_3^3 - u_1u_2^2u_3)x_1 + u_1^2u_3^3 + u_1^2u_2^2u_3)x_2 \neq 0$; Line BC is non-isotropic.

Example 437 (A-586: 1.23s, 21). *The Lemoine point of a triangle is the point of intersection of the lines joining the midpoints of the sides of the triangle to the midpoints of the corresponding altitudes.*

Points B, C, A are arbitrarily chosen. Points M_1, M_2, I, D, M are constructed (in order) as follows: M_1 is the midpoint of B and C; M_2 is the midpoint of A and C; $\tan(CBI) = \tan(M_2BA)$; $\tan(BAI) = \tan(M_1AC)$; $DA \perp BC$; D is on line BC; M is the midpoint of A and D. The **conclusion**: Points M_1, M and I are collinear.

$B = (0,0)$, $C = (u_1,0)$, $A = (u_2,u_3)$, $M_1 = (x_1,0)$, $M_2 = (x_2,x_3)$, $I = (x_5,x_4)$, $D = (u_2,0)$, $M = (u_2,x_6)$.

The **nondegenerate** conditions: $(((-u_1u_2 + u_1^2)u_3^2 - u_1u_2^3 + u_1^2u_2^2)x_1 + u_1u_3^4 + (2u_1u_2^2 - u_1^2u_2)u_3^2 + u_1u_2^4 - u_1^2u_2^3)x_3 + ((-u_1u_3^3 - u_1u_2^2u_3)x_1 + u_1^2u_3^3 + u_1^2u_2^2u_3)x_2 \neq 0$; Line BC is non-isotropic.

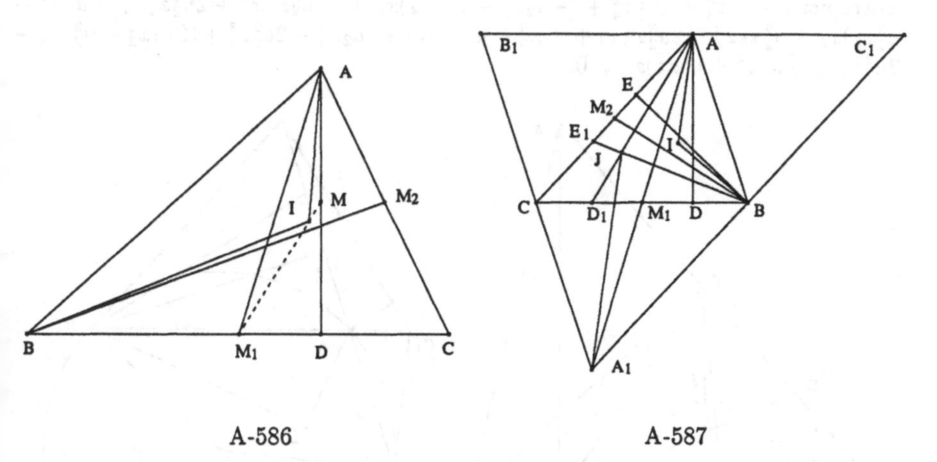

A-586 A-587

Example 438 (A-587: 4.37s, 33). *The isotomic of the orthocenter of a triangle is the Lemoine point of the anticomplementary triangle.*

Points C, B, A are arbitrarily chosen. Points M_1, M_2, I, D, E, E_1, D_1, J, B_1, C_1, A_1 are constructed (in order) as follows: M_1 is the midpoint of B and C; M_2

is the midpoint of A and C; $\tan(CBI) = \tan(M_2BA)$; $\tan(BAI) = \tan(M_1AC)$; $DA \perp BC$; D is on line BC; E is on line AC; $EB \perp AC$; M_2 is the midpoint of E and E_1; M_1 is the midpoint of D and D_1; J is on line AD_1; J is on line BE_1; $B_1C \parallel AB$; $B_1A \parallel BC$; $C_1B \parallel AC$; C_1 is on line B_1A; A_1 is on line B_1C; A_1 is on line C_1B. The **conclusion**: $\tan(BA_1A) = \tan(JA_1C)$.

$C = (0,0)$, $B = (u_1, 0)$, $A = (u_2, u_3)$, $M_1 = (x_1, 0)$, $M_2 = (x_2, x_3)$, $I = (x_5, x_4)$, $D = (u_2, 0)$, $E = (x_7, x_6)$, $E_1 = (x_8, x_9)$, $D_1 = (x_{10}, 0)$, $J = (x_{12}, x_{11})$, $B_1 = (x_{13}, u_3)$, $C_1 = (x_{14}, u_3)$, $A_1 = (x_{16}, x_{15})$.

The **nondegenerate** conditions: $((u_1u_2u_3^2 + u_1u_2^3 - 2u_1^2u_2^2 + u_1^3u_2)x_1 - u_1u_3^4 + (-2u_1u_2^2 + 2u_1^2u_2 - u_1^3)u_3^2 - u_1u_2^4 + 2u_1^2u_2^3 - u_1^3u_2^2)x_3 + ((u_1u_3^3 + (u_1u_2^2 - 2u_1^2u_2 + u_1^3)u_3)x_1)x_2 + (-u_1^2u_3^3 + (-u_1^2u_2^2 + 2u_1^3u_2 - u_1^4)u_3)x_1 \neq 0$; Line BC is non-isotropic; Line AC is non-isotropic; Line BE_1 intersects line AD_1; Points B, C and A are not collinear; Points B_1, A and C are not collinear; Line C_1B intersects line B_1C.

Example 439 (A256-1: 7.77s, 33; 7.67s, 138). *If DEF is the orthic triangle of ABC, show that the symmedian points of the triangles AEF, BFD, CDE lie on the medians of ABC.*

Points C, B, A are arbitrarily chosen. Points M_3, D, E, D_1, E_1, J are constructed (in order) as follows: M_3 is the midpoint of A and B; $DA \perp BC$; D is on line BC; E is on line AC; $EB \perp AC$; D_1 is the midpoint of C and E; E_1 is the midpoint of C and D; $\tan(EDD_1) = \tan(JDC)$; $\tan(CEJ) = \tan(E_1ED)$. The **conclusion**: Points C, M_3 and J are collinear.

$C = (0,0)$, $B = (u_1, 0)$, $A = (u_2, u_3)$, $M_3 = (x_1, x_2)$, $D = (u_2, 0)$, $E = (x_4, x_3)$, $D_1 = (x_5, x_6)$, $E_1 = (x_7, 0)$, $J = (x_9, x_8)$.

The **nondegenerate** conditions: Line BC is non-isotropic; Line AC is non-isotropic; $((-u_2x_4^3 + 2u_2^2x_4^2 + (-u_2x_3^2 - u_2^3)x_4)x_6 + (-u_2x_3x_4^2 + 2u_2^2x_3x_4 - u_2x_3^3 - u_2^3x_3)x_5 + u_2^2x_3x_4^2 - 2u_2^3x_3x_4 + u_2^2x_3^3 + u_2^4x_3)x_7 + (u_2x_4^4 - 2u_2^2x_4^3 + (2u_2x_3^2 + u_2^3)x_4^2 - 2u_2^2x_3^2x_4 + u_2x_3^4 + u_2^3x_3^2)x_6 \neq 0$.

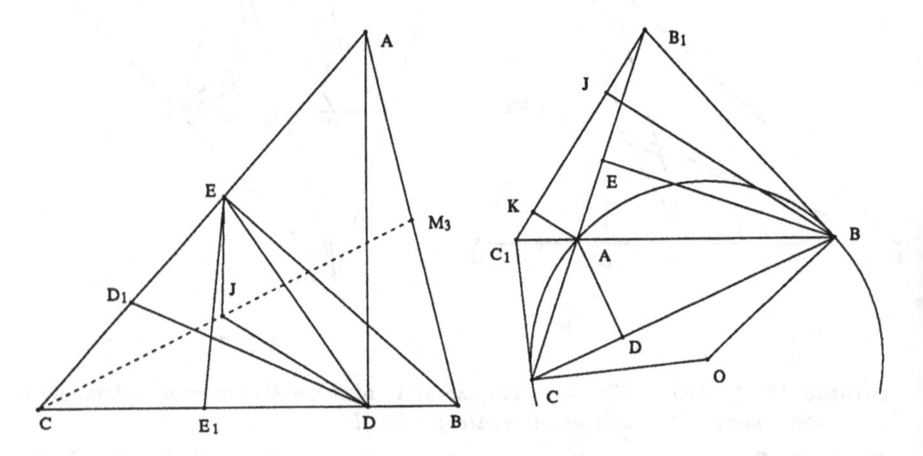

A256-1 A256-3

Example 440 (A256-3: 277.02s, 7472). *Show that the distances of the vertices of a triangle from the Lemoine axis are proportional to the squares of the respective altitudes.*

Points A, B, C are arbitrarily chosen. Points O, B_1, C_1, E, D, K, J are constructed (in order) as follows: $OA \equiv OC$; $OA \equiv OB$; B_1 is on line AC; $B_1B \perp OB$; C_1 is on line AB; $C_1C \perp OC$; $EB \perp AC$; E is on line AC; D is on line BC; $DA \perp BC$; K is on line B_1C_1; $KA \perp B_1C_1$; J is on line B_1C_1; $JB \perp B_1C_1$. The **conclusion**: $\overline{AK}^2\ \overline{BE}^2\ \overline{BE}^2 - \overline{BJ}^2\ \overline{AD}^2\ \overline{AD}^2 = 0$.

$A = (0,0)$, $B = (u_1,0)$, $C = (u_2,u_3)$, $O = (x_2,x_1)$, $B_1 = (x_4,x_3)$, $C_1 = (x_5,0)$, $E = (x_7,x_6)$, $D = (x_9,x_8)$, $K = (x_{11},x_{10})$, $J = (x_{13},x_{12})$.

The **nondegenerate** conditions: Points A, B and C are not collinear; Line OB is not perpendicular to line AC; Line OC is not perpendicular to line AB; Line AC is non-isotropic; Line BC is non-isotropic; Line B_1C_1 is non-isotropic; Line B_1C_1 is non-isotropic.

Example 441 (A257-7: 12.77s, 157). *Show that the sides of the pedal triangle of the symmedian point of a given triangle are proportional to the medians of the given triangle.*

Points C, B, A are arbitrarily chosen. Points M_1, M_2, I, B_1, C_1, A_1 are constructed (in order) as follows: M_1 is the midpoint of B and C; M_2 is the midpoint of A and C; $\tan(BAI) = \tan(M_1AC)$; $\tan(CBI) = \tan(M_2BA)$; B_1 is on line CA; $B_1I \perp CA$; C_1 is on line AB; $C_1I \perp AB$; A_1 is on line BC; $A_1I \perp BC$. The **conclusion**: $A_1C_1 \cdot M_1A = B_1C_1 \cdot M_2B$.

$C = (0,0)$, $B = (u_1,0)$, $A = (u_2,u_3)$, $M_1 = (x_1,0)$, $M_2 = (x_2,x_3)$, $I = (x_5,x_4)$, $B_1 = (x_7,x_6)$, $C_1 = (x_9,x_8)$, $A_1 = (x_5,0)$.

The **nondegenerate** conditions: $((u_1u_2u_3^2 + u_1u_2^3 - 2u_1^2u_2^2 + u_1^3u_2)x_1 - u_1u_3^4 + (-2u_1u_2^2 + 2u_1^2u_2 - u_1^3)u_3^2 - u_1u_2^4 + 2u_1^2u_2^3 - u_1^3u_2^2)x_3 + ((u_1u_3^3 + (u_1u_2^2 - 2u_1^2u_2 + u_1^3)u_3)x_1)x_2 + (-u_1^2u_3^3 + (-u_1^2u_2^2 + 2u_1^3u_2 - u_1^4)u_3)x_1 \neq 0$; Line CA is non-isotropic; Line AB is non-isotropic; Line BC is non-isotropic.

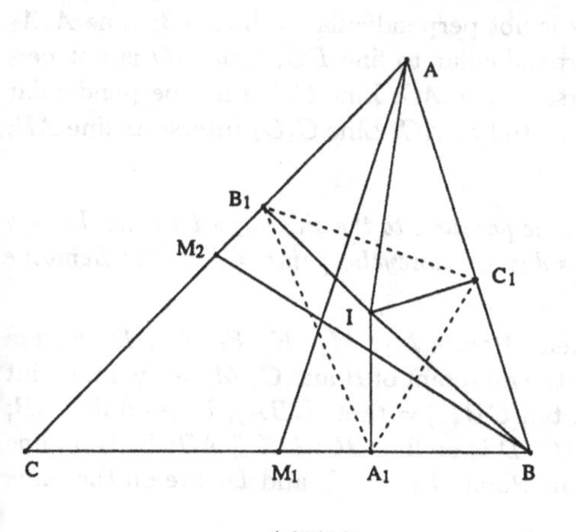

A257-7

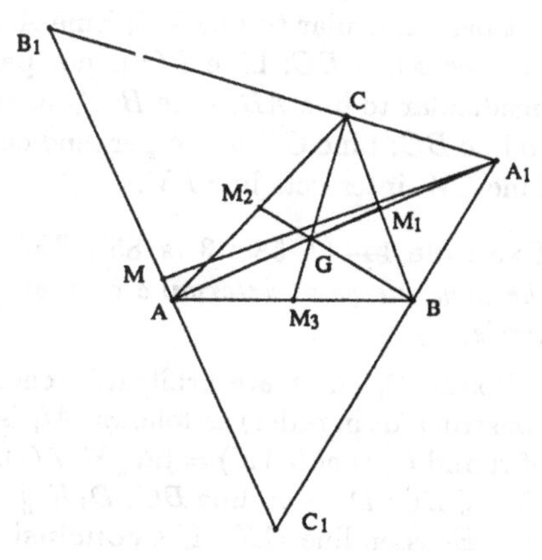

A257-8

Example 442 (A257-8: 15.15s, 534; 8.22s, 288). *If through the vertices of a triangle perpendiculars are drawn to the medians of the triangle, show that the symmedian point of the triangle thus formed coincides with the centroid of the given triangle.*

Points A, B, C are arbitrarily chosen. Points M_1, M_2, M_3, A_1, B_1, C_1, G, M are constructed (in order) as follows: M_1 is the midpoint of B and C; M_2 is the midpoint of A and C; M_3 is the midpoint of B and A; $A_1B \perp BM_2$; $A_1C \perp CM_3$; $B_1A \perp AM_1$; B_1 is on line A_1C; C_1 is on line B_1A; C_1 is on line BA_1; G is on line BM_2; G is on line AM_1; M is the midpoint of B_1 and C_1. The **conclusion**: $\tan(C_1A_1M) = \tan(GA_1B_1)$.

$A = (0,0)$, $B = (u_1,0)$, $C = (u_2,u_3)$, $M_1 = (x_1,x_2)$, $M_2 = (x_3,x_4)$, $M_3 = (x_5,0)$, $A_1 = (x_7,x_6)$, $B_1 = (x_9,x_8)$, $C_1 = (x_{11},x_{10})$, $G = (x_{13},x_{12})$, $M = (x_{14},x_{15})$.

The **nondegenerate** conditions: Line CM_3 intersects line BM_2; Line A_1C is not perpendicular to line AM_1; Line BA_1 intersects line B_1A; Line AM_1 intersects line BM_2.

Example 443 (A257-9: 5.75s, 68). *Through the vertices B, C of the triangle ABC parallels are drawn to the tangent to the circumcircle (O) at A, meeting AC, AB in A_1, A_2. The line A_1A_2 meets BC in U. Show that AU and its analogous BV, CW are concurrent.*

Points A, B, C are arbitrarily chosen. Points O, A_1, A_2, U, B_1, B_2, V, C_1, C_2, W, I are constructed (in order) as follows: $OA \equiv OC$; $OA \equiv OB$; A_1 is on line AC; $A_1B \perp AO$; A_2 is on line AB; $A_2C \perp AO$; U is on line BC; U is on line A_1A_2; B_1 is on line BC; $B_1A \perp BO$; B_2 is on line AB; $B_2C \perp BO$; V is on line AC; V is on line B_1B_2; C_1 is on line BC; $C_1A \perp CO$; C_2 is on line AC; $C_2B \perp CO$; W is on line AB; W is on line C_1C_2; I is on line BV; I is on line CW. The **conclusion**: Points A, U and I are collinear.

$A = (0,0)$, $B = (u_1,0)$, $C = (u_2,u_3)$, $O = (x_2,x_1)$, $A_1 = (x_4,x_3)$, $A_2 = (x_5,0)$, $U = (x_7,x_6)$, $B_1 = (x_9,x_8)$, $B_2 = (x_{10},0)$, $V = (x_{12},x_{11})$, $C_1 = (x_{14},x_{13})$, $C_2 = (x_{16},x_{15})$, $W = (x_{17},0)$, $I = (x_{19},x_{18})$.

The **nondegenerate** conditions: Points A, B and C are not collinear; Line AO is not perpendicular to line AC; Line AO is not perpendicular to line AB; Line A_1A_2 intersects line BC; Line BO is not perpendicular to line BC; Line BO is not perpendicular to line AB; Line B_1B_2 intersects line AC; Line CO is not perpendicular to line BC; Line CO is not perpendicular to line AC; Line C_1C_2 intersects line AB; Line CW intersects line BV.

Example 444 (A-588: 3.7s, 85). *The three parallels to the sides of a triangle through the Lemoine point determine on these sides six concyclic points (The first Lemoine circle).*

Points B, C, A are arbitrarily chosen. Points M_1, M_2, K, F_1, D_1, D, E_1 are constructed (in order) as follows: M_1 is the midpoint of B and C; M_2 is the midpoint of A and C; $\tan(BAK) = \tan(M_1AC)$; $\tan(CBK) = \tan(M_2BA)$; F_1 is on line AB; $F_1K \parallel BC$; D_1 is on line BC; $D_1K \parallel AC$; D is on line BC; $DK \parallel AB$; E_1 is on line AC; E_1 is on line DK. The **conclusion**: Points E_1, F_1, D and D_1 are on the same

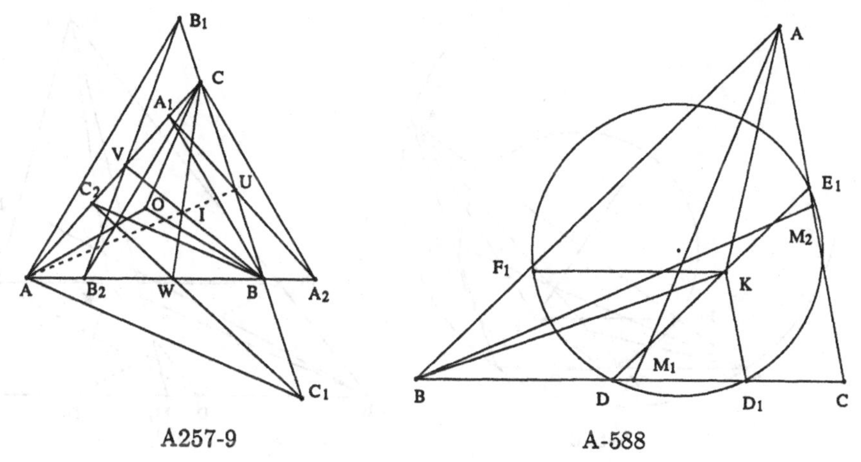

A257-9 A-588

circle.

$B = (0,0)$, $C = (u_1, 0)$, $A = (u_2, u_3)$, $M_1 = (x_1, 0)$, $M_2 = (x_2, x_3)$, $K = (x_5, x_4)$, $F_1 = (x_6, x_4)$, $D_1 = (x_7, 0)$, $D = (x_8, 0)$, $E_1 = (x_{10}, x_9)$.

The **nondegenerate** conditions: $(((-u_1u_2 + u_1^2)u_3^2 - u_1u_2^3 + u_1^2u_2^2)x_1 + u_1u_3^4 + (2u_1u_2^2 - u_1^2u_2)u_3^2 + u_1u_2^4 - u_1^2u_2^3)x_3 + ((-u_1u_3^3 - u_1u_2^2u_3)x_1 + u_1^2u_3^3 + u_1^2u_2^2u_3)x_2 \neq 0$; Points B, C and A are not collinear; Points A, C and B are not collinear; Points A, B and C are not collinear; Line DK intersects line AC.

Example 445 (A-590: 2.27s, 21). *The center of the first Lemoine circle of a triangle lies midway between the circumcenter and the Lemoine point of the triangle.*

Points B, C, A are arbitrarily chosen. Points M_1, M_2, K, F_1, D_1, D, O, L are constructed (in order) as follows: M_1 is the midpoint of B and C; M_2 is the midpoint of A and C; $\tan(BAK) = \tan(M_1AC)$; $\tan(CBK) = \tan(M_2BA)$; F_1 is on line AB; $F_1K \parallel BC$; D_1 is on line BC; $D_1K \parallel AC$; D is on line BC; $DK \parallel AB$; $OB \equiv OA$; $OB \equiv OC$; $LD \equiv LF_1$; $LD \equiv LD_1$. The **conclusion**: L is the midpoint of O and K.

$B = (0,0)$, $C = (u_1, 0)$, $A = (u_2, u_3)$, $M_1 = (x_1, 0)$, $M_2 = (x_2, x_3)$, $K = (x_5, x_4)$, $F_1 = (x_6, x_4)$, $D_1 = (x_7, 0)$, $D = (x_8, 0)$, $O = (x_{10}, x_9)$, $L = (x_{12}, x_{11})$.

The **nondegenerate** conditions: $(((-u_1u_2 + u_1^2)u_3^2 - u_1u_2^3 + u_1^2u_2^2)x_1 + u_1u_3^4 + (2u_1u_2^2 - u_1^2u_2)u_3^2 + u_1u_2^4 - u_1^2u_2^3)x_3 + ((-u_1u_3^3 - u_1u_2^2u_3)x_1 + u_1^2u_3^3 + u_1^2u_2^2u_3)x_2 \neq 0$; Points B, C and A are not collinear; Points A, C and B are not collinear; Points A, B and C are not collinear; Points B, C and A are not collinear; Points D, D_1 and F_1 are not collinear.

Example 446 (A-591: 2.05s, 21).

Points B, C, A are arbitrarily chosen. Points M_1, M_2, K, F_1, D_1, D, E are constructed (in order) as follows: M_1 is the midpoint of B and C; M_2 is the midpoint of A and C; $\tan(BAK) = \tan(M_1AC)$; $\tan(CBK) = \tan(M_2BA)$; F_1 is on line AB; $F_1K \parallel BC$; D_1 is on line BC; $D_1K \parallel AC$; D is on line BC; $DK \parallel AB$; E is on line AC; E is on line F_1K. The **conclusion**: $F_1D_1 \equiv DE$.

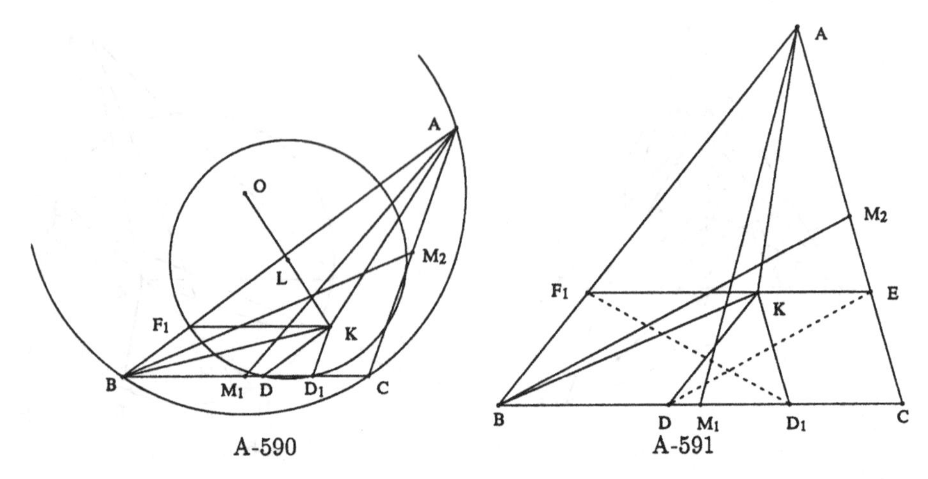

A-590 A-591

$B = (0,0)$, $C = (u_1,0)$, $A = (u_2,u_3)$, $M_1 = (x_1,0)$, $M_2 = (x_2,x_3)$, $K = (x_5,x_4)$, $F_1 = (x_6,x_4)$, $D_1 = (x_7,0)$, $D = (x_8,0)$, $E = (x_9,x_4)$.

The **nondegenerate** conditions: $(((-u_1u_2+u_1^2)u_3^2 - u_1u_2^3 + u_1^2u_2^2)x_1 + u_1u_3^4 + (2u_1u_2^2 - u_1^2u_2)u_3^2 + u_1u_2^4 - u_1^2u_2^3)x_3 + ((-u_1u_3^3 - u_1u_2^2u_3)x_1 + u_1^2u_3^3 + u_1^2u_2^2u_3)x_2 \neq 0$; Points B, C and A are not collinear; Points A, C and B are not collinear; Points A, B and C are not collinear; Line F_1K intersects line AC.

Example 447 (A260-1: 1.75s, 21).

Points B, C, A are arbitrarily chosen. Points M_1, M_2, K, F_1, D_1, D, E are constructed (in order) as follows: M_1 is the midpoint of B and C; M_2 is the midpoint of A and C; $\tan(BAK) = \tan(M_1AC)$; $\tan(CBK) = \tan(M_2BA)$; F_1 is on line AB; $F_1K \parallel BC$; D_1 is on line BC; $D_1K \parallel AC$; D is on line BC; $DK \parallel AB$; E is on line AC; E is on line F_1K. The **conclusion**: $F_1D \equiv D_1E$.

$B = (0,0)$, $C = (u_1,0)$, $A = (u_2,u_3)$, $M_1 = (x_1,0)$, $M_2 = (x_2,x_3)$, $K = (x_5,x_4)$, $F_1 = (x_6,x_4)$, $D_1 = (x_7,0)$, $D = (x_8,0)$, $E = (x_9,x_4)$.

The **nondegenerate** conditions: $(((-u_1u_2+u_1^2)u_3^2 - u_1u_2^3 + u_1^2u_2^2)x_1 + u_1u_3^4 + (2u_1u_2^2 - u_1^2u_2)u_3^2 + u_1u_2^4 - u_1^2u_2^3)x_3 + ((-u_1u_3^3 - u_1u_2^2u_3)x_1 + u_1^2u_3^3 + u_1^2u_2^2u_3)x_2 \neq 0$; Points B, C and A are not collinear; Points A, C and B are not collinear; Points A, B and C are not collinear; Line F_1K intersects line AC.

Example 448 (A260-2: 73.98s, 3102; 46.55s, 160).

Points B, C, A are arbitrarily chosen. Points M_1, M_2, K, F_1, D_1, D, E, L, I, J are constructed (in order) as follows: M_1 is the midpoint of B and C; M_2 is the midpoint of A and C; $\tan(BAK) = \tan(M_1AC)$; $\tan(CBK) = \tan(M_2BA)$; F_1 is on line AB; $F_1K \parallel BC$; D_1 is on line BC; $D_1K \parallel AC$; D is on line BC; $DK \parallel AB$; E is on line AC; E is on line F_1K; $LD \equiv LF_1$; $LD \equiv LD_1$; I is on line D_1E; $IL \perp D_1E$; J is on line F_1D; $JL \perp F_1D$. The **conclusion**: $LI \equiv LJ$.

$B = (0,0)$, $C = (u_1,0)$, $A = (u_2,u_3)$, $M_1 = (x_1,0)$, $M_2 = (x_2,x_3)$, $K = (x_5,x_4)$, $F_1 = (x_6,x_4)$, $D_1 = (x_7,0)$, $D = (x_8,0)$, $E = (x_9,x_4)$, $L = (x_{11},x_{10})$, $I = (x_{13},x_{12})$, $J = (x_{15},x_{14})$.

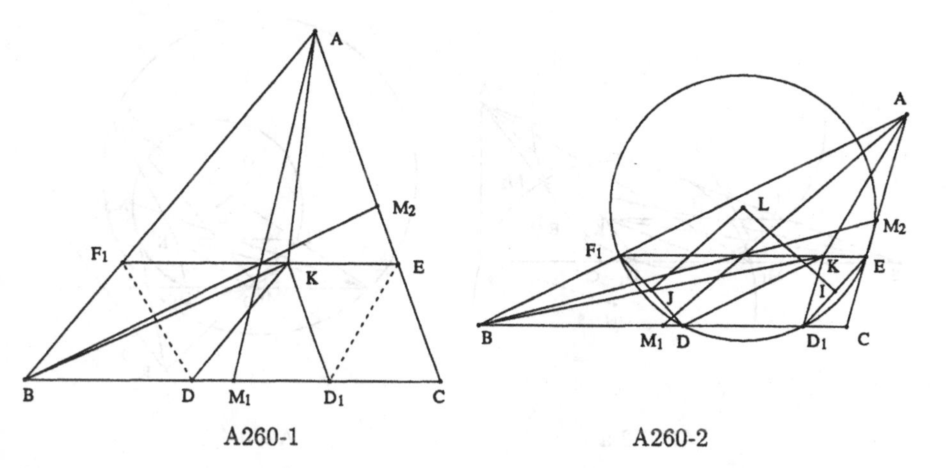

A260-1 A260-2

The **nondegenerate** conditions: $(((-u_1u_2+u_1^2)u_3^2-u_1u_2^3+u_1^2u_2^2)x_1+u_1u_3^4+(2u_1u_2^2-u_1^2u_2)u_3^2+u_1u_2^4-u_1^2u_2^3)x_3+((-u_1u_3^3-u_1u_2^2u_3)x_1+u_1^2u_3^3+u_1^2u_2^2u_3)x_2 \neq 0$; Points B, C and A are not collinear; Points A, C and B are not collinear; Points A, B and C are not collinear; Line F_1K intersects line AC; Points D, D_1 and F_1 are not collinear; Line D_1E is non-isotropic; Line F_1D is non-isotropic.

Example 449 (A260-2-a: 21.62s, 340).

Points B, C, A are arbitrarily chosen. Points M_1, M_2, K, F_1, D_1, D, E, L, J, S, N are constructed (in order) as follows: M_1 is the midpoint of B and C; M_2 is the midpoint of A and C; $\tan(BAK) = \tan(M_1AC)$; $\tan(CBK) = \tan(M_2BA)$; F_1 is on line AB; $F_1K \parallel BC$; D_1 is on line BC; $D_1K \parallel AC$; D is on line BC; $DK \parallel AB$; E is on line AC; E is on line F_1K; $LD \equiv LF_1$; $LD \equiv LD_1$; J is on line F_1D; $JL \perp F_1D$; S is on line BC; $SA \perp BC$; $NS \equiv NM_2$; $NS \equiv NM_1$. The **conclusion**: $NS \equiv LJ$.

$B = (0,0)$, $C = (u_1,0)$, $A = (u_2,u_3)$, $M_1 = (x_1,0)$, $M_2 = (x_2,x_3)$, $K = (x_5,x_4)$, $F_1 = (x_6,x_4)$, $D_1 = (x_7,0)$, $D = (x_8,0)$, $E = (x_9,x_4)$, $L = (x_{11},x_{10})$, $J = (x_{13},x_{12})$, $S = (u_2,0)$, $N = (x_{15},x_{14})$.

The **nondegenerate** conditions: $(((-u_1u_2+u_1^2)u_3^2-u_1u_2^3+u_1^2u_2^2)x_1+u_1u_3^4+(2u_1u_2^2-u_1^2u_2)u_3^2+u_1u_2^4-u_1^2u_2^3)x_3+((-u_1u_3^3-u_1u_2^2u_3)x_1+u_1^2u_3^3+u_1^2u_2^2u_3)x_2 \neq 0$; Points B, C and A are not collinear; Points A, C and B are not collinear; Points A, B and C are not collinear; Line F_1K intersects line AC; Points D, D_1 and F_1 are not collinear; Line F_1D is non-isotropic; Line BC is non-isotropic; Points S, M_1 and M_2 are not collinear.

Example 450 (A260-4: 3.12s, 28).

Points B, C, A are arbitrarily chosen. Points M_1, M_2, K, F_1, D_1, D, O, L, R are constructed (in order) as follows: M_1 is the midpoint of B and C; M_2 is the midpoint of A and C; $\tan(BAK) = \tan(M_1AC)$; $\tan(CBK) = \tan(M_2BA)$; F_1 is on line AB; $F_1K \parallel BC$; D_1 is on line BC; $D_1K \parallel AC$; D is on line BC; $DK \parallel AB$; $OB \equiv OA$; $OB \equiv OC$; $LD \equiv LF_1$; $LD \equiv LD_1$; R is on line CA; R is on line F_1D.

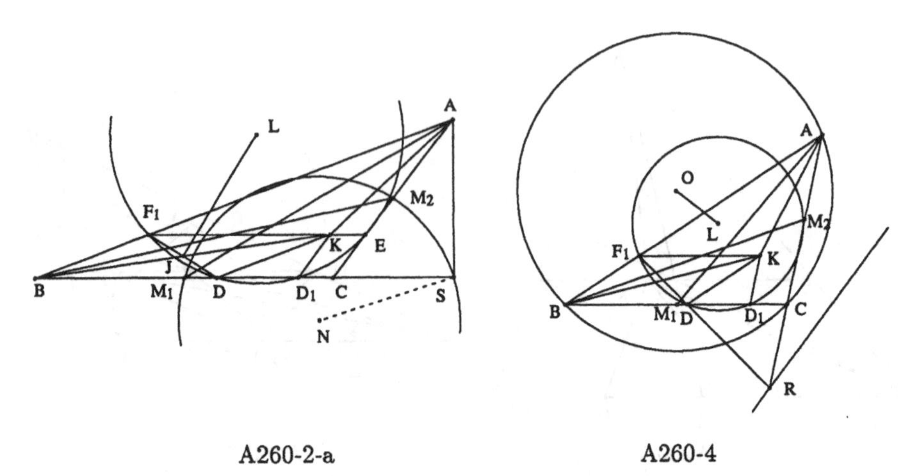

A260-2-a A260-4

The **conclusion**: Point R is on the radical axises of circle (O, OB) and circle (L, LD).

$B = (0,0)$, $C = (u_1,0)$, $A = (u_2,u_3)$, $M_1 = (x_1,0)$, $M_2 = (x_2,x_3)$, $K = (x_5,x_4)$, $F_1 = (x_6,x_4)$, $D_1 = (x_7,0)$, $D = (x_8,0)$, $O = (x_{10},x_9)$, $L = (x_{12},x_{11})$, $R = (x_{14},x_{13})$.

The **nondegenerate** conditions: $(((-u_1u_2+u_1^2)u_3^2 - u_1u_2^3 + u_1^2u_2^2)x_1 + u_1u_3^4 + (2u_1u_2^2 - u_1^2u_2)u_3^2 + u_1u_2^4 - u_1^2u_2^3)x_3 + ((-u_1u_3^3 - u_1u_2^2u_3)x_1 + u_1^2u_3^3 + u_1^2u_2^2u_3)x_2 \neq 0$; Points B, C and A are not collinear; Points A, C and B are not collinear; Points A, B and C are not collinear; Points B, C and A are not collinear; Points D, D_1 and F_1 are not collinear; Line F_1D intersects line CA.

Example 451 (A260-5: 5.73s, 32). *Show that the radical axis of the first Lemoine circle and the nine-point circle of a triangle passes through the points in which the Lemoine parallels meet the corresponding sides of the orthic triangle.*

Points B, C, A are arbitrarily chosen. Points M_1, M_2, K, D_1, D, F, L, S, N, Q, R are constructed (in order) as follows: M_1 is the midpoint of B and C; M_2 is the midpoint of A and C; $\tan(BAK) = \tan(M_1AC)$; $\tan(CBK) = \tan(M_2BA)$; D_1 is on line BC; $D_1K \parallel AC$; D is on line BC; $DK \parallel AB$; F is on line AB; F is on line D_1K; $LD \equiv LF$; $LD \equiv LD_1$; S is on line BC; $SA \perp BC$; $NS \equiv NM_2$; $NS \equiv NM_1$; Q is on line BA; $QC \perp BA$; R is on line SQ; R is on line FD_1. The **conclusion**: Point R is on the radical axises of circle (N, NS) and circle (L, LD).

$B = (0,0)$, $C = (u_1,0)$, $A = (u_2,u_3)$, $M_1 = (x_1,0)$, $M_2 = (x_2,x_3)$, $K = (x_5,x_4)$, $D_1 = (x_6,0)$, $D = (x_7,0)$, $F = (x_9,x_8)$, $L = (x_{11},x_{10})$, $S = (u_2,0)$, $N = (x_{13},x_{12})$, $Q = (x_{15},x_{14})$, $R = (x_{17},x_{16})$.

The **nondegenerate** conditions: $(((-u_1u_2+u_1^2)u_3^2 - u_1u_2^3 + u_1^2u_2^2)x_1 + u_1u_3^4 + (2u_1u_2^2 - u_1^2u_2)u_3^2 + u_1u_2^4 - u_1^2u_2^3)x_3 + ((-u_1u_3^3 - u_1u_2^2u_3)x_1 + u_1^2u_3^3 + u_1^2u_2^2u_3)x_2 \neq 0$; Points A, C and B are not collinear; Points A, B and C are not collinear; Line D_1K intersects line AB; Points D, D_1 and F are not collinear; Line BC is non-isotropic; Points S, M_1 and M_2 are not collinear; Line BA is non-isotropic; Line FD_1 intersects line SQ.

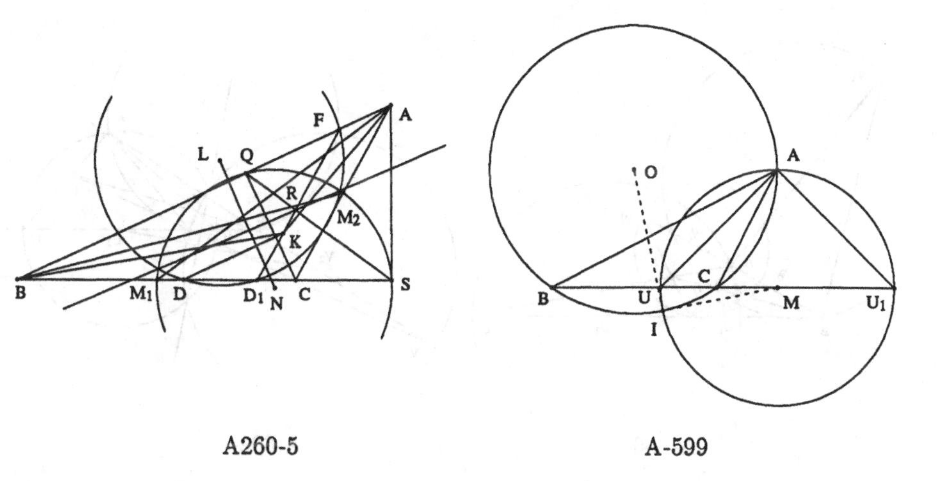

A260-5 A-599

Example 452 (A-599: 7.03s, 333; 5.03s, 133). *The circumcircle of a triangle is orthogonal to the three Apollonian circles of the triangle.*

Points B, C, A are arbitrarily chosen. Points U, U_1, M, O, I are constructed (in order) as follows: U is on line BC; $\tan(BAU) = \tan(UAC)$; $U_1A \perp UA$; U_1 is on line BC; M is the midpoint of U and U_1; $OB \equiv OC$; $OB \equiv OA$; $IM \equiv MU$; $IO \equiv OB$. The **conclusion**: $OI \perp MI$.

$B = (0,0)$, $C = (u_1,0)$, $A = (u_2,u_3)$, $U = (x_1,0)$, $U_1 = (x_2,0)$, $M = (x_3,0)$, $O = (x_5,x_4)$, $I = (x_7,x_6)$.

The **nondegenerate** conditions: $(2u_2 - u_1)u_3 \neq 0$; Line BC is not perpendicular to line UA; Points B, A and C are not collinear; Line OM is non-isotropic.

Example 453 (A-600: 3041.67s, 18290). *The three Apollonian circles of a triangle have two points in common (the isodynamic points).*

Points B, C, I are arbitrarily chosen. Points A, U, U_1, V, V_1, W, W_1, U_2, V_2, W_2, O are constructed (in order) as follows: $\tan(CBI) = \tan(IBA)$; $\tan(BCI) = \tan(ICA)$; U is on line BC; U is on line AI; $U_1A \perp UA$; U_1 is on line BC; V is on line AC; V is on line BI; V_1 is on line AC; $V_1B \perp BI$; W is on line AB; W is on line CI; W_1 is on line BA; $W_1C \perp CW$; U_2 is the midpoint of U and U_1; V_2 is the midpoint of V and V_1; W_2 is the midpoint of W and W_1; $OU_2 \equiv U_2U$; $OV_2 \equiv V_2V$. The **conclusion**: $W_2C \equiv W_2O$.

$B = (0,0)$, $C = (u_1,0)$, $I = (u_2,u_3)$, $A = (x_2,x_1)$, $U = (x_3,0)$, $U_1 = (x_4,0)$, $V = (x_6,x_5)$, $V_1 = (x_8,x_7)$, $W = (x_{10},x_9)$, $W_1 = (x_{12},x_{11})$, $U_2 = (x_{13},0)$, $V_2 = (x_{14},x_{15})$, $W_2 = (x_{16},x_{17})$, $O = (x_{19},x_{18})$.

The **nondegenerate** conditions: Points I, C and B are not collinear; $\angle CIB$ is not right; Line AI intersects line BC; Line BC is not perpendicular to line UA; Line BI intersects line AC; Line BI is not perpendicular to line AC; Line CI intersects line AB; Line CW is not perpendicular to line BA; Line V_2U_2 is non-isotropic.

Example 454 (A-602-a: 9.1s, 103). *The isodynamic points of a triangle lie on the*

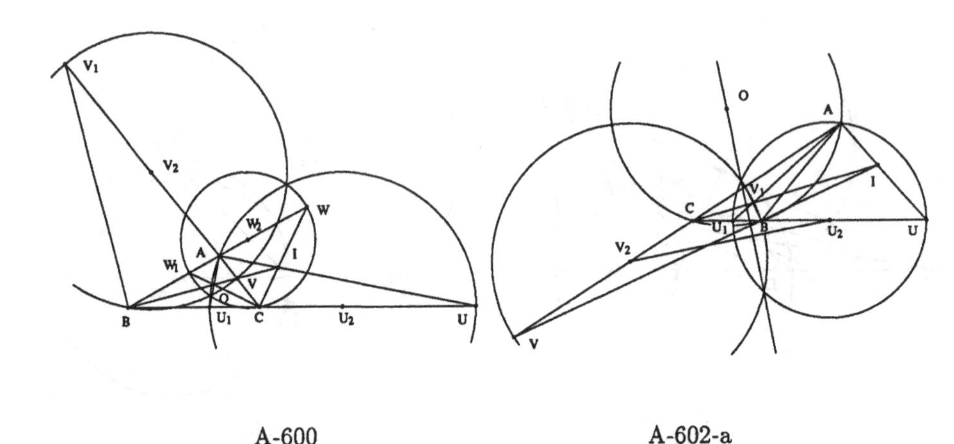

A-600 A-602-a

Brocard diameters of the triangle.

Points C, B, I are arbitrarily chosen. Points A, O, U, U_1, V, V_1, U_2, V_2 are constructed (in order) as follows: $\tan(CBI) = \tan(IBA)$; $\tan(BCI) = \tan(ICA)$; $OC \equiv OB$; $OC \equiv OA$; U is on line BC; U is on line AI; $U_1A \perp UA$; U_1 is on line BC; V is on line AC; V is on line BI; V_1 is on line AC; $V_1B \perp BI$; U_2 is the midpoint of U and U_1; V_2 is the midpoint of V and V_1. The **conclusion**: Point O is on the radical axises of circle (V_2, V_2V) and circle (U_2, U_2U).

$C = (0,0)$, $B = (u_1,0)$, $I = (u_2,u_3)$, $A = (x_2,x_1)$, $O = (x_4,x_3)$, $U = (x_5,0)$, $U_1 = (x_6,0)$, $V = (x_8,x_7)$, $V_1 = (x_{10},x_9)$, $U_2 = (x_{11},0)$, $V_2 = (x_{12},x_{13})$.

The **nondegenerate** conditions: Points I, C and B are not collinear; $\angle CIB$ is not right; Points C, A and B are not collinear; Line AI intersects line BC; Line BC is not perpendicular to line UA; Line BI intersects line AC; Line BI is not perpendicular to line AC.

Example 455 (A-602-b: 5.02s, 142).

Points B, C, I are arbitrarily chosen. Points A, A_1, B_1, K, U, U_1, V, V_1, U_2, V_2 are constructed (in order) as follows: $\tan(CBI) = \tan(IBA)$; $\tan(BCI) = \tan(ICA)$; A_1 is the midpoint of C and B; B_1 is the midpoint of A and C; $\tan(CBK) = \tan(B_1BA)$; $\tan(BAK) = \tan(A_1AC)$; U is on line BC; U is on line AI; $U_1A \perp UA$; U_1 is on line BC; V is on line AC; V is on line BI; V_1 is on line AC; $V_1B \perp BI$; U_2 is the midpoint of U and U_1; V_2 is the midpoint of V and V_1. The **conclusion**: Point K is on the radical axises of circle (V_2, V_2V) and circle (U_2, U_2U).

$B = (0,0)$, $C = (u_1,0)$, $I = (u_2,u_3)$, $A = (x_2,x_1)$, $A_1 = (x_3,0)$, $B_1 = (x_4,x_5)$, $K = (x_7,x_6)$, $U = (x_8,0)$, $U_1 = (x_9,0)$, $V = (x_{11},x_{10})$, $V_1 = (x_{13},x_{12})$, $U_2 = (x_{14},0)$, $V_2 = (x_{15},x_{16})$.

The **nondegenerate** conditions: Points I, C and B are not collinear; $\angle CIB$ is not right; $((-u_1x_2^3 + u_1^2x_2^2 - u_1x_1^2x_2 + u_1^2x_1^2)x_3 + u_1x_2^4 - u_1^2x_2^3 + 2u_1x_1^2x_2^2 - u_1^2x_1^2x_2 + u_1x_1^4)x_5 + ((-u_1x_1x_2^2 - u_1x_1^3)x_3 + u_1^2x_1x_2^2 + u_1^2x_1^3)x_4 \neq 0$; Line AI intersects line BC;

Line BC is not perpendicular to line UA; Line BI intersects line AC; Line BI is not perpendicular to line AC.

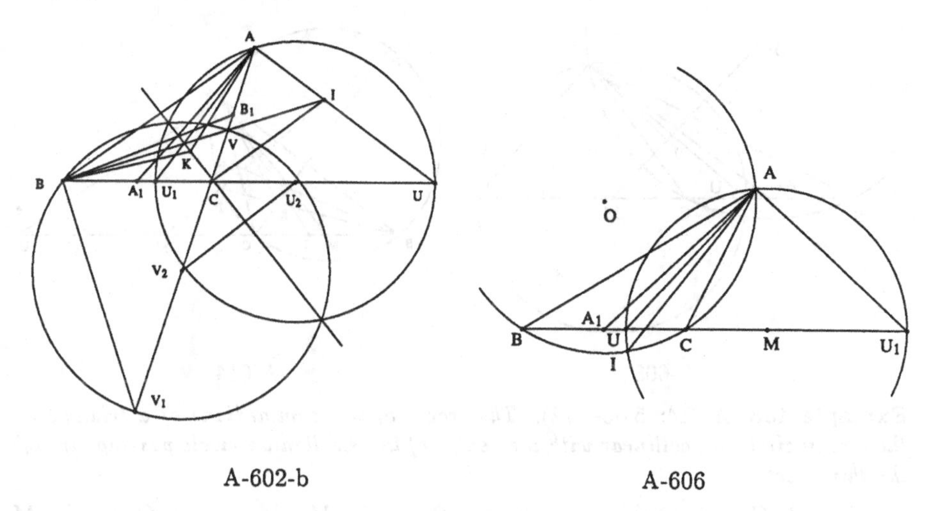

A-602-b A-606

Example 456 (A-606: 1.83s, 29). *The common chord of the circumcircle and a Apollonian circle of a triangle coincides with the corresponding symmedian of the triangle.*

Points B, C, A are arbitrarily chosen. Points U, U_1, M, O, I, A_1 are constructed (in order) as follows: U is on line BC; $\tan(BAU) = \tan(UAC)$; $U_1A \perp UA$; U_1 is on line BC; M is the midpoint of U and U_1; $OB \equiv OC$; $OB \equiv OA$; $IM \equiv MU$; $IO \equiv OB$; A_1 is the midpoint of B and C. The **conclusion**: $\tan(BAI) = \tan(A_1AC)$.

$B = (0,0)$, $C = (u_1,0)$, $A = (u_2,u_3)$, $U = (x_1,0)$, $U_1 = (x_2,0)$, $M = (x_3,0)$, $O = (x_5,x_4)$, $I = (x_7,x_6)$, $A_1 = (x_8,0)$.

The **nondegenerate** conditions: $(2u_2 - u_1)u_3 \neq 0$; Line BC is not perpendicular to line UA; Points B, A and C are not collinear; Line OM is non-isotropic.

Example 457 (A-608: 3.5s, 87). *The polar of the circumcenter of a triangle with respect to an Apollonian circle coincides with the corresponding symmedian of the triangle.*

Points B, C, A are arbitrarily chosen. Points U, U_1, M, O, D, I, A_1 are constructed (in order) as follows: U is on line BC; $\tan(BAU) = \tan(UAC)$; $U_1A \perp UA$; U_1 is on line BC; M is the midpoint of U and U_1; $OB \equiv OC$; $OB \equiv OA$; D is on line OM; $DM \equiv MU$; O and I are inversive wrpt circle (M, MD); I is on line OM; A_1 is the midpoint of B and C. The **conclusion**: $\tan(BAI) = \tan(A_1AC)$.

$B = (0,0)$, $C = (u_1,0)$, $A = (u_2,u_3)$, $U = (x_1,0)$, $U_1 = (x_2,0)$, $M = (x_3,0)$, $O = (x_5,x_4)$, $D = (x_7,x_6)$, $I = (x_9,x_8)$, $A_1 = (x_{10},0)$.

The **nondegenerate** conditions: $(2u_2 - u_1)u_3 \neq 0$; Line BC is not perpendicular to line UA; Points B, A and C are not collinear; Line OM is non-isotropic; $-x_4 \neq 0$; $-x_4 \neq 0$.

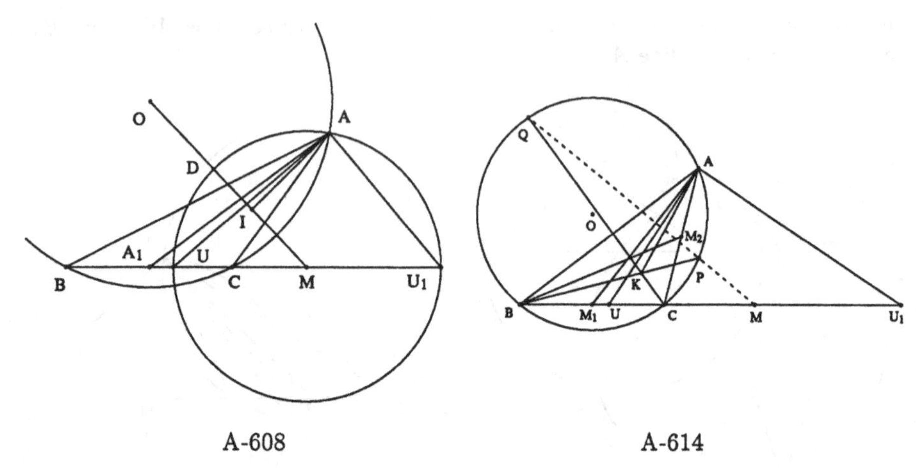

A-608 A-614

Example 458 (A-614: 5.33s, 72). *The traces of two symmedians of a triangle on the circumcircle are collinear with the center of the Apollonian circle passing through the third vertex.*

Points B, C, A are arbitrarily chosen. Points O, M_1, M_2, K, P, Q, U, U_1, M are constructed (in order) as follows: $OB \equiv OA$; $OB \equiv OC$; M_1 is the midpoint of B and C; M_2 is the midpoint of A and C; $\tan(BAK) = \tan(M_1AC)$; $\tan(CBK) = \tan(M_2BA)$; $PO \equiv OB$; P is on line BK; $QO \equiv OB$; Q is on line CK; U is on line BC; $\tan(BAU) = \tan(UAC)$; $U_1A \perp UA$; U_1 is on line BC; M is the midpoint of U and U_1. The **conclusion**: Points M, P and Q are collinear.

$B = (0,0)$, $C = (u_1, 0)$, $A = (u_2, u_3)$, $O = (x_2, x_1)$, $M_1 = (x_3, 0)$, $M_2 = (x_4, x_5)$, $K = (x_7, x_6)$, $P = (x_9, x_8)$, $Q = (x_{11}, x_{10})$, $U = (x_{12}, 0)$, $U_1 = (x_{13}, 0)$, $M = (x_{14}, 0)$.

The **nondegenerate** conditions: Points B, C and A are not collinear; $(((-u_1u_2 + u_1^2)u_3^2 - u_1u_2^3 + u_1^2u_2^2)x_3 + u_1u_3^4 + (2u_1u_2^2 - u_1^2u_2)u_3^2 + u_1u_2^4 - u_1^2u_2^3)x_5 + ((-u_1u_3^3 - u_1u_2^2u_3)x_3 + u_1^2u_3^3 + u_1^2u_2^2u_3)x_4 \neq 0$; Line BK is non-isotropic; Line CK is non-isotropic; $(2u_2 - u_1)u_3 \neq 0$; Line BC is not perpendicular to line UA. In addition, the following nondegenerate conditions, which come from reducibility and have been detected by our prover, should be also added: $Q \neq C$; $P \neq B$.

Example 459 (A-615: 7.22s, 72). *The two lines joining two vertices of a triangle to the traces of their symmedians on the circumcircle meet at the trace of the third symmedian on the Lemoine axis of the triangle.*

Points B, C, A are arbitrarily chosen. Points O, M_1, M_2, K, P, Q, I, B_1, A_1 are constructed (in order) as follows: $OB \equiv OA$; $OB \equiv OC$; M_1 is the midpoint of B and C; M_2 is the midpoint of A and C; $\tan(BAK) = \tan(M_1AC)$; $\tan(CBK) = \tan(M_2BA)$; $PO \equiv OB$; P is on line BK; $QO \equiv OB$; Q is on line CK; I is on line CP; I is on line BQ; B_1 is on line AC; $B_1B \perp OB$; A_1 is on line BC; $A_1A \perp OA$. The **conclusions**: (1) Points K, A and I are collinear; (2) Points I, A_1 and B_1 are collinear.

$B = (0,0)$, $C = (u_1, 0)$, $A = (u_2, u_3)$, $O = (x_2, x_1)$, $M_1 = (x_3, 0)$, $M_2 = (x_4, x_5)$, $K = (x_7, x_6)$, $P = (x_9, x_8)$, $Q = (x_{11}, x_{10})$, $I = (x_{13}, x_{12})$, $B_1 = (x_{15}, x_{14})$, $A_1 =$

$(x_{16}, 0)$.

The **nondegenerate** conditions: Points B, C and A are not collinear; $(((-u_1u_2+u_1^2)u_3^2 - u_1u_2^3 + u_1^2u_2^2)x_3 + u_1u_3^4 + (2u_1u_2^2 - u_1^2u_2)u_3^2 + u_1u_2^4 - u_1^2u_2^3)x_5 + ((-u_1u_3^3 - u_1u_2^2u_3)x_3 + u_1^2u_3^3 + u_1^2u_2^2u_3)x_4 \neq 0$; Line BK is non-isotropic; Line CK is non-isotropic; Line BQ intersects line CP; Line OB is not perpendicular to line AC; Line OA is not perpendicular to line BC. In addition, the following nondegenerate conditions, which come from reducibility and have been detected by our prover, should be also added: $Q \neq C$; $P \neq B$.

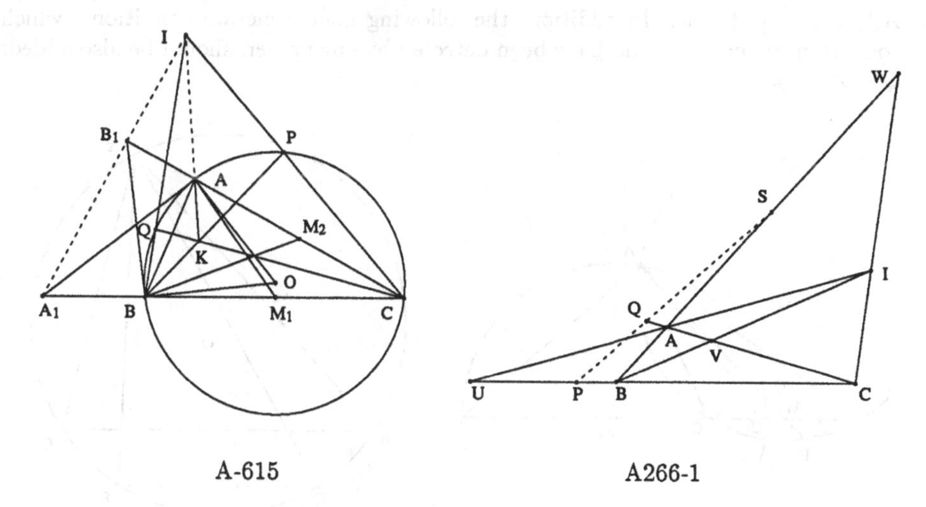

A-615 A266-1

Example 460 (A266-1: 6.4s, 189). *Show that the mediators of the internal bisectors of the angles of a triangle meet the respective sides of the triangle in three collinear points.*

Points B, C, I are arbitrarily chosen. Points A, U, V, W, P, Q, S are constructed (in order) as follows: $\tan(CBI) = \tan(IBA)$; $\tan(BCI) = \tan(ICA)$; U is on line BC; U is on line AI; V is on line AC; V is on line BI; W is on line AB; W is on line CI; P is on line BC; $PA \equiv PU$; Q is on line AC; $QB \equiv QV$; S is on line AB; $SC \equiv SW$. The **conclusion**: Points S, P and Q are collinear.

$B = (0,0)$, $C = (u_1, 0)$, $I = (u_2, u_3)$, $A = (x_2, x_1)$, $U = (x_3, 0)$, $V = (x_5, x_4)$, $W = (x_7, x_6)$, $P = (x_8, 0)$, $Q = (x_{10}, x_9)$, $S = (x_{12}, x_{11})$.

The **nondegenerate** conditions: Points I, C and B are not collinear; $\angle CIB$ is not right; Line AI intersects line BC; Line BI intersects line AC; Line CI intersects line AB; Line AU is not perpendicular to line BC; Line BV is not perpendicular to line AC; Line CW is not perpendicular to line AB.

Example 461 (A267-5: 6.68s, 24; 2.65s, 25). *Show that the median through a given vertex of a triangle meets the Apollonian circle passing through the vertex considered and the circumcircle in two points which, with the other two vertices of the triangle, determine a parallelogram.*

Points B, C, A are arbitrarily chosen. Points U, U_1, M, O, A_1, P, Q are constructured (in order) as follows: U is on line BC; $\tan(BAU) = \tan(UAC)$; $U_1A \perp UA$; U_1 is on line BC; M is the midpoint of U and U_1; $OB \equiv OC$; $OB \equiv OA$; A_1 is the midpoint of B and C; $PO \equiv OB$; P is on line AA_1; $QM \equiv MA$; Q is on line AA_1. The **conclusion**: $BP \parallel CQ$.

$B = (0,0)$, $C = (u_1,0)$, $A = (u_2,u_3)$, $U = (x_1,0)$, $U_1 = (x_2,0)$, $M = (x_3,0)$, $O = (x_5,x_4)$, $A_1 = (x_6,0)$, $P = (x_8,x_7)$, $Q = (x_{10},x_9)$.

The **nondegenerate** conditions: $(2u_2 - u_1)u_3 \neq 0$; Line BC is not perpendicular to line UA; Points B, A and C are not collinear; Line AA_1 is non-isotropic; Line AA_1 is non-isotropic. In addition, the following nondegenerate conditions, which come from reducibility and have been detected by our prover, should be also added: $Q \neq A$; $P \neq A$.

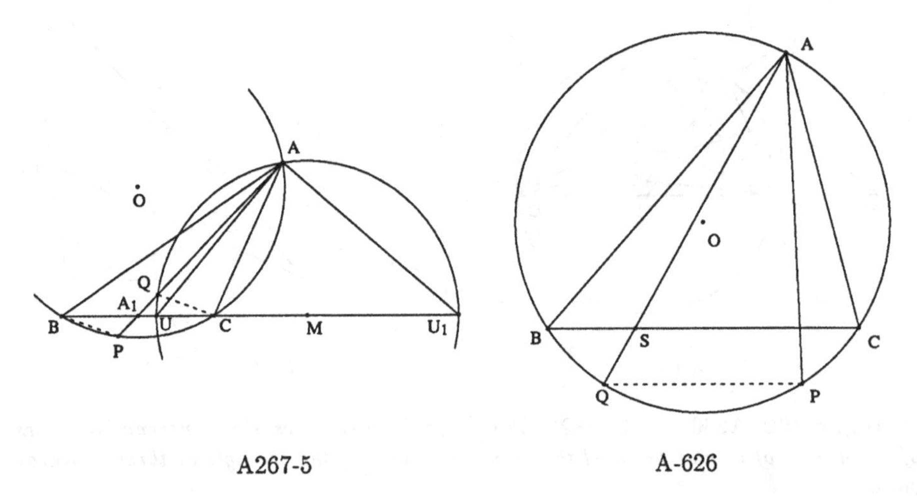

A267-5 A-626

Example 462 (A-626: 3.08s, 92). *The line joining the traces, on the circumcircle of a triangle, of two isogonal lines of an angle of the triangle is parallel to the side opposite the vertex considered.*

Points B, C, A are arbitrarily chosen. Points O, P, S, Q are constructed (in order) as follows: $OA \equiv OB$; $OC \equiv OB$; $PO \equiv OB$; S is on line BC; $\tan(BAS) = \tan(PAC)$; $QO \equiv OB$; Q is on line AS. The **conclusion**: $PQ \parallel BC$.

$B = (0,0)$, $C = (u_1,0)$, $A = (u_2,u_3)$, $O = (x_2,x_1)$, $P = (x_3,u_4)$, $S = (x_4,0)$, $Q = (x_6,x_5)$.

The **nondegenerate** conditions: Points C, B and A are not collinear; $((2u_2 - u_1)u_3)x_3 + (u_3^2 - u_2^2 + u_1u_2)u_4 - u_3^3 - u_2^2u_3 \neq 0$; Line AS is non-isotropic. In addition, the following nondegenerate conditions, which come from reducibility and have been detected by our prover, should be also added: $Q \neq A$.

Example 463 (A-627: 0.73s, 4). *The line joining the two projections of a given point upon the sides of an angle is perpendicular to the isogonal conjugate of the line joining the given point to the vertex of the angle.*

Points A, B, C, M are arbitrarily chosen. Points N, Q, P are constructed (in order) as follows: $\tan(BAM) = \tan(NAC)$; Q is on line AB; $QM \perp AB$; P is on line AC; $PM \perp AC$. The **conclusion**: $NA \perp PQ$.

$A = (0,0)$, $B = (u_1,0)$, $C = (u_2,u_3)$, $M = (u_4,u_5)$, $N = (x_1,u_6)$, $Q = (u_4,0)$, $P = (x_3,x_2)$.

The **nondegenerate** conditions: $A \neq C$, $A \neq B$, $A \neq M$; Line AB is non-isotropic; Line AC is non-isotropic.

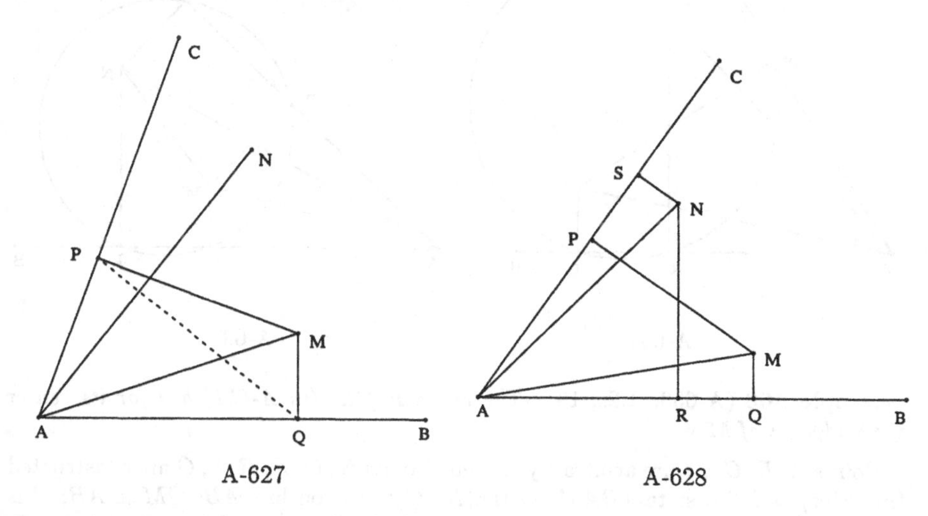

A-627 A-628

Example 464 (A-628: 1.75s, 47). *The distances, from the sides, of an angle, of two points on two isogonal lines are inversely proportional.*

Points A, B, C, M are arbitrarily chosen. Points N, Q, P, R, S are constructed (in order) as follows: $\tan(BAM) = \tan(NAC)$; Q is on line AB; $QM \perp AB$; P is on line AC; $PM \perp AC$; R is on line AB; $RN \perp AB$; S is on line AC; $SN \perp AC$. The **conclusion**: $MQ \cdot NR = MP \cdot NS$.

$A = (0,0)$, $B = (u_1,0)$, $C = (u_2,u_3)$, $M = (u_4,u_5)$, $N = (x_1,u_6)$, $Q = (u_4,0)$, $P = (x_3,x_2)$, $R = (x_1,0)$, $S = (x_5,x_4)$.

The **nondegenerate** conditions: $A \neq C$, $A \neq B$, $A \neq M$; Line AB is non-isotropic; Line AC is non-isotropic; Line AB is non-isotropic; Line AC is non-isotropic.

Example 465 (A-631: 3.72s, 81; 2.9s, 101). *The four projections upon the two sides of an angle of two points on two isogonal conjugate lines are concyclic.*

Points A, B, C, M are arbitrarily chosen. Points N, Q, P, R, S are constructed (in order) as follows: $\tan(BAM) = \tan(NAC)$; Q is on line AB; $QM \perp AB$; P is on line AC; $PM \perp AC$; R is on line AB; $RN \perp AB$; S is on line AC; $SN \perp AC$. The **conclusion**: Points R, Q, S and P are on the same circle.

$A = (0,0)$, $B = (u_1,0)$, $C = (u_2,u_3)$, $M = (u_4,u_5)$, $N = (x_1,u_6)$, $Q = (u_4,0)$, $P = (x_3,x_2)$, $R = (x_1,0)$, $S = (x_5,x_4)$.

The **nondegenerate** conditions: $A \neq C$, $A \neq B$, $A \neq M$; Line AB is non-isotropic; Line AC is non-isotropic; Line AB is non-isotropic; Line AC is non-isotropic.

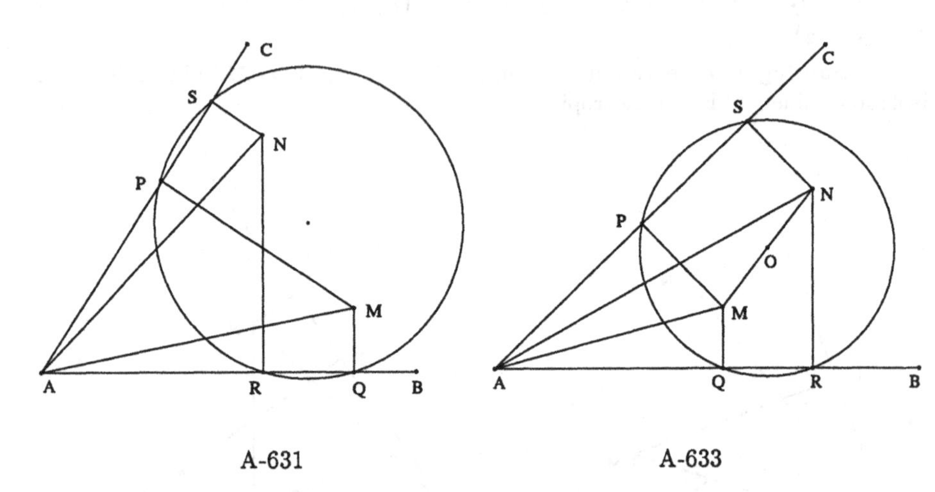

A-631 A-633

Example 466 (A-633: 1.2s, 13). *The circle PQRS (in A-631) has for its center the midpoint of MN.*

Points A, B, C, M are arbitrarily chosen. Points N, Q, P, R, S, O are constructed (in order) as follows: $\tan(BAM) = \tan(NAC)$; Q is on line AB; $QM \perp AB$; P is on line AC; $PM \perp AC$; R is on line AB; $RN \perp AB$; S is on line AC; $SN \perp AC$; O is the midpoint of N and M. The **conclusion**: $OQ \equiv OS$.

$A = (0,0)$, $B = (u_1, 0)$, $C = (u_2, u_3)$, $M = (u_4, u_5)$, $N = (x_1, u_6)$, $Q = (u_4, 0)$, $P = (x_3, x_2)$, $R = (x_1, 0)$, $S = (x_5, x_4)$, $O = (x_6, x_7)$.

The **nondegenerate** conditions: $A \neq C$, $A \neq B$, $A \neq M$; Line AB is non-isotropic; Line AC is non-isotropic; Line AB is non-isotropic; Line AC is non-isotropic.

Example 467 (A-634: 1.83s, 48). *If two lines are antiparallel with respect to an angle, the perpendiculars dropped upon them from the vertex are isogonal in the angle considered.*

Points B, C, A are arbitrarily chosen. Points R, O, S, M, N are constructed (in order) as follows: R is on line AB; $OB \equiv OR$; $OB \equiv OC$; $SO \equiv OB$; S is on line AC; $MA \perp RS$; N is on line BC; $NA \perp BC$. The **conclusion**: $\tan(BAM) = \tan(NAC)$.

$B = (0,0)$, $C = (u_1, 0)$, $A = (u_2, u_3)$, $R = (x_1, u_4)$, $O = (x_3, x_2)$, $S = (x_5, x_4)$, $M = (x_6, u_5)$, $N = (u_2, 0)$.

The **nondegenerate** conditions: $A \neq B$; Points B, C and R are not collinear; Line AC is non-isotropic; $R \neq S$; Line BC is non-isotropic. In addition, the following nondegenerate conditions, which come from reducibility and have been detected by our prover, should be also added: $S \neq C$.

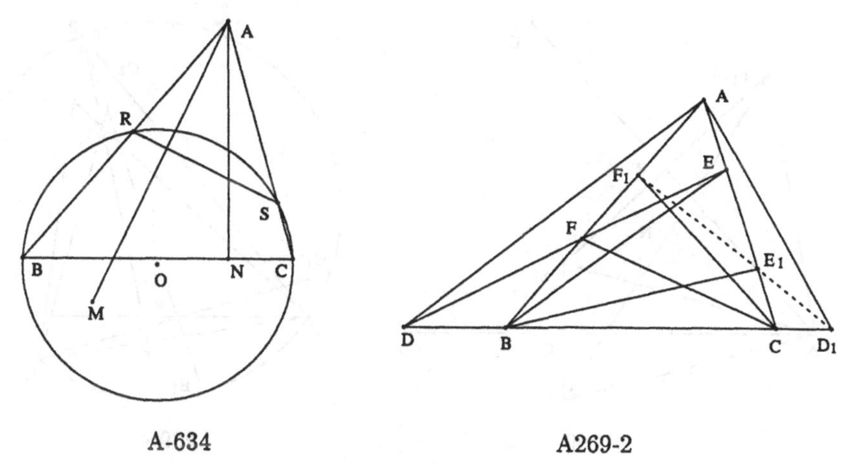

A-634 A269-2

Example 468 (A269-2: 2.73s, 45). *The isogonal conjugate of the three lines joining the vertices of a triangle to the points of intersection of the respectively opposite sides with a transversal meet these sides in three collinear points.*

Points B, C, A are arbitrarily chosen. Points D, E, F, D_1, E_1, F_1 are constructed (in order) as follows: D is on line BC; E is on line AC; F is on line ED; F is on line AB; D_1 is on line BC; $\tan(BAD) = \tan(D_1AC)$; E_1 is on line CA; $\tan(CBE) = \tan(E_1BA)$; F_1 is on line AB; $\tan(BCF) = \tan(F_1CA)$. The **conclusion**: Points D_1, E_1 and F_1 are collinear.

$B = (0,0)$, $C = (u_1,0)$, $A = (u_2,u_3)$, $D = (u_4,0)$, $E = (x_1,u_5)$, $F = (x_3,x_2)$, $D_1 = (x_4,0)$, $E_1 = (x_6,x_5)$, $F_1 = (x_8,x_7)$.

The **nondegenerate** conditions: $B \neq C$; $A \neq C$; Line AB intersects line ED; $((2u_2-u_1)u_3)u_4-u_3^3-u_2^2u_3 \neq 0$; $-u_1^2u_3x_1+(-u_1u_3^2-u_1u_2^2+u_1^2u_2)u_5 \neq 0$; $u_1^2u_3x_3+(-u_1u_3^2-u_1u_2^2+u_1^2u_2)x_2-u_1^3u_3 \neq 0$.

Example 469 (A269-3: 1.33s, 7). *Show that the line joining a given point to the vertex of a given angle has for its isogonal line the mediator of the segment determined by the symmetrics of the given point with respect to the sides of the angle.*

Points A, B, C, M are arbitrarily chosen. Points P, Q, P_1, Q_1, N are constructed (in order) as follows: P is on line AC; $PM \perp AC$; Q is on line AB; $QM \perp AB$; P is the midpoint of M and P_1; Q is the midpoint of M and Q_1; N is the midpoint of P_1 and Q_1. The **conclusion**: $AN \perp P_1Q_1$.

$A = (0,0)$, $B = (u_1,0)$, $C = (u_2,u_3)$, $M = (u_4,u_5)$, $P = (x_2,x_1)$, $Q = (u_4,0)$, $P_1 = (x_3,x_4)$, $Q_1 = (u_4,x_5)$, $N = (x_6,x_7)$.

The **nondegenerate** conditions: Line AC is non-isotropic; Line AB is non-isotropic.

Example 470 (A269-4: 55.8s, 1630). *Two triangles ABC, $A_1B_1C_1$ are perspective and inscribed in the same circle (O); if A_2, B_2, C_2 are the traces, on BC, CA, AB of the isogonal conjugates of the line A_1A, B_1B, C_1C, show that the lines AA_2,*

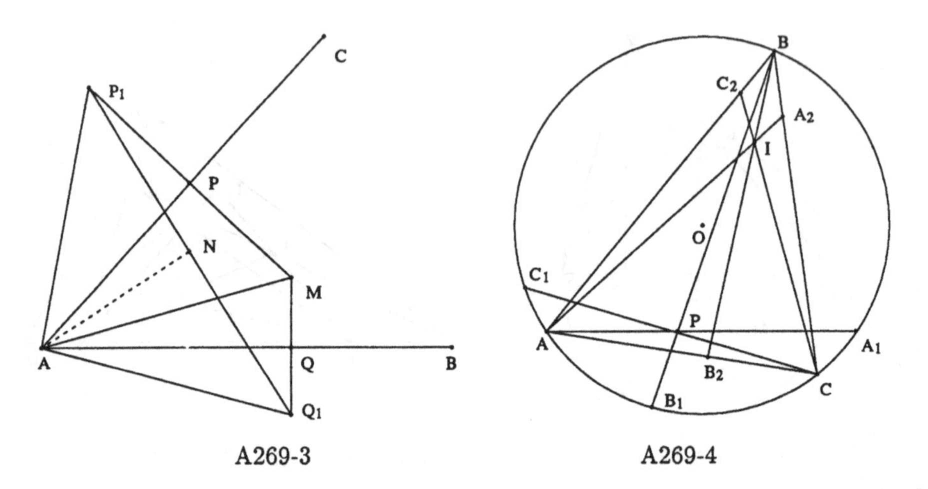

A269-3

A269-4

BB_2, CC_2 are concurrent.

Points A, A_1, B are arbitrarily chosen. Points O, C, P, B_1, C_1, A_2, B_2, C_2, I are constructed (in order) as follows: $OA \equiv OA_1$; $OA \equiv OB$; $CO \equiv OA$; P is on line AA_1; $B_1O \equiv OA$; B_1 is on line PB; $C_1O \equiv OA$; C_1 is on line PC; $\tan(BAA_1) = \tan(A_2AC)$; A_2 is on line BC; $\tan(ABB_1) = \tan(B_2BC)$; B_2 is on line AC; $\tan(ACC_1) = \tan(C_2CB)$; C_2 is on line AB; I is on line BB_2; I is on line AA_2. The **conclusion**: Points I, C and C_2 are collinear.

$A = (0,0)$, $A_1 = (u_1, 0)$, $B = (u_2, u_3)$, $O = (x_2, x_1)$, $C = (x_3, u_4)$, $P = (u_5, 0)$, $B_1 = (x_5, x_4)$, $C_1 = (x_7, x_6)$, $A_2 = (x_9, x_8)$, $B_2 = (x_{11}, x_{10})$, $C_2 = (x_{13}, x_{12})$, $I = (x_{15}, x_{14})$.

The **nondegenerate** conditions: Points A, B and A_1 are not collinear; $A \neq A_1$; Line PB is non-isotropic; Line PC is non-isotropic; $u_1u_3x_3^2 + u_1u_3u_4^2 + (-u_1u_3^2 - u_1u_2^2)u_4 \neq 0$; $(-u_3x_3^2 + 2u_2u_3x_3 - u_3u_4^2 + (u_3^2 - u_2^2)u_4)x_5 + (u_2x_3^2 + (u_3^2 - u_2^2)x_3 + u_2u_4^2 - 2u_2u_3u_4)x_4 + (-u_3^3 - u_2^2u_3)x_3 + (u_2u_3^2 + u_2^3)u_4 \neq 0$; $(-u_3x_3^2 + 2u_2u_4x_3 + u_3u_4^2 + (-u_3^2 - u_2^2)u_4)x_7 + (-u_2x_3^2 + (-2u_3u_4 + u_3^2 + u_2^2)x_3 + u_2u_4^2)x_6 + u_3x_3^3 - u_2u_4x_3^2 + u_3u_4^2x_3 - u_2u_4^3 \neq 0$; Line AA_2 intersects line BB_2. In addition, the following nondegenerate conditions, which come from reducibility and have been detected by our prover, should be also added: $C_1 \neq C$; $B_1 \neq B$.

Example 471 (A269-7: 1.62s, 38). *Show that the four perpendiculars to the sides of an angle at four concyclic points form a parallelogram whose opposite vertices lie on isogonal conjugate lines with respect to the given angle.*

Points A, B, C are arbitrarily chosen. Points O, D, E, F, I are constructed (in order) as follows: $OA \equiv OC$; $OA \equiv OB$; $DO \equiv OA$; $ED \perp DC$; $EA \perp AB$; $FC \perp CD$; $FB \perp BA$; I is on line AB; I is on line CD. The **conclusion**: $\tan(AIE) = \tan(FIC)$.

$A = (0,0)$, $B = (u_1, 0)$, $C = (u_2, u_3)$, $O = (x_2, x_1)$, $D = (x_3, u_4)$, $E = (0, x_4)$, $F = (u_1, x_5)$, $I = (x_6, 0)$.

The **nondegenerate** conditions: Points A, B and C are not collinear; Line AB

intersects line DC; Line BA intersects line CD; Line CD intersects line AB.

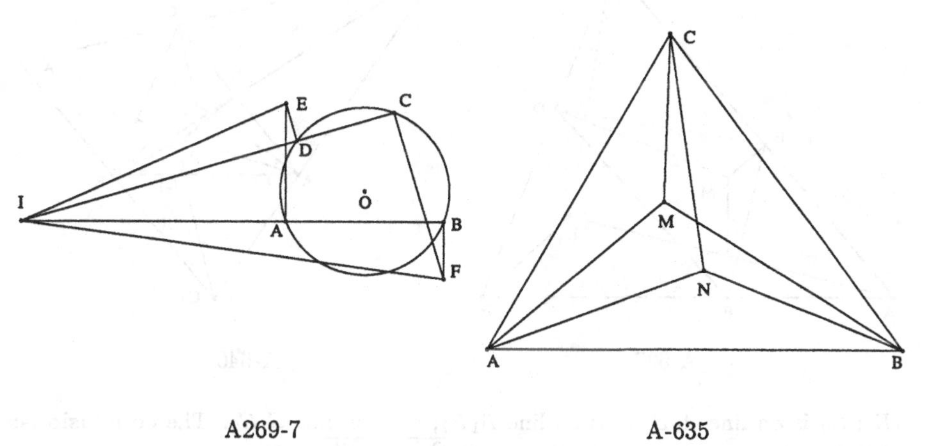

A269-7 A-635

Example 472 (A-635: 1.58s, 25). *The isogonal conjugates of the three lines joining a given point to the vertices of a given triangle are concurrent (The isogonal conjugate point).*

Points A, B, C, M are arbitrarily chosen. Points N are constructed (in order) as follows: $\tan(CBM) = \tan(NBA)$; $\tan(BAM) = \tan(NAC)$. The **conclusion**: $\tan(ACM) = \tan(NCB)$.

$A = (0,0)$, $B = (u_1,0)$, $C = (u_2,u_3)$, $M = (u_4,u_5)$, $N = (x_2,x_1)$.

The **nondegenerate** conditions: $-u_1^3u_3u_5^2 + (u_1^3u_3^2 + u_1^3u_2^2 - u_1^4u_2)u_5 - u_1^3u_3u_4^2 + u_1^4u_3u_4 \neq 0$.

Example 473 (A-637: 3.1s, 80). *The perpendiculars dropped from the vertices of a triangle upon the corresponding sides of the pedal triangle of a given point are concurrent.*

Points A, B, C, M are arbitrarily chosen. Points D, E, F, I are constructed (in order) as follows: D is on line BC; $DM \perp BC$; E is on line AC; $EM \perp AC$; F is on line AB; $FM \perp AB$; $IB \perp DF$; $IA \perp EF$. The **conclusion**: $CI \perp ED$.

$A = (0,0)$, $B = (u_1,0)$, $C = (u_2,u_3)$, $M = (u_4,u_5)$, $D = (x_2,x_1)$, $E = (x_4,x_3)$, $F = (u_4,0)$, $I = (x_6,x_5)$.

The **nondegenerate** conditions: Line BC is non-isotropic; Line AC is non-isotropic; Line AB is non-isotropic; Points E, F and D are not collinear.

Example 474 (A-640: 245.2s, 4403). *If two points are isogonal for a triangle, the pedal triangle of one is homothetic to the antipedal triangle of the other.*

Points A, B, C, Q are arbitrarily chosen. Points P, A_1, B_1, C_1, B_2, A_2, C_2, O are constructed (in order) as follows: $\tan(ABP) = \tan(QBC)$; $\tan(BAP) = \tan(QAC)$; A_1 is on line BC; $A_1P \perp BC$; B_1 is on line AC; $B_1P \perp AC$; C_1 is on line AB; $C_1P \perp AB$; $B_2C \perp CQ$; $B_2A \perp AQ$; A_2 is on line B_2C; $A_2B \perp QB$; C_2 is on line

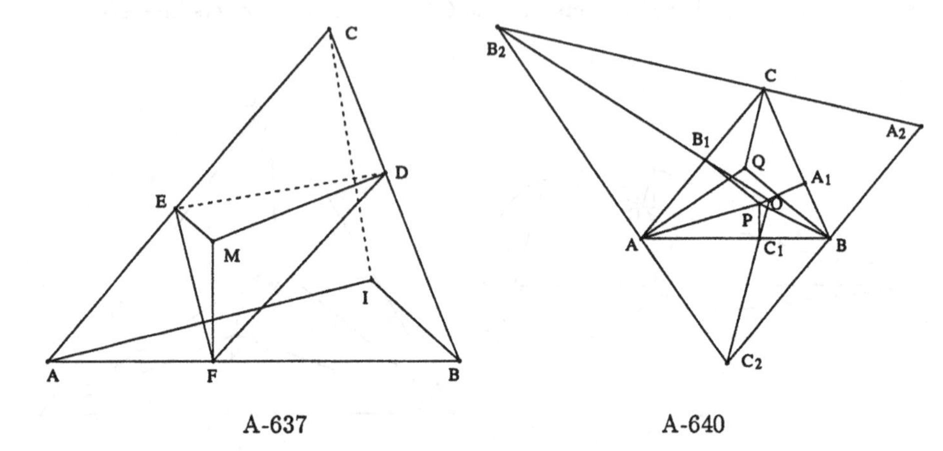

A-637 A-640

AB_2; C_2 is on line A_2B; O is on line B_1B_2; O is on line C_1C_2. The **conclusions**: (1) Points O, A_1 and A_2 are collinear; (2) $\frac{\overline{OC_1}}{\overline{OC_2}} - \frac{\overline{OB_1}}{\overline{OB_2}} = 0$.

$A = (0,0)$, $B = (u_1, 0)$, $C = (u_2, u_3)$, $Q = (u_4, u_5)$, $P = (x_2, x_1)$, $A_1 = (x_4, x_3)$, $B_1 = (x_6, x_5)$, $C_1 = (x_2, 0)$, $B_2 = (x_8, x_7)$, $A_2 = (x_{10}, x_9)$, $C_2 = (x_{12}, x_{11})$, $O = (x_{14}, x_{13})$.

The **nondegenerate** conditions: $-u_1^3u_3u_5^2 + (u_1^3u_3^2 + u_1^3u_2^2 - u_1^4u_2)u_5 - u_1^3u_3u_4^2 + u_1^4u_3u_4 \neq 0$; Line BC is non-isotropic; Line AC is non-isotropic; Line AB is non-isotropic; Points A, Q and C are not collinear; Line QB is not perpendicular to line B_2C; Line A_2B intersects line AB_2; Line C_1C_2 intersects line B_1B_2.

Example 475 (A-641: 38.63s, 1371). *The six projections of two isogonal conjugate points upon the sides of a triangle are concyclic.*

Points A, B, C, M are arbitrarily chosen. Points N, D, E, F, F_1 are constructed (in order) as follows: $\tan(CBM) = \tan(NBA)$; $\tan(BAM) = \tan(NAC)$; D is on line BC; $DM \perp BC$; E is on line AC; $EM \perp AC$; F is on line AB; $FM \perp AB$; F_1 is on line BC; $F_1N \perp BC$. The **conclusion**: Points E, F, D and F_1 are on the same circle.

$A = (0,0)$, $B = (u_1, 0)$, $C = (u_2, u_3)$, $M = (u_4, u_5)$, $N = (x_2, x_1)$, $D = (x_4, x_3)$, $E = (x_6, x_5)$, $F = (u_4, 0)$, $F_1 = (x_8, x_7)$.

The **nondegenerate** conditions: $-u_1^3u_3u_5^2 + (u_1^3u_3^2 + u_1^3u_2^2 - u_1^4u_2)u_5 - u_1^3u_3u_4^2 + u_1^4u_3u_4 \neq 0$; Line BC is non-isotropic; Line AC is non-isotropic; Line AB is non-isotropic; Line BC is non-isotropic.

Example 476 (A-643: 7.5s, 148). *The circumcircle of the pedal triangle of a point for a given triangle cuts the sides of the given triangle again in the vertices of the pedal triangle of a second point, which point is the isogonal conjugate of the first point with respect to the given triangle.*

Points A, B, C, P are arbitrarily chosen. Points A_1, B_1, C_1, O, A_2, C_2, B_2, Q are constructed (in order) as follows: A_1 is on line BC; $A_1P \perp BC$; B_1 is on

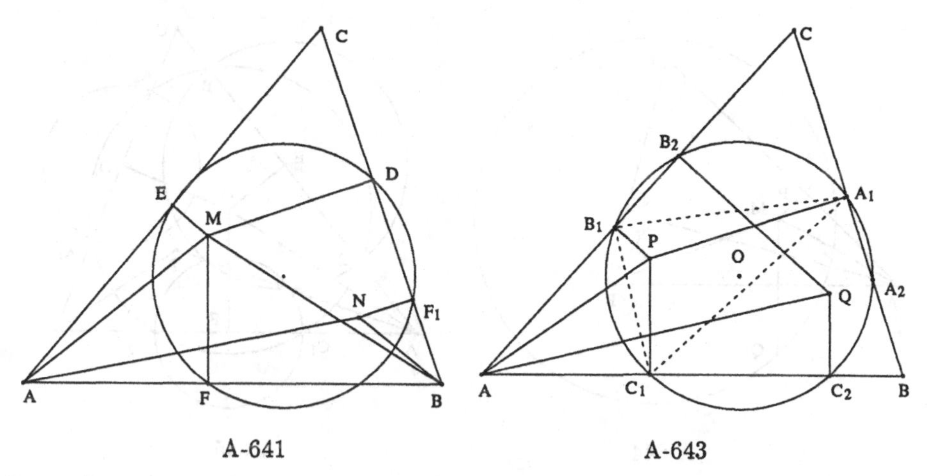

A-641 A-643

line AC; $B_1P \perp AC$; C_1 is on line AB; $C_1P \perp AB$; $OC_1 \equiv OB_1$; $OC_1 \equiv OA_1$; $A_2O \equiv OC_1$; A_2 is on line BC; $C_2O \equiv OC_1$; C_2 is on line AB; $B_2O \equiv OC_1$; B_2 is on line CA; $QB_2 \perp AC$; $QC_2 \perp AB$. The **conclusions**: (1) $\tan(BAP) = \tan(QAC)$; (2) $QA_2 \perp BC$.

$A = (0,0)$, $B = (u_1,0)$, $C = (u_2,u_3)$, $P = (u_4,u_5)$, $A_1 = (x_2,x_1)$, $B_1 = (x_4,x_3)$, $C_1 = (u_4,0)$, $O = (x_6,x_5)$, $A_2 = (x_8,x_7)$, $C_2 = (x_9,0)$, $B_2 = (x_{11},x_{10})$, $Q = (x_9,x_{12})$.

The **nondegenerate** conditions: Line BC is non-isotropic; Line AC is non-isotropic; Line AB is non-isotropic; Points C_1, A_1 and B_1 are not collinear; Line BC is non-isotropic; Line AB is non-isotropic; Line CA is non-isotropic; Points A, B and C are not collinear. In addition, the following nondegenerate conditions, which come from reducibility and have been detected by our prover, should be also added: $B_2 \neq B_1$; $C_2 \neq C_1$; $A_2 \neq A_1$.

Example 477 (A-644: 2.73s, 13). *If two points are isogonal with respect to a triangle, each is the center of the circle determined by the symmetrics of the other with respect to the sides of the triangle.*

Points A, B, C, M are arbitrarily chosen. Points N, P, Q, P_1, Q_1 are constructed (in order) as follows: $\tan(CBM) = \tan(NBA)$; $\tan(BAM) = \tan(NAC)$; P is on line AC; $PM \perp AC$; Q is on line AB; $QM \perp AB$; P is the midpoint of M and P_1; Q is the midpoint of M and Q_1. The **conclusion**: $NP_1 \equiv NQ_1$.

$A = (0,0)$, $B = (u_1,0)$, $C = (u_2,u_3)$, $M = (u_4,u_5)$, $N = (x_2,x_1)$, $P = (x_4,x_3)$, $Q = (u_4,0)$, $P_1 = (x_5,x_6)$, $Q_1 = (u_4,x_7)$.

The **nondegenerate** conditions: $-u_1^3u_3u_5^2 + (u_1^3u_3^2 + u_1^3u_2^2 - u_1^4u_2)u_5 - u_1^3u_3u_4^2 + u_1^4u_3u_4 \neq 0$; Line AC is non-isotropic; Line AB is non-isotropic.

Example 478 (A-645: 4.78s, 120). *The radical center of the three circles having for diameters the segments intercepted by a given circle on the sides of a triangle is the isogonal conjugate of the center of the given circle with respect to the triangle.*

Points A, B, C, O are arbitrarily chosen. Points C_1, B_1, A_1, D, E, F, C_2, B_2,

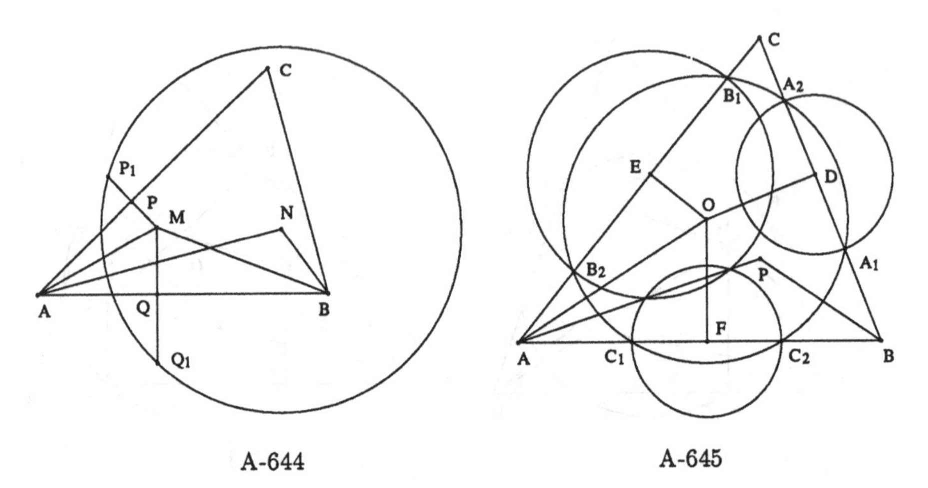

A-644 A-645

A_2, P are constructed (in order) as follows: C_1 is on line AB; B_1 is on line AC; $B_1O \equiv OC_1$; A_1 is on line BC; $A_1O \equiv OC_1$; D is on line BC; $DO \perp BC$; E is on line AC; $EO \perp AC$; F is on line AB; $FO \perp AB$; F is the midpoint of C_1 and C_2; E is the midpoint of B_1 and B_2; D is the midpoint of A_1 and A_2; Point P is on the radical axises of circle (F, FC_1) and circle (D, DA_1); Point P is on the radical axises of circle (F, FC_1) and circle (E, EB_1). The **conclusion**: $\tan(BAP) = \tan(OAC)$.

$A = (0,0)$, $B = (u_1, 0)$, $C = (u_2, u_3)$, $O = (u_4, u_5)$, $C_1 = (u_6, 0)$, $B_1 = (x_2, x_1)$, $A_1 = (x_4, x_3)$, $D = (x_6, x_5)$, $E = (x_8, x_7)$, $F = (u_4, 0)$, $C_2 = (x_9, 0)$, $B_2 = (x_{10}, x_{11})$, $A_2 = (x_{12}, x_{13})$, $P = (x_{15}, x_{14})$.

The **nondegenerate** conditions: $A \neq B$; Line AC is non-isotropic; Line BC is non-isotropic; Line BC is non-isotropic; Line AC is non-isotropic; Line AB is non-isotropic; Points F, E and D are not collinear.

Example 479 (A273-1: 2.77s, 47). *Given two isogonal points for a triangle, show that the product of their distances from a side of the triangle is equal to the analogous product for each of the other two sides.*

Points A, B, C, M are arbitrarily chosen. Points N, E, F, E_1, F_1 are constructed (in order) as follows: $\tan(CBM) = \tan(NBA)$; $\tan(BAM) = \tan(NAC)$; E is on line AC; $EM \perp AC$; F is on line AB; $FM \perp AB$; E_1 is on line AC; $E_1N \perp AC$; F_1 is on line AB; $F_1N \perp AB$. The **conclusion**: $ME \cdot NE_1 = MF \cdot NF_1$.

$A = (0,0)$, $B = (u_1, 0)$, $C = (u_2, u_3)$, $M = (u_4, u_5)$, $N = (x_2, x_1)$, $E = (x_4, x_3)$, $F = (u_4, 0)$, $E_1 = (x_6, x_5)$, $F_1 = (x_2, 0)$.

The **nondegenerate** conditions: $-u_1^3u_3u_5^2 + (u_1^3u_3^2 + u_1^3u_2^2 - u_1^4u_2)u_5 - u_1^3u_3u_4^2 + u_1^4u_3u_4 \neq 0$; Line AC is non-isotropic; Line AB is non-isotropic; Line AC is non-isotropic; Line AB is non-isotropic.

Example 480 (A274-3: 1.05s, 19). *Show that the vertices of the tangential triangle of ABC are the isogonal conjugates of the vertices of the anticomplementary triangle of ABC.*

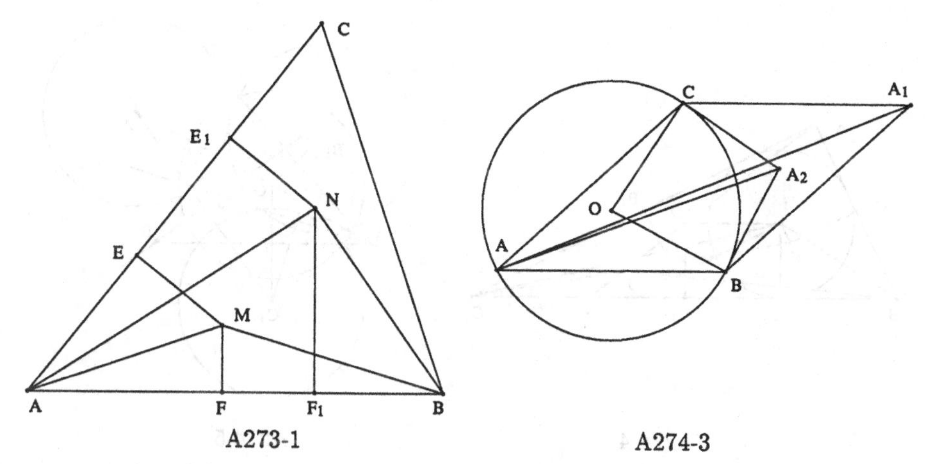

A273-1 A274-3

Points A, B, C are arbitrarily chosen. Points O, A_1, A_2 are constructed (in order) as follows: $OA \equiv OC$; $OA \equiv OB$; $A_1B \parallel AC$; $A_1C \parallel AB$; $A_2B \perp BO$; $A_2C \perp CO$. The **conclusion**: $\tan(BAA_2) = \tan(A_1AC)$.

$A = (0,0)$, $B = (u_1, 0)$, $C = (u_2, u_3)$, $O = (x_2, x_1)$, $A_1 = (x_3, u_3)$, $A_2 = (x_5, x_4)$.

The **nondegenerate** conditions: Points A, B and C are not collinear; Points A, B and C are not collinear; Points C, O and B are not collinear.

Example 481 (A274-4: 4.75s, 41). *Show that the lines joining the vertices of a triangle to the projections of the incenter upon the mediators of the respectively opposite sides meet in a point – the isotomic conjugate of the Gergonne point of the triangle.*

Points B, C, I are arbitrarily chosen. Points A, X, A_1, B_1, C_1, D, E, F, J, Y are constructed (in order) as follows: $\tan(CBI) = \tan(IBA)$; $\tan(BCI) = \tan(ICA)$; X is on line BC; $XI \perp BC$; A_1 is the midpoint of B and C; B_1 is the midpoint of C and A; C_1 is the midpoint of A and B; $DI \parallel BC$; $DA_1 \perp BC$; $EI \parallel AC$; $EB_1 \perp AC$; $FI \parallel AB$; $FC_1 \perp AB$; J is on line BE; J is on line AD; Y is on line BC; Y is on line AD. The **conclusions**: (1) Points C, F and J are collinear; (2) A_1 is the midpoint of X and Y.

$B = (0,0)$, $C = (u_1, 0)$, $I = (u_2, u_3)$, $A = (x_2, x_1)$, $X = (u_2, 0)$, $A_1 = (x_3, 0)$, $B_1 = (x_4, x_5)$, $C_1 = (x_6, x_7)$, $D = (x_3, u_3)$, $E = (x_9, x_8)$, $F = (x_{11}, x_{10})$, $J = (x_{13}, x_{12})$, $Y = (x_{14}, 0)$.

The **nondegenerate** conditions: Points I, C and B are not collinear; $\angle CIB$ is not right; Line BC is non-isotropic; Line BC is non-isotropic; Line AC is non-isotropic; Line AB is non-isotropic; Line AD intersects line BE; Line AD intersects line BC.

Example 482 (A274-5: 2.72s, 66). *Show that the three symmetrics of a given circle with respect to the sides of a triangle have for radical center the isogonal point of the center of the given circle with respect to the triangle.*

Points A, B, C, O, R are arbitrarily chosen. Points D, E, F, A_1, B_1, C_1, P are constructed (in order) as follows: D is on line BC; $DO \perp BC$; E is on line AC;

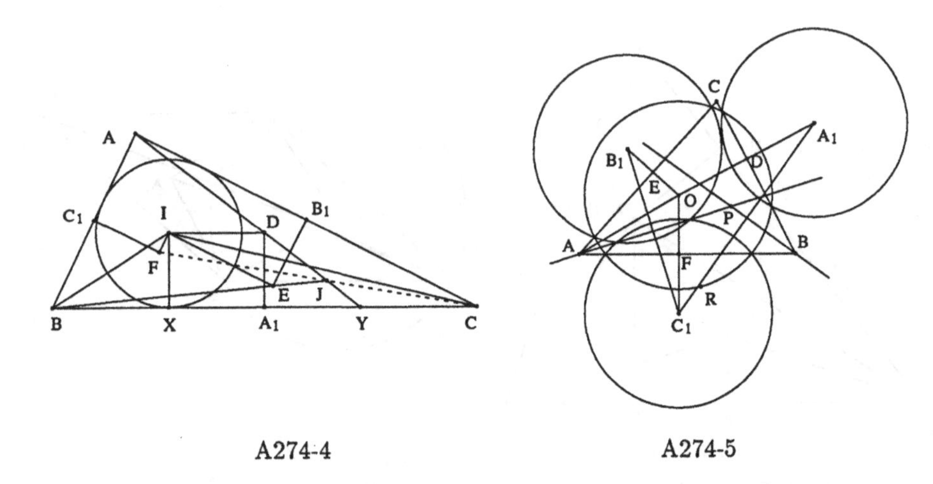

A274-4 A274-5

$EO \perp AC$; F is on line AB; $FO \perp AB$; D is the midpoint of O and A_1; E is the midpoint of O and B_1; F is the midpoint of O and C_1; Point P is on the radical axises of circle (C_1, OR) and circle (A_1, OR); Point P is on the radical axises of circle (C_1, OR) and circle (B_1, OR). The **conclusion**: $\tan(BAP) = \tan(OAC)$.

$A = (0,0)$, $B = (u_1,0)$, $C = (u_2,u_3)$, $O = (u_4,u_5)$, $R = (u_6,u_7)$, $D = (x_2,x_1)$, $E = (x_4,x_3)$, $F = (u_4,0)$, $A_1 = (x_5,x_6)$, $B_1 = (x_7,x_8)$, $C_1 = (u_4,x_9)$, $P = (x_{11},x_{10})$.

The **nondegenerate** conditions: Line BC is non-isotropic; Line AC is non-isotropic; Line AB is non-isotropic; Points C_1, B_1 and A_1 are not collinear.

Example 483 (A-650: 1.57s, 70). *The three adjoint circles of the direct group have a point in common.*

Points B, C, A are arbitrarily chosen. Points A_1, B_1, C_1, N are constructed (in order) as follows: $A_1C \perp CA$; $A_1B \equiv A_1C$; $B_1A \perp AB$; $B_1A \equiv B_1C$; $C_1B \perp BC$; $C_1A \equiv C_1B$; $NC_1 \equiv C_1B$; $NA_1 \equiv A_1B$. The **conclusion**: $B_1C \equiv B_1N$.

$B = (0,0)$, $C = (u_1,0)$, $A = (u_2,u_3)$, $A_1 = (x_2,x_1)$, $B_1 = (x_4,x_3)$, $C_1 = (0,x_5)$, $N = (x_7,x_6)$.

The **nondegenerate** conditions: Points B, C and A are not collinear; Points A, C and B are not collinear; Points A, B and C are not collinear; Line A_1C_1 is non-isotropic. In addition, the following nondegenerate conditions, which come from reducibility and have been detected by our prover, should be also added: $N \neq B$.

Remark. For a triangle ABC, let (AB) be the circle passing through vertices A and B and tangent at B to the side B, and similarly for the circles (BC) and (CA). The three circles (AB), (BC) and (CA) is called the direct group of adjoint circles. Similarly, we have other three circles (BA), (AC) and (CB) called the indirect group of adjoint circles.

Example 484 (A-651: 9.28s, 543). *The three adjoint circles of the indirect group have a point in common.*

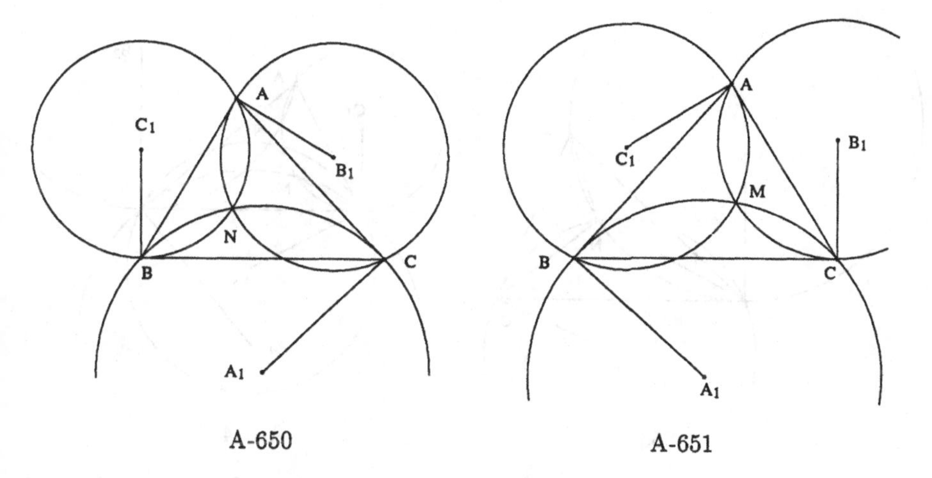

A-650 A-651

Points B, C, A are arbitrarily chosen. Points A_1, B_1, C_1, M are constructed (in order) as follows: $A_1B \perp BA$; $A_1C \equiv A_1B$; $B_1C \perp CB$; $B_1A \equiv B_1C$; $C_1A \perp AC$; $C_1B \equiv C_1A$; $MC_1 \equiv C_1B$; $MA_1 \equiv A_1B$. The **conclusion**: $B_1C \equiv B_1M$.

$B = (0,0)$, $C = (u_1,0)$, $A = (u_2,u_3)$, $A_1 = (x_2,x_1)$, $B_1 = (u_1,x_3)$, $C_1 = (x_5,x_4)$, $M = (x_7,x_6)$.

The **nondegenerate** conditions: Points C, B and A are not collinear; Points A, C and B are not collinear; Points B, A and C are not collinear; Line A_1C_1 is non-isotropic. In addition, the following nondegenerate conditions, which come from reducibility and have been detected by our prover, should be also added: $M \neq B$.

Remark. The two points in A-650 and A-652 are called the Brocard points of the triangle.

Example 485 (A-653: 0.97s, 12). *For the Brocard point in A-650 we have $\angle NAB = \angle NBC = \angle NCA$. Similar for the Brocard point in A-651.*

Points B, C, A are arbitrarily chosen. Points A_1, C_1, N are constructed (in order) as follows: $A_1C \perp CA$; $A_1B \equiv A_1C$; $C_1B \perp BC$; $C_1A \equiv C_1B$; $NC_1 \equiv C_1B$; $NA_1 \equiv A_1B$. The **conclusion**: $\tan(NAB) = \tan(NBC)$.

$B = (0,0)$, $C = (u_1,0)$, $A = (u_2,u_3)$, $A_1 = (x_2,x_1)$, $C_1 = (0,x_3)$, $N = (x_5,x_4)$.

The **nondegenerate** conditions: Points B, C and A are not collinear; Points A, B and C are not collinear; Line A_1C_1 is non-isotropic. In addition, the following nondegenerate conditions, which come from reducibility and have been detected by our prover, should be also added: $N \neq B$.

Example 486 (A-654: 7.05s, 29; 4.5s, 77). *The Brocard points are a pair of isogonal points of the triangle.*

Points B, C, A are arbitrarily chosen. Points A_1, C_1, A_2, B_2, M, N are constructed (in order) as follows: $A_1C \perp CA$; $A_1B \equiv A_1C$; $C_1B \perp BC$; $C_1A \equiv C_1B$; $A_2B \perp BA$; $A_2C \equiv A_2B$; $B_2C \perp CB$; $B_2A \equiv B_2C$; $MC_1 \equiv C_1B$; $MA_1 \equiv A_1B$; $NB_2 \equiv B_2C$; $NA_2 \equiv A_2B$. The **conclusion**: $\tan(ABM) = \tan(NBC)$.

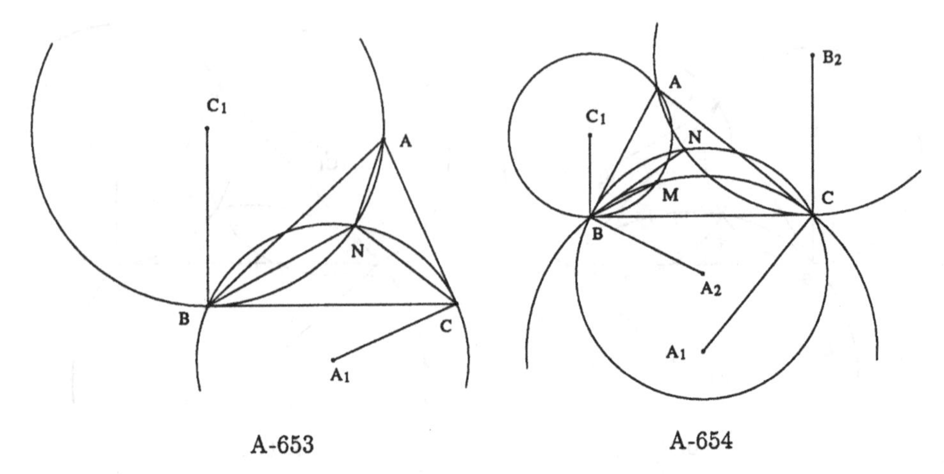

A-653 A-654

$B = (0,0)$, $C = (u_1, 0)$, $A = (u_2, u_3)$, $A_1 = (x_2, x_1)$, $C_1 = (0, x_3)$, $A_2 = (x_5, x_4)$, $B_2 = (u_1, x_6)$, $M = (x_8, x_7)$, $N = (x_{10}, x_9)$.

The **nondegenerate** conditions: Points B, C and A are not collinear; Points A, B and C are not collinear; Points C, B and A are not collinear; Points A, C and B are not collinear; Line A_1C_1 is non-isotropic; Line A_2B_2 is non-isotropic. In addition, the following nondegenerate conditions, which come from reducibility and have been detected by our prover, should be also added: $N \neq C$; $M \neq B$.

Example 487 (A-661: 396.95s, 9405; 82.27s, 501). *The traces on the circumcircle of a triangle of the lines joining the vertices to a Brocard point form a triangle congruent to the given triangle.*

Points B, C, A are arbitrarily chosen. Points A_1, C_1, N, O, B_2, A_2 are constructed (in order) as follows: $A_1C \perp CA$; $A_1B \equiv A_1C$; $C_1B \perp BC$; $C_1A \equiv C_1B$; $NC_1 \equiv C_1B$; $NA_1 \equiv A_1B$; $OB \equiv OC$; $OB \equiv OA$; B_2 is on line AN; $B_2O \equiv OB$; A_2 is on line CN; $A_2O \equiv OB$. The **conclusion**: $B_2A_2 \equiv AB$.

$B = (0,0)$, $C = (u_1, 0)$, $A = (u_2, u_3)$, $A_1 = (x_2, x_1)$, $C_1 = (0, x_3)$, $N = (x_5, x_4)$, $O = (x_7, x_6)$, $B_2 = (x_9, x_8)$, $A_2 = (x_{11}, x_{10})$.

The **nondegenerate** conditions: Points B, C and A are not collinear; Points A, B and C are not collinear; Line A_1C_1 is non-isotropic; Points B, A and C are not collinear; Line AN is non-isotropic; Line CN is non-isotropic. In addition, the following nondegenerate conditions, which come from reducibility and have been detected by our prover, should be also added: $A_2 \neq C$; $B_2 \neq A$; $N \neq B$.

Example 488 (A-663: 17.25s, 805). *The two Brocard points of a triangle are equidistant from the circumcenter of the triangle.*

Points B, C, A are arbitrarily chosen. Points A_1, C_1, A_2, B_2, M, N, O are constructed (in order) as follows: $A_1C \perp CA$; $A_1B \equiv A_1C$; $C_1B \perp BC$; $C_1A \equiv C_1B$; $A_2B \perp BA$; $A_2C \equiv A_2B$; $B_2C \perp CB$; $B_2A \equiv B_2C$; $MC_1 \equiv C_1B$; $MA_1 \equiv A_1B$; $NB_2 \equiv B_2C$; $NA_2 \equiv A_2B$; $OA \equiv OB$; $OB \equiv OC$. The **conclusion**: $ON \equiv OM$.

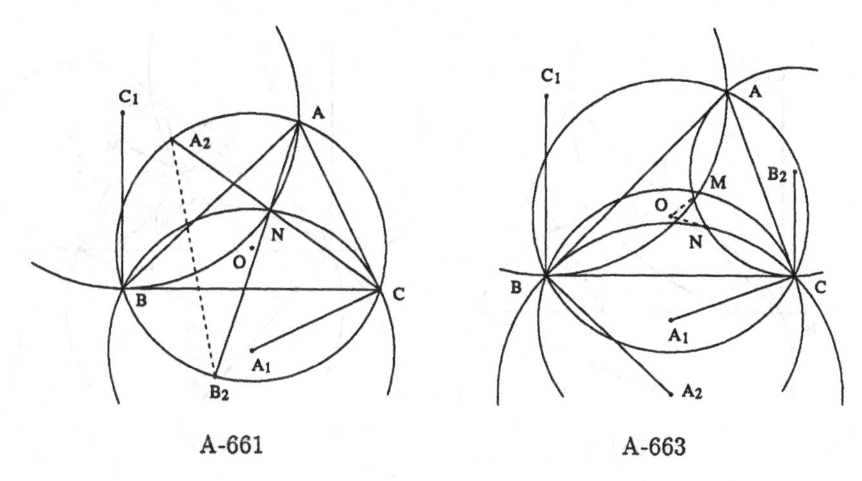

A-661 A-663

$B = (0,0)$, $C = (u_1, 0)$, $A = (u_2, u_3)$, $A_1 = (x_2, x_1)$, $C_1 = (0, x_3)$, $A_2 = (x_5, x_4)$, $B_2 = (u_1, x_6)$, $M = (x_8, x_7)$, $N = (x_{10}, x_9)$, $O = (x_{12}, x_{11})$.

The **nondegenerate** conditions: Points B, C and A are not collinear; Points A, B and C are not collinear; Points C, B and A are not collinear; Points A, C and B are not collinear; Line A_1C_1 is non-isotropic; Line A_2B_2 is non-isotropic; Points B, C and A are not collinear. In addition, the following nondegenerate conditions, which come from reducibility and have been detected by our prover, should be also added: $N \neq C$; $M \neq B$.

Example 489 (A-664: 4.57s, 170). *The pedal triangle of a Brocard point is similar to the given triangle.*

Points B, C, A are arbitrarily chosen. Points A_1, C_1, G, N, L, M are constructed (in order) as follows: $A_1C \perp CA$; $A_1B \equiv A_1C$; $C_1B \perp BC$; $C_1A \equiv C_1B$; $GC_1 \equiv C_1B$; $GA_1 \equiv A_1B$; N is on line AB; $NG \perp AB$; L is on line BC; $LG \perp BC$; M is on line AC; $MG \perp AC$. The **conclusion**: $NL \cdot BC = LM \cdot AB$.

$B = (0,0)$, $C = (u_1, 0)$, $A = (u_2, u_3)$, $A_1 = (x_2, x_1)$, $C_1 = (0, x_3)$, $G = (x_5, x_4)$, $N = (x_7, x_6)$, $L = (x_5, 0)$, $M = (x_9, x_8)$.

The **nondegenerate** conditions: Points B, C and A are not collinear; Points A, B and C are not collinear; Line A_1C_1 is non-isotropic; Line AB is non-isotropic; Line BC is non-isotropic; Line AC is non-isotropic. In addition, the following nondegenerate conditions, which come from reducibility and have been detected by our prover, should be also added: $G \neq B$.

Example 490 (A-665: 47.88s, 2293). *The pedal triangle of the two Brocard points are congruent.*

Points B, C, A are arbitrarily chosen. Points A_1, C_1, A_2, B_2, G, G_1, L, M, N, L_1, M_1, N_1 are constructed (in order) as follows: $A_1C \perp CA$; $A_1B \equiv A_1C$; $C_1B \perp BC$; $C_1A \equiv C_1B$; $A_2B \perp BA$; $A_2C \equiv A_2B$; $B_2C \perp CB$; $B_2A \equiv B_2C$; $GC_1 \equiv C_1B$; $GA_1 \equiv A_1B$; $G_1B_2 \equiv B_2C$; $G_1A_2 \equiv A_2B$; L is on line BC; $LG \perp BC$; M is on line AC; $MG \perp AC$; N is on line AB; $NG \perp AB$; L_1 is on line BC; $L_1G_1 \perp BC$;

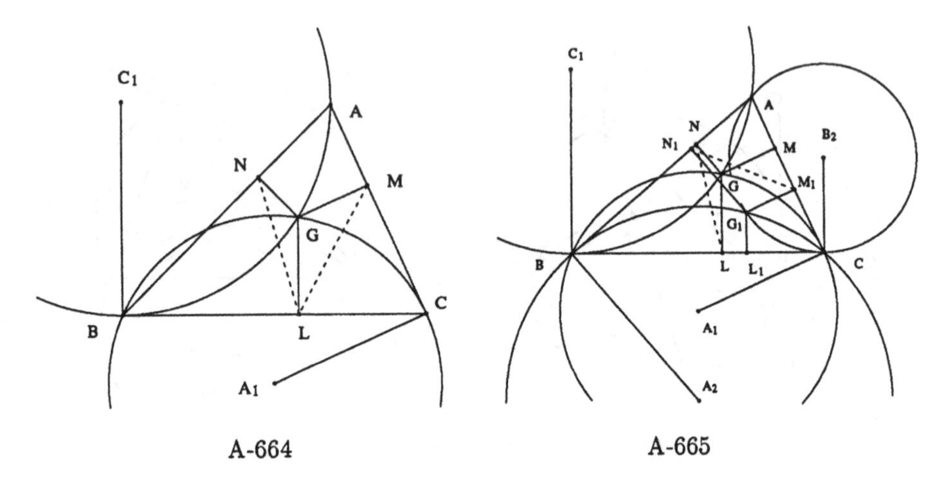

A-664 A-665

M_1 is on line AC; $M_1G_1 \perp AC$; N_1 is on line AB; $N_1G_1 \perp AB$. The **conclusion:** $LN \equiv N_1M_1$.

$B = (0,0)$, $C = (u_1,0)$, $A = (u_2,u_3)$, $A_1 = (x_2,x_1)$, $C_1 = (0,x_3)$, $A_2 = (x_5,x_4)$, $B_2 = (u_1,x_6)$, $G = (x_8,x_7)$, $G_1 = (x_{10},x_9)$, $L = (x_8,0)$, $M = (x_{12},x_{11})$, $N = (x_{14},x_{13})$, $L_1 = (x_{10},0)$, $M_1 = (x_{16},x_{15})$, $N_1 = (x_{18},x_{17})$.

The **nondegenerate** conditions: Points B, C and A are not collinear; Points A, B and C are not collinear; Points C, B and A are not collinear; Points A, C and B are not collinear; Line A_1C_1 is non-isotropic; Line A_2B_2 is non-isotropic; Line BC is non-isotropic; Line AC is non-isotropic; Line AB is non-isotropic; Line BC is non-isotropic; Line AC is non-isotropic; Line AB is non-isotropic. In addition, the following nondegenerate conditions, which come from reducibility and have been detected by our prover, should be also added: $G_1 \neq C$; $G \neq B$.

Example 491 (A-667: 26.28s, 889). *The Brocard points of a triangle lie on the Brocard circle.*

Points A, B, C are arbitrarily chosen. Points M_1, M_3, K, O, A_1, C_1, M, R are constructed (in order) as follows: M_1 is the midpoint of B and C; M_3 is the midpoint of A and B; $\tan(BAK) = \tan(M_1AC)$; $\tan(ACK) = \tan(M_3CB)$; $OB \equiv OA$; $OB \equiv OC$; $A_1C \perp CA$; $A_1B \equiv A_1C$; $C_1B \perp BC$; $C_1A \equiv C_1B$; $MC_1 \equiv C_1B$; $MA_1 \equiv A_1B$; R is the midpoint of O and K. The **conclusion:** $RO \equiv RM$.

$A = (0,0)$, $B = (u_1,0)$, $C = (u_2,u_3)$, $M_1 = (x_1,x_2)$, $M_3 = (x_3,0)$, $K = (x_5,x_4)$, $O = (x_7,x_6)$, $A_1 = (x_9,x_8)$, $C_1 = (x_{11},x_{10})$, $M = (x_{13},x_{12})$, $R = (x_{14},x_{15})$.

The **nondegenerate** conditions: $(((u_1u_2 - u_1^2)u_3^2 + u_1u_2^3 - u_1^2u_2^2)x_2 + (u_1u_3^3 + u_1u_2^2u_3)x_1)x_3 + (-u_1u_3^4 + (-2u_1u_2^2 + u_1^2u_2)u_3^2 - u_1u_2^4 + u_1^2u_2^3)x_2 + (-u_1^2u_3^3 - u_1^2u_2^2u_3)x_1 \neq 0$; Points B, C and A are not collinear; Points B, C and A are not collinear; Points A, B and C are not collinear; Line A_1C_1 is non-isotropic. In addition, the following nondegenerate conditions, which come from reducibility and have been detected by our prover, should be also added: $M \neq B$.

Example 492 (A-670-a: 1936.77s, 26294; 237.82s, 3067). *The Brocard circle of a*

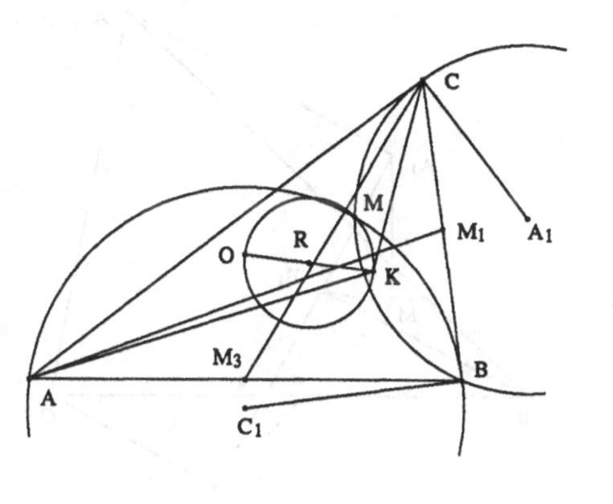

A-667

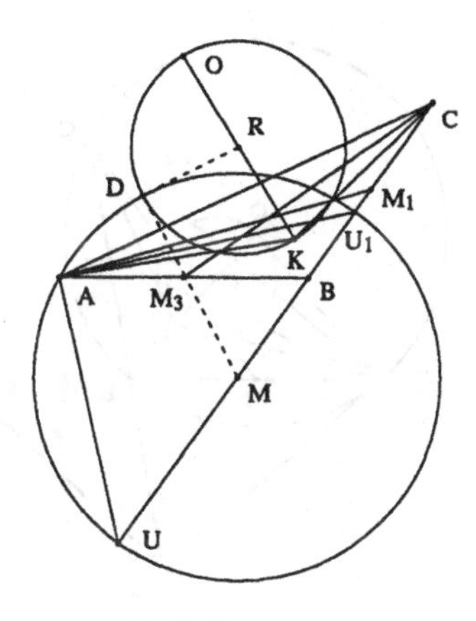

A-670-a

triangle is orthogonal to the Apollonian circles.

Points A, B, C are arbitrarily chosen. Points M_1, M_3, K, O, R, U, U_1, M, D are constructed (in order) as follows: M_1 is the midpoint of B and C; M_3 is the midpoint of A and B; $\tan(BAK) = \tan(M_1AC)$; $\tan(ACK) = \tan(M_3CB)$; $OB \equiv OA$; $OB \equiv OC$; R is the midpoint of O and K; U is on line BC; $\tan(BAU) = \tan(UAC)$; $U_1A \perp UA$; U_1 is on line BC; M is the midpoint of U and U_1; $DR \equiv RO$; $DM \equiv MU$. The **conclusion**: $MD \perp RD$.

$A = (0,0)$, $B = (u_1,0)$, $C = (u_2,u_3)$, $M_1 = (x_1,x_2)$, $M_3 = (x_3,0)$, $K = (x_5,x_4)$, $O = (x_7,x_6)$, $R = (x_8,x_9)$, $U = (x_{11},x_{10})$, $U_1 = (x_{13},x_{12})$, $M = (x_{14},x_{15})$, $D = (x_{17},x_{16})$.

The **nondegenerate** conditions: $(((u_1u_2 - u_1^2)u_3^2 + u_1u_2^3 - u_1^2u_2^2)x_2 + (u_1u_3^3 + u_1u_2^2u_3)x_1)x_3 + (-u_1u_3^4 + (-2u_1u_2^2 + u_1^2u_2)u_3^2 - u_1u_2^4 + u_1^2u_2^3)x_2 + (-u_1^2u_3^3 - u_1^2u_2^2u_3)x_1 \neq 0$; Points B, C and A are not collinear; $-u_1u_3^3 + (-u_1u_2^2 + u_1^3)u_3 \neq 0$; $u_3 \neq 0$; Line BC is not perpendicular to line UA; Line MR is non-isotropic.

Example 493 (A-670-b: 7.8s, 21; 5.32s, 83). *The radical axis of the Brocard circle and the circumcircle coincides with the Lemoine axis.*

Points A, B, C are arbitrarily chosen. Points M_1, M_3, K, O, R, X are constructed (in order) as follows: M_1 is the midpoint of B and C; M_3 is the midpoint of A and B; $\tan(BAK) = \tan(M_1AC)$; $\tan(ACK) = \tan(M_3CB)$; $OB \equiv OA$; $OB \equiv OC$; R is the midpoint of O and K; X is on line BC; $XA \perp OA$. The **conclusion**: Point X is on the radical axises of circle (R, RO) and circle (O, OB).

$A = (0,0)$, $B = (u_1,0)$, $C = (u_2,u_3)$, $M_1 = (x_1,x_2)$, $M_3 = (x_3,0)$, $K = (x_5,x_4)$, $O = (x_7,x_6)$, $R = (x_8,x_9)$, $X = (x_{11},x_{10})$.

The **nondegenerate** conditions: $(((u_1u_2 - u_1^2)u_3^2 + u_1u_2^3 - u_1^2u_2^2)x_2 + (u_1u_3^3 + u_1u_2^2u_3)x_1)x_3 + (-u_1u_3^4 + (-2u_1u_2^2 + u_1^2u_2)u_3^2 - u_1u_2^4 + u_1^2u_2^3)x_2 + (-u_1^2u_3^3 - u_1^2u_2^2u_3)x_1 \neq 0$; Points B, C and A are not collinear; Line OA is not perpendicular to line BC.

Example 494 (A-691: 1.85s, 42). *The perpendiculars dropped upon the sides of a triangle from the projections of the opposite vertices upon a given line are concurrent*

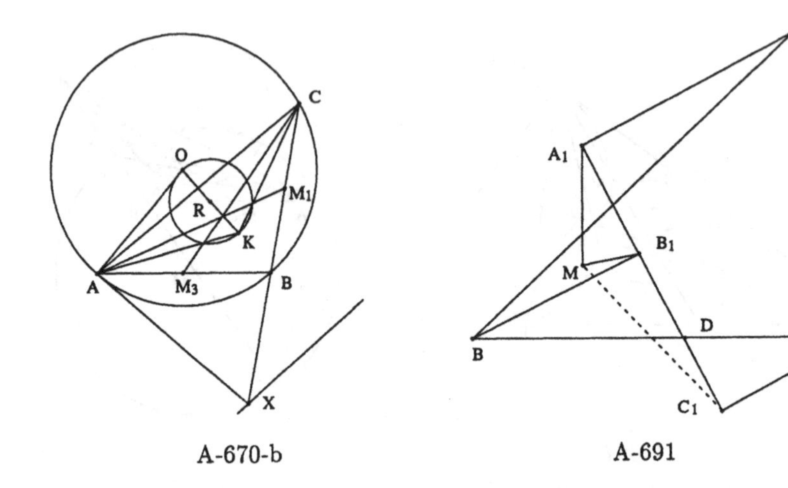

A-670-b A-691

(The Orthopole of the given line).

Points B, C, A, A_1 are arbitrarily chosen. Points D, B_1, C_1, M are constructed (in order) as follows: D is on line BC; $DA_1 \perp A_1A$; B_1 is on line DA_1; $B_1B \perp A_1D$; C_1 is on line DA_1; $C_1C \perp DA_1$; $MB_1 \perp AC$; $MA_1 \perp BC$. The **conclusion**: $C_1M \perp AB$.

$B = (0,0)$, $C = (u_1,0)$, $A = (u_2,u_3)$, $A_1 = (u_4,u_5)$, $D = (x_1,0)$, $B_1 = (x_3,x_2)$, $C_1 = (x_5,x_4)$, $M = (u_4,x_6)$.

The **nondegenerate** conditions: Line A_1A is not perpendicular to line BC; Line A_1D is non-isotropic; Line DA_1 is non-isotropic; Points B, C and A are not collinear.

Example 495 (A-694: 6.92s, 381). *The orthopole, for a given triangle, of a circumdiameter of that triangle lies on the nine-point circle of the triangle.*

Points B, C, A are arbitrarily chosen. Points O, B_1, C_1, M, M_1, M_2, M_3, N are constructed (in order) as follows: $OB \equiv OA$; $OB \equiv OC$; B_1 is on the circle with diagonal BO; C_1 is on line OB_1; $C_1C \perp OB_1$; $MC_1 \perp AB$; $MB_1 \perp AC$; M_1 is the midpoint of B and C; M_2 is the midpoint of A and C; M_3 is the midpoint of A and B; $NM_3 \equiv NM_2$; $NM_3 \equiv NM_1$. The **conclusion**: $NM \equiv NM_3$.

$B = (0,0)$, $C = (u_1,0)$, $A = (u_2,u_3)$, $O = (x_2,x_1)$, $B_1 = (x_3,u_4)$, $C_1 = (x_5,x_4)$, $M = (x_7,x_6)$, $M_1 = (x_8,0)$, $M_2 = (x_9,x_{10})$, $M_3 = (x_{11},x_{12})$, $N = (x_{14},x_{13})$.

The **nondegenerate** conditions: Points B, C and A are not collinear; Line OB_1 is non-isotropic; Points A, C and B are not collinear; Points M_3, M_1 and M_2 are not collinear.

Example 496 (A-695: 3.87s, 140). *The orthopole of a line for a given triangle lies on the Simson line perpendicular to the given line.*

Points B, C, A are arbitrarily chosen. Points O, P, D, F, C_1, B_1, M are constructed (in order) as follows: $OB \equiv OC$; $OB \equiv OA$; $PO \equiv OB$; D is on line BC; $DP \perp BC$; F is on line AB; $FP \perp AB$; $C_1C \parallel DF$; $B_1C_1 \perp FD$; $B_1B \parallel FD$; $MC_1 \perp AB$; $MB_1 \perp AC$. The **conclusion**: Points F, D and M are collinear.

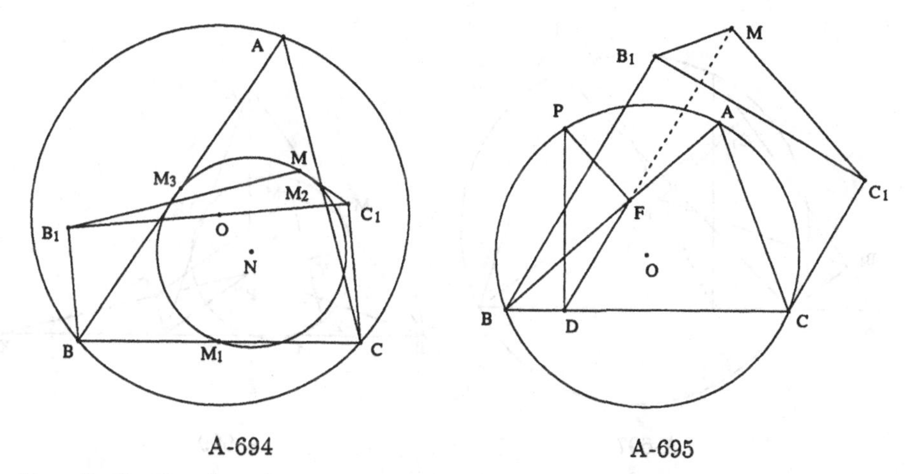

A-694 A-695

$B = (0,0)$, $C = (u_1,0)$, $A = (u_2,u_3)$, $O = (x_2,x_1)$, $P = (x_3,u_4)$, $D = (x_3,0)$, $F = (x_5,x_4)$, $C_1 = (x_6,u_5)$, $B_1 = (x_8,x_7)$, $M = (x_{10},x_9)$.

The **nondegenerate** conditions: Points B, A and C are not collinear; Line BC is non-isotropic; Line AB is non-isotropic; $D \neq F$; Line FD is non-isotropic; Points A, C and B are not collinear.

Example 497 (A-697: 10.45s, 159). *If a line meets the circumcircle of a triangle, the Simson lines of the points of intersection with the circle meet in the orthopole of the line for the triangle.*

Points B, C, A are arbitrarily chosen. Points O, P, D, F, X, C_1, B_1, M are constructed (in order) as follows: $OB \equiv OA$; $OB \equiv OC$; $PO \equiv OB$; D is on line BC; $DP \perp BC$; F is on line AC; $FP \perp AC$; X is on line BC; C_1 is on line PX; $C_1C \perp PX$; B_1 is on line PX; $B_1B \perp PX$; $MC_1 \perp AB$; $MB_1 \perp AC$. The **conclusion**: Points F, D and M are collinear.

$B = (0,0)$, $C = (u_1,0)$, $A = (u_2,u_3)$, $O = (x_2,x_1)$, $P = (x_3,u_4)$, $D = (x_3,0)$, $F = (x_5,x_4)$, $X = (u_5,0)$, $C_1 = (x_7,x_6)$, $B_1 = (x_9,x_8)$, $M = (x_{11},x_{10})$.

The **nondegenerate** conditions: Points B, C and A are not collinear; Line BC is non-isotropic; Line AC is non-isotropic; $B \neq C$; Line PX is non-isotropic; Line PX is non-isotropic; Points A, C and B are not collinear.

Example 498 (A-699: 33.1s, 1317). *If a line passes through the orthocenter of a triangle, the symmetric of the orthopole of this line with respect to the line lies on the nine-point circle.*

Points B, C, A are arbitrarily chosen. Points P, X, B_1, C_1, M, Y, Z, M_1, M_2, M_3, N are constructed (in order) as follows: $PB \perp AC$; $PA \perp BC$; X is on line BC; B_1 is on line PX; $B_1B \perp PX$; C_1 is on line PX; $C_1C \perp PX$; $MC_1 \perp AB$; $MB_1 \perp AC$; Y is on line XP; $YM \perp XP$; Y is the midpoint of Z and M; M_1 is the midpoint of B and C; M_2 is the midpoint of C and A; M_3 is the midpoint of A and B; $NM_3 \equiv NM_2$; $NM_3 \equiv NM_1$. The **conclusion**: $NM_3 \equiv NZ$.

$B = (0,0)$, $C = (u_1,0)$, $A = (u_2,u_3)$, $P = (u_2,x_1)$, $X = (u_4,0)$, $B_1 = (x_3,x_2)$,

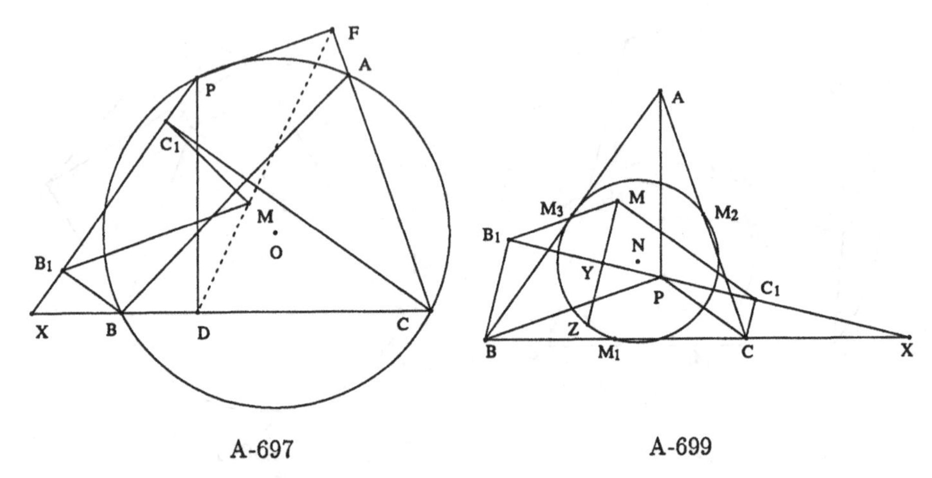

A-697 A-699

$C_1 = (x_5, x_4)$, $M = (x_7, x_6)$, $Y = (x_9, x_8)$, $Z = (x_{10}, x_{11})$, $M_1 = (x_{12}, 0)$, $M_2 = (x_{13}, x_{14})$, $M_3 = (x_{15}, x_{16})$, $N = (x_{18}, x_{17})$.

The **nondegenerate** conditions: Points B, C and A are not collinear; $B \neq C$; Line PX is non-isotropic; Line PX is non-isotropic; Points A, C and B are not collinear; Line XP is non-isotropic; Points M_3, M_1 and M_2 are not collinear.

Example 499 (A291-1: 3.95s, 113). *P and Q are any pair of isogonal conjugates with respect to a triangle ABC. If AQ cuts the circumcircle in O, and OP cuts BC in R, show that QR is parallel to AP.*

Points A, B, C, P are arbitrarily chosen. Points Q, N, O, R are constructed (in order) as follows: $\tan(CBP) = \tan(QBA)$; $\tan(BAP) = \tan(QAC)$; $NA \equiv NC$; $NA \equiv NB$; $ON \equiv NA$; O is on line AQ; R is on line BC; R is on line OP. The **conclusion**: $QR \parallel AP$.

$A = (0,0)$, $B = (u_1, 0)$, $C = (u_2, u_3)$, $P = (u_4, u_5)$, $Q = (x_2, x_1)$, $N = (x_4, x_3)$, $O = (x_6, x_5)$, $R = (x_8, x_7)$.

The **nondegenerate** conditions: $-u_1^3 u_3 u_5^2 + (u_1^3 u_3^2 + u_1^3 u_2^2 - u_1^4 u_2) u_5 - u_1^3 u_3 u_4^2 + u_1^4 u_3 u_4 \neq 0$; Points A, B and C are not collinear; Line AQ is non-isotropic; Line OP intersects line BC. In addition, the following nondegenerate conditions, which come from reducibility and have been detected by our prover, should be also added: $O \neq A$.

Example 500 (A291-2-0: 376.3s, 5906; 109.77s, 1065). *Through the orthogonal projections of the vertices of a triangle ABC upon a given line m parallels are dawn to the respective sides of ABC forming a triangle $A_1B_1C_1$. Show that the orthopole of m for ABC is concyclic with the points A_1, B_1, C_1.*

Points X, Y are arbitrarily chosen. Points Z, A, B, C, M, C_1, A_1, B_1, O are constructed (in order) as follows: Z is on line XY; $AX \perp XY$; $BY \perp XY$; $CZ \perp XY$; $MY \perp AC$; $MX \perp BC$; $C_1Y \parallel AC$; $C_1X \parallel BC$; A_1 is on line YC_1; $A_1Z \parallel AB$; B_1 is on line ZA_1; B_1 is on line XC_1; $OA_1 \equiv OC_1$; $OA_1 \equiv OB_1$. The **conclusion**: $OA_1 \equiv OM$.

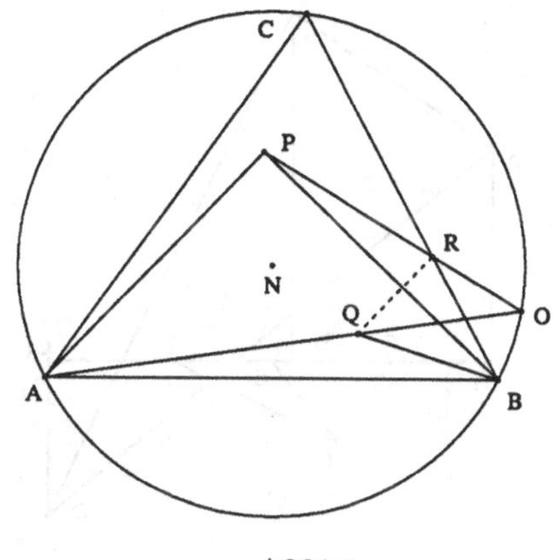

A291-1

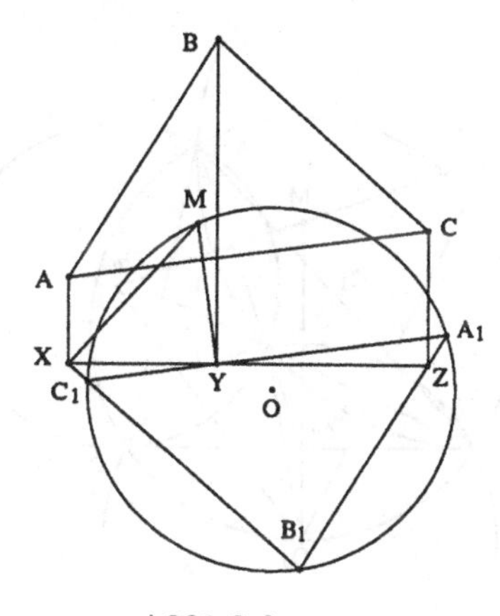

A291-2-0

$X = (0,0)$, $Y = (u_1,0)$, $Z = (u_2,0)$, $A = (0,u_3)$, $B = (u_1,u_4)$, $C = (u_2,u_5)$, $M = (x_2,x_1)$, $C_1 = (x_4,x_3)$, $A_1 = (x_6,x_5)$, $B_1 = (x_8,x_7)$, $O = (x_{10},x_9)$.

The **nondegenerate** conditions: $X \neq Y$; $X \neq Y$; $X \neq Y$; $X \neq Y$; Points B, C and A are not collinear; Points B, C and A are not collinear; Line AB intersects line YC_1; Line XC_1 intersects line ZA_1; Points A_1, B_1 and C_1 are not collinear.

Example 501 (A291-2: 0.0s, 5906). *If O and G are the circumcenter and the centroid of the triangle ABC, and O_a, O_b, O_c the circumcenters of the triangle GBC, GCA, GAB, show that the points O and G are respectively the centroid and the Lemoine point of the triangle $O_aO_bO_c$.*

Points B, C, A are arbitrarily chosen. Points A_1, B_1, C_1, G, O_A, O_B, O_C, O, M are constructed (in order) as follows: A_1 is the midpoint of C and B; B_1 is the midpoint of A and C; C_1 is the midpoint of B and A; G is on line BB_1; G is on line A_1A; $O_AB \equiv O_AG$; $O_AB \equiv O_AC$; $O_BC \equiv O_BG$; $O_BC \equiv O_BA$; $O_CB \equiv O_CG$; $O_CB \equiv O_CA$; $OB \equiv OA$; $OB \equiv OC$; M is the midpoint of O_B and O_C. The **conclusions**: (1) $\tan(O_BO_AG) = \tan(MO_AO_C)$; (2) Points O_A, O and M are collinear.

$B = (0,0)$, $C = (u_1,0)$, $A = (u_2,u_3)$, $A_1 = (x_1,0)$, $B_1 = (x_2,x_3)$, $C_1 = (x_4,x_5)$, $G = (x_7,x_6)$, $O_A = (x_9,x_8)$, $O_B = (x_{11},x_{10})$, $O_C = (x_{13},x_{12})$, $O = (x_{15},x_{14})$, $M = (x_{16},x_{17})$.

The **nondegenerate** conditions: Line A_1A intersects line BB_1; Points B, C and G are not collinear; Points C, A and G are not collinear; Points B, A and G are not collinear; Points B, C and A are not collinear.

Example 502 (A291-4-0: 0.82s, 8). *If A_1, B_1, C_1 are the projections of the vertices of a triangle ABC upon a given line, show that the perpendiculars from the midpoints of the segments B_1C_1, C_1A_1, A_1B_1 upon the respective sides of ABC are concurrent.*

Points A_1, B_1 are arbitrarily chosen. Points C_1, A, B, C, D, E, F, O are constructed (in order) as follows: C_1 is on line A_1B_1; $AA_1 \perp A_1B_1$; $BB_1 \perp A_1B_1$; $CC_1 \perp A_1B_1$; D is the midpoint of B_1 and C_1; E is the midpoint of A_1 and C_1; F

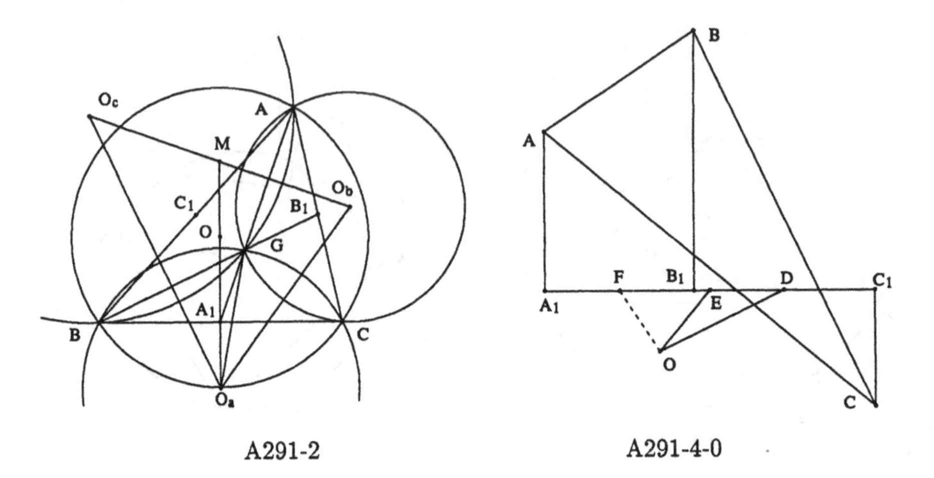

A291-2 A291-4-0

is the midpoint of A_1 and B_1; $OE \perp AC$; $OD \perp BC$. The **conclusion**: $OF \perp AB$.

$A_1 = (0,0)$, $B_1 = (u_1,0)$, $C_1 = (u_2,0)$, $A = (0,u_3)$, $B = (u_1,u_4)$, $C = (u_2,u_5)$, $D = (x_1,0)$, $E = (x_2,0)$, $F = (x_3,0)$, $O = (x_5,x_4)$.

The **nondegenerate** conditions: $A_1 \neq B_1$; $A_1 \neq B_1$; $A_1 \neq B_1$; $A_1 \neq B_1$; Points B, C and A are not collinear.

Example 503 (A291-4: 3.58s, 21). *If EDF is the orthic triangle of the triangle ABC, and X_1, Y_1, Z_1 are the symmetrics, with respect to the symmedian point K of ABC, of the projections X, Y, Z of K upon BC, CA, AB, show that X_1, Y_1, Z_1 are the Lemoine points of the triangles AEF, BFD, CDE.*

Points A, B, C are arbitrarily chosen. Points M_1, M_3, K, X, X_1, E, F, M are constructed (in order) as follows: M_1 is the midpoint of B and C; M_3 is the midpoint of A and B; $\tan(BAK) = \tan(M_1AC)$; $\tan(ACK) = \tan(M_3CB)$; X is on line BC; $XK \perp BC$; K is the midpoint of X_1 and X; E is on line AC; $EB \perp AC$; F is on line AB; $FC \perp AB$; M is the midpoint of E and F. The **conclusion**: $\tan(BAM) = \tan(X_1AC)$.

$A = (0,0)$, $B = (u_1,0)$, $C = (u_2,u_3)$, $M_1 = (x_1,x_2)$, $M_3 = (x_3,0)$, $K = (x_5,x_4)$, $X = (x_7,x_6)$, $X_1 = (x_8,x_9)$, $E = (x_{11},x_{10})$, $F = (u_2,0)$, $M = (x_{12},x_{13})$.

The **nondegenerate** conditions: $(((u_1u_2 - u_1^2)u_3^2 + u_1u_2^3 - u_1^2u_2^2)x_2 + (u_1u_3^3 + u_1u_2^2u_3)x_1)x_3 + (-u_1u_3^4 + (-2u_1u_2^2 + u_1^2u_2)u_3^2 - u_1u_2^4 + u_1^2u_2^3)x_2 + (-u_1^2u_3^3 - u_1^2u_2^2u_3)x_1 \neq 0$; Line BC is non-isotropic; Line AC is non-isotropic; Line AB is non-isotropic.

Example 504 (A291-5-a: 3.3s, 76). *The distance between the orthopoles of the two bisectors of the same angle of a triangle is equal to the circumdiameter.*

Points A, B, C are arbitrarily chosen. Points E, D, D_1, M, M_1, O are constructed (in order) as follows: $\tan(BAE) = \tan(EAC)$; D is on line AE; $DB \perp EA$; D_1 is on the circle with diagonal BA; $D_1A \perp AD$; $MD \perp AC$; $MA \perp BC$; $M_1D_1 \perp AC$; $M_1A \perp BC$; $OA \equiv OC$; $OA \equiv OB$. The **conclusion**: $2OA = MM_1$.

$A = (0,0)$, $B = (u_1,0)$, $C = (u_2,u_3)$, $E = (x_1,u_4)$, $D = (x_3,x_2)$, $D_1 = (x_5,x_4)$,

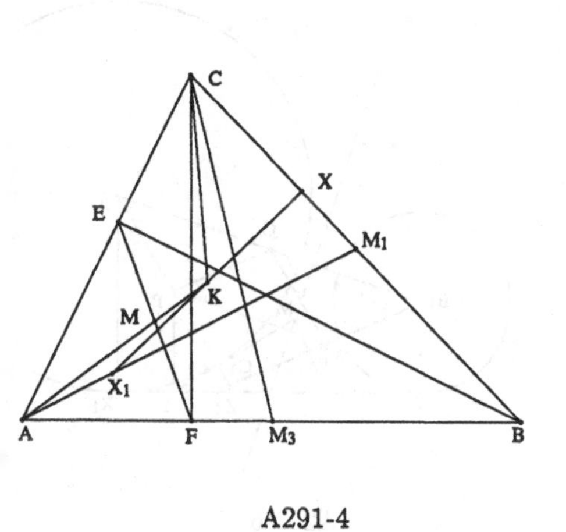

A291-4

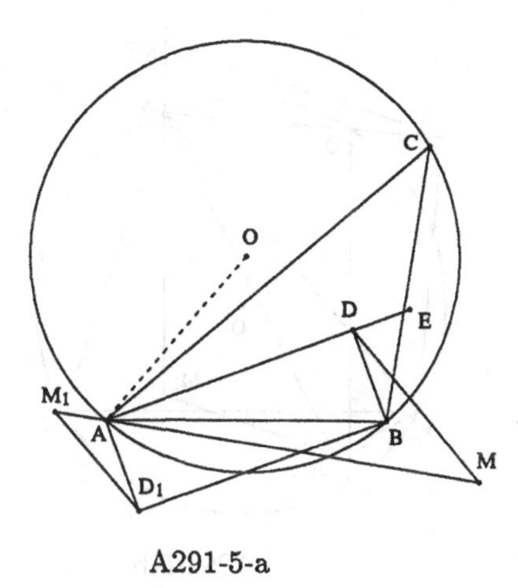

A291-5-a

$M = (x_7, x_6)$, $M_1 = (x_9, x_8)$, $O = (x_{11}, x_{10})$.

The **nondegenerate** conditions: Points A, C and B are not collinear; Line EA is non-isotropic; Line AD is non-isotropic; Points B, C and A are not collinear; Points B, C and A are not collinear; Points A, B and C are not collinear. In addition, the following nondegenerate conditions, which come from reducibility and have been detected by our prover, should be also added: $D_1 \neq A$.

Example 505 (A291-5-b: 133.87s, 3826; 12.9s, 192). *If a chord PQ of the circumcircle (O) of a triangle ABC is perpendicular to BC, the distance between the orthopoles of the lines AP, AQ is equal to PQ.*

Points B, C, A are arbitrarily chosen. Points O, P, Q, B_1, B_2, M_1, M_2 are constructed (in order) as follows: $OB \equiv OA$; $OB \equiv OC$; $PO \equiv OB$; $QP \perp BC$; $QO \equiv OB$; B_1 is on line AP; $B_1B \perp PA$; B_2 is on line AQ; $B_2B \perp QA$; $M_1B_1 \perp AC$; $M_1A \perp BC$; $M_2B_2 \perp AC$; $M_2A \perp BC$. The **conclusion**: $PQ \equiv M_1M_2$.

$B = (0,0)$, $C = (u_1, 0)$, $A = (u_2, u_3)$, $O = (x_2, x_1)$, $P = (x_3, u_4)$, $Q = (x_3, x_4)$, $B_1 = (x_6, x_5)$, $B_2 = (x_8, x_7)$, $M_1 = (u_2, x_9)$, $M_2 = (u_2, x_{10})$.

The **nondegenerate** conditions: Points B, C and A are not collinear; Line BC is non-isotropic; Line PA is non-isotropic; Line QA is non-isotropic; Points B, C and A are not collinear; Points B, C and A are not collinear. In addition, the following nondegenerate conditions, which come from reducibility and have been detected by our prover, should be also added: $Q \neq P$.

Example 506 (A291-6: 4.38s, 16; 2.45s, 16). *Show that the orthopole of a line for a triangle is the radical center of the three circles tangent to the given line and having for centers the vertices of the anticomplementary triangle of the given triangle.*

Points X, Y, C are arbitrarily chosen. Points A, B, M, C_1, B_1, A_1, X_1, Y_1, Z_1 are constructed (in order) as follows: $AX \perp XY$; $BY \perp XY$; $MY \perp AC$; $MX \perp BC$; $C_1B \parallel AC$; $C_1A \parallel BC$; B_1 is on line C_1A; $B_1C \parallel AB$; A_1 is on line B_1C; A_1 is on line C_1B; $X_1A_1 \perp XY$; X_1 is on line XY; $Y_1B_1 \perp XY$; Y_1 is on line XY; $Z_1C_1 \perp XY$; Z_1 is on line XY. The **conclusion**: Point M is on the radical axises

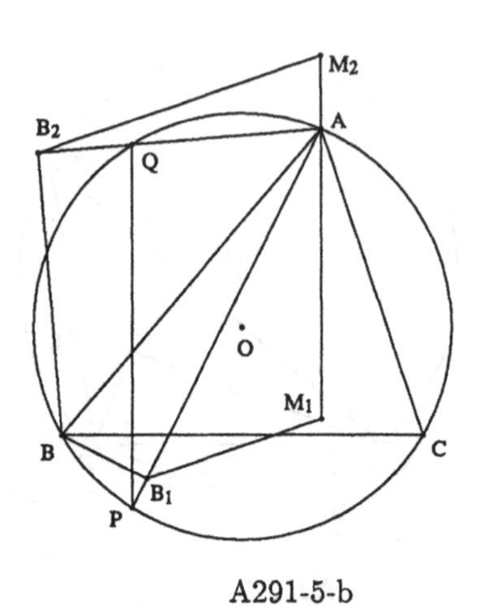

A291-5-b

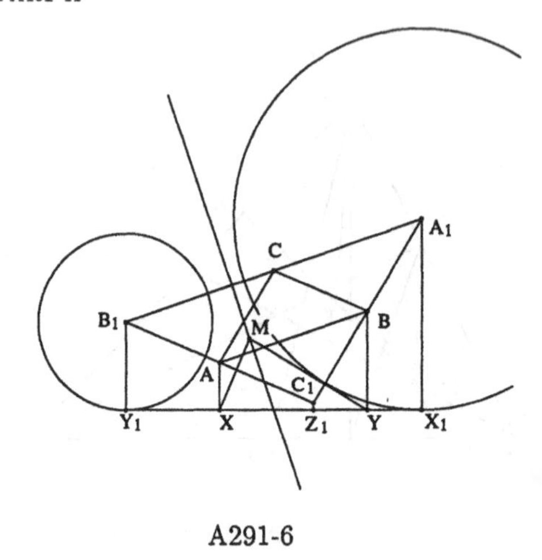

A291-6

of circle (A_1, A_1X_1) and circle (B_1, B_1Y_1).

$X = (0,0)$, $Y = (u_1,0)$, $C = (u_2,u_3)$, $A = (0,u_4)$, $B = (u_1,u_5)$, $M = (x_2,x_1)$, $C_1 = (x_4,x_3)$, $B_1 = (x_6,x_5)$, $A_1 = (x_8,x_7)$, $X_1 = (x_8,0)$, $Y_1 = (x_6,0)$, $Z_1 = (x_4,0)$.

The **nondegenerate** conditions: $X \neq Y$; $X \neq Y$; Points B, C and A are not collinear; Points B, C and A are not collinear; Points A, B and C_1 are not collinear; Line C_1B intersects line B_1C; Line XY is non-isotropic; Line XY is non-isotropic; Line XY is non-isotropic.

Example 507 (A292-7: 32.23s, 2582). *If P and P_1, Q and Q_1 are pairs of isogonal conjugate points, show that the points $R = PQ \cap P_1Q_1$, $R_1 = PQ_1 \cap P_1Q$ are also isogonal conjugates.*

Points A, B, C, Q, P are arbitrarily chosen. Points P_1, Q_1, R, R_1 are constructed (in order) as follows: $\tan(CBP) = \tan(P_1BA)$; $\tan(BAP) = \tan(P_1AC)$; $\tan(CBQ) = \tan(Q_1BA)$; $\tan(BAQ) = \tan(Q_1AC)$; R is on line P_1Q_1; R is on line PQ; R_1 is on line P_1Q; R_1 is on line PQ_1. The **conclusion**: $\tan(BAR) = \tan(R_1AC)$.

$A = (0,0)$, $B = (u_1,0)$, $C = (u_2,u_3)$, $Q = (u_4,u_5)$, $P = (u_6,u_7)$, $P_1 = (x_2,x_1)$, $Q_1 = (x_4,x_3)$, $R = (x_6,x_5)$, $R_1 = (x_8,x_7)$.

The **nondegenerate** conditions: $-u_1^3u_3u_7^2 + (u_1^3u_3^2 + u_1^3u_2^2 - u_1^4u_2)u_7 - u_1^3u_3u_6^2 + u_1^4u_3u_6 \neq 0$; $-u_1^3u_3u_5^2 + (u_1^3u_3^2 + u_1^3u_2^2 - u_1^4u_2)u_5 - u_1^3u_3u_4^2 + u_1^4u_3u_4 \neq 0$; Line PQ intersects line P_1Q_1; Line PQ_1 intersects line P_1Q.

Example 508 (A292-8: 29.33s, 681; 18.63s, 119). *Show that the isogonal conjugate of the Nagel point is collinear with the circumcenter and the incenter of the triangle.*

Points B, C, I are arbitrarily chosen. Points A, I_A, I_B, X_A, Y_B, N, N_1, O are constructed (in order) as follows: $\tan(CBI) = \tan(IBA)$; $\tan(BCI) = \tan(ICA)$; $I_AB \perp IB$; I_A is on line AI; I_B is on line BI; I_B is on line CI_A; X_A is on line BC; $X_AI_A \perp BC$; Y_B is on line CA; $Y_BI_B \perp AC$; N is on line AX_A; N is on line BY_B; $\tan(BCN) = \tan(N_1CA)$; $\tan(CBN) = \tan(N_1BA)$; $OC \equiv OA$; $OB \equiv OA$. The **conclusion**: Points O, N_1 and I are collinear.

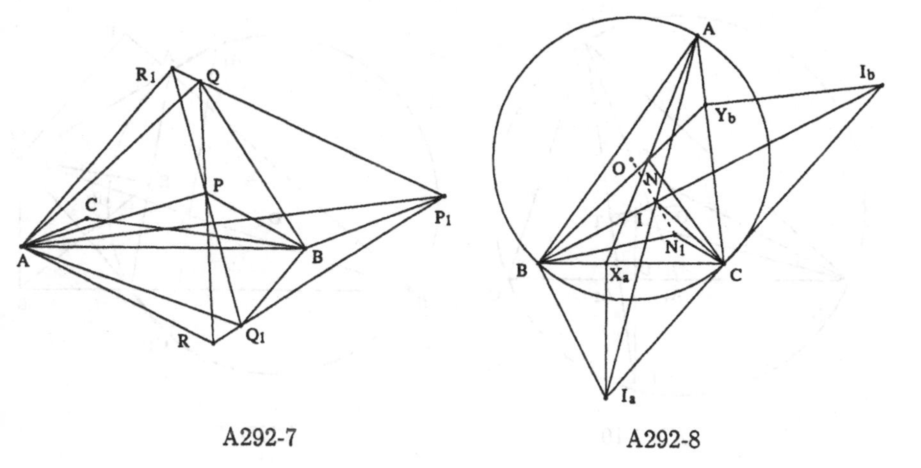

A292-7 A292-8

$B = (0,0)$, $C = (u_1,0)$, $I = (u_2,u_3)$, $A = (x_2,x_1)$, $I_A = (x_4,x_3)$, $I_B = (x_6,x_5)$, $X_A = (x_4,0)$, $Y_B = (x_8,x_7)$, $N = (x_{10},x_9)$, $N_1 = (x_{12},x_{11})$, $O = (x_{14},x_{13})$.

The **nondegenerate** conditions: Points I, C and B are not collinear; $\angle CIB$ is not right; Line AI is not perpendicular to line IB; Line CI_A intersects line BI; Line BC is non-isotropic; Line AC is non-isotropic; Line BY_B intersects line AX_A; $-u_1^3x_1x_{10}^2 + u_1^4x_1x_{10} - u_1^3x_1x_9^2 + (u_1^3x_2^2 - u_1^4x_2 + u_1^3x_1^2)x_9 \neq 0$; Points B, A and C are not collinear.

Example 509 (A292-10: 5.07s, 80). *Show that the symmetric, with respect to a side of a given triangle, of the corresponding vertex of the cosymmedian triangle is the isogonal conjugate of the foot of the perpendicular from the circumcenter upon the symmedian considered.*

Points B, C, A are arbitrarily chosen. Points A_1, A_2, O, P, D, Q, T are constructed (in order) as follows: A_1 is the midpoint of B and C; $\tan(BAA_1) = \tan(A_2AC)$; A_2 is on line BC; $OB \equiv OA$; $OB \equiv OC$; $PO \equiv OB$; P is on line AA_2; D is on line BC; $DP \perp BC$; D is the midpoint of P and Q; T is on line AA_2; $TO \perp AA_2$. The **conclusions**: (1) $\tan(CBT) = \tan(QBA)$; (2) $\tan(BAT) = \tan(QAC)$.

$B = (0,0)$, $C = (u_1,0)$, $A = (u_2,u_3)$, $A_1 = (x_1,0)$, $A_2 = (x_2,0)$, $O = (x_4,x_3)$, $P = (x_6,x_5)$, $D = (x_6,0)$, $Q = (x_6,x_7)$, $T = (x_9,x_8)$.

The **nondegenerate** conditions: $((2u_2 - u_1)u_3)x_1 - u_3^3 - u_2^2u_3 \neq 0$; Points B, C and A are not collinear; Line AA_2 is non-isotropic; Line BC is non-isotropic; Line AA_2 is non-isotropic. In addition, the following nondegenerate conditions, which come from reducibility and have been detected by our prover, should be also added: $P \neq A$.

Example 510 (A292-15: 57.1s, 1692). *M is any point in the plane of a triangle ABC. If MA, MB, MC cut the circumcircle again in A_1, B_1, C_1, and D, E, F are the projections of M upon BC, CA, AB, show that triangles $A_1B_1C_1$, DEF are directly similar and that M in $A_1B_1C_1$ corresponds with the isogonal conjugate*

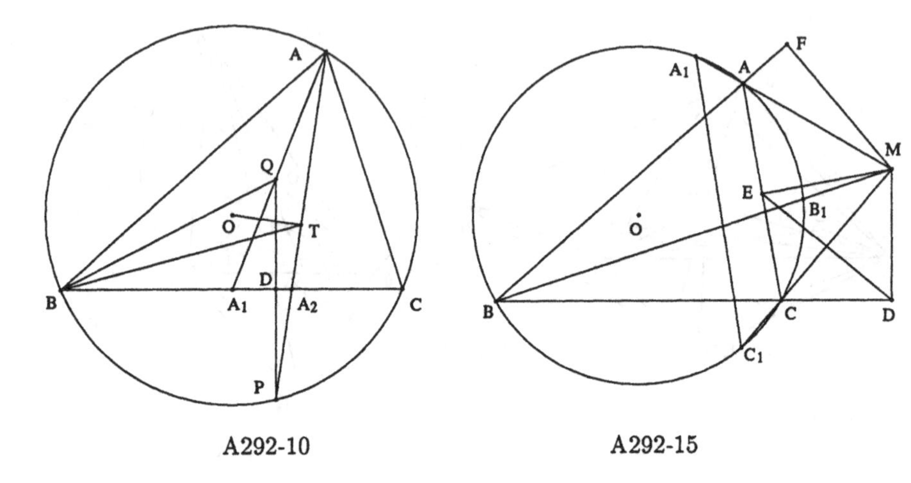

A292-10 A292-15

of M in DEF.

Points B, C, A, M are arbitrarily chosen. Points O, A_1, B_1, C_1, D, E, F are constructed (in order) as follows: $OB \equiv OA$; $OB \equiv OC$; $A_1O \equiv OB$; A_1 is on line AM; $B_1O \equiv OB$; B_1 is on line BM; $C_1O \equiv OB$; C_1 is on line CM; D is on line BC; $DM \perp BC$; E is on line AC; $EM \perp AC$; F is on line AB; $FM \perp AB$. The **conclusions**: (1) $\tan(EDM) = \tan(MA_1C_1)$; (2) $\tan(EDF) = \tan(B_1A_1C_1)$.

$B = (0,0)$, $C = (u_1,0)$, $A = (u_2,u_3)$, $M = (u_4,u_5)$, $O = (x_2,x_1)$, $A_1 = (x_4,x_3)$, $B_1 = (x_6,x_5)$, $C_1 = (x_8,x_7)$, $D = (u_4,0)$, $E = (x_{10},x_9)$, $F = (x_{12},x_{11})$.

The **nondegenerate** conditions: Points B, C and A are not collinear; Line AM is non-isotropic; Line BM is non-isotropic; Line CM is non-isotropic; Line BC is non-isotropic; Line AC is non-isotropic; Line AB is non-isotropic. In addition, the following nondegenerate conditions, which come from reducibility and have been detected by our prover, should be also added: $C_1 \neq C$; $B_1 \neq B$; $A_1 \neq A$.

Example 511 (A292-16: 40.93s, 1387; 40.42s, 190). *Show that the circumcenter of a triangle is the centroid of the antipedal triangle of the Lemoine point of the triangle.*

Points B, C, A are arbitrarily chosen. Points A_1, B_1, S, A_2, C_2, B_2, O, D are constructed (in order) as follows: A_1 is the midpoint of B and C; B_1 is the midpoint of C and A; $\tan(CBS) = \tan(B_1BA)$; $\tan(BAS) = \tan(A_1AC)$; $A_2C \perp CS$; $A_2B \perp BS$; C_2 is on line BA_2; $C_2A \perp AS$; B_2 is on line AC_2; B_2 is on line CA_2; $OB \equiv OC$; $OA \equiv OB$; D is on line B_2C_2; D is on line OA_2. The **conclusion**: D is the midpoint of B_2 and C_2.

$B = (0,0)$, $C = (u_1,0)$, $A = (u_2,u_3)$, $A_1 = (x_1,0)$, $B_1 = (x_2,x_3)$, $S = (x_5,x_4)$, $A_2 = (x_7,x_6)$, $C_2 = (x_9,x_8)$, $B_2 = (x_{11},x_{10})$, $O = (x_{13},x_{12})$, $D = (x_{15},x_{14})$.

The **nondegenerate** conditions: $(((-u_1u_2 + u_1^2)u_3^2 - u_1u_2^3 + u_1^2u_2^2)x_1 + u_1u_3^4 + (2u_1u_2^2 - u_1^2u_2)u_3^2 + u_1u_2^4 - u_1^2u_2^3)x_3 + ((-u_1u_3^3 - u_1u_2^2u_3)x_1 + u_1^2u_3^3 + u_1^2u_2^2u_3)x_2 \neq 0$; Points B, S and C are not collinear; Line AS is not perpendicular to line BA_2; Line CA_2 intersects line AC_2; Points A, B and C are not collinear; Line OA_2 intersects line B_2C_2.

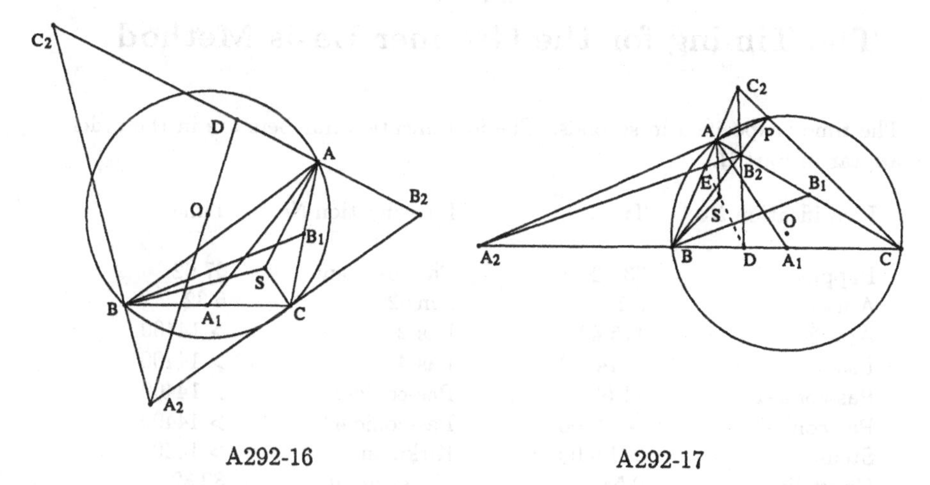

A292-16 A292-17

Example 512 (A292-17: 3.17s, 25). *Show that the trilinear polar of a point on the circumcircle of a triangle passes through the symmedian point of the given triangle.*

Points B, C, A are arbitrarily chosen. Points O, P, A_1, B_1, S, A_2, B_2, C_2, D, E are constructed (in order) as follows: $OB \equiv OA$; $OB \equiv OC$; $PO \equiv OB$; A_1 is the midpoint of B and C; B_1 is the midpoint of C and A; $\tan(CBS) = \tan(B_1BA)$; $\tan(BAS) = \tan(A_1AC)$; A_2 is on line BC; A_2 is on line AP; B_2 is on line AC; B_2 is on line BP; C_2 is on line AB; C_2 is on line CP; D is on line BC; D is on line B_2C_2; E is on line B_2A_2; E is on line BA. The **conclusion**: Points E, D and S are collinear.

$B = (0,0)$, $C = (u_1,0)$, $A = (u_2,u_3)$, $O = (x_2,x_1)$, $P = (x_3,u_4)$, $A_1 = (x_4,0)$, $B_1 = (x_5,x_6)$, $S = (x_8,x_7)$, $A_2 = (x_9,0)$, $B_2 = (x_{11},x_{10})$, $C_2 = (x_{13},x_{12})$, $D = (x_{14},0)$, $E = (x_{16},x_{15})$.

The **nondegenerate** conditions: Points B, C and A are not collinear; $(((-u_1u_2+u_1^2)u_3^2 - u_1u_2^3 + u_1^2u_2^2)x_4 + u_1u_3^4 + (2u_1u_2^2 - u_1^2u_2)u_3^2 + u_1u_2^4 - u_1^2u_2^3)x_6 + ((-u_1u_3^3 - u_1u_2^2u_3)x_4 + u_1^2u_3^3 + u_1^2u_2^2u_3)x_5 \neq 0$; Line AP intersects line BC; Line BP intersects line AC; Line CP intersects line AB; Line B_2C_2 intersects line BC; Line BA intersects line B_2A_2.

Appendix
The Timing for the Gröbner Basis Method

The time is specified in seconds. The identification numbers are in the order they appear in Part II.

Identification No	Time	Identification No	Time
Pappus	33.32	Pappus-point	67.62
Ams-1	9.1	Ams-2	5.32
Ams-3	115.43	Pas-2	> 14400
Pas-3	> 14400	Pas-4	> 14400
Pas-conic-1	71.63	Pas-conic-2	> 14400
Pas-conic-3	> 14400	Pas-conic-4	> 14400
Steiner	> 14400	Kirkman	> 14400
Gauss-line	9.58	Gauss-point	32.82
Gauss-conic	5255.4	Pappus-dual	25.53
Brianchon	> 14400	Altitude-dual	4.67
Simson-dual	> 14400	E1.5-19	12.17
E2.3-21	7.0	E2.3-22	33.02
E2.3-25	36.77	E2.3-26	> 14400
E2.3-27	159.3	E2.4-16	435.13
C3-4	1.58	C18-4	1.13
C20-2	1.03	C22-5	2.68
C2.45	2.95	C25-3	2.65
C45-2	412.17	C2.81	323.47
C46-2	1.5	C47-3	3.75
Morley	308.07	C49-1	> 14400
C3.11	6.9	C3.14	1.37
C3.36	9.37	C65-2-a	10.73
C65-2-b	> 14400	C69-3	37.65
C73-1	7.98	C76-1	> 14400
Y-2	2.28	Y-4-a	5.65
Y-4-b	14.45	Y-6	9.95
Y-8	7.95	Y-8-c	81.33
Y-9-b	2.12	Y-11-b	2.18
Y-15-a	4.58	Y-15-b	3.82
Y-25	25.25	Y-27	213.32
Y-38-b	82.55	Y-page-83	63.3
Y-57	> 14400	Y-68	116.13
Ogilvy-1	340.68	Ogilvy-2	215.6
Harmonic-set	4.37	Quadrangular-set	40.0
H-28	6.03	H-64	21.2
H-67	2.82	H-68	> 14400
H-69	521.57	H-102	14.87

H-104	4.07	H189-200	11.07
H199-212a	4.62	H-379	41.35
Ptolemy	3.33	Pratt-4	> 14400
Pratt-5	14.22	M-3	15.92
M-8	38.88	M10-26	6.0
M10-32	3.05	M10-34	0.78
M10-40	3.07	M12-46	4.02
M12-47	5.92	M-13	55.3
M-16	4.12	M-18	1.62
M-19	73.45	M21-63	> 14400
M21-67	1.03	M21-68	> 14400
M22-77	3.82	M23-80	2.95
M23-81	> 14400	M24-95	7.17
M25-96	4.12	M25-98	2.6
M-24	1.35	M-26	5.68
M-27	19.18	M-28	> 14400
M34-121	3.45	M34-122	4.48
M35-127	> 14400	M37-149	7.7
M38-156	22.07	M38-157	14.55
M39-158	20.63	M39-161	1.23
M40-167	> 14400	M40-168	72.13
M40-169	5.53	A30-28	17.93
A30-29	4.25	A30-30	2.82
A30-32	5.82	A31-35	7.12
A31-37	23.32	A31-39	29.48
A31-40	6.13	A31-41	5.05
A31-43	6.38	A31-44	503.73
A31-45	3.67	A31-47	5.13
A32-49	11.62	A32-56	8.6
A32-58	2.52	A52-1	15.23
A57-2	187.7	A57-3	11.8
A-73	1.58	A-79	1.28
A-81	2.53	A-85	0.93
A-86	1.0	A-88	3.35
A64-10-a	25.82	A65-4	111.9
A68-8	13.23	A68-9	6.32
A70-9	3.47	A70-10	21.08
A70-12	4.27	A-109	3.78
A-110	5.98	A-113	16.03
A73-2	5.57	A-116	3.03
A-120	5.48	A-122	5.72
A-123	6.97	A-126	13.97
A-127	5.42	A78-1	19.55
A-134	8.95	A-152	110.85
A-154	27.97	A-160	3.97
A-162	13.52	A-164	45.15

A87-4-a	29.07	A87-4b	18.9
A93-14	17.8	A-168	26.58
A93-17	18.87	A93-20	12.38
A-175	0.47	A-178	4.98
A-180	5.65	A-181	6.08
A97-5	3.92	A97-6	22.57
A97-8	5.47	A-186	6.43
A-188	1.97	A-189	4.37
A-191	16.22	A99-1	30.78
A99-2	4.02	A99-3	97.83
A99-4	3.32	A99-5	4.15
A-195	1.28	A-197	7.37
A-201	4.87	A-203	2.58
A-204	1.33	A-205	24.18
A103-5	20.15	A-207	13.18
A-208	6.12	A-209-a	14.78
A-209-b	18.17	A-210	25.98
A-211	38.08	A-213	39.18
A-214	11.88	A-215	29.28
A108-2	2.77	A108-3	4.47
A108-4	5.77	A108-5	15.83
A108-6	3.78	A108-7	10.37
A109-18	12.85	A-224	7.32
A-227	24.02	A-228	43.1
A111-1	7.55	A111-2	9.28
A111-5	8.65	A111-9	41.3
A115-4	8.38	A115-5	4.67
A116-11	4.05	A116-12	14.2
A116-13	80.3	A116-14	9.03
A117-23	27.67	A117-40	30.23
A118-42	10.05	A118-43	10.75
A118-44	15.72	A118-47	3.67
A118-49	59.52	A118-50	3.33
A118-53	11.93	A119-54	3.72
A119-55	6.52	A119-56	17.18
A119-61	11.37	A119-63	77.55
A119-67	10.85	A120-68-a	> 14400
A120-69	35.82	A120-70	16.2
A120-71	85.98	A120-72	13.07
A120-73	78.88	A120-76	12.82
A120-78	5.3	A120-79	35.48
A120-80	49.92	A121-81	6.28
A121-82	11.02	A121-84	10.58
A121-86	11.87	A121-88	51.57
A121-89	2.0	A121-91	27.83
A122-92	9.32	A122-93	1.52

A122-96	37.25	A-242	11.58
A-244	18.07	A-246	1.75
A-248	4.03	A127-2	77.27
A-258	9.37	A-259	12.87
A-261	7.57	A-262	8.6
A-263	20.62	A-264	3.58
A134-1	213.77	A134-2	4.93
A134-4	12.28	A134-5	9.42
A134-6	13.83	A135-7	6.08
A135-8	1334.2	A135-9	93.5
A-273	5.35	A-274	5.9
A-275	9.52	A-276	5.4
A-277	14.95	A-278	4.02
A-279	2.5	A138-1	4.12
A138-2	50.4	A-282	5.02
A-284	49.17	A-286	31.85
A-287	5.57	A-288	4.52
A-290	9.28	A-291	21.85
A145-2	10.25	A145-4	6.4
A145-5	7.43	A145-6	6.38
A145-7	3.58	A145-10	11.83
A145-11	6.98	A-294	284.38
A-294-a	335.25	A-295	7.23
A-293	2894.05	A-297	65.4
A-298	80.85	A-299	90.33
A-302	137.73	A-303	> 14400
A-304	> 14400	A-305	6.33
A149-1	19.95	A149-3	67.02
A149-4	> 14400	A149-7	> 14400
A150-11	344.92	A150-12	73.02
A153-3	26.22	A-310	3.02
A-313	1.67	A-314	12.15
A-316	11.28	A-321	12.52
A-318	5.23	A-324	9.87
A158-1	3.12	A158-2	3.82
A158-4	38.17	A158-5	18.13
A158-6	38.93	A158-7	242.52
A158-8	5.73	A-326	3.47
A-328	3.95	A-330	18.32
A-333	38.37	A161-2	2.98
A161-4	31.72	A-335	37.52
A-338	38.23	A-339	51.07
A-340	3.63	A-341	3.75
A-342	3.7	A-343	11.73
A164-1	725.63	A164-3	7.15
A164-4	21.95	A164-5	10.87

A165-5	5.83	A165-6	14.05
A165-7	34.58	A165-8	30.83
A165-9	82.38	A165-10	36.9
A168-2	2.98	A-351	3.48
A-353	11.68	A171-7	3.67
A171-8	5.7	A171-10	8.0
A171-12	18.62	A171-13	12.5
A171-4	2.12	A171-5	3.18
A-362	> 14400	A174-1	15.92
A174-3	16.68	A174-5a	2.18
A174-6	2.03	A-368	4.2
A-372	135.18	A177-8-a	2.95
A177-8-b	8.75	A177-10	5.6
A177-12a	38.48	A-386	22.13
A182-3-a	> 14400	A183-1b	22.05
A184-2	7.27	A184-3	13.37
A184-4	17.27	A184-5	9.88
A184-7	11.38	A184-8	5.98
A189-8	3.48	A189-9	4.0
A190-12	5.13	A-410	11975.2
A193-7	12.9	A193-8	10.67
A195-3	1929.8	A194-5	6.78
A194-7	> 14400	A194-8	> 14400
A-425	1.93	A-428a	1.3
A-429	51.87	A196-5	17.75
A196-2	> 14400	A197-9	4.2
A197-12	12.53	A197-13	1.47
A221-1	10.92	A227-1	3.4
A227-3	3.17	A227-6	12.87
A229-24	720.93	A242-9	125.05
A-547-a	21.55	A-547-b	10.33
A246-2	9.2	A247-4	34.07
A247-5	61.2	A247-7	47.6
A-559	10.97	A-560	4.83
A-562	2.67	A-564	23.52
A-565	12.28	A-566	6.5
A-568-a	5.42	A252-5	2.78
A252-7b	17.82	A252-8	5.82
A252-9	8.62	A252-10	122.18
A252-11-a	17.68	A-572	6.23
A-574	11.07	A-577	11.43
A-579	15.22	A-581	18.47
A-583	17.37	A-584	3.97
A-586	1.62	A-587	23.27
A256-1	19.2	A256-3	901.9
A257-7	55.18	A257-8	25.77

A257-9	38.53	A-588	27.23
A-590	40.82	A-591	11.88
A260-1	12.9	A260-2	81.78
A260-2-a	65.53	A260-4	49.22
A260-5	73.12	A-599	187.28
A-600	271.15	A-602-a	44.18
A-602-b	99.48	A-606	194.45
A-608	19.32	A-614	24.9
A-615	87.33	A266-1	43.15
A267-5	24.68	A-626	24.33
A-627	1.93	A-628	6.57
A-631	6.35	A-633	6.17
A-634	20.88	A269-2	18.98
A269-3	4.45	A269-4	12526.4
A269-7	7.98	A-635	6.82
A-637	14.55	A-640	160.02
A-641	337.55	A-643	82.97
A-644	18.58	A-645	24.25
A273-1	370.92	A274-3	2.47
A274-4	56.63	A274-5	28.25
A-650	10.38	A-651	7.67
A-653	6.27	A-654	14.97
A-661	837.47	A-663	35.37
A-664	56.67	A-665	206.18
A-667	41.45	A-670-a	> 14400
A-670-b	29.12	A-691	13.55
A-694	40.68	A-695	74.72
A-697	127.8	A-699	136.05
A291-1	149.92	A291-2-0	225.33
A291-2	30.1	A291-4-0	3.1
A291-4	18.13	A291-5-a	66.55
A291-5-b	4.73	A291-6	5.13
A292-7	> 14400	A292-8	65.88
A292-10	10.18	A292-15	104.17
A292-16	55.9	A292-17	116.37

Subject Index

Page numbers in boldface represent the definition or the main source of items indexed.

Index of Examples

Numbers are referred to *example numbers*. Numbers in boldface represent the definition or the main source of items indexed. This index is helpful for finding a particular theorem in the 512 examples. The terminologies representing the main feature of an example are selected to be indexed. The names of persons are connected with the theorems named after them.

9 781402 00330